MODERN QUATERNARY RESEARCH IN SOUTHEAST ASIA

VOLUME 17
Series editors: Juliette Pasveer & Ken Aplin

Lake Ayamaru, central Bird's Head, Papua, Indonesia. Photograph by Johan Jelsma.

The Djief Hunters
26,000 Years of Rainforest Exploitation on the Bird's Head of Papua, Indonesia

JULIETTE M. PASVEER
Department of Archaeology and Natural History, The Australian National University, Canberra, Australia

MODERN QUATERNARY RESEARCH IN SOUTHEAST ASIA

VOLUME 17

CRC Press
Taylor & Francis Group
Boca Raton London New York

CRC Press is an imprint of the
Taylor & Francis Group, an **informa** business
A BALKEMA BOOK

Subsidy for the publication of this book was supplied by the Netherlands Organisation for Scientific Research (NWO).

Cover illustration: Karst formation on the Ayamaru Plateau. Courtesy Johan Jelsema

CRC Press
Taylor & Francis Group
6000 Broken Sound Parkway NW, Suite 300
Boca Raton, FL 33487-2742

First issued in hardback 2019

ISBN-13: 978-90-5809-663-0 (hbk)

Visit the Taylor & Francis Web site at
http://www.taylorandfrancis.com

and the CRC Press Web site at
http://www.crcpress.com

Library of Congress Cataloging-in-Publication Data

A Catalogue record for this book is available from the Library of Congress

ISSN 0168-6151

Contents

List of Figures

List of Tables

Preface

The present volume of *Modern Quaternary Research in Southeast Asia* contains the results of archaeological research by Pasveer on the Bird's Head Peninsula of the island of New Guinea. The research was submitted for a PhD dissertation at the University of Groningen, the Netherlands, and funded within the Dutch-Indonesian 'ISIR' project. Many outcomes of the ISIR project, including some preliminary results from the archaeological research, were reported in volume 15 of this series (*Bird's Head Approaches*, edited by G.-J. Bartstra).

Technically, the study area falls outside of the geographic region of Southeast Asia and within Oceania. However, for two reasons, we consider the subject matter appropriate for publication in this series. Firstly, the Bird's Head is politically part of the Indonesian province of Papua (formerly Irian Jaya) and thus is likely to attract increasing attention from Indonesian archaeologists. And secondly, the Bird's Head represents a region of biological and cultural interaction between the two great geographic regions and thus warrants the attention of all regional scholars.

The Djief Hunters contains a detailed account of cultural and biological remains excavated from two prehistoric sites in the central Bird's Head, with a combined archaeological record spanning c. 26,000 radiocarbon years. The work is pioneering in the sense that the sites are the first to be systematically excavated and analysed in the entire western half of New Guinea. However, it is also pioneering in the sense that it contains new methods of analysis and interpretation of the abundant faunal remains from the two sites, and some new observations on the nature and extent of environmental change and human adaptations in a tropical lowland environment. We hope that the range of material included in the volume will make it a valuable reference for Quaternary researchers all of persuasions.

A few 'domestic' issues concerning *Modern Quaternary Research in Southeast Asia* need to be mentioned. First of all, the circumstances of publication changed upon the retirement in 2000 of Mr A.T. Balkema. Although A.A. Balkema Publishers continued to exist (and still does today), it was now managed within the broader umbrella of another Netherlands-based publisher, Swets & Zeitlinger Publishers. Subsequently, Swets & Zeitlinger Publishers was sold in November 2003 to Taylor & Francis, a UK-based scientific publisher. The production of new 'MQR' volumes was somewhat delayed due to these events and also by the relocation of the series editors from the western side of Australia to the eastern side. However, the delay also gave time for reflection, one consequence of which is the changed size and 'modernised' cover of this and all future volumes in the series.

The production of *The Djief Hunters* was made possible through financial support kindly provided by the Netherlands Organisation for Scientific Research (NWO). We would like to thank Janjaap Blom of A.A. Balkema Publishers for his patience and support during the production of this initial volume of the 'new look' *Modern Quaternary Research in Southeast Asia.*

Juliette Pasveer
Department of Archaeology and Natural History
The Australian National University
Canberra
Australia

Ken Aplin
Community Ecology Group
CSIRO Sustainable Ecosystems
Canberra
Australia

Acknowledgements

There are numerous people who have had part in this project, all of whom provided an invaluable contribution and to whom I would like to express my gratitude. First of all, the fieldwork in Ayamaru would not have been such a great success without the inexhaustible help and energy from our Indonesian fieldwork team: Sapri Hadiwisastra and Yono Silalahi (Indonesian Institute of Sciences, R&D Center for Geotechnology, Bandung), Elimas Kambuaya (Sekolah Menengah Pertama, Ayamaru) and Johnny Naa (Kartapura). Mr Kambuaya has been an invaluable source of information both before and after the fieldwork. In recognition of his efforts, a newly discovered Striped Possum (*Dactylopsila kambuayai*) found amongst the excavated fauna remains has been named after him. I would also like to thank the Kepala Desas and the people of Ayamaru (Mefkajim), Men/Semoe/Kartapura and Suwiam/Mapura for their hospitality and help during our stay and during the fieldwork, and for their permission to access their sites. D.D. Bintarti (National Research Centre of Archaeology, Jakarta) assisted and accompanied us during the reconnaissance trip to the Bird's Head in 1994.

Much of my inspiration for this work came from my extensive discussions with Ken Aplin (Commonwealth Scientific and Industrial Research Organisation, Canberra), who has given me invaluable insights in the field of archaeozoology and Australasian-Pacific archaeology, and who helped with faunal identifications. Reinder Reinders, Wietske Prummel, Gert-Jan Bartstra, Marcel Niekus [all associated (or used to be) with the Groningen Institute of Archaeology (GIA), University of Groningen], and Matthew Spriggs [School of Archaeology and Anthropology, Australian National University (ANU)] gave valuable comments on the draft of this thesis.

My collaboration with Gifford Miller (Center for Geochronological Research/Dept. of Geological Sciences, University of Colorado) and Simon Clarke (School of Geosciences, University of Wollongong) has been particularly fruitful. As part of their programme involving amino acid racemisation of ratite eggshell from the Australian-Pacific region, they used the eggshell material from both Bird's Head sites and provided me with a large proportion of the radiocarbon dates from the caves.

Rien Dam (CICAT, Delft University of Technology) and Sander van der Kaars (School of Geography & Environmental Science, Monash University) allowed me to use their unpublished data obtained from the pollen cores taken from the Ayamaru Lakes. Geoff Hope (Dept. of Archaeology and Natural History, ANU) helped me to interpret these data. I am grateful for their help and collaboration.

Many people helped with specific problems or particular materials; of these I would like to specifically mention (in no particular order): Tim Flannery and John Scanlon (South Australian Museum, Adelaide); Walter Boles (Australian Museum, Sydney); Peter Hiscock (School of Archaeology and Anthropology, ANU); Marcel Niekus (GIA, University of Groningen); Rob Moolenbeek (Zoological Museum Amsterdam), Katherine Szabó, Glenn Summerhayes and Wallace Ambrose (Dept. of Archaeology and Natural History, ANU); Rainer Grün (Environmental Geochemistry and Geochronology, ANU); Lyn Craven (Australian National Herbarium); Gerry Allen (formerly Western Australian Museum); Jan Delvigne (Dept. of Physical Geography, University of Groningen); Rob Braam (Dutch Topographical Service); and David Bulbeck and Kathleen McConnell (see their contributions in Appendix 1 and 2, respectively). I am grateful for their time, energy, and equipment that they invested for my purpose.

Many thanks furthermore to Truman Simanjuntak (National Research Centre of Archaeology, Jakarta) for translating the summary of this thesis in Bahasa Indonesia, and to Nadine Guthrie (formerly Edith Cowan University, Perth), Mirjam Weijns and Hans Zwier (GIA, University of Groningen), for providing some of the excellent drawings of stone tools, bone points and cartographic illustrations. Some of the excavated materials have been photographed by Peter van der Syde (Laboratory of Cell Biology and Electron Microscopy, University of Groningen). Ietse Stokroos and Freerk Dijk (same department) assisted with Scanning Electron Microscopy.

I thank New Holland Publishers and the photographers of the illustrations in Appendix 3 for their permission to reproduce these images from Tim Flannery's book *Mammals of New Guinea*. I also thank my friend and colleague Johan Jelsma (De Steekproef, the Netherlands; formerly GIA, University of Groningen), who gave me full access to all material he excavated from Toé Cave, as well as to his photographic material from the area.

Various departments and institutes have offered me office space or access to their reference collections. Their hospitality has made it a pleasure to work there: The Groningen Institute of Archaeology, University of Groningen; The Laboratory for Anatomy and Embryology, University of Groningen; The Department of Archaeology and Natural History, Research School of Pacific and Asian Studies, the Australian National University in Canberra; the Western Australian Museum in Perth; the Mammal Department at the Australian Museum in Sydney; The Australian National Wildlife Collection at the Commonwealth Scientific and Industrial Research Organisation (CSIRO) in Canberra.

The radiocarbon dates were provided by the National Science Foundation Radiocarbon Facility, University of Arizona; the Australian Nuclear Science and Technology Organisation (ANSTO); the Centre for Isotope Research, Radiocarbon Laboratory, University of Groningen; and the Radiocarbon Accelerator Unit, Research Laboratory for Archaeology and the History of Art, Oxford University. The dated sample 'OZE 542' was funded by AINSE grant 99/020 to Wallace Ambrose and Matthew Spriggs (Dept. of Archaeology and Natural History / School of Archaeology and Anthropology, ANU). The OZF/OZG dates on eggshell, generated as part of the project by Giff Miller and Simon Clarke mentioned above, were funded by AINSE grant 01/107 and 02/091 to Colin V. Murray-Wallace, Gifford H. Miller, Allan R. Chivas and Simon J. Clarke. Amino acid racemisation analyses were made by Charles P. Hart at the University of Colorado, with support provided by the U.S. NSF grant (ATM-9709806).

This research was carried out as part of the Dutch-Indonesian ISIR project (Irian Jaya Studies, a Programme for Interdisciplinary Research), funded by the Netherlands Foundation for the Advancement of Tropical Research (WOTRO). I thank the ISIR members for their encouragement and support during this study. The publication of this thesis was made possible through financial support from the Netherlands Organisation for Scientific Research (NWO).

All who contributed to this project and gave their valuable comments on the drafts of this thesis no doubt influenced my views in many ways; however, I alone am responsible for the contents of this work.

1. Introduction

1.1. ORIGIN OF THE PROJECT

In 1993, a multidisciplinary research project was set up to investigate the cultural and natural diversity of Indonesia's easternmost province of Papua[1], at that time still known as Irian Jaya. The province of Papua, forming the western half of the equatorial island of New Guinea, has attracted relatively little scientific attention and is widely regarded as one of the least explored regions on Earth. However, it is also known as a region of exceptional biological and cultural diversity, and hence of enormous scientific potential.

The research project, known as 'The Irian Jaya Studies: A programme for Interdisciplinary Research' or ISIR, was designed to fill some of the gaps in our knowledge of this vast and complex tropical region. The project was funded by the Netherlands Foundation for the Advancement of Tropical Research (WOTRO), a governmental institution operating within the framework of the Netherlands Organisation for Scientific Research (NWO), and was developed in collaboration with the Indonesian Institute of Sciences (LIPI).

The geographic focus of ISIR was narrowed to the westernmost peninsula of the province, the Bird's Head[2] (see Fig. 1.1). The main reason behind this choice was related to the fact that the Bird's Head is situated on the boundary between Island Southeast Asia and Oceania and thus represents an interesting cultural and biological transition zone. Moreover, some recent anthropological and linguistic studies had been conducted in this area from the Netherlands. Unfortunately, this advantage did not apply in the case of archaeological research – in this respect the Bird's Head was an almost complete

[1] The general term 'New Guinea' will be used to refer to the mainland as well as the surrounding islands that currently make up the Indonesian province of Papua and the independent country Papua New Guinea (see Fig. 1.1). 'Papua' is used as shorthand for the Indonesian province of Papua, formerly known as Irian Barat, West Irian and Irian Jaya. Papua New Guinea includes the eastern half of the main island of New Guinea along with the islands of the Bismarck Archipelago (New Britain, New Ireland, etc.), the Admiralty Group (including Manus Island), the northern Solomon Islands, and the islands of Milne Bay. 'Mainland New Guinea' refers to the main landmass without its associated satellite islands. 'Melanesia' includes New Guinea and the extended island chain consisting of the Solomon Islands, Vanuatu, New Caledonia and Fiji, while 'Island Melanesia' refers to the islands of Melanesia described above, excluding the New Guinea mainland.

[2] See footnote on page 2.

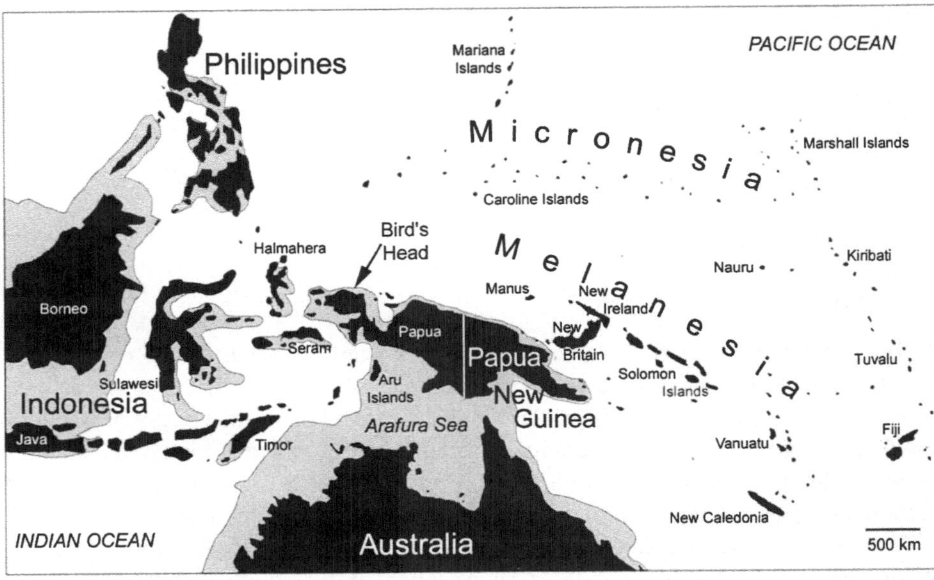

Figure 1.1. Location of the Indonesian province of Papua, and other places mentioned in the text. Continental shelf is indicated in grey. The continental shelf supporting the islands of Java and Borneo is part of the larger Pleistocene landmass of Sunda. The enlarged landmass that united New Guinea and Australia during Pleistocene glacial episodes is known as Sahul.

unknown. It had, however, a history of early colonial and biological exploration, which could guide the selection of research topics and study sites.

In 1994, a first reconnaissance trip was undertaken to several areas within the Bird's Head to locate a suitable location for archaeological research. The area showing greatest promise was a region of limestone karst around the Ayamaru Lakes in the central Bird's Head (see Fig. 1.2 and Chapter 2), and it was this area that was selected for detailed site survey and excavation. In 1995, I undertook a longer expedition to the Ayamaru area, together with fellow ISIR project member Johan Jelsma (formerly Groningen Institute of Archaeology).

This thesis presents the results of the fieldwork and subsequent analysis of the excavated material. In the remainder of this introduction, first the history of the region will be outlined, including the early explorations and major scientific expeditions. Then previous archaeological investigations in the wider Indo-Pacific region will be reviewed and an overview will be given of relevant archaeological knowledge and issues; this provides a context for introducing the objectives of this study.

[2] The westernmost peninsula of Papua is known under a variety of names (e.g. Vogelkop, Bird's Head, Kepala Burung, or Doberai Peninsula). For simplicity's sake, this peninsula (which excludes the more southern Fakfak or Bomberai Peninsula) will be referred to in this thesis as the Bird's Head, as this is the term by which it is known to most Pacific archaeologists and other scientists.

1.2. DISCOVERY, EXPLORATION AND ARCHAEOLOGICAL RESEARCH IN THE REGION

1.2.1. *European discovery, exploration and colonial history*

From early explorers' accounts and the discovery of non-native artefacts through the region, it is clear that Papua, and the Bird's Head region in particular, have had a long history of visitation from and contact with Southeast Asian neighbours. As Spriggs (1998: 931) recently observed

> 'A 16th century European account from the early period of European contact in the region describes people with rice, buffalo, goats, many wearing gold jewellery, possessing slaves and fleets of 160 ships. Those researchers working in the eastern half of the island of New Guinea, in present-day Papua New Guinea or indeed in other parts of Melanesia would be amazed to hear that this description refers not to somewhere in Island Southeast Asia, but to areas either on or adjacent to the Bird's Head.'

Spriggs derived this information from a report by the Portuguese Miguel Roxo de Brito, translated by Sollewijn Gelpke (1994), about his voyage to the Raja Ampat Islands, the MacCluer Gulf and Seram in 1581-82. Other historical information about the Bird's Head's economic position in Southeast Asia before Europeans appeared on the stage include reports by other early explorers and later colonial officials, and from archaeological studies in surrounding areas.

The first Europeans to sight the coast of New Guinea may have been the Portuguese when they reached the island of Banda in the early 16th century (Whittaker et al. 1975: 182). In 1526, the Portuguese Governor-elect of the Moluccas, Dom Jorge de Meneses, was driven by the northwest monsoon either onto the northwest coast of the Bird's Head, possibly at Warsai (Galis 1953: 8), or into the island of Waigeo (Whittaker et al. 1975: 182). The name Ilhas dos Papuas, used to indicate either Waigeo (Whittaker et al. 1975: 182) or the general group of islands in this area (Galis 1953: 9), probably derives from the Malay word *papuwah*, meaning fuzzy-haired, a reference to one of the most characteristic features of its inhabitants (Whittaker et al. 1975: 183). It was not until almost 20 years later, in 1545, that the mainland of New Guinea was first visited and given its current name by the Spaniard De Retes, who landed at the mouth of the Mamberamo River and claimed the land in the name of the King of Spain.

Initial European interest in the area focussed on rumours of gold and possibilities for trade and religious conversions. The first Dutch ship to visit the area was the 'Duyfken' in 1602. Although the Dutch saw limited possibilities for trade in the area, they regarded the western half of the island as important to safeguarding their trading empire and shipping routes in the adjacent Indonesian Archipelago. Through the 17th century various Dutch missions to New Guinea attempted to obtain knowledge about the land, people and potential for gold (e.g. Jansz in 1605, Carstensz in 1623, Pool in 1636, Keyts in 1678; see Whittaker et al. 1975: 196-202). These generally led to hostile encounters with the native inhabitants and little progress was made towards wider exploration of the island. Other nations also showed little interest in New Guinea, presumably discouraged by its inaccessible shores, uneconomical position in relation to the established trade routes, and reputedly violent inhabitants. Indeed, it was not until the 18th century that British settlement of Australia sparked a renewal of colonial interest in New Guinea. Even so, it took until 1875 before European traders, explorers, missionaries and researchers found the motivation and courage to penetrate a mere fifteen miles or so into the hostile hinterland of both mainland New Guinea and New Britain (Whittaker et al. 1975: 181).

After false rumours that the British had established a trading post on the southwest coast of New Guinea, the Dutch formally proclaimed sovereignty of the western half of the island in 1828 and established a settlement at Triton Bay. Beset by diseases and by attacks from the native population, this first attempt at settlement was short-lived. In 1875 the eastern border of West New Guinea was established at the 141st meridian, corresponding to the eastern shore of Humboldt Bay, east of present-day Jayapura.

The eastern half of New Guinea received little attention from the western world until it was annexed by Queensland in 1883. This annexation, although repudiated, brought to the fore the question of intervention in the eastern half of the island. A year later, east New Guinea was subdivided into two Protectorates, the northeastern half under German control ('Kaiser Wilhelmsland') and the southeastern half under Great Britain. This was a strategic move, motivated by the political situation in Europe at the time, rather than by economic considerations. The British Protectorate became the colony of 'British New Guinea' in 1888 (Whittaker et al. 1975: 437); it later became the 'Territory of Papua' after it fell under Australian control in 1905 (Essai 1961: 8). German New Guinea remained under German protection until 1914 when World War I broke out, at which time it fell under Australian Military Administration until it became Australia's mandated 'Territory of New Guinea' in 1920. During and after World War II the two territories were administered as a single unit by the Commonwealth of Australia, as recognised by the name 'Territory of Papua and New Guinea'. Finally, in 1975 the eastern half of New Guinea became the independent country of Papua New Guinea.

Dutch New Guinea remained a colony of the Netherlands until it was handed over, via the UN, to Indonesia in 1962-63, whence it became known as Irian Barat or West Irian. Later, the province was renamed Irian Jaya, as it is still known by many today. However, in the context of political unrest amongst the native Irian people, encouraged in recent years by East Timor's struggle for independence, the province was given semi-autonomous status under the new name of Papua in January 2002. The title 'Nueva Guinea', employed by the Spanish in 1545 for the main island, is the only original name that is still current today.

1.2.2. *Early scientific explorations*

Many nations have contributed to the scientific exploration of New Guinea [reviewed, among others, by Schumacher (1954) and Frodin & Gressitt (1982)]. The early French scientific expeditions, led by Louis de Saulses de Freycinet in 1817, Louis Isidor

Duperrey in 1823, and Dumont d'Urville (1823) mapped several locations around the island of Waigeo, northwest of the Bird's Head. Duperrey was the first to estimate the altitude of the Arfak Mountains in the Bird's Head – the first altitude estimation of any mountains in New Guinea. Coastal areas around the Bird's Head were the focus of botanical and ethnological collecting in 1858, with visits by Alfred Russell Wallace and by the first Dutch scientific expedition to New Guinea on the ship 'Etna'.

The Italians D'Albertis and Beccari collected birds and insects in the Arfak Mountains of the Bird's Head in 1872 and were the first explorers to venture any distance inland from the coast. In 1876 and 1877, d'Albertis travelled to eastern New Guinea and sailed up the Fly River in an attempt to discover its source; he reached a point some 275 km inland and obtained an admirable collection of birds and plants. On his second expedition to eastern New Guinea he discovered the mouth of the Strickland River. The Russian explorer Miklucho Maclay conducted anthropological studies in Astrolabe Bay in eastern New Guinea in 1874. Around the same time Captain John Moresby was exploring the eastern part of New Guinea, naming numerous bays, rivers and islands, and essentially finished off the exploration of the east coast.

Following d'Albertis' lead, other expeditions attempted to penetrate the central mountain ranges via the major river systems in the late 1800s-early 1900s. Among the most successful and best known are the English and Dutch expeditions to the Carstensz Mountains, and the Lorentz expedition to the 4750-m summit of Mt Wilhelmina. These expeditions were mostly designed to obtain scientific material (although the second Lorentz Expedition had the sole purpose of beating the British to the New Guinean eternal snow, long visible from the coast). Some of the most important scientific expeditions in the eastern half of New Guinea were those led by Macgregor between 1889 and 1898. In 1889 he reached the summit of Mt Victoria (4075 m).

From the 1930s on, biological exploration was dominated by the Archbold Expeditions, most of which took place in the eastern half of the island, but one, in 1938-39, a joint American-Dutch expedition, visited the Baliem Valley and the Mt Wilhelmina area. Dutch expeditions also visited the Wissel Lakes. The six Archbold expeditions in the eastern half of New Guinea collected in various areas including the Owen Stanley Range in 1933-34, the source of the Fly in 1936-37, and most recently in 1964 on the Huon Peninsula. All these expeditions resulted in large collections of botanical, zoological, geological and ethnological materials, but very little effort was made to investigate New Guinea's past.

Further Dutch exploration of western New Guinea after World War II was mainly focussed on economic resources, especially oil, and on extending political control. Missionaries penetrated various areas in pursuit of religious conversion. This gradual spread of colonial influence led to a steady increase in knowledge of native cultures and languages, and, to lesser extent, of the natural environment.

1.2.3. *Previous archaeological research in the region*

Early archaeological field reports from the Dutch-controlled territory mention historic or prehistoric sites or objects at various locations, including Waigeo and the Bomberai

Peninsula (e.g. Galis 1957a,b; 1964) and also report finds of Dongson drums and other bronze or brass objects at Lake Sentani (De Bruyn 1962; see also Agogino 1979), on the islands of Asei and Kwadeware (Simanjuntak 1998: 947), and in the Ayamaru District (Elmberg 1959; Galis 1964). Röder (1940) reports a site near Fakfak on the Bomberai Peninsula, the cave of Dudumunir, where ceramics and stone artefacts have been found in stratified context. Unfortunately, none of these early reports led to any systematic archaeological research. The province was effectively closed to archaeological research over much of the last three decades (Spriggs 1998).

The first archaeological excavation in western New Guinea was carried out as a 'side activity' by members of the 1971-1973 Australian-Indonesian Carstenz Glaciers Expeditions (Hope et al. 1976). Hope & Hope (1976) reported that Mapala Rockshelter on Mt Jaya (c. 4000 m; see Fig. 1.2) contained bones, ash, a few stone artefacts, and shell remains. A radiocarbon date of 5440 BP was obtained from an exposed charcoal-rich layer (Hope & Hope 1976: 234), but most of the deposit was left *in situ*. Other indications of early human activity in the region came from research at Ijomba Bog at 3600 m on Mt Jaya, where a pollen record suggests that firing may have led to an opening up of forest cover by 10,500 BP (Hope & Hope 1976: 236). Similar, indirect evidence of early human activity in the highlands of Papua is reported by Haberle et al. (1991: 36), who provide evidence for deliberate or accidental firing of vegetation in the Baliem Valley by 32,000 BP.

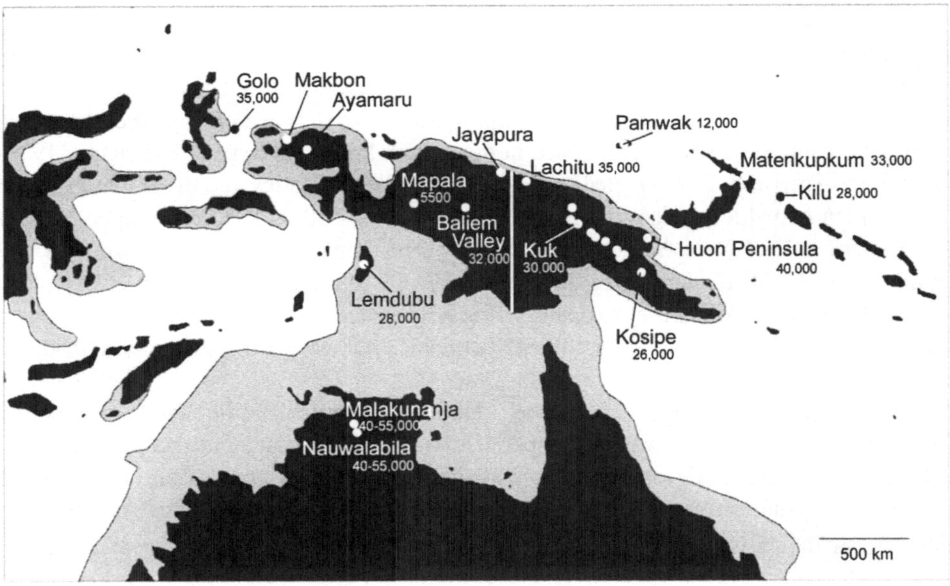

Figure 1.2. Location of some of the archaeological and palynological sites mentioned in the text, with the earliest associated radiometric dates (uncalibrated ^{14}C or OSL)in years BP. The series of unnamed white dots in the Papua New Guinean Highlands all date to younger than 30,000 BP. The grey area that unites New Guinea and Australia represents the Pleistocene landmass of Sahul.

The American archaeologist Wilhelm G. Solheim II was one of the first to conduct a systematic regional survey and test excavations in western New Guinea. He surveyed several coastal areas around the Bird's Head and the Bomberai Peninsula in 1975-76 and again in 1990 (Solheim II 1978, 1979, 1985, 1998; Solheim II & Ap 1977; Solheim II & Mansoben 1977; and pers. comm.). Shell middens, abundant pottery (including specimens with decorations similar to Lapita pottery, see below) and porcelain were found at several sites around the village of Makbon on the north coast of the Bird's Head, east of Sorong (Solheim II 1998: 30, 39). No further details of these sites have been published.

Finally, in 1995, the National Research Centre of Archaeology in Jakarta, in cooperation with ORSTOM in France, investigated some sites in the area around Papua's provincial capital Jayapura (Simanjuntak 1998), the first results of which were published by Simanjuntak (1996).

Prior to the 1950s only occasional prehistoric finds were reported from the area now united as Papua New Guinea (e.g. Meyer 1907; Pöch 1907; Leask 1943). The first professional excavations in this area were conducted in 1959 by Sue and Ralph Bulmer (Bulmer & Bulmer 1964). Since that time, many more sites have been excavated in Papua New Guinea and elsewhere in Island Melanesia, mainly by Australian and Papua New Guinean researchers (see Lilley 1992).

The oldest site currently known on the mainland of New Guinea is an open site on the north coastal Huon Peninsula, dating back to around 40,000 BP (Groube et al. 1986; see Fig. 1.2). At present, the oldest site in the main highlands region, dating around 30,000 BP, is Kuk Swamp, located near Mount Hagen; this site has yielded sparse evidence for human activity at this early date, in the form of humanly transported fire-cracked rocks and some charcoal (Golson 1977a; White & O'Connell 1982: 42). Overlying deposits at Kuk contain far more substantial evidence for human activity in the form of early drainage systems, the oldest phase dating to c. 9000 BP (Golson 1977a; Denham et al. 2003). No other sites in New Guinea have yet produced evidence of such early agricultural activity, although several show elaborate drainage systems dating to around 5000 BP or later (Harris & Hughes 1978; Gillieson et al. 1985).

Other sites of Pleistocene age in New Guinea's interior include an open site at Kosipe, situated high in the Owen Stanley Range, with stone tools including the distinctive 'waisted' blades dating back to 26,000 BP (White et al. 1970), and Nombe Rockshelter, with a basal layer containing the bones of now-extinct animals found in association with stone artefacts and probably dating to around 20-25,000 BP (Mountain 1983). Other well-known sites in the New Guinea Highlands are Batari (8000 BP), Aibura (4000 BP), and Kafiavana (11,000 BP), analysed by White (1972). Contact between the highlands and the coast goes back to at least c. 9000 years ago, as is evident from the presence of marine shells at Kafiavana around that time (White 1972: 93). Wañelek, dating to 15,000 BP (Bulmer 1977, 1991), and NFX, dating possibly up to 18,000 BP (Watson & Cole 1977) are both open sites in the Highlands. Yuku and Kiowa are rockshelter sites excavated by Sue Bulmer in 1959; Yuku is dated to 12,000 BP but waisted tools, similar to those from Kosipe, are still present in levels from below this date and the initial occupation is therefore thought to be older (White & O'Connell 1982: 57).

Long pollen records from montane localities also document early human presence within the landscape (Haberle et al. 2001). However, human impact is sometimes difficult to disentangle from climate-driven environmental changes, which saw a major expansion of permanent snow across many of the higher mountains and an associated depression of major vegetation zones by as much as 2000 m (Hope 1996).

Besides the Huon Peninsula locality, only one other Pleistocene site is known from a near coastal context; this is Lachitu Rockshelter (Gorecki et al. 1991; Gorecki 1993) in a tectonically active location on the north coast, with a basal age of 35,000 BP. Other coastal sites, mostly situated along the south coast of Papua New Guinea, are of mid- to late Holocene age (e.g. Vanderwal 1978; Egloff 1979; Rhoads 1980; Irwin 1985). Many of the coastal sites have been excavated to discover the prehistory of historically documented trading systems, mainly involving the exchange of pottery and sago. It is likely that many Pleistocene sites on the southern side of Papua New Guinea were drowned by the rising sea level (Groube 1989: 295).

Archaeological research in Island Melanesia started with excavations in the Bismarck Archipelago during the 1960s. Most work has been carried out on New Ireland and New Britain (e.g. Downie & White 1978; Allen et al. 1988; Allen & Gosden 1991; Gosden & Robertson 1991; Marshall & Allen 1991; Pavlides & Gosden 1994; Leavesley & Allen 1998). The site of Matenkupkum on New Ireland is currently the oldest site in Island Melanesia, with a date of 33,000 BP (Gosden & Robertson 1991). The site of Kilu on Buka, the northernmost island of the Solomon Group, was occupied by 28,000 BP, proving that people were able to cross open sea distances of 130-180 km (Wickler & Spriggs 1988: 705). Occupation of Pamwak rockshelter on Manus commenced more than 12,000 BP, in this case requiring an open water crossing of 200 km or more of open sea (Fredericksen et al. 1993: 151). The prehistory of Island Melanesia also gives indications of the early domestication of plants (Yen 1995), translocation of animals (Flannery & White 1991), and the movement of obsidian over some 350 km, beginning in the Late Pleistocene and continuing into the Holocene (Gosden 1993: 133). Interestingly, this region provides evidence of island exploitation at an earlier time than anywhere else in the world (Keegan & Diamond 1987).

A dominant feature of the later Holocene prehistory of Island Melanesia is the widespread appearance of the Lapita culture complex, characterised by a highly distinctive pottery style, and by evidence for horticulture, domesticated animals, highly developed maritime skills and extensive trade connections (Spriggs 1995: 116-118). The distribution of the Lapita culture and its derivatives corresponds quite well with the modern distribution of Austronesian languages (Bellwood 1991; Diamond 2000), and this fact, together with evidence from human genetics (e.g. Hill & Serjeantson 1989; Cann 2001), suggests that the spread of the Lapita culture marks a phase of active migration. This most likely originated in Taiwan, with people moving eastwards into the Philippines and from there into Island Melanesia and the wider Pacific region (Bellwood 1991; Kirch 1997: 15), although genetic evidence also suggests an origin in eastern Indonesia (Oppenheimer & Richards 2001). It must be noted, however, that not all archaeologists accept the migration model for Lapita origins. Alternative views see the Lapita tradition as a more local development (e.g. White & Allen 1980; J. Allen 1996, 2000), or one that was perhaps influenced by both Austronesian immigrants and the

Melanesian descendants of the Pleistocene colonisers of New Guinea and the Bismarck Archipelago (Kirch 1997; Allen 2000), or possibly by distant events via the rapid dispersal of technological innovations (e.g. Spriggs 2000a). One site on the north coast of mainland New Guinea (Spriggs 1997), and one on a small island in the same area (Terrell & Welsch 1997: 558) have produced Lapita pottery, while other sites in this area contain pottery that is similar (Swadling et al. 1989; Gorecki et al. 1991; Simunjuntak 1998). Pottery from these northern New Guinean sites might also be earlier than any genuine Lapita sites (Swadling et al. 1989: 109; Gorecki et al. 1991: 121), although the dating of these sites is contested (Spriggs 1996a: 329-335, 1996b: 43-44; Lilley 2000: 182).

The descendants of Lapita peoples were the first colonists of many of the smaller islands of Island Melanesia and beyond into the remote Pacific. Various models have been proposed for the colonisation process, including the 'fast train' model (e.g. Diamond 1988: 307), the 'slow boat' model (e.g. Gibbons 2001: 1735), the 'triple-I' model (Green 1991; Kirch 1997: 46), and the 'entangled bank' or 'voyaging corridor' model (Terrell & Welsch 1997). Most of these models allow for extensive culture contact and intermarriage along the way between 'new' and 'old' peoples, especially on the large islands of Melanesia with their long histories of occupation. Genetic evidence supports the idea of extensive interbreeding between recent migrants and Papuan-speaking Melanesians (Lum & Cann 1998).

The area immediately to the west of New Guinea was until recently an archaeological unknown. However, the area has long been of central interest as an arena of early and later human movements (Spriggs 2000a). In recent years, archaeological research has been conducted along two possible migration routes into New Guinea (Birdsell 1977), one passing through the northern Moluccas (Bellwood et al. 1998), and another passing through the Aru Islands (Veth et al. 1998; O'Connor et al. 2002). In the northern Moluccas there is evidence for human occupation from 35,000 BP in Golo Cave, with indications of animal translocation from New Guinea around 8000 BP (Bellwood et al. 1998: 261). Lemdubu Cave on the Aru Islands recently provided evidence for occupation dating back to before 28,000 BP (O'Connor et al. 2002).

Through much of the Late Pleistocene, New Guinea was united with Australia because of lowered sea levels, to comprise a large landmass known as Sahul (White & O'Connell 1982; see Fig. 1.1). The Pleistocene and later archaeology of Australia is thus directly relevant to that of New Guinea and *vice versa*, although they are more often than not seen as separate areas of investigation.

Systematic archaeological research in Australia did not commence until the late 1950s and early 1960s, but the number of excavated sites now vastly exceeds that in the wider New Guinea region. Based on radiocarbon dating, many Australian sites have produced ages in the range 35-40,000 BP, but recent re-dating of some sites using a variety of alternative methods indicates that some of these sites may be considerably older, perhaps exceeding the 'radiocarbon barrier' of c. 45,000 BP. Potentially the earliest sites, based on combinations of Optically Stimulated Luminescence (OSL), Thermoluminescence (TL), Electron Spin Resonance (ESR) and Uranium-series dates, are the Mungo 3 skeleton in southeastern Australia, with an estimated age of 62,000 BP (Thorne et al. 1999), and Nauwalabila and Malakunanja Rockshelters in northern

Australia, with estimated ages for first occupation of between 53 and 60,000 BP (Roberts et al. 1993, 1994). These early dates for human occupation of Australia are by no means universally accepted, partly on the grounds that the radiocarbon dates obtained so far for the oldest occupation of the wider Melanesian and Australian region show a clear pattern for first colonisation around 35-40,000 years ago. This is consistent with the current theories about the first appearance and spread of modern humans elsewhere outside Africa, whereas the older ages are not (Allen 1994; Allen & Holdaway 1995; O'Connell & Allen 1998). However, the evidence for occupation of Australia at least by 48,000 BP appears increasingly secure (Turney et al. 2001).

1.3. RESEARCH CONTEXT AND OBJECTIVES

The Bird's Head is clearly a region of great interest for New Guinean and greater Australasian prehistory. Beginning with the most remote periods of time, the Bird's Head represents one of the more obvious landfalls for the earliest colonists of Sahul, particularly for people coming in via the 'northern route' through the northern Moluccas (Birdsell 1977). As an area with complex topography and geology (Ratman 1998), the Bird's Head would have presented an attractive range of habitat types to early colonists, including mangrove forest, rocky coastlines, a variety of lowland forest types, montane forests and both lowland and upland lakes. Some of these habitats probably experienced only minor changes over the last 50,000 years or so, but others may have shown dramatic environmental changes. In particular the highland areas of the Bird's Head have experienced dramatic elevational shifts in plant and animal communities (Van der Kaars 1991), while its southern margin has been completely reconfigured through exposure of the broad continental shelf between the Bird's Head and Misool Island to the southwest (Dam 1998: 774-776). These environmental changes may have also brought about significant alterations in the nature and distribution of human activity within the landscape.

During later periods of prehistory various outside influences came to bear once again on the Bird's Head region. At some point in time the region came under the influence of agricultural traditions, with the introduction of new crops and domestic animals and a probable shift from essentially nomadic to a more sedentary lifestyle. Possibly at the same time, or perhaps later again, a pottery-making tradition made its mark on the region. And perhaps even more recently, the area came to fall within the expanding sphere of influence of larger regional powers, first those of Southeast and South Asia, and later those of the European colonial period.

This brief sketch of the history of the Bird's Head of course represents a very minimalist view of its past. The region today supports a considerable diversity of languages, the majority of which belong to the West Papuan Phylum (Reesink 1996, 1998; also represented on islands of the northern Moluccas), but with isolated languages of the Trans New Guinea Phylum (Pawley 1998) including one that may be related to languages of the Marind stock (De Vries 1998). Linguistic heterogeneity of this kind may be due to past movement of groups of people, which in turn may point to periods of social or environmental upheaval. Such events might be registered archaeologically, but

Figure 1.3. The Ayamaru Lakes surrounded by karst hills in the centre of the Bird's Head of Papua. Courtesy J.Jelsma.

are entirely unpredictable in terms of general regional patterns. Austronesian languages as such are not spoken anywhere on the Bird's Head, but they are spoken on the islands of Salawatti, Batanta and Waigeo to the immediate west of the Bird's Head and in the Bird's 'neck' region to the east. Reesink (1998) mentions 'heavy Austronesian influence' in Bird's Head languages, and posits a long history of both peaceful and antagonistic interactions, the latter often motivated by the taking of slaves, with women being taken as wives.

The primary research objective of the project was to locate and excavate stratified sites to document changes in prehistoric activities and cultural traditions. Because so little previous work had been undertaken in western New Guinea, there was no way of predicting in advance what particular issues of Australasian-Pacific prehistory might be more or less relevant to the Bird's Head region.

The fieldwork was successful and two sites with contrasting but complementary stratigraphic sequences were excavated in the Ayamaru region of the central Bird's Head (see Fig. 1.2 and 1.3). Both produced abundant stone artefacts and organic remains. Initial study of the excavated materials generated a suite of more specific research issues. These included the impact of climate change on the distribution of animal species in a lowland environment; the pattern of exploitation of the dominant prey animal (a species of wallaby, locally known as *Djief*) through time; the pattern of exploitation of raw materials for stone artefact production; and the mode of manufacture and function of bone artefacts. Stratigraphic analysis and radiocarbon dating became a vital issue, not only as a means of connecting the two sequences, but also as means of demonstrating the integrity of each deposit. The results of the study proved to have important bearing on various issues of wider significance, including the extent of environmental change in lowland tropical habitats during glacial episodes; the timing and circumstances of occupation of lowland rainforest habitats; and the possibility of sustainable exploitation of large mammals in rainforest habitat.

The following chapter describes the research area, the fieldwork design, and the methods used to analyse the stratigraphy. Chapter 3 and 4 give a full description of the two investigated sites, their stratigraphy and dating, and each ends with a comprehensive account of the occupation history. Chapters 5 and 6 are devoted to more detailed analyses of the excavated stone and bone artefacts. Chapter 7, 8 and 9 are devoted to the abundant faunal remains: Chapter 7 analyses the changing pattern of exploitation of food resources through time; Chapter 8 examines the impact of sustained hunting of the forest wallaby through time; and Chapter 9 investigates the environmental significance of the changing species composition of the mammal fauna. All of these threads are drawn together and discussed in the concluding Chapter 10.

Not all categories of excavated material are analysed in detail; however, specialised reports on the limited human and botanical remains are given in Appendices 1 and 2, respectively. Appendix 3 presents information on the modern occurrence of animal species in the study area, obtained on our behalf by our right hand in the field, Mr Elimas Kambuaya. Raw data on which the faunal analyses are based are provided in my unpublished PhD thesis (Pasveer 2003).

Some of the results described in these chapters have been published earlier in preliminary form (Pasveer 1995; Pasveer & Bartstra 1996; Pasveer 1998; Pasveer & Aplin 1999; Aplin et al. 1999; Pasveer et al. 2002). Although most of the information is still regarded as valid, it must be emphasised that the results and interpretations given in this volume supersede those given in previous publications.

2. The Research Area, Fieldwork Design and Methods of Stratigraphic Analysis

2.1. THE RESEARCH AREA: NATURAL ENVIRONMENT

The research area is located in the centre of the Bird's Head Peninsula, on the Ayamaru Plateau (Fig. 2.1), centred on the three connected Ayamaru Lakes [officially named Lake Ayamaru (the westernmost) and Lake Hain (comprising the two smaller eastern basins); Pigram & Sukanta 1989]. Today, the Meybrat[1] people live around the lakes (see section 2.2). In Meybrat language Lake Ayamaru is called *Maru Jaw* (Miedema 1986: 1) or *Maru Tä* (Elmberg 1968: 16), and the two smaller basins of Lake Hain are *Semetu* and *Maru Jate* (Miedema 1986: 1) or *Maru Jow* (Elmberg 1968: 16). The lakes occupy an elongate depression in the limestone running roughly northwest-southeast and measuring some 15 km long and 1-3 km wide. The margins of the lakes support a zone of marshy grassland and the lakes themselves are covered with dense aquatic vegetation (see Fig. 2.2 and 2.3). Water levels fluctuate seasonally, and large areas are flooded especially to the northeast (Miedema 1986). Water depth varies between 20 and 200 cm; some deeper channels are visible on aerial photographs (Fig. 2.4). The westernmost lake is mainly fed by rain- and groundwater sources; springs occur along the northernmost edge of Lake Ayamaru near Sekior (or Segior) village (Reynders 1961: 19; Miedema 1986: 3). In contrast, the eastern basins are fed by the river Isme that comes from the southeast and bends sharply towards the northeast before entering the lakes (Reynders 1961; see Fig. 2.1). The lake complex drains to the east into the river Kais. Unpublished research by Dam and Van der Kaars indicates that the present-day Ayamaru Lakes – or at least the westernmost lake – were formed in mid-Holocene times (see Chapter 9).

The lakes are positioned centrally on the Ayamaru Plateau, a heavily karstified region of Miocene limestone, calcareous and non-calcareous mudstone and conglomerate. The average elevation of the plateau is reported by Visser & Hermes (1962) as 350 m above sea level. The plateau is dominated by sinus karst (called 'polygonal or cockpit karst' by Pigram & Sukanta 1989; see Fig. 2.5) with a local relief of c. 50-100 m; many of the low, beehive-shaped hills contain caves.

The Ayamaru Plateau extends for 20-30 km to the south and southwest of the lakes before giving way to a broad zone of relict alluvial landforms dissected by wide flooded river valleys (Pigram & Sukanta 1989). Fifty kilometres to the east, a mountain block with isolated peaks above 2000 m altitude flanks the plateau. More distant are the

[1] The spelling of Meybrat (both the language and the people) varies among authors (e.g. Mejbrat, Mejprat, Meybrat, Maybrat).

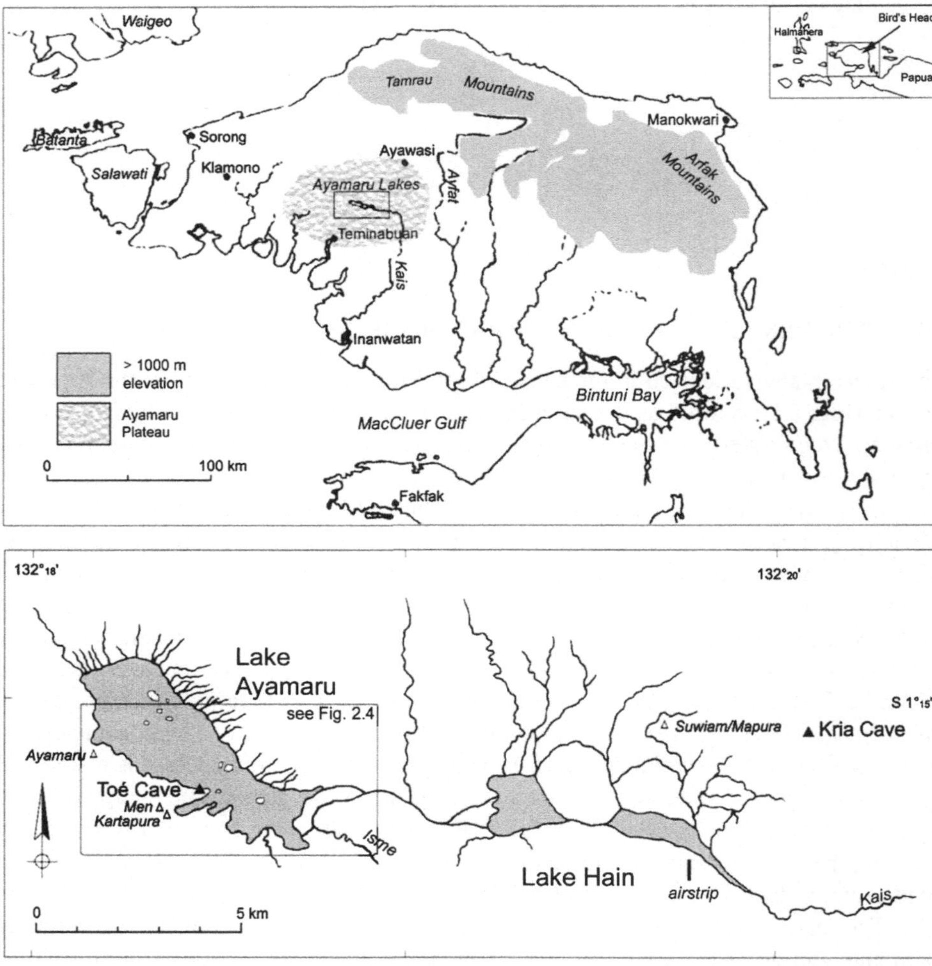

Figure 2.1. Map of the Bird's Head with the research area and location of the sites. The boundaries of the lakes are subject to change due to seasonal flooding. Lake Hain consists of two small basins. Based on a drawing by Hans Zwier.

Tamrau Mountains, 65 km to the north, and the Arfak Mountains, more than 100 km to the northeast, both with substantial areas above 2000 m.

Average annual precipitation over the plateau is very high, with 5591 mm recorded by Elmberg (1968: 16) in 1954, about half of which fell in June, July and August. Berlage (1949) reported annual precipitation of c. 3800 mm at Teminabuan (see Fig. 2.1) and c. 4850 mm at Aitinjo, both located to the southwest of the Ayamaru Lakes. Low clouds frequently blanket the area, especially in the mornings.

Except where removed by human activities, lowland rainforest blankets the entire region. Polak (1998) provided structural and floristic information on the rainforest community in the Ayawasi region, located at 500 m above sea level and about 40 km

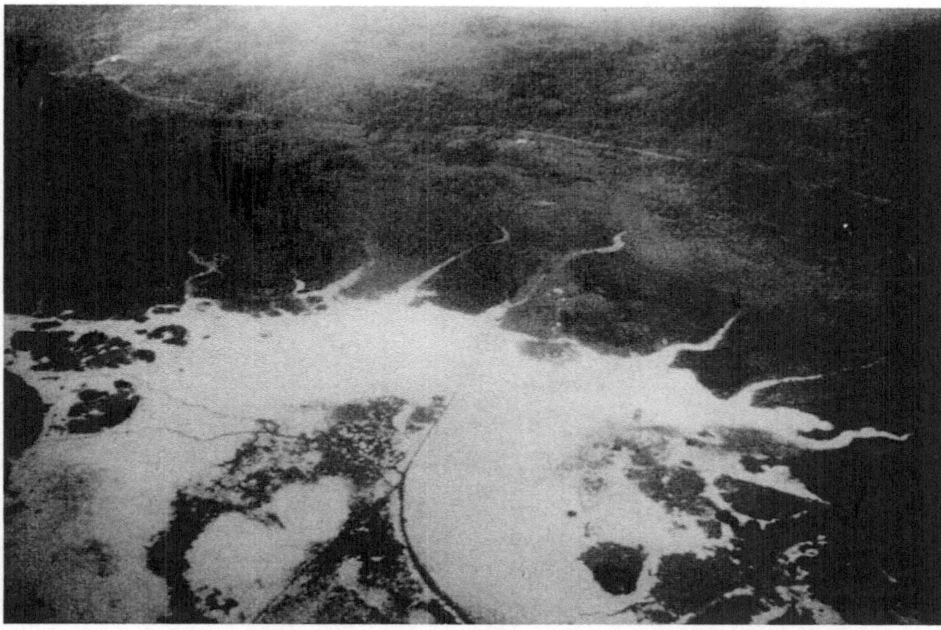

Figure 2.2. One of the Ayamaru Lakes viewed from the air. Note the canoe tracks through the aquatic vegetation, and the sealed road that runs south of the lake (top of the photograph). Courtesy J.Jelsma.

northeast of Lake Ayamaru. In this area, the limestone hills and associated skeletal soils support a low (c. 10 m), relatively dense but floristically non-diverse forest dominated by *Syzygium* spp. and other Myrtaceae. The intervening, lower-lying areas are frequently disturbed by gardening and other activities, but can support a taller and more diverse forest with Myrtaceae dominant and Guttiferae, Dipterocarpaceae and Burseraceae subdominant. One or a few tree species are dominant in both hill and non-hill forest types.

The Ayamaru Plateau has received little systematic zoological attention and its contemporary fauna remains poorly documented. Aplin (1998) compared the fauna of the upper levels of the two archaeological deposits described here with a list of typical lowland species recorded from elsewhere on the Bird's Head Peninsula. The two lists are very similar, indicating that the Ayamaru Plateau supports a typical lowland rainforest fauna, albeit with a number of significant additions (see Chapter 9).

Mr Elimas Kambuaya, junior-highschool (SMP1) teacher at Mefkajim/Ayamaru, compiled some information on contemporary animal populations on my behalf. Kambuaya interviewed local inhabitants using photographs of various distinctive species (see Appendix 3). His reports suggest that game is now relatively scarce in the immediate vicinity of the major settlements, although forest wallabies (*Dorcopsis muelleri*[2]), Spotted Cuscus (*Spilocuscus maculatus*) and striped possums (*Dactylopsila*

[2] See footnote on page 16.

Figure 2.3. View across Lake Ayamaru, looking towards the southern margin. Courtesy J.Jelsma.

sp.) can still be obtained in the vicinity of the lakes. Tree-kangaroos (*Dendrolagus inustus*) and the Long-beaked Echidna (*Zaglossus bruijnii*) are said to be absent from the immediate area, but still present in the Suwiam/Mapura and Ayawasi areas to the northeast of the lakes. Specimens of Northern Common Cuscus (*Phalanger orientalis*) and Emerald Python (*Morelia viridis*) were consumed in the village of Suwiam/Mapura during the period of archaeological fieldwork, and a juvenile *Phalanger orientalis* and a young cassowary were kept locally as pets.

The Ayamaru Lakes support only a small number of native fish species and most of these are of very small size. These include indigenous species of goby, eels, a catfish-eel, and rainbow fishes. The Dutch introduced some larger fish species, such as the carp (*Cyprinus carpio*) and labyrinth fishes (Anabantidae; Miedema 1986: Table 1) into the lakes in the mid-1930s to provide new sources of animal protein (Brongersma 1958: 51). Crayfish are abundant in the lakes.

---

[2] Unless stated otherwise, the taxonomic nomenclature for vertebrate species mentioned in this thesis follows Flannery (1995a) for mammals; Cogger (2000) for reptiles; Coates (1985) for birds; and G.R. Allen (1996) for fish. Please refer to these references for the authority of species names. Authorities for other groups are given in the text.

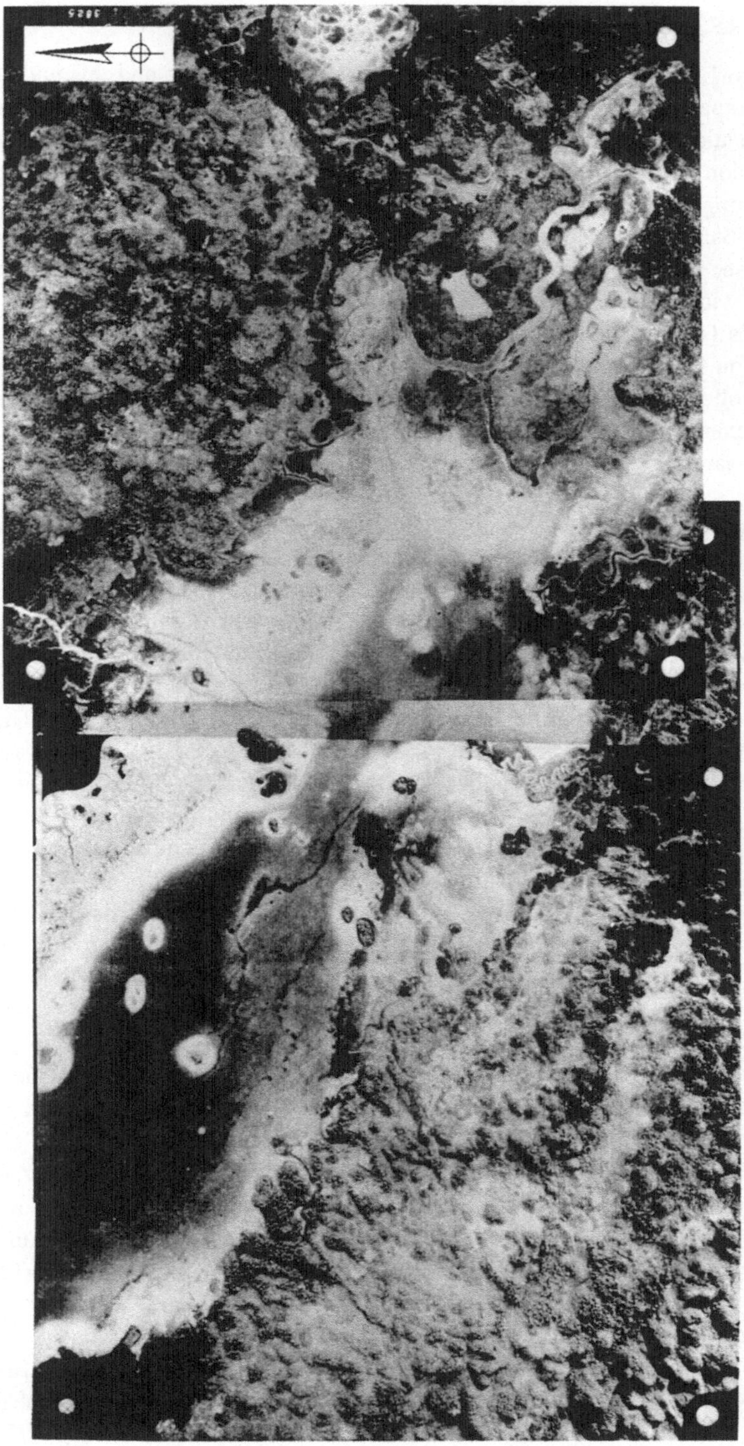

Figure 2.4. Aerial photographs of Lake Ayamaru (see Fig. 2.1). Provided by the Topographical Service in the Netherlands; photo numbers from left to right: 3828, 3827, 3825.

2.2. THE RESEARCH AREA: CULTURAL ENVIRONMENT

The Ayamaru Lakes area has a long history of colonial influence and this undoubtedly had a large impact on settlement patterns and elements of economic activity. The present human population of the area is concentrated around the lakes and in the vicinity of several mission settlements to the east and northeast. However, small villages are scattered throughout the region and the rainforest canopy is interrupted by numerous current and abandoned gardens (see Fig. 2.5). One sealed road runs south of the lakes to the airstrip (see Fig. 2.2) and connects the area with Teminabuan, the largest population centre in the wider region, located to the southwest.

Elmberg's (1955, 1959, 1965, 1968) ethnographic studies provide a rich source of information on the contemporary inhabitants of the Ayamaru Lakes area. These people are known collectively as the Meybrat, but they are divided into various dialectic sub-groups (for example, Mej Brat, Mej Marú, Mej Sawiet), who do not always understand each other's language (Elmberg 1955: 2). Today, Bahasa Indonesia is taught at local schools.

The Meybrat people have been heavily influenced by the presence of Christian missions, and more recently by the Indonesian administration. Many traditional beliefs and practices have probably disappeared, while others have become mixed with newly introduced beliefs. This was illustrated during the fieldwork period in relation to requests to enter particular caves. In one case the owner of a cave brought everyone of the survey team together to seek approval with a Christian prayer before disturbing the cave deposit. In another instance, a request to enter a particular cave was refused. This cave was said to have been the place of exile for people affected by leprosy, and it was feared that the souls of the deceased, together with the disease itself, would still reside there. Elmberg (1959: 74) reported that Meybrat people traditionally held strong beliefs in the power of departed souls to affect the well-being of the living and that they conducted '...a number of ceremonies for the deceased to ensure prosperity'.

The present-day economy consists of horticulture, fishing and occasional hunting. Sago palms are present in limited numbers but the production of sago has not been reported or observed in the area, although sago is available at the markets. The thin and generally poor limestone soils represent a major limitation to agricultural productivity throughout the region (Elmberg 1955: 8; Miedema 1986: 1).

Elmberg (1968: 16) reported that hunting was not of any great significance in Meybrat economy, but further noted that kangaroos and other game were valued for ceremonial exchange. Indeed, greater emphasis was placed on collecting and much of the daily protein was obtained from sago grubs and butterfly larvae, snails, lizards and frogs, with occasional eggs, birds and mice. Fishing was also a relatively minor source of protein (Elmberg 1955: 59; 1968: 16); this is no surprise, considering the small size of the native fish of the lakes. However, all fish foods were considered delicacies (Elmberg 1955: 60).

Traps and spears were the main hunting tools, the spear made of wood with either one (for hunting) or five (for fishing) iron or bamboo points (Elmberg 1955: 14). Bow and arrows were not manufactured locally and, apart from a lucky occasional hit, the weapon was more used for amusement than for actual hunting (Elmberg 1955: 61).

Interestingly, blowpipes made of short sections of bamboo were also used among the northern Meybrat; Elmberg (1959: 78) notes their use 'to a good effect for hunting bats in grottos during periods of food shortage'.

Various organic materials, such as wood, bamboo or bone, were used in the manufacture of tools. Some tools were disposable and only used once (e.g. digging stick), others were carried around for as long as they lasted (Elmberg 1955: 12). Elmberg (1955: 12, 56) did not observe the use of stone tools, but noted that they were referred to in old tales. Steel axes and adzes were already in wide use during the 1950s; the earliest imports reputedly came via the Kei Islands (Elmberg 1955: 56).

Paintings were made on bark walls of sacred houses, rather than on rocks or cave walls. Body painting was applied infrequently, mainly during initiations or periods of mourning, or sometimes as a remedy against illness (Elmberg 1955: 43, 49, 78, 83).

Traditional mortuary practices involved laying the body on a scaffold and collecting the bones when they were clean. The skull, lower jaw and thighbones were distributed amongst the children of the deceased and the remaining bones were put away in a small cave or rock crevice (Elmberg 1955: 69-70).

Contacts with the coast were plentiful; many forefathers and clans were reported to have migrated between the Ayamaru area and coastal settlements (e.g. the village of Konda or Teminabuan), and some were even said to have come from the island of Salawati (Elmberg 1955: 10).

Figure 2.5. Karst formation on the Ayamaru Plateau. Courtesy J.Jelsma.

2.3. FIELD SURVEY

After examination of geological maps and aerial photographs (the latter probably of American origin taken between World War II and 1960), the Ayamaru Lakes region was selected for an archaeological survey because of the presence of a large karst area surrounding the lakes. Moreover, the Ayamaru area is relatively easy to access by light aircraft from Sorong and there are mopeds and an occasional car or truck available for transport.

Archaeological survey of a karst area was favoured for several reasons. Karst areas commonly feature numerous caves or rockshelters (Jennings 1985: 135) and these represent the major source of comparative archaeological information for both Papua New Guinea and Australia (Smith & Sharp 1993). Caves tend to serve as a focal point for human activities, providing good prospects for recovery of stratified archaeological sequences. They are relatively easy to locate, especially with the guidance of local people, even in a forested landscape. In contrast, open sites may be almost impossible to locate under thick forest cover and, unless semi-permanent village sites are present, they generally do not accumulate stratified deposits. Rockshelter or cave deposits may also have a better chance of remaining relatively undisturbed through time (Smith & Sharp 1993), although their physical configuration is by no means absolutely stable (Straus 1979).

The recovery of organic remains was of high priority in the selection of a field area, both for absolute dating and for investigation of prehistoric economic activities. Preservation of organic material is often quite good in limestone caves (Jennings 1985), while destruction of organic remains in open sites is generally very rapid in tropical environments.

Many places in Papua New Guinea show evidence of prehistoric occupation in caves, and ethnological sources (e.g. Gorecki 1979) document the use of caves for mortuary practices and other activities. Although ethnographic sources are limited for the western part of New Guinea, some ethnographies and historical documents report the existence and use of caves in certain areas, including the Meybrat area around the Ayamaru Lakes (Elmberg 1955, 1965; Galis 1964).

Caves were located with the assistance of local villagers, who also provided information about the traditional usage of specific caves. The local schoolteacher Elimas Kambuaya, who established the location of the caves, their owners and their accessibility, considerably facilitated this process. Where permission to enter a cave was granted, notes were made on the cave entrances and chambers, and on the surrounding natural environment. The thickness and contents of any potential archaeological deposit were determined by means of augering with an 'Edelman' drill, generally at or near the entrance of the cave.

The area turned out to have high archaeological potential. Despite difficulties of accessing many areas, we managed to visit eight localities containing 15 cave systems or rockshelters, seven of which were found to contain archaeological material ranging from isolated finds through to substantial deposits. Two site locations lay on the southern margin of the western lake, three on a peninsula, and two on islands in the westernmost basin. Three other sites were located within a maximum distance of 2-3 km of the high water level of the lake complex. Details of the location, form, context and content of the investigated sites are provided in my unpublished PhD thesis (Pasveer 2003).

Based on accessibility and the apparent depth and richness of the archaeological deposit, two sites were selected for test excavation. These were Toé Cave and Kria Cave, separated by a distance of c. 12 km. Test excavations took three weeks in each cave, and were completed in Toé Cave. Unfortunately, plans to continue excavation in Kria Cave were not realised owing to political disturbances in Irian Jaya (as it was then known) during 1996.

2.4. EXCAVATION METHODS

The same method of excavation was employed for both sites. Excavation was carried out in horizontal units of 1 m^2 (where possible) and, with few exceptions (see footnote 3), in vertical spits of 5 cm depth. The use of arbitrary spits reflected a lack of visible stratigraphy and the poor light conditions that became worse as the excavations progressed; a kerosene lamp was used to supplement light in the trenches.

All finds were drawn per spit on scale 1:10, and, in some cases, photographed *in situ*. Stone artefacts, pieces of ochre and charcoal, human bones, and exceptional finds, such as large, intact animal bones or large shells, were recorded three-dimensionally. Horizontal measurements were recorded in centimetres and related back to a common horizontal datum at the southwestern corner of Square 0N0E (0 North, 0 East). Vertical measurements were recorded in relation to an arbitrary fixed datum in each site.

Excavated sediment was searched for archaeological materials by hand. This was carried out for three main reasons: because a local water supply was unavailable for wet-sieving in the case of Kria Cave; because the sediment was rich in clay in both sites and could not be dry-sieved; and because the excavated soil was needed for backfilling and in the case of Toé, could not be lost by wet-sieving in the lake. Nevertheless, the rate of recovery of even small items is thought to be very high. A sediment sample taken from every spit was sieved in a net with 2 mm mesh width to check for smaller elements such as fish bones or small mammal teeth. Two persons carried out the excavation, and two others searched through the excavated soil; in both sites the same persons performed the same tasks.

Samples for radiocarbon dating (bone and/or charcoal) were collected and kept separate without cleaning during excavation. Samples for pollen analyses were taken at every 5 cm depth from the surface to the bottom, using empty film containers as a standard measure of volume.

[3] In Squares 1N1E and 0N0E in Toé Cave, the surface showed a slight depression, hence there is no deposit in the top 5 cm. The first two spits in these squares are 10-cm spits (5-15 and 15-25 cm depth); the first spit in both Square 2N1E and 1S0E is also a 10-cm spit (0-10 cm depth). In the graphs in the following chapters these spits have been evenly split and presented as two 5-cm spits as to give a more consistent representation of distributions of material and material condition. The same applies to one spit in Kria Cave where the sediment between 180 and 190 cm depth in Square 0N0E was excavated as one 10-cm spit, due to a mix up of deposit.

Before refilling the trench in each site, a latex peel was made of the stratigraphical profile in one square, following the method described by Orliac (1975). This peel was subsequently studied under better natural or artificial light conditions. The excavation pits were refilled after excavation to prevent excessive erosion and personal accidents.

Neither of the cave deposits was excavated completely. The excavation in Toé Cave covered roughly 40% of the horizontal entrance area (see Fig. 4.3 in Chapter 4). In Kria Cave only about 12% of the rockshelter was excavated (see Fig. 3.2). For Toé, with excavation squares running from wall to wall, the excavated sample is likely to be representative of the whole cave entrance. However, for Kria, the two test pits may not be fully representative of the spatial variation across the entire rockshelter. Nevertheless, it is assumed that vertical trends shared between each of the two test pits in Kria Cave are representative for the site as a whole.

2.5. ANALYTICAL METHODS

The occupation history of both sites in the Bird's Head, to be discussed in Chapters 3 and 4, was reconstructed from the following sources:

- the visible stratigraphy of the deposits, especially as observed on the latex peels;
- the vertical distribution of the various categories of cultural materials;
- the pattern of weathering, erosion and fragmentation of bone and stone; and
- radiocarbon dates and A/I ratios from amino acid racemisations (AAR) of cassowary eggshell.

By combining the information from these different sources it was possible to assess the integrity of the deposits in each cave and to reconstruct their occupational histories. Some general comments on each of these methods are provided below.

2.5.1. *Stratigraphy*

Although caves are often considered stable containers for prehistoric remains (Glover 1979: 304), the sedimentation and erosion processes are potentially quite complex. Roof fall, sink action and general erosion and re-deposition of sediment are not uncommon in caves, and these processes can literally turn a stratigraphic sequence upside down.

Because of the primitive excavation circumstances and the minimal light conditions in the caves, no obvious primary strata or features were recognised during excavation or during inspection of the *in situ* profiles. However, the latex peel taken from one trench wall in each site provided a means of studying the stratigraphy away from the actual excavation. The depth range of the identified strata is provided in centimetres; these are meant as a rough indication of major boundaries, rather than as a precise vertical division of the deposit; some distortion of the latex peel has undoubtedly occurred.

2.5.2. *Distribution of cultural material*

Following sorting, the various categories of cultural material were counted and/or weighed on an electronic balance. For each square the vertical distribution was plotted for each category, generally using weight as a measure of abundance. Where the excavated area was less than 1 m^2, either because of an encroaching cave wall (Squares 2N1E and 1S0E in Toé) or the presence of large immovable rocks (certain spits in Squares 1N1E and 0N0E in Toé and in Square 1N1E in Kria Cave), an appropriate correction factor was applied to the abundance measure for each category. These corrected values are presented in all weight/volume graphs (except for human bone) in Chapters 3 and 4. For certain classes of remains, such as bone points and pottery sherds, a count of individual specimens is more meaningful than weight. The vertical distributions were compared between adjacent squares in order to find common stratigraphic patterns and to assess the nature of horizontal patterning within the sites.

Detailed analyses of the stone and bone artefacts, and of the fauna can be found in subsequent chapters; human bone and carbonised fruits are described in Appendices 1 and 2, respectively.

2.5.3. *Taphonomy*

Origin of the faunal remains

The bulk of the faunal remains is considered to be of anthropogenic origin. No large predators that might transport their prey into caves are known to occur in New Guinea. The dog-sized, extinct Tasmanian Tiger (*Thylacinus cynocephalus*) has been recorded from Late Pleistocene sites in the Highlands of New Guinea (Flannery 1995a), but never in a lowland context. The recently extinct population of this species in Tasmania is not reported to have accumulated bone from its prey in caves.

Minor quantities of small rodent bones may be prey remains of a raptorial bird or a small mammalian carnivore. However, these bones are scarce and are mainly found in the upper levels of the deposits. They are distinguished from the anthropogenic remains by their typically unfragmented and unburnt state.

Natural deaths in the caves may account for some of the remains. However, the majority of species identified in the sites are not cave dwellers, except for some bats, some larger rodents and the cuscus *Phalanger gymnotis*; remains of these animals were typically burnt and fragmented. The occasional remains of a cave-roosting, insectivorous bat, *Hipposideros diadema*, may be the result of natural deaths.

Preservation of organic and inorganic material

Fluctuations in the quantity of archaeological material through the deposits may reflect changes in the intensity of occupation of the sites, provided that taphonomic factors and the effects of soil conditions and other destructive agents can be accounted for. As Hiscock (1990) has clearly demonstrated, post-depositional alteration of archaeological collections, both organic and inorganic, can be of sufficient magnitude as to lead to spurious interpretations of stratigraphic sequences.

Preservation of animal bone

Animal bone in archaeological deposits can be affected by a variety of physical and biological processes. Inorganic or mineral components can be removed under acidic groundwater conditions (corrosion) or altered by chemical replacement (mineralisation). Organic components are affected by heat and moisture changes, and can, if oxygen is present, be attacked and broken down by microbial activity. Bone can also be subject to direct attrition as a result of scavenging or other mechanical processes (e.g. trampling, water movement).

The preservational state of animal bones was assessed using two different methods: visual inspection for signs of post-depositional alteration, in particular to the unburnt material (see below); and examination of changes in the proportion of unburnt to burnt fractions in each sample.

Three states of burning of bone are recognised, drawing on previous classifications by Ubelaker (1978), Shipman et al. (1984) and Pearce & Luff (1994):

- unburnt to lightly burnt (bone is pale, yellow to light brown and porous to moderately dense in texture). The lightly burnt condition is usually the result of cooking of bone inside a fleshed carcass rather than direct contact with flames; in the Kria and Toé collections staining of unburnt bone blurred the distinction between the unburnt and lightly burnt categories, hence they are grouped under the heading 'unburnt';
- burnt (bone is very dark brown or dark grey to black and very dense). The burnt condition usually results from direct contact with flame or intense heat; and
- calcined (bone is grey to white, very dense and sometimes crazed or warped). All organic material has been lost from calcined bone, resulting in shrinkage and increased brittleness.

For the visual assessment of the post-depositional damage of unburnt bone, each sample was scored on three criteria:

- reticulate surface corrosion, probably caused by plant roots;
- tooth marks indicative of animal scavenging; and
- general corrosion of the bone surface, presumably caused a combination of acidic water solution and microbial breakdown.

For each criterion the sample of unburnt bone from each spit was rated on the following scale:

1: 0% of the total unburnt bone surface is affected
2: less than 10% of the total unburnt bone surface is affected
3: 10-30% of the total unburnt bone surface is affected
4: 30-50% of the total unburnt bone surface is affected
5: more than 50% of the total unburnt bone surface is affected
0: no unburnt bone present (NB: because the categories 'unburnt' and 'lightly burnt' were merged at a later stage, a score of 0 means that no bone was present that was truly unburnt; any lightly burnt bone was excluded from this assessment).

Burning of bone alters its chemical composition, primarily by oxidation of the organic content (Lyman 1994). This tends to make it less susceptible to microbial attack and possibly also to scavenging. However, burning also results in a denser structure that makes burnt bone more resistant to most physical and chemical processes. This is especially true of calcined bone that has lost its entire organic component and has passed through a process of structural change that involves appreciable shrinkage and a significant increase in density. Aplin (1981) has noted that calcined bone is extremely resistant to both chemical and microbial breakdown and is sometimes the only bone material remaining in archaeological sites.

Because of these considerations, the visual criteria for post-depositional alteration were scored on the unburnt bone sample only. However, the total bone sample was used to assess post-depositional alterations by plotting the changing proportions of various burning categories: unburnt, burnt and calcined bone. The expectation is that, under conditions that are destructive to organic remains, unburnt bone will show a proportional decrease over time, as it undergoes preferential destruction by microbial and chemical processes, while burnt bone and calcined bone will proportionally increase in the assemblage. However, this rather simplistic scenario may obviously be complicated by aspects of soil chemistry and by variations in the proportions of unburnt to burnt bone in the original assemblages, the latter possibly reflecting variation in placement of hearths and other aspects of human behaviour. For this reason, it is essential that patterns of burning frequency are not used alone to infer post-depositional alteration, but always in combination with other visual criteria.

Preservation of stone

Hiscock (1985) emphasised that stone artefacts are subject to post-depositional alteration of the kind that affects bone, albeit under the operation of different processes. For this study, the preservation condition of stone artefacts was assessed according to two criteria: the extent of weathering and the degree of fragmentation.

Weathering was scored according to Hiscock's (1985) definitions, but with the addition of one extra stage to a total of five stages. For reasons that will be explained below, only unheated chert was included in the analysis. The stages are as follows:

- stage 0 (no patination): no sign of weathering, the colour and texture of the chert being the same on the outside and inside of the artefact, the latter as observed on freshly broken surfaces;
- stage 1 (lightly patinated): an irregular faint white film of patination on the surface, but with the original colour still visible in many places;
- stage 2 (patinated): a dense white to creamy patination covering the entire surface, the original colour only visible on the more recently broken surfaces; patination is limited to the outer one or two millimetres from the surface and does not extend through the body of the artefact;
- stage 3 (heavily weathered): patination extends through the body of the artefact, showing the typical white colour even on freshly broken surfaces; thinner parts of the artefact, such as along the edges, have become powdery, but most of the artefact still retains a relatively hard and smooth surface;
- stage 4 (very heavily weathered): the specimen is completely white and most of its outer surface has become porous and powdery; the artefact is very easy to break or damage.

As with bone, heating of chert appears to make it more resistant to chemical weathering (contra Purdy & Clark 1979, and Sheppard & Pavlish 1992: 46). Heating of chert effectively reduces its porosity, as the microcrystals fit closer together as intercrystalline water is removed (Purdy 1974). Burnt chert is thus more dense and cemented in its structure. However, excessive heating may also cause the artefact to crack, causing fragmentation. Burnt chert shows various characteristic features including the presence of potlids or their scars, crazing or crenated fracturing of the surface, and irregular fragmentation (Hiscock 1990). The presence of these features was recorded to assess the amount of burnt chert through the deposit.

Fragmentation itself will also promote weathering of a stone artefact assemblage, for as a rule, smaller and thinner items will weather more rapidly than larger items, due to its larger surface area in relation to volume. Two measures are used here for chert artefact fragmentation. The first is simply the average weight of the chert fragments in each spit, determined by dividing the total weight of chert by the total number of fragments (TNF). The second measure focuses on the recognisable chert flakes. Following the method of Hiscock (1985), the Minimum Number of Flakes (MNF) in each spit was calculated by recording the portion of each flake that was present: proximal, distal, marginal, medial, or mid-section, or a combination of these: proximo-marginal indicates that the flake had lost its distal section plus one of the margins. Medial stands for the absence of both proximal and distal ends; mid-section means the absence of both margins; a longitudinal conesplit indicates that the flake was broken exactly through the bulb, basically splitting the flake in half. From these data it is possible to compare the MNF with the total number of fragments (TNF); samples that have suffered more intense fragmentation will have higher values of TNF/MNF.

Density of bone and stone

The density (weight/volume) of animal bone and stone was used previously as an indicator of preservational state for each of these materials in Kria and Toé Caves (Jelsma 1998; Pasveer 1998). Bones and stone artefacts that have been subject to removal of their original inorganic or organic matrix might be expected to have a reduced density. However, fossilisation of bone might be expected to result in increased densities; the same effect may also occur with burning of bone and stone.

To estimate densities, the total sample of animal bone material per spit was weighed on an electronic balance; for bone samples weighing more than 10 g the volume of the material was then determined in a glass measuring jug filled with water. The same procedure was applied to the chert material (Jelsma 1998; Pasveer 1998).

This method may be considered fairly crude, as the processes of preservation and fossilisation of bone as well as stone are complex (Piepenbrink 1989; Sheppard & Pavlish 1992). Moreover, the effects of burning on both stone and bone ideally need to be taken into account; and different species and even different bones from one species have different densities (Binford & Bertram 1977; Nicholson 1996). Similarly, for stone the composition and density will vary according to its original location in the nodule (Purdy 1974: 44); the presence of impurities and the presence of cortex on a particular specimen can also be expected to influence its overall density. A number of specimens (both stone and bone) also had a carbonate crust that was too thick and strongly attached to be removed.

2.5.4. *Dating*

In the New Guinea region absolute dating methods represent the most important means of establishing archaeological chronology. In spite of major changes that have occurred in economies over time, the basic Melanesian stone tool kit has changed little in morphology over long periods of time (Golson 1977b; White 1972: 148), hence relative dating based on typochronology is generally of limited or no use. Items made of more perishable material such as wood or bamboo may have shown greater stylistic variation in both time and space (Peterson 1971); unfortunately, these materials rarely survive in archaeological deposits.

The earliest use of pottery in the region is generally accepted to be around 3500 BP on Kayoa Island in the northern Moluccas (Bellwood et al. 1998: 263-264), despite claims for ceramic manufacture and use back to c. 5000 BP in the Sepik area of Papua New Guinea (Swadling et al. 1989; Gorecki et al. 1991). In other coastal areas of Papua New Guinea, pottery has been in use for at least 2000 years, with more recent usage in the highlands (Watson 1979; White & O'Connell 1982: 173, 176). The first use of pottery in the Bird's Head is thus likely to date to the last 2-3000 years at most. The dates of introduction of various domestic animals (pigs, dogs) might also provide useful chronological markers, but these are not yet well established. Previous studies of faunal remains have shown signs of changes related both to climate and human land use (e.g. Aplin 1981; Hope et al. 1993; Mountain 1993; Flannery 1999; Pasveer & Aplin 1998), but these studies are too few to provide any regional synthesis.

The last decade has seen a revolution in the use of absolute dating methods in archaeology, both in terms of the variety of methods that can now be applied, and in terms of the understanding of the phenomena of reliability and accuracy. However, many of the techniques are still regarded as experimental and, when applied to common problems, they often seem to provide variable results. The current debate on the age of the first Australians clearly illustrates this problem and shows the continuing limitations of absolute dating methods in archaeology (e.g. Allen 1994; Roberts et al. 1994; Allen & Holdaway 1995; Chappell et al. 1996; O'Connell & Allen 1998; Van Huet et al. 1998).

To a lesser extent the same caveats still apply in regard to radiocarbon dates. Many site reports show inconsistencies among sets of radiocarbon dates and between these dates and the stratigraphic and archaeological evidence. In many cases, these problems probably derive from a lack of stratigraphic integrity of the dated samples, but there are also situations in which the dated materials may have been subject to unrecognised diagenesis or sample contamination. These problems are well illustrated in the present study by inconsistencies in radiocarbon dates obtained from both Toé Cave and Kria Cave, but especially from the former site. Nevertheless, close attention to stratigraphic issues, combined with the use of several different dating materials and methods, allowed a satisfactory resolution of these problems.

The archaeological sequences in both Kria and Toé Caves were dated primarily using the Accelerator Mass Spectometry (AMS) radiocarbon method. AMS dates were provided by the National Science Foundation Radiocarbon Facility, University of Arizona (code AA in Table 3.2 and 4.2 in Chapters 3 and 4, respectively); Centre for Isotope Research, Radiocarbon Laboratory, Groningen University (code GrA); Australian Nuclear Science and Technology Organisation (ANSTO – code OZE, OZF and OZG); and The Radiocarbon Accelerator Unit, Research Laboratory for Archaeology and the History of Art, Oxford University (code OxA). The material used for AMS consisted of charcoal, animal bone and cassowary eggshell.

Avian eggshell, especially that of large birds such as the cassowaries (*Casuarius* spp.), represents a particularly good material for archaeological radiocarbon dating for three reasons. Firstly, the dense calcite structure is resistant to diagenetic alteration and therefore preserves indigenous inorganic carbon. Secondly, the extent of amino acid racemisation (AAR) can serve as an independent check on sample age (Pasveer et al. 2002; Clarke forthc.). And thirdly, the presence of eggshell is often directly linked to human activity, especially in cave sites. The eggshell dates were performed on the carbon dioxide evolved by acid digestion of eggshell calcite. Prior to submission of the samples to the accelerator, at least 50% of the eggshell carbonate was removed by stoichiometric addition of 2N HCl to minimize contamination by exchange with younger CO_2. The amino acid data comprises the extent of isoleucine epimerisation determined by the ratio of D-alloisoleucine to L-isoleucine (A/I ratio) in the total acid hydrolysate (Clarke forthc.). The AAR determinations were made either by Gifford Miller of the Center for Geochronological Research, Institute of Arctic and Alpine Research, and Department of Geological Sciences, University of Colorado, or by Simon Clarke, School of Geosciences, University of Wollongong.

Unless stated otherwise, all radiocarbon ages in this thesis are presented as uncalibrated ages. The calibrated dates in Tables 3.2 and 4.2 (Chapters 3 and 4 respectively) are calculated with CALIB 4.3 (Stuiver & Reimer 1993) using the dataset of Stuiver et al. (1998). No correction for the Southern Hemisphere offset was applied because the sites are located only just below the Equator. The correction factor for this offset is still under study (McCormac et al. 1998).

Besides AMS, an attempt was made to date the lower levels of Toé Cave using Electron Spin Resonance (ESR), performed on a tooth of the New Guinean rat *Mallomys* sp. cf. *M. aroaensis*. ESR was performed by the Quaternary Dating Research Centre at the Australian National University in Canberra.

3. Occupation History of Kria Cave

3.1. DESCRIPTION OF THE CAVE

Kria Cave lies about 2 km northeast of the eastern basin of the Ayamaru Lakes, and c. 3 km east of the village of Suwiam/Mapura (see Fig. 2.1 in Chapter 2). The location of the cave, as determined by means of a Global Positioning System, is 01°15'711" S and 132°20'064" E, and about 325 m above mean sea level.

The cave has several entrances leading to a deeper cavern system within the low hill in which it is located. Two entrances, adjacent to each other at the eastern side of the hill, face east-northeast and east-southeast, while a third, located on the opposite side of the hill, faces west. All have the shape of a rockshelter, with passages leading back into the actual cave. From any one of the entrances the daylight coming in through the other two is visible. The cave itself has several larger chambers, some of which are sparsely sunlit by small holes in the ceiling or walls. Large numbers of small, insectivorous bats inhabit these inner chambers, and thick layers of bat guano are present, in some places covered by thin layers of deposited calcite.

In the main gallery behind the entrances large rocks have fallen down from the ceiling and these support many *in situ* stalactites and stalagmites. Considering the high humidity in the cave, and the presence of calcite layers enclosing recent guano deposits, it is likely that the speleothem formation in the cave is continuing at a rapid pace today, with major formations formed in a matter of decades or centuries (S. Hadiwisastra pers. comm.). The walls of the cave are covered in moss and there is a constant dripping of water from the ceiling. Despite some effort to locate possible rock paintings or carvings, none were detected.

The western entrance is the largest, measuring c. 9 m wide and c. 5 m high. Coring revealed a deposit of some 75 cm depth, contained between large blocks of roof fall. However, with the possible exception of some charcoal, there was no obvious cultural material. The eastern entrance facing east-southeast measures c. 3 m wide and c. 3 m high and contains c. 30 cm of deposit, with some shell fragments and charcoal recovered by coring. The entrance facing east-northeast is c. 6 m wide, c. 2.5 m deep and c. 3 m high (Fig. 3.1 and 3.2). With 105 cm of archaeologically productive sediment indicated by initial coring (on excavation this turned out to be over 2 m), it was obvious that this entrance was the main focus for human occupation of the cave. It was therefore selected for test excavation. The interior of the cave itself did not contain any obvious archaeological deposit.

Figure 3.1. Kria Cave. Top: rockshelter where the excavation took place. Bottom: early stage of the excavation; note the passage into the cave in the back wall (compare Fig. 3.2).

3.2. EXCAVATION IN KRIA CAVE

The excavation grid was set up over most of the surface of the rockshelter, directed at 65° from north. Two test pits were excavated, one in a more or less central position (Square 1N1E; see Fig. 3.1 and 3.2), and one near the back wall of the rockshelter (Square 0N0E). These locations represented the optimal conditions for excavation: the northern part of the surface is covered with large, immovable rocks; the southern wall is formed by a very low-hanging, tilted ceiling and the regular dripping of rainwater has produced a pothole in the surface; and in front of the shelter, the tree-covered surface slopes away steeply and has probably been subjected to considerable erosion.

All depth measurements were initially related to a datum point at the back of the rockshelter, approximately 31 cm above the surface of the excavated squares. The surface rises slightly towards the entrance, but is roughly horizontal at the location of the two test pits. All measurements of depth were eventually converted into centimetres below the surface of each square. Horizontal measurements were taken from the zero-point established at the southwest corner of Square 0N0E.

The excavation reached a depth of approximately 195 cm below the surface in Square 0N0E and 170 cm in Square 1N1E. Unfortunately, because of time constraints, the bedrock was not reached in either square. Coring and probing in the bottom of Square 0N0E indicated that a further 40 cm of archaeological deposit (producing burnt bones and stone flakes) remained above bedrock or massive roof fall. Probing in Square 1N1E indicated a further 20 cm of remaining deposit. This indicates that the bedrock lies at depths 235 cm and 190 cm respectively in Squares 0N0E and 1N1E, i.e. deeper at the rear of the shelter and rising towards the front.

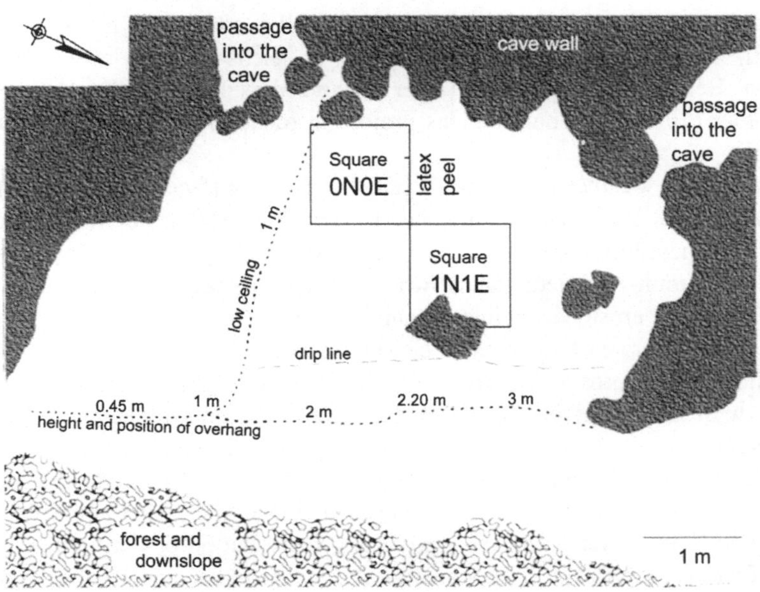

Figure 3.2. Plan of the excavated rockshelter of Kria Cave with the location of the excavation squares.

3.3. OCCUPATION HISTORY OF KRIA CAVE

Methods used to reconstruct the occupation history of Kria Cave are described in Chapter 2.

3.3.1. *Stratigraphy*

The latex peel represents a part of the northern wall in Square 0N0E (see Fig. 3.2): a central section of 30 cm width, covering the total length of the profile from the surface down to the lowest excavated levels (see Fig. 3.3). Held in full natural light some different textures and shades of colour are visible.

The upper 10-12 cm are light grey (10YR 7/1; layer 1 of Fig. 3.3). This is underlain on the western side (towards the back wall of the cave) by a thin, slightly darker layer (10YR 6/1; layer 2) a few centimetres thick. Layers 1 and 2 give way to a light brownish grey layer (10YR 6/2; layer 3a) that continues until some 25 cm below the surface. A projecting piece of human bone at 15 cm depth corresponds with a concentration of human bones in this square (see below). Underlying layer 2 is a small layer or feature (this is not clear from the narrow peel), which reaches a maximum thickness of 7 cm on the western side (layer 3b). This layer is light brownish grey in colour (10YR 6/2) and contains pieces of (human?) bone and a speck of charcoal. In depth it corresponds to the concentration of human bones; however, the projecting bone fragment at 15 cm is clearly located outside of this possible feature.

Below 25 cm depth, the soil grades into a darker grey layer (10YR 5/1; layer 4), which contains some specks of charcoal and shell. This layer extends until about the 40-45 cm level. A small specimen of a *Melanoides* shell is attached to the west side of the latex peel, corresponding to a concentration of shells at this level noted during excavation. Between 40 and 80 cm (layer 5) the sediment returns to a lighter greyish brown colour (10YR 5/2), but contains projecting rocks, fragments of bone and small stone flakes.

The soil has a distinctly granular texture from the surface down to 80 cm but is smooth below that level. The significance of this change is not currently understood. However one possibility suggested by Jan Delvigne (Dept. of Physical Geography, University of Groningen, Netherlands) who examined the peel in March 1997, is that it reflects changes in erosional regime within the site catchment, with the recent periods seeing an increased rate of soil erosion causing mobilisation of soil aggregates or 'peds' from within the soil profile. Alternatively, the contrast might reflect post-depositional alteration with the smoother texture of the lower sediments reflecting a leaching of calcium carbonate with consequent loss of any original pedal structure.

Between 80 and 105 cm (layer 6) below the surface the soil has a smooth texture and a darker greyish brown colour (10YR 4/2), presumably due to a higher humic content than the overlying layer. Layer 6 has some projecting rocks but no obvious cultural remains. From c. 105 to 125 cm (layer 7) the soil is similar in texture but somewhat paler than layer 6, and it also contains projecting bone fragments. Below this depth and down to 160 cm (layer 8) the sediment is brown (10YR 4/3), smooth in texture, and contains a large projecting stone flake, fragments of burnt bone and a piece of red ochre. The lowest

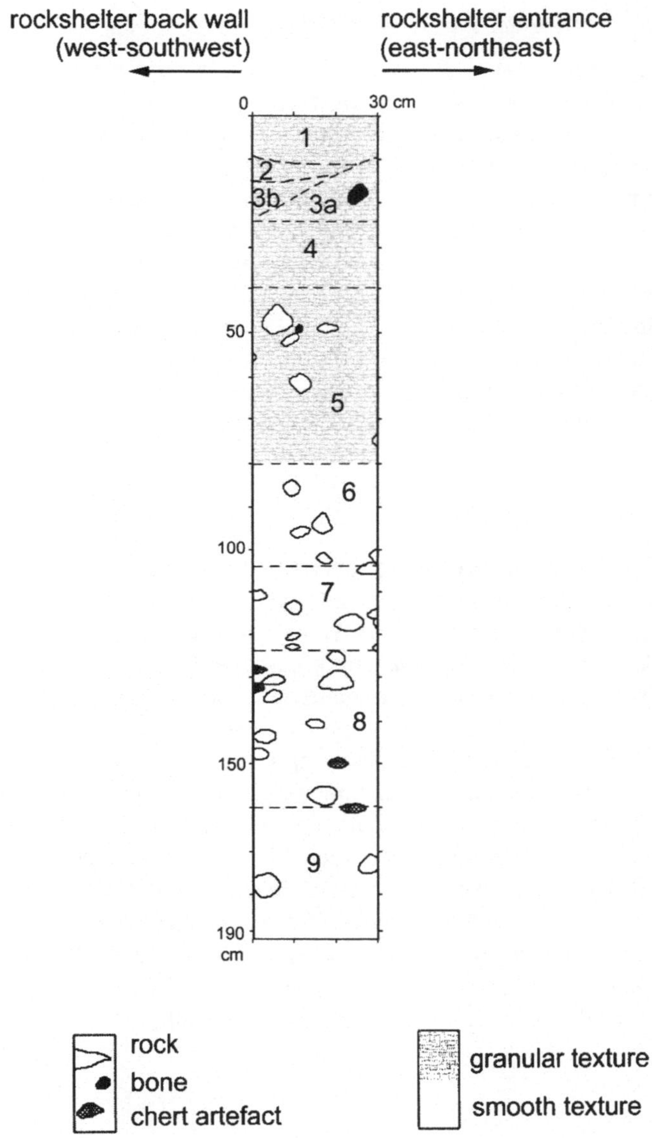

Figure 3.3. Schematic image of the latex peel from Square 0N0E in Kria Cave. Layer 1: light grey; 2: grey; 3a: light brownish grey with animal and human bone; 3b: light brownish grey with animal (and human?) bone and charcoal; 4: grey with charcoal and shell; 5: greyish brown with bone and stone flakes; 6: dark greyish brown; 7: dark greyish brown with bone; 8: brown with stone flakes, burnt bone and red ochre; 9: very dark greyish brown with burnt bone.

part of the peel (layer 9, 160-190 cm) is a dark greyish brown colour (10YR 3/2) with projecting fragments of burnt bone. Coring of the unexcavated deposit in this square encountered bedrock or massive roof fall at 235 cm.

3.3.2. *Vertical distribution of cultural material*

Among the cultural remains discovered in Kria Cave are mammal bones, including human, bones of birds and avian eggshell fragments, fish and reptile bones, crab claws, molluscan shell fragments, ochre, charcoal, and stone and bone artefacts. Pottery is represented by only four sherds from the top levels.

The pattern of occurrence of each of these major categories through the sequence in each square is shown in Figures. 3.4 to 3.12 (note that the weights of all categories except human bone have been corrected for excavated area in Square 1N1E; see also Chapter 2, section 2.5.2. Values for 10-cm spits have been evenly split and presented as two 5-cm spits; see note 3 on page 21). Animal bone and stone flakes comprise the bulk of the total excavated material, with lesser quantities of shell and other cultural remains. Further analyses of chert and bone artefacts, fauna, human, and botanic remains are presented in the following chapters and in the appendices.

Faunal remains

The total weight of animal bone is just over 12.5 kilograms (12,691 g). The bulk of the bones (estimated to be around 80%) come from a single species of forest wallaby, the Brown Dorcopsis (*Dorcopsis muelleri*), known locally as *Djief.* Other species present include tree-kangaroos (*Dendrolagus* spp.), cuscuses (*Spilocuscus* spp. and *Phalanger* spp.), echidna (*Zaglossus bruijnii*), ringtails (*Pseudochirulus* spp. and *Pseudochirops* spp.) and other possums (*Dactylopsila* spp. and *Petaurus* sp.), bandicoots (*Echymipera* spp.), rodents, birds (including *Casuarius* spp.), snakes, lizards, and both megachiropteran and microchiropteran bats. Pig is represented only in one of the uppermost levels.

The quantity of animal bones in Square 1N1E (Fig. 3.4) shows three clear peaks, one between 30 and 55 cm and another between 105 and 125 cm and a third between 135 and 160 cm. These are flanked by zones of generally lower abundance above 30 cm, between 55 and 105 cm depth and below 160 cm. The animal bone material in Square 0N0E shows an even stronger bimodality, with peak values at 60-80 cm and again below 120 cm, and low quantities above 60 cm, between 90 and 120 cm and at 150-155 cm depth. A minor peak occurs at 110-115 cm in Square 0N0E.

Human bone

A total of 789 g of human bone was recovered. The skeletal remains are concentrated in the upper levels of Square 0N0E (with minor quantities in Square 1N1E), the bulk between 15 and 25 cm depth (Fig. 3.5). The remains belong to two individuals, one adult and a juvenile, both probably male. While the burials appear to be primary, the bodies are incomplete (partly due to the limited area excavated) and the bones did not lie in anatomical position. Further details of these remains are discussed by Bulbeck in Appendix 1.

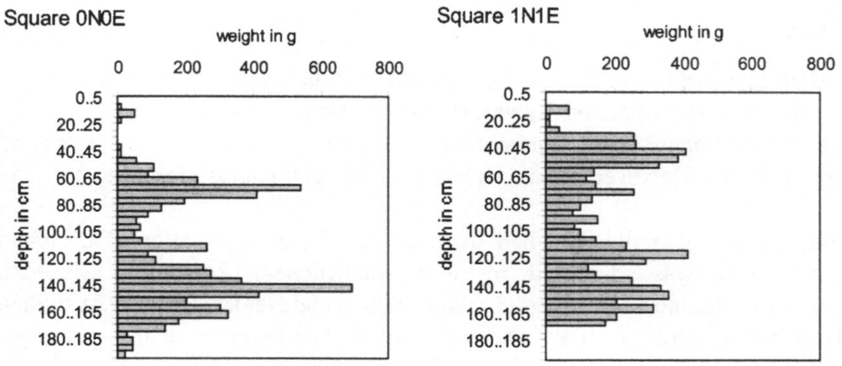

Figure 3.4. Kria Cave, distribution of animal bone through the deposit. The weights shown in Square 1N1E (in this and all following graphs, except human bone) are corrected for excavated area.

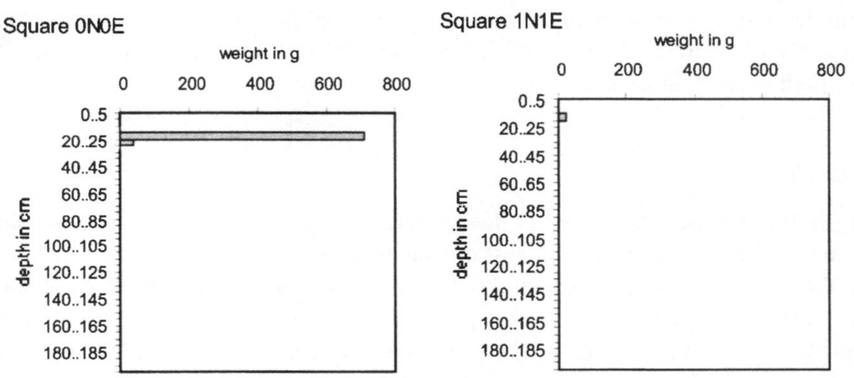

Figure 3.5. Kria Cave, distribution of human remains through the deposit.

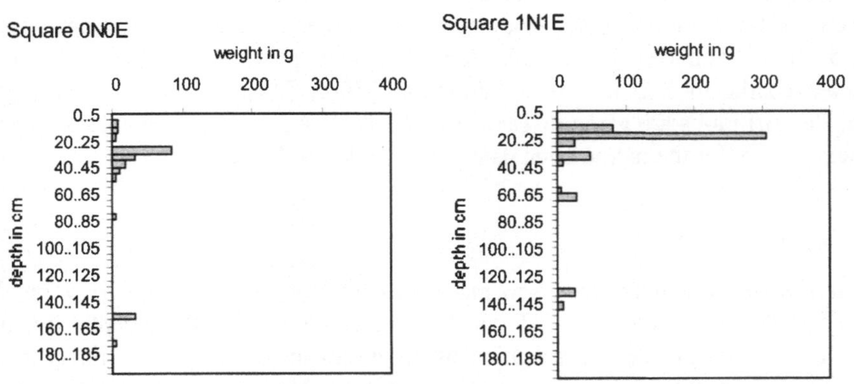

Figure 3.6. Kria Cave, distribution of molluscan shell through the deposit. See Chapter 7 for a more detailed analysis.

Molluscan shell

Molluscan shell is present in relatively large quantities in the upper levels of the deposit in Square 1N1E, but only occasionally and in low quantities below c. 35 cm (Fig. 3.6). The same is true for Square 0N0E, where the bulk of shell fragments occurs above c. 55 cm, peaking at 25 cm. However, in this square a small peak is also present at 150-155 cm depth.

The total quantity of molluscan shell fragments is 712 g. A selection of the more complete specimens was identified by Rob Moolenbeek (Zoological Museum, Amsterdam) and Katherine Szabó (Dept. of Archaeology and Natural History, Australian National University), either to genus or species level. The majority of the specimens belong to a fresh water species of *Melanoides*. Other species are: *Ellobium* sp.; *Bellamya* sp. cf. *B. kowiayiensis* (Brazier 1886); *Batissa* sp.; *Polymesoda viridescens* (Tapparone Canefri 1883); *Hyridella* sp. cf. *H. misoolensis* (Schepman, 1897); *Anadara granosa* (Linnaeus 1758); and *Saccostrea cucullata* (Born 1778). The latter two and *Ellobium* sp. are mangrove species.

Various land snails are also represented including species of *Cyclotus*, *Pupinella*, *Palaina* and *Papuina* along with unidentified Helicarionidae. A more detailed analysis of molluscan shell is given in Chapter 7.

Avian eggshell

Avian eggshell fragments are present in most levels through the deposit, but with a total weight of only 57 g. In general the highest quantities are found in the bottom half of the deposit, although there is a large peak at 65-70 cm in Square 0N0E, and one at 35-40 cm in Square 1N1E (see Fig. 3.7). The lowest quantities are found in Square 0N0E between c. 80 and 105 cm, and in Square 1N1E between c. 65 and 80 cm. The vertical distribution corresponds reasonably well with that of animal bone, with two peaks evident in both squares.

The majority of the fragments come from eggs of a cassowary (*Casuarius* sp.), although there are also some much thinner fragments that probably represent one or more species of brush turkeys (Megapodidae). Both classes of eggshell fragments have been used in an amino acid racemisation (AAR) project by the University of Wollongong (Clarke forthc.) in collaboration with Gifford Miller of the University of Colorado. The A/I ratios are given in Table 3.2 and Table 4.2 and have been used as an independent control for the radiocarbon dates obtained for both sites.

Botanic remains

Among the charcoal fragments are a few pieces of carbonised fruits. These were found only in Square 0N0E, between 110-115 cm, 140-145 cm and 165-170 cm depth. It was not possible to obtain precise identification of these remains, although it is likely that they derive from a member of the family Elaeocarpaceae. The remains are described by Kathleen McConnell in Appendix 2.

Fragments of *Celtis* seeds and banana seeds were found in Toé Cave, but not in Kria.

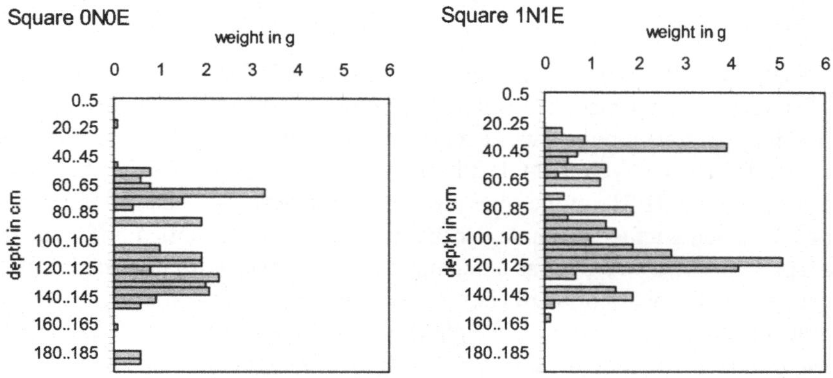

Figure 3.7. Kria Cave, distribution of avian eggshell through the deposit.

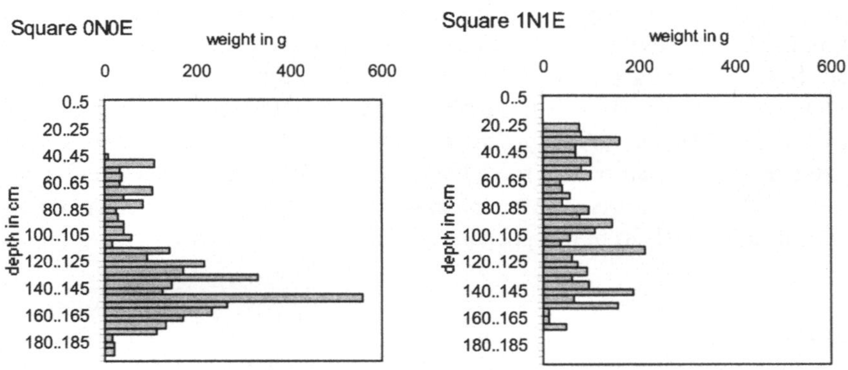

Figure 3.8. Kria Cave, distribution of chert (in weight) through the deposit.

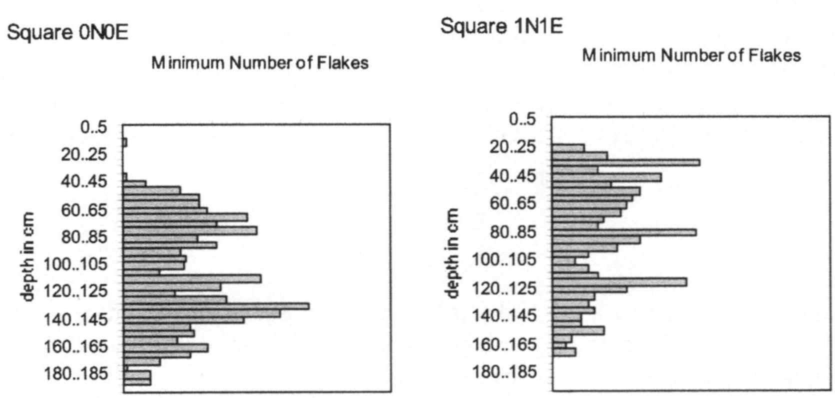

Figure 3.9. Kria Cave, distribution of Minimum Number of chert Flakes (MNF) through the deposit.

Stone artefacts

A detailed analysis of the stone collection of both Kria and Toé Cave is presented in Chapter 5. These results supersede the preliminary information given by Pasveer (1998).

Almost 7.5 kg of stone artefacts was excavated from Kria Cave. The majority of the artefacts are made of a relatively light, brittle chert, which comprises 75.4% (or 5570 g) of the total weight, and 91.8% of the total number of artefacts (1402 out of 1528). The remainder of the stone artefact assemblage consists of limestone, silicified tuff, quartzite (including a pebble which was used as a hammerstone) and metamorphic material. Here I will concentrate on the distribution of chert artefacts within the deposit.

The vertical distribution of chert by weight (Fig. 3.8) shows that most of the material was deposited in the lower section of Square 0N0E, with a sharp fall in quantities above 110 cm depth. The distribution of chert in Square 1N1E is more stable through time, with slight peaks between 20 and 60 cm, between 60 and 110 cm, and again from 110 cm down until the lowest excavated levels.

When the quantity of stone artefacts is expressed as the Minimum Number of Flakes (MNF: Hiscock 1985; see Chapter 2, p.26), a somewhat different pattern emerges (see Fig. 3.9). In particular, MNF values are almost equal above and below 110 cm in Square 0N0E and there is an extra peak between 110 and 125 cm depth. The MNF values in Square 1N1E show a clear decline from the top levels to the bottom, but three minor peaks are visible albeit with slightly shifted boundaries, the lowermost one focussed between 80 and 105 cm, rather than spread over 60 to 110 cm depth. These differences are presumably due to a decrease in the average weight of the stone artefacts through time; a more detailed analysis of this phenomenon is given in Chapter 5.

Bone artefacts

A total of 92 bone points, or fragments thereof, were found in Kria Cave. Most of these are between 2 and 4 cm long, with occasional examples of 1.5 to over 5 cm. Twenty-eight of these are worked at both ends (i.e. bipoints), while the remainder have single points and could either be original unipoints, or damaged and/or incomplete bipoints.

With one possible exception (a modified bone fragment) bone artefacts do not occur above 45 cm in Square 0N0E or above 35 cm in Square 1N1E (see Fig. 3.10). They otherwise occur throughout the deposit, with no obvious systematic variation in abundance through either square. A detailed analysis of the bone artefacts is given in Chapter 6.

Pottery

Pottery is confined to the upper levels and was found only in Square 0N0E. Three sherds were found in the upper 10 cm and a fourth at a depth of 20-25 cm of this square. Three of the four pottery sherds are undecorated; one has incised cross lines (see Fig. 3.11). One undecorated piece comes from the rim of a pot, the others are body sherds. On macroscopic inspection, the pottery has a temper that includes quartz grains and a feldspar-like mineral (G. Summerhayes pers. comm.). This indicates a source area for this material where volcanic or metamorphic weathering products are available. The nearest potential source for such materials is found in the mountain ranges of Tamrau and Arfak to the north and east of the lakes, and on the islands of Waigeo, Batanta and Salawati to the west of the Bird's Head (Ratman 1998: figure 7).

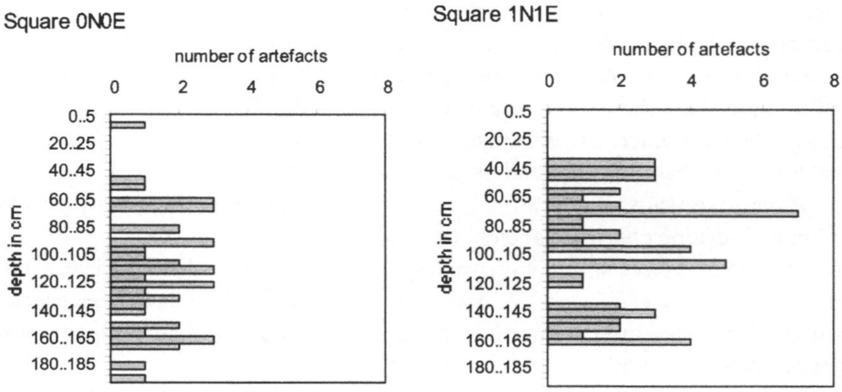

Figure 3.10. Kria Cave, distribution of bone artefacts through the deposit.

Figure 3.11. The four pottery sherds from Kria Cave. Specimens 628, 630, 635, 671.

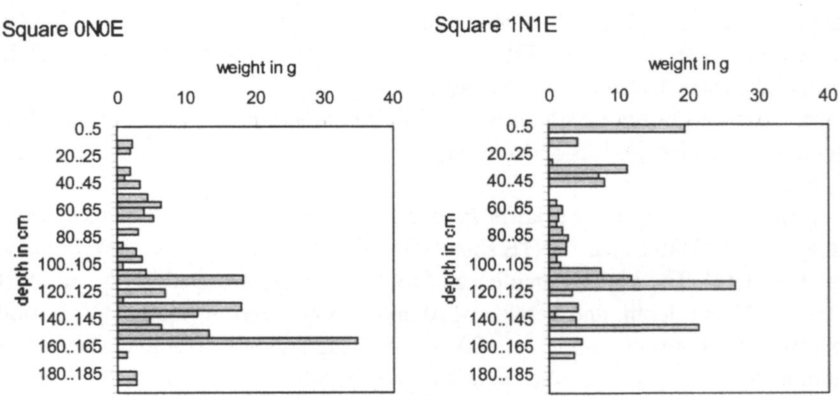

Figure 3.12. Kria Cave, distribution of ochre through the deposit.

Ochre

Fragments of red and yellow ochre are found in both squares in virtually all levels (Fig. 3.12), though especially in the lower half of the deposit of Square 0N0E. In Square 1N1E some peaks in material are seen between c. 30-45 cm as well as between c. 105-120 cm. The total weight is 312 g.

The ochre is highly variable in colour and degree of purity. Some fragments are an intense red or rich orange colour and have a fine, clayey texture. However, many more are paler and mottled, and contain various inclusions, including quartz sand grains and pockets of unpigmented clay. Some fragments have a hard, iron-rich veneer over part of their surface. The presence of embedded quartz sand grains in many of the ochre fragments indicates that this material is not derived from weathering of the parent limestone of the cave walls. Rather, it is presumably derived from a weathering profile, either within a sandstone unit or an ancient lateritic soil.

None of the ochre pieces shows any grinding facets or other evidence of use, such as was found in Toé Cave (see Chapter 4). Alternatively, the ochre may have been prepared in a way that might leave little archaeological trace, for example pounded up and mixed with water rather than ground or used as a 'crayon'.

When Elmberg visited the Ayamaru area in the 1950s, the Meybrat people, who inhabit the region, were not particularly focussed on the use of colour in decoration. Although rock art is known from the southern edge of the MacCluer Gulf (e.g. Röder 1939), Elmberg (1955: 43, 49, 78, 83) mentions that paintings were only seen on bark walls of sacred houses. Body painting was only applied on particular occasions or events, but was generally rare. However, it is possible, if not likely considering the occurrence of ochre throughout the deposit in both Kria and Toé Caves, that the use of pigments for decoration was more important in the past.

3.3.3. *Weathering, corrosion and fragmentation of bone and stone*

Post-depositional damage of bone

Root damage on bone is relatively minor in most samples from Kria Cave. It is most noticeable in remains from Square 1N1E at the front of the rockshelter and especially above 40 cm and between 95 and 130 cm. Square 0N0E shows less root damage overall, but some was observed in samples from around 40-45 cm and in the zone between 110-130 cm with a peak also at 155-160 cm depth (see Fig. 3.13; for methods used see Chapter 2, p. 24-25).

Signs of animal scavenging are fairly randomly distributed through each sequence, although the remains from Square 1N1E show somewhat higher numbers of tooth marks overall (see Fig. 3.14). The highest incidence of tooth marks is observed in Square 0N0E between 25 and 40 cm depth, and between 110 and 135 cm, and in Square 1N1E tooth marks are generally increasing above c. 100 cm. Although some specimens show more or less random tooth marks over the bone surface, in other specimens the damage is concentrated around muscle insertion features, i.e. those places where muscle or connective tissue is likely to have survived human consumption. From the size and nature of the tooth marks, the most likely scavenger of bones in the Kria site was a marsupial carnivore, perhaps *Dasyurus albopunctatus* or *Myoictis melas*; the latter species is reported to scavenge around villages within Papua New Guinea (Flannery 1995a).

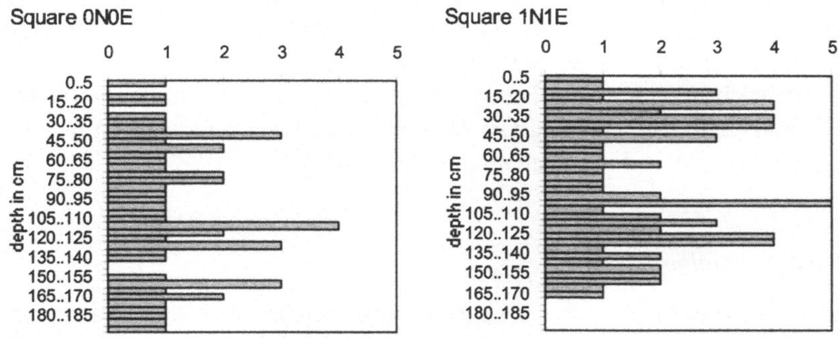

Figure 3.13. Kria Cave, degree of root damage on unburnt bone through the deposit. 0 = no unburnt bone present; 1 = 0% root damage; 2 = <10% root damage; 3 = 10-30% root damage; 4 = 30-50% root damage; 5 = >50% root damage. See Chapter 2 for methods used.

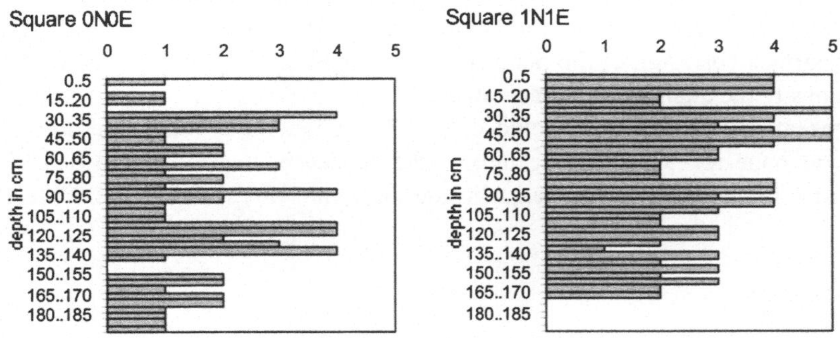

Figure 3.14. Kria Cave, abundance of tooth marks on unburnt bone through the deposit. 0 = no unburnt bone present; 1 = 0% tooth marks; 2 = <10% tooth marks; 3 = 10-30% tooth marks; 4 = 30-50% tooth marks; 5 = >50% tooth marks. See Chapter 2 for methods used.

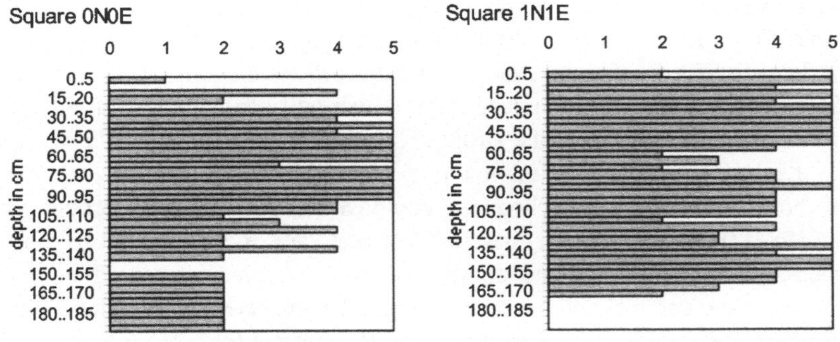

Figure 3.15. Kria Cave, degree of corrosion of unburnt bone through the deposit. 0 = no unburnt bone present; 1 = 0% corrosion; 2 = <10% corrosion; 3 = 10-30% corrosion; 4 = 30-50% corrosion; 5 = > 50% corrosion. See Chapter 2 for methods used.

The degree of general corrosion of the unburnt bones shows more marked variation through the deposit (see Fig. 3.15). In Square 0N0E, the unburnt bone is quite badly corroded down to 105 cm, but shows little sign of corrosion in the lower levels. In Square 1N1E corrosion is overall quite strong, but particularly so in the upper 60 cm, and below c. 130 cm. Bone from 60 to 75 cm is visibly less corroded in this square. The situation in Square 1N1E is thus less clear than in Square 0N0E; this may be related to its position closer to the drip line of the cave. The generally better condition of the unburnt bone material in the lower levels (especially in Square 0N0E) might be related to more rapid burial under a slightly higher sedimentation rate or perhaps to a more stable long-term temperature and moisture regime (see below).

Proportion of burnt vs. unburnt bone

The proportion of unburnt bone shows a slight overall increase with depth in Square 1N1E, interrupted by lower values between 60-80 cm (Fig 3.16; see Chapter 2, p.24-25 for definitions of burning categories). Square 0N0E, in contrast, shows a series of major fluctuations in the proportions of unburnt to burnt bone, with peak values of unburnt bone at 20-35 cm, 70-80 cm, 115-135 cm, and 160 cm. These fluctuations are clearly not random, but they show no obvious correlation with peaks in the visible criteria of scavenging or surface corrosion of the bone. They also show no clear correlation to the major fluctuations in the quantity of animal bone.

For the present, it is therefore concluded that the variations in the proportion of unburnt to burnt bone are a primary feature of the sequence, most likely caused by changing patterns of human behaviour (e.g. shifting location of fireplaces across the rear of the shelter).

Bone density

Bone densities fluctuate somewhat through the sequence in Square 1N1E, but there is no overall trend (Fig. 3.17). In Square 0N0E, bone density appears to increase slightly towards the base of the deposit. Bone from the uppermost levels of this square varies greatly in density. While it seems reasonable to relate this trend to the lesser degree of corrosion of bones from the lower levels of this square, the increase in density might also be explained, at least in part, by the presence of calcium carbonate encrustation on some of the bones. This crust was most pronounced in the lower half of the deposit in Square 0N0E and was in some cases too thick and strongly attached to be removed. The impact of this factor on the density estimates is unknown.

Burning of bone alters both its chemical composition and physical properties (Shipman et al. 1984), making it denser than unburnt bone and more resistant to most destructive processes. Variation in bone density between samples, measured on the combined unburnt, burnt and calcined remains, might be expected to correlate with variation in burning composition. However, there is little evidence for any association in either square (compare Figures 3.16 and 3.17), possibly because of the crude nature of the density estimate. The low values in the upper levels of Square 1N1E may be related

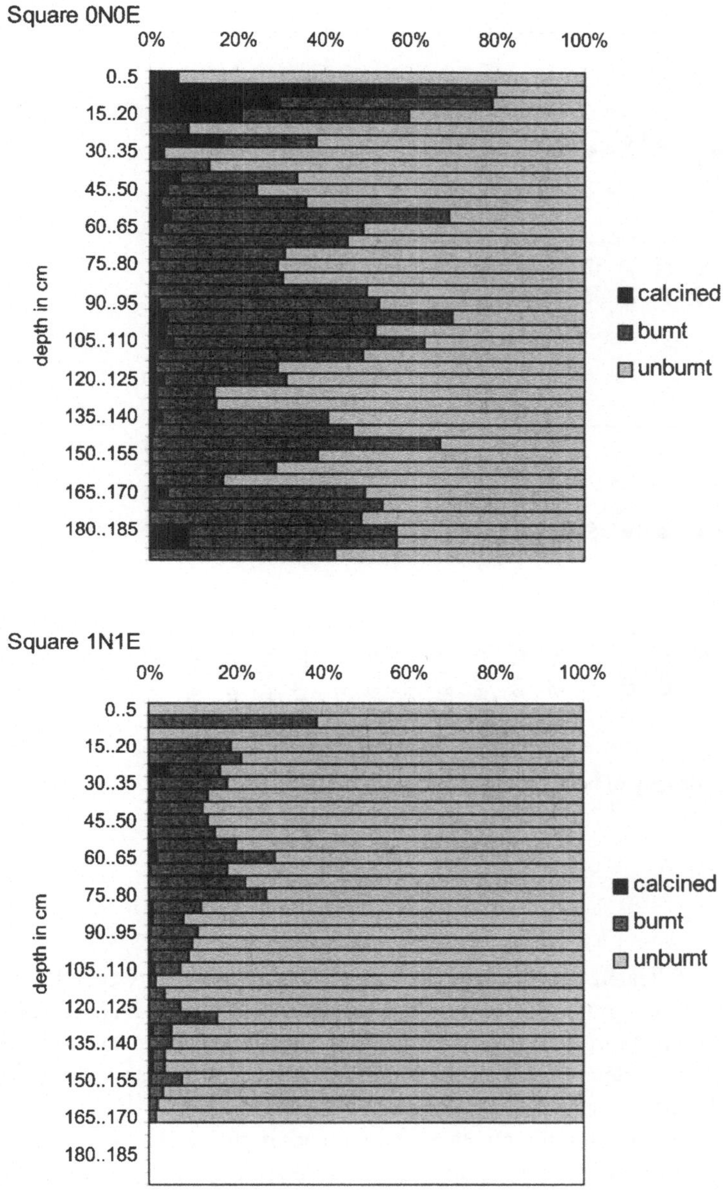

Figure 3.16. Kria Cave, proportions of burning categories of animal bone through the deposit.

to the observed very poor state of preservation of bone in these levels. However, it might also reflect the presence in these samples of pig bone, or unrecognised fragments of human bone, which are generally less dense than other animal bone.

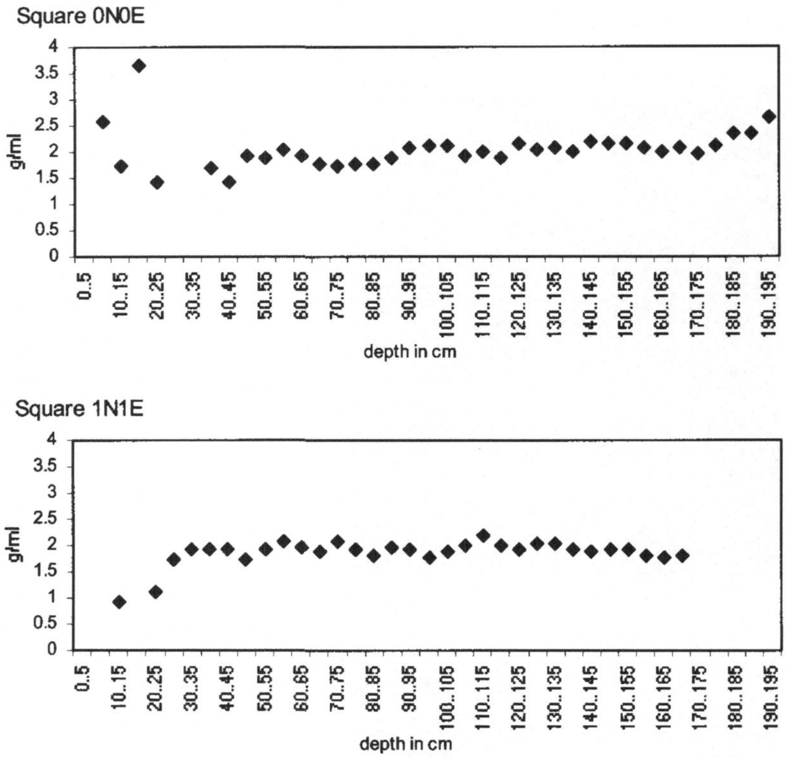

Figure 3.17. Kria Cave, density of bone through the deposit.

Heating of chert

The vertical pattern of heated chert fragments varies considerably between the two squares. In Square 0N0E the proportion of burning fluctuates markedly, with at least five major peaks rising to values in the range 40% to 95%, mostly separated by 'lows' of 15% or less (Fig. 3.18). In Square 1N1E the proportion of burning is also variable but remains lower overall, with only two levels exceeding 40% burnt.

The higher incidence of burning of chert at the rear of the shelter parallels the burning pattern already described for animal bone. More detailed comparison of the burning patterns for animal bone and chert also reveals a relatively good fit in the location of major peaks for Square 0N0E, but a less convincing association for Square 1N1E. As indicated earlier, this most likely reflects the preferential location of fireplaces close to the back wall of the cave, especially during the earlier phases of occupation. The association further suggests that chert was not heated in a deliberate attempt to improve its flaking properties, both rather that bone and stone alike were unintentionally burnt in, under or around fireplaces.

Square 0N0E

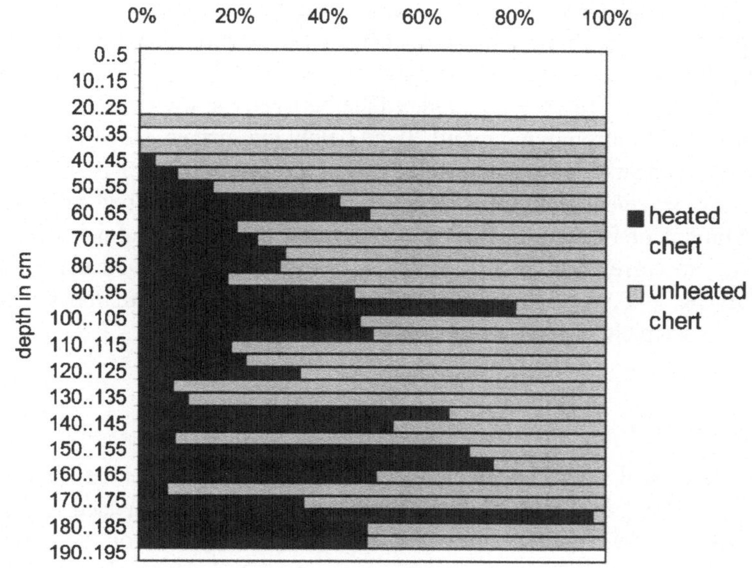

Square 1N1E

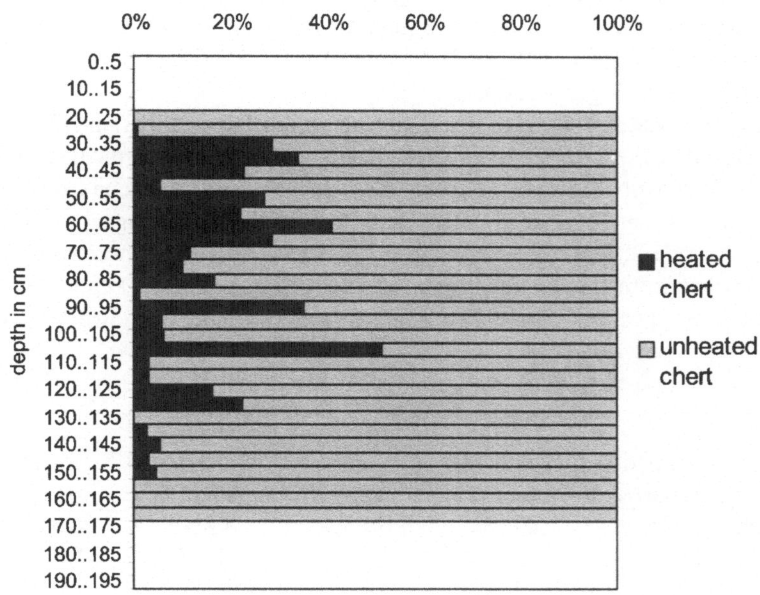

Figure 3.18. Kria Cave, proportions of unheated vs. heated chert through the deposit (based on weight).

Fragmentation of chert

The fluctuations in the fragmentation index in both squares appear random, although there are some overall trends visible. As expressed by the total number of fragments divided by the Minimum Number of Flakes (Fig. 3.19, bottom), fragmentation decreases in Square 0N0E from the bottom of the site through to 45 cm, with little or no chert above that level. In Square 1N1E the index falls between the lowermost levels and 100 cm, then rises slightly through to around 20 cm below the surface. However, the average fragment weight appears higher in the lower half of the deposit (Fig. 3.19, top).

Plotting the fragmentation (as determined by the total number of fragments / Minimum Number of Flakes), against the proportion of heated fragments through the deposit shows no correlation in Square 1N1E (R^2 = 0.005), but a clear correlation in Square 0N0E (R^2 = 0.35). This is, again, most likely the result of the much higher proportions of heated chert in the latter square.

Weathering of chert

In most levels heated chert is visibly less weathered than unheated chert. This is illustrated very clearly in Table 3.1 that shows the distribution of heated and unheated material across the five weathering stages (for full descriptions of these stages see Chapter 2). The negative association between heating and weathering is remarkably strong (Chi-square 365.53777, d.f. 4, $p < 0.00000$). All fragments with weathering stage 0 (no patination) are heated. At the other end of the spectrum, 95% of the most heavily weathered fragments (stage 4) show no signs of heating. Specimens with weathering stage 4 do not occur above 105 cm in Square 0N0E or 85 cm in Square 1N1E. This shows, contra e.g. Purdy & Clark (1979), and Sheppard & Pavlish (1992), that heating of chert slows the weathering process down.

Excluding the heated fragments from consideration, the distribution of weathering classes for unheated chert artefacts shows interesting patterns through the deposit in Kria (Fig 3.20). In Square 0N0E lightly patinated chert is confined to the zone from 100 to 45 cm. Below 100 cm there is a steady increase in the proportion of stage 3 weathering at the expense of stage 2. Stage 4 weathering is observed in a zone between 115 and 155 cm, but is absent from the lowest levels. Square 1N1E, produced almost no stage 1 material, and shows peaks of stage 4 specimens in both the upper and lower levels.

Table 3.1. Kria Cave, weathering stages of heated and unheated chert.

Kria Cave	no patination stage 0	lightly patinated stage 1	patinated stage 2	heavily weathered stage 3	very heavily weathered stage 4
heated chert	100%	80.5%	39.3%	8.8%	5.6%
	N = 20	N = 99	N = 242	N = 50	N = 4
unheated	0%	19.5%	60.7%	91.2%	94.4%
chert	N = 0	N = 24	N = 374	N = 519	N = 67

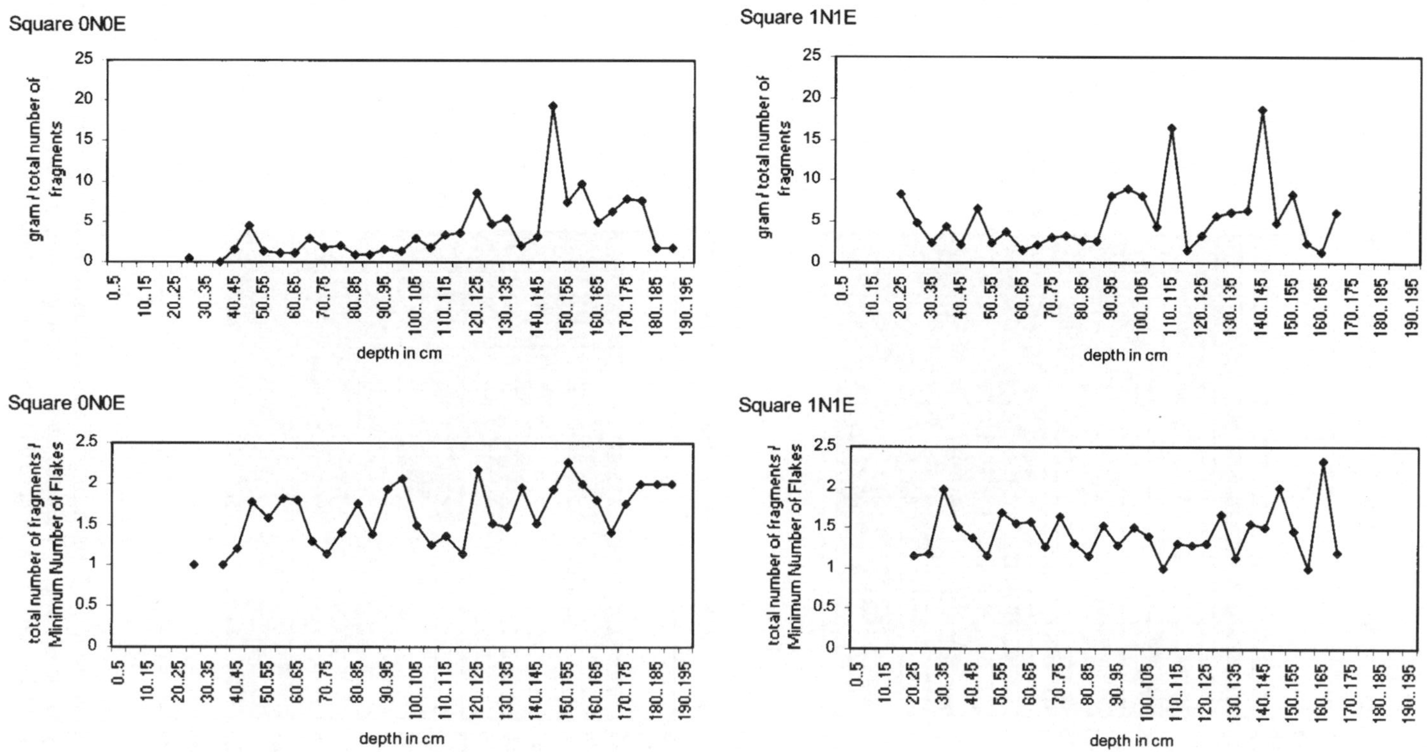

Figure 3.19. Kria Cave, fragmentation of chert through the deposit. Top left and right: total weight / total number of fragments; bottom left and right: total number of fragments / Minimum Number of Flakes.

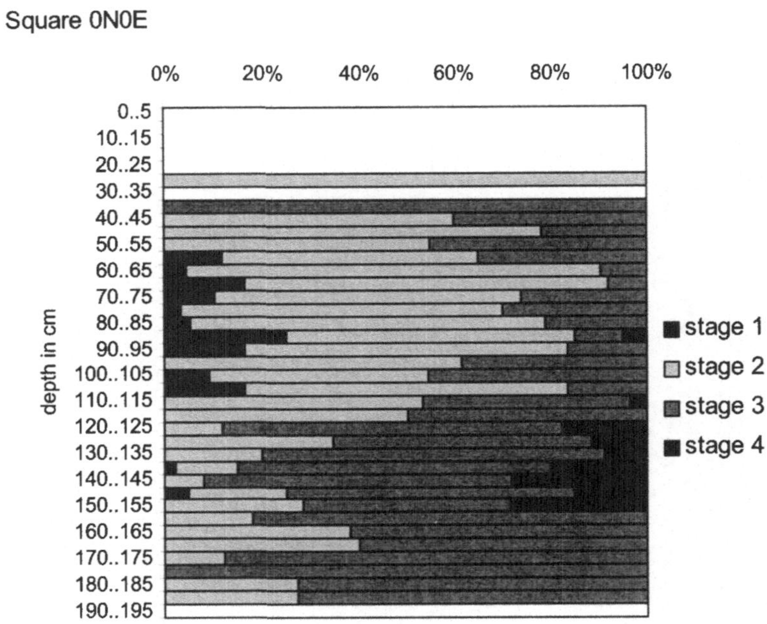

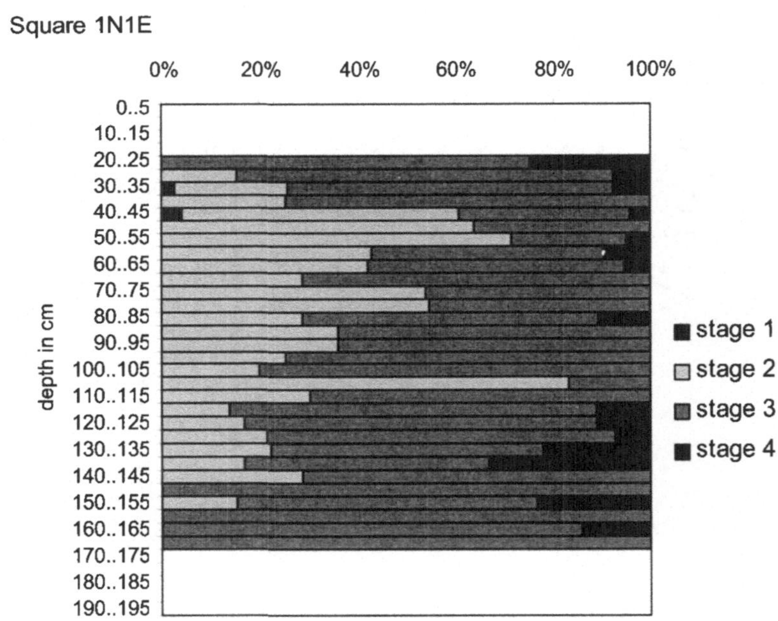

Figure 3.20. Kria Cave, degree of weathering of unheated chert through the deposit. Stage 1: lightly patinated; stage 2: patinated; stage 3: heavily weathered; stage 4: very heavily weathered. No artefacts with stage 0 (no patination) are present.

As a general rule, all weathering processes tend to proceed most rapidly under fluctuating environmental conditions. Square 1N1E is located towards the entrance of the rockshelter, close to the drip line, and is under the influence of incoming sun and rain; its contents have probably been subjected to considerable fluctuations in temperature and moisture. Square 0N0E is positioned against the back wall of the cave and is, in that respect, less influenced by the effects of the weather. It is thus not surprising that chert recovered from Square 1N1E is more weathered than that from Square 0N0E.

The absence of stage 4 materials below 155 cm in Square 0N0E appears anomalous. However, it is possible that soil moisture conditions below this level have been more stable as a result of the sloping bedrock that would tend to direct incoming moisture back towards the rear of the rockshelter. Further support for this notion is found in the contrasting preservation of bone in the two squares. As described earlier, Square 0N0E indeed shows remarkably little corrosion of unburnt bone in the lower levels and the bone is often protected by a carbonate crust. Square 1N1E shows an overall higher degree of corrosion of unburnt bone, corresponding with the generally higher weathering of chert in this square, and there is no indication that bone from the deeper levels has been subject to any reduced weathering activity. Despite these observations, it must be emphasised that bone and stone weather in very different ways and at potentially quite different rates, hence one must be careful not to over-interpret any similarities between the patterns of bone and stone preservation.

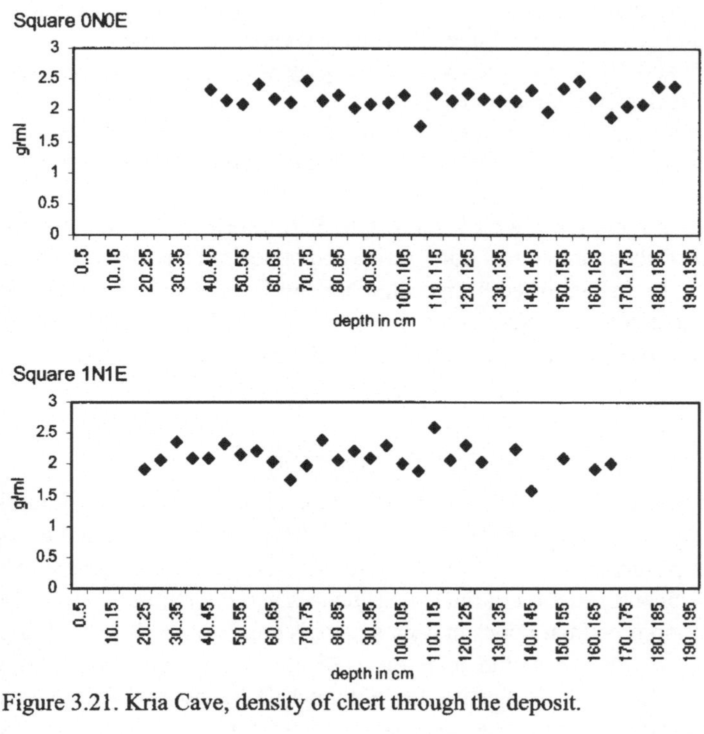

Figure 3.21. Kria Cave, density of chert through the deposit.

Chert density

In both squares the density of chert shows some variation but with no overall trend (Fig. 3.21). This variation may partly reflect other factors such as variable representation of cortex, or of remains of calcium carbonate crust. The density values are not significantly correlated with the proportions of heated or weathered chert.

3.3.4. *Identification of stratigraphic units*

The upper levels of the sequence, from the surface down to c. 40 cm in Square 0N0E, and to c. 20 cm in Square 1N1E, are clearly distinguished from the underlying levels. Animal bone, eggshell, and stone and bone artefacts are all virtually absent above 40 cm, while molluscan shell and human bones reach peak abundance in these levels. Pig bones and pottery are also only found in the upper spits.

The distributions of cultural material found in Square 0N0E correspond well with the projecting finds and sedimentary layers observed in the latex peel from this square (Fig. 3.3). Combining the evidence from both sources, five stratigraphic 'Units' are distinguished in this square:

- Unit I, from the surface down to about 40-45 cm (layer 1-4), a light grey to grey-brown granular sediment, characterised by a low density of stone artefacts and animal bones, an absence of bone artefacts, but the presence of pottery, pig and human bones and a concentration of shell fragments (predominantly *Melanoides* sp.);
- Unit II, between 40-45 and 80 cm (layer 5), a light grey-brown granular sediment with abundant animal bone, stone and bone artefacts, and eggshell fragments;
- Unit III, between 80 and 105 cm (layer 6), a brown-grey smooth-textured sediment with only minor quantities of cultural remains;
- Unit IV, between 105 and 125 cm (layer 7), a slightly paler brown-grey, smooth-textured sediment with higher quantities of animal bone, stone and bone artefacts, and eggshell; and
- Unit V, between 125 cm and the lowest levels (layer 8-9), a yellow-brown, smooth-textured sediment with abundant animal bone, stone and bone artefacts, and eggshell.

Units I and II are clearly distinguished from each other and from underlying levels on sedimentological criteria and/or on their cultural content. Unit III, IV and V differ predominantly in quantity of material and in the intensity of weathering of the soil.

3.3.5. *Correlation between squares*

Square 1N1E shows the same basic changes in the nature and abundance of cultural materials as 0N0E, albeit with some differences. Some of these are clearly related to its position closer to the drip line; however, others are not so readily explained. To resolve these contrasts, some vertical translation of the two sequences was attempted.

The vertical distribution of animal bones in each square is compared in Fig. 3.22 (top left). Apart from a near perfect fit in the uppermost 25 cm, the patterns show little obvious correspondence. However, by shifting the Square 1N1E sequence downwards by 20 cm (Fig. 3.22, top right), the two sequences attain a very good alignment of peaks and lowpoints. A similar result is obtained with the vertical distributions of stone artefacts (Fig. 3.22, bottom left and right). This congruent result suggests that, at least for the greater parts of their history, the cave sediments accumulated with a downward slope of some 10 degrees away from the entrance. This is of course consistent with the observed difference in depth to bedrock in the two squares, being 235 cm in Square 0N0E and 190 cm in Square 1N1E. Infilling at the rear of the cave, thereby levelling the present rockshelter floor in the area of the two squares, evidently occurred only recently during the accumulation of the deposit, perhaps through deposition of the shell-rich layer 4 (see Fig. 3.23) in Square 0N0E (which corresponds to the observed peak in shell quantities at 15-20 cm in Square 1N1E). The very good correspondence in the vertical distribution of both animal bones and stone flakes between the two squares, once allowance is made for the slight dip in the bulk of the sediments below layer 4 in Square 0N0E, further indicates that the stratigraphic sequences in both squares are of high integrity and thus have not been subject to significant local disturbance through human activities or through natural processes such as water erosion or slumping.

3.3.6. *Horizontal distribution*

Even though only 2 m^2 of the rockshelter has been excavated, there are clear indications that the horizontal distribution of cultural materials is not random within the site. In the upper levels, the four pottery sherds were found exclusively in Square 0N0E, at the rear of the cave. Human bones occurred in both squares, but predominantly in Square 0N0E and adjacent squares; the bones themselves did not lie in anatomical position. The significance of this distribution is not clear (see Appendix 1).

In contrast, the densest concentration of molluscan shell was found in the upper levels of Square 1N1E, with lower densities spread through a greater vertical range in Square 0N0E. As noted above, the dense concentration corresponds to layer 4 of the latex peel (see Fig. 3.23), and appears to represent an expanded 'infill' unit that effectively levelled the cave floor.

Although the total quantity of animal bone recovered from each square is very similar, there are remarkable differences in the proportion of burnt and unburnt material (Fig. 3.16). In Square 1N1E, 70 to 95 % of the material is unburnt, and the proportion of burnt material remains fairly constant through the deposit. In Square 0N0E, the proportion of burnt bone fluctuates between 20 and 70 %. Square 0N0E also contains relatively more calcined bone than Square 1N1E. As demonstrated earlier, this pattern is unlikely to reflect the differential preservation of unburnt vs. burnt bone between the squares or layers. A reasonable explanation in terms of human behaviour concerns the shifting location of the fireplace: across the back wall of the rockshelter (Square 0N0E), and away from the drip line (Square 1N1E).

Stone artefacts were found in larger quantities in Square 0N0E, especially in the lower half, and the highest peak in animal bone also occurs in this square. Bone points and eggshell, on the other hand, occur in slightly larger quantities in Square 1N1E. These contrasts might reflect a functional difference between a 'discard' area along the

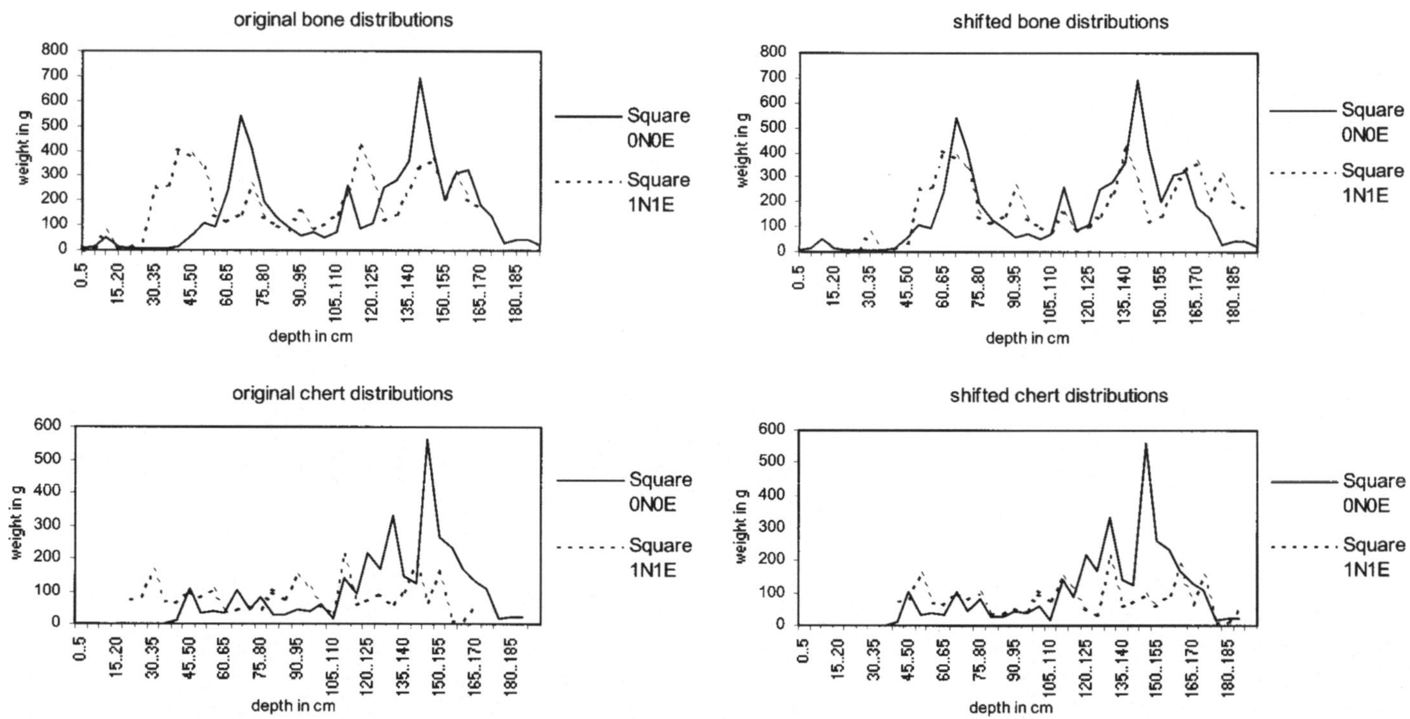

Figure 3.22. Kria Cave, correlation between squares of distributions of animal bone (top left and right) and chert (bottom left and right) through the deposit.

back wall of the rockshelter, receiving larger quantities of stone flaking debris and animal bones, and a 'living' area nearer to the drip line, receiving smaller items of food refuse (e.g. fragments of eggshell) and small, functional items such as bone artefacts. However, it is also possible that the concentration of material along the back wall is at least partly a result of gradual, down-slope movement of discarded materials under the force of gravity.

3.3.7. *Dating*

Ten samples of animal bone, cassowary eggshell or charcoal from Kria Cave were submitted for AMS dating (see Table 3.2 on page 56). Seven of these produced internally consistent results, falling into almost perfect succession (Fig. 3.23).

Two charcoal samples (GrA 6312 and 6313) gave anomalously young dates. These dates were both obtained from small fragments of charcoal that were not found in association with any special cultural or stratigraphic feature. Because other charcoal samples (GrA 6310 and OxA 6043) from both higher and lower stratigraphic levels gave seemingly reliable results, the anomalous dates are probably not the result of contamination, but more likely reflect the introduction of small fragments of charcoal from higher levels during excavation under dark, cramped conditions.

A sample of animal bone from 175-180 cm in Square 0N0E (processed in Oxford) did not yield enough organic material for dating. However, the alkaline fraction (Mook & Streurman 1983) of a sample of animal bone fragments from 5 cm higher in the sequence (GrA 9103) did yield a plausible date. Because of the inferred slope in the deposit, the radiocarbon date of 6900 BP at 155-160 cm in Square 1N1E is equivalent to the 175-180 cm level in Square 0N0E. The A/I ratios of the two dated fragments of cassowary eggshell are consistent with their age as determined by AMS (S.J. Clarke pers. comm.).

Examination of the age-depth relations indicates that the deposit accumulated at a fairly even rate through Unit V to Unit II, but with a decline in sedimentation in Unit I. This decline points either to a dramatic reduction in sedimentation rate or to a depositional hiatus, followed by a resumption of sediment accumulation.

These alternatives can be modelled statistically from the age-depth relations. Under a scenario of continuous but variable deposition (Fig. 3.24 top), the age-depth relationship can be expressed in the form of $y = 1981.1\mathrm{Ln}(x) - 348.5$ (R^2=0.99). This model implies a steadily declining sedimentation rate for each successive unit and an extrapolated date of initial occupation and sediment accumulation of 7258 years BP. Under the alternative scenario of interrupted deposition (Fig. 3.24 bottom), the age-depth relationship through Units V to II gives an extremely strong association under a simple linear model as expressed by $y = 92.303x + 3582.7$ (R^2=0.97), with a sedimentation rate of 0.054 cm per year. This is only minimally improved even under a complex polynomial model (sedimentation rate in Unit I is a maximum of 0.022 cm per year). The initial occupation of the site is then revised to 7875 years BP. Under either model, initial occupation of the site probably dates to some time between 7000 and 8000 uncalibrated years BP.

A pig bone in Square 1N1E, dated to a minimum[1] of 1840 ± 40 BP, was positioned only 10 cm above the top of Unit II. This indicates that any depositional hiatus between Units I and II might have lasted for a maximum of about 2000 years. The sharp boundary between Units I and II, and the fact that Unit I is qualitatively as well as quantitatively different from all underlying units, both support the notion that the site was abandoned for some time, followed by a more recent, less intensive and possibly different use of the site. It is possible that some of the top levels of Unit II were removed before Unit I was deposited on top, although there is no direct evidence that this occurred.

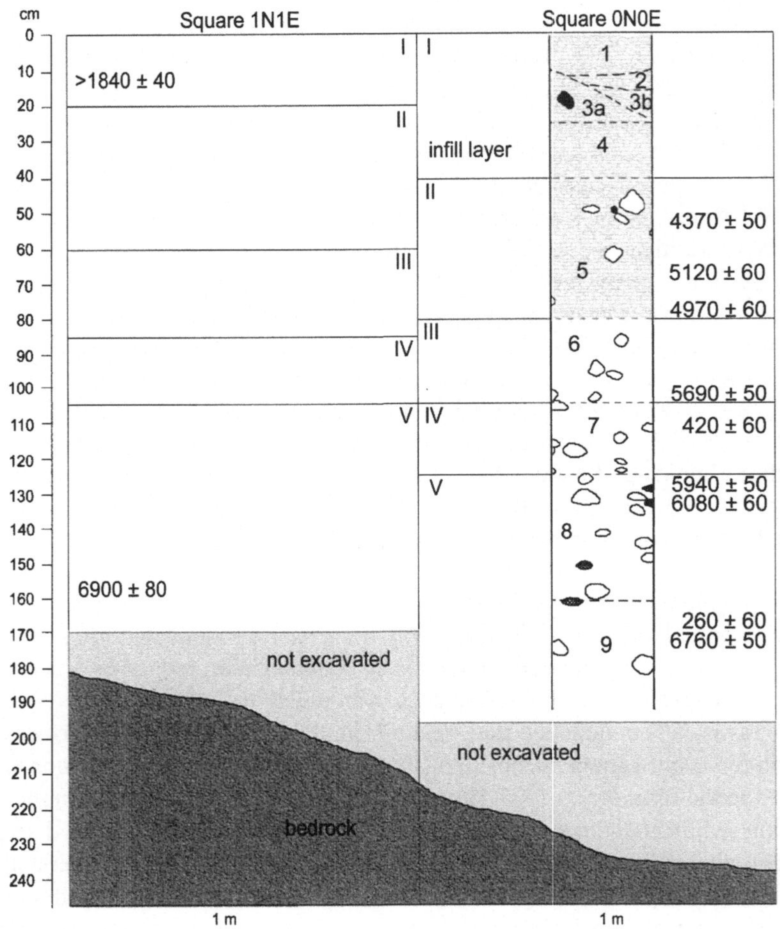

Figure 3.23. Kria Cave, schematic image of the stratigraphy, with position of uncalibrated radiocarbon dates (in years BP) and layers as identified on the latex peel (compare Fig. 3.3).

[1] This sample has a C/N ratio of 4.2, 2.8% collagen, and 4.1% carbon content. The high value for C/N suggests that the sample is probably contaminated and the date should be regarded as a minimum age (Q. Hua, ANSTO, pers. comm.).

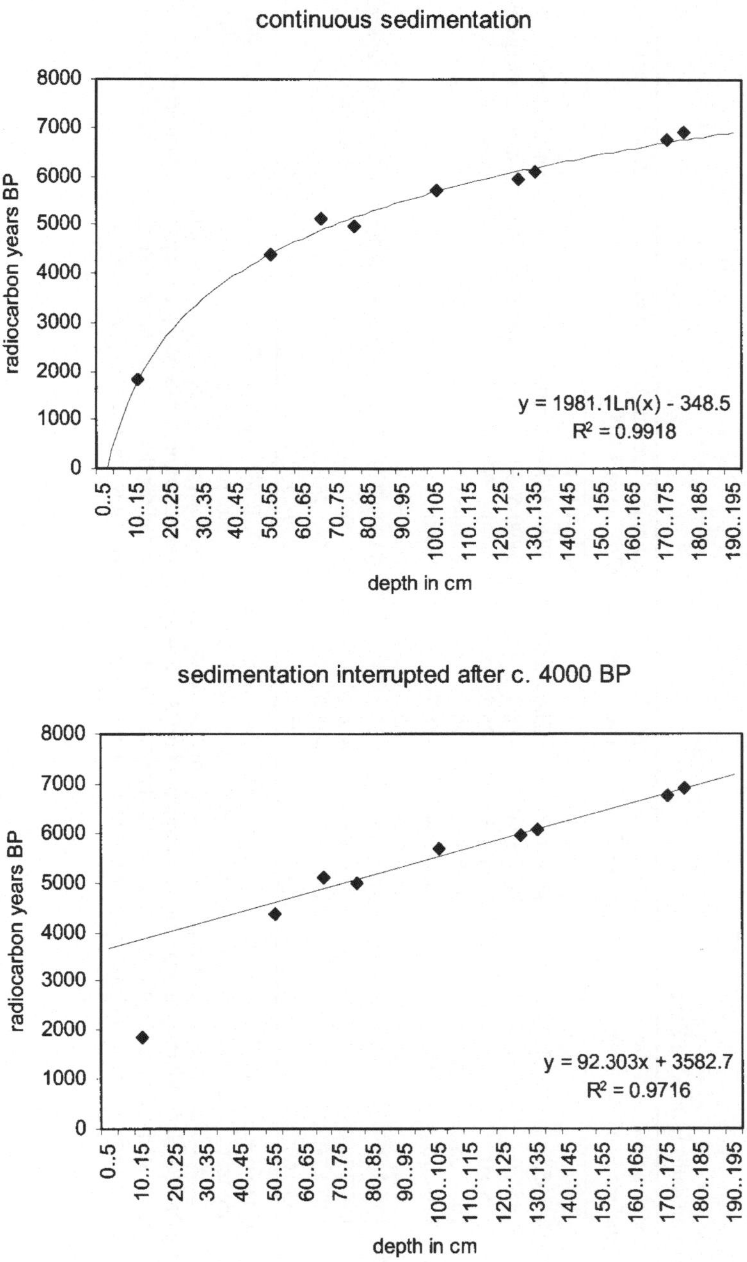

Figure 3.24. Kria Cave, sedimentation rate as determined by age-depth relation. Top: continuous sedimentation; bottom: assumed hiatus between Unit I and II.

Table 3.2. Radiocarbon dates from Kria Cave.

laboratory ref. no.	depth below surface (cm)	Unit and level	radiocarbon years BP	$\delta^{13}C$ (‰, PDB)	calibrated age range (2σ) in years BP[4]	A/I ratios	dated material
OZE 542	10-15[1]	I, 6	>1840 ± 40[3]	-23.3	>1870-1630		pig bone
GrA 9100	50-55	II, 3	4370 ± 50	-25.8	5210-4840		animal bone[5]
OZF 245	65-70	II, 6	5120 ± 60	-12.2	5990-5730	0.165 ± 0.001	*Casuarius* eggshell
GrA 6310	75-80	II, 8	4970 ± 60	-25.7	5890-5590		charcoal
GrA 9101	100-105	III, 5	5690 ± 50	-26.5	6640-6320		animal bone[5]
GrA 6312	110-115	IV, 2	420 ± 60	-25.7	540-315		charcoal
GrA 9102	125-130	V, 1	5940 ± 50	-25.8	6890-6660		animal bone[5]
OZF 246	130-135	V, 2	6080 ± 60	-13.5	7180-6750	0.225 ± 0.009	*Casuarius* eggshell
GrA 6313	165-170	V, 9	260 ± 60	-26.2	470-2		charcoal
GrA 9103	170-175	V, 10	6760 ± 50	-25.0	7680-7510		animal bone[5]
OxA 6043	155-160[2]	V, 11	6900 ± 80	-26.4	7930-7590		charcoal

Notes: The data in this table supersede those given in Pasveer (2003). All samples, except OZE 542 and OxA 6043 were taken from Square 0N0E. For interpretation see text. [1]The top levels in Unit I run virtually horizontal, hence OZE 542 from Square 1N1E is equivalent to 10-15 cm depth in Square 0N0E. [2]OxA 6043 comes from a depth of 155-160 cm Square 1N1E; due to a slope in the deposit, this sample is equivalent to 175-180 cm in Square 0N0E. [3]See footnote on page 54. [4]Calibrated with CALIB 4.3 (Stuiver & Reimer 1993) using the dataset of Stuiver et al. (1998). As the sites are located only just below the Equator, the dates have not been corrected for Southern Hemisphere offset. [5]These dates are based on the alkaline fraction of a bulk bone sample, consisting predominantly of remains of a wallaby (*Dorcopsis muelleri*).

3.4. CONCLUSION

In Kria Cave, five occupation units were identified on the basis of combined sedimentological and cultural criteria. Unit I, forming the upper 20 to 40 cm of the deposit, is distinguished from the underlying units by the presence of pottery and pig bone, albeit both in small quantities, and the presence of discrete concentrations of freshwater gastropod shells (*Melanoides* sp.) and human bones. This unit produced relatively small quantities of animal bone, very few stone artefacts and no bone tools. The sharp stratigraphic and cultural boundary between Units I and II suggests either a period of abandonment of the site from around 4000 BP and/or that some of Unit II may have been removed before Unit I was deposited. The presence of pottery and pig remains suggest a maximum age for Unit I of c. 3000 BP (Bellwood et al. 1998). Based on a radiocarbon date of >1840 BP on pig bone from this unit, the maximum length of any hiatus between Units I and II is estimated to be around 2000 years.

Units II-V produced considerably larger quantities of animal bone, predominantly derived from the forest wallaby *Dorcopsis muelleri* and other lowland rainforest species, and larger numbers of stone and bone artefacts, with the main difference between these units being the quantity of various materials. No pottery or pig remains and no discrete cultural features such as middens or burials were found in these units. Units I and II are distinguished from the underlying units III to V by a change from granular to smooth sediment. The significance of this textural change is not known. Unit II produced especially large quantities of stone and animal bone. Units III and IV both produced smaller quantities of bone and stone, but with no trend in reduction in other cultural materials. Unit III is especially humic and shows evidence of more pronounced weathering of bone and stone; it is possible that sedimentation slowed slightly during this period, perhaps coupled with a reduced rate of visitation of the site.

Unit V is distinguished from the overlying units by the browner, more mottled appearance of the sediment, presumably a result of significant *in situ* weathering. A return to large quantities of stone and animal bones and a slight change in the composition of the fauna, now including minor quantities of montane elements (see Chapter 9), also distinguishes this unit from those above. Core samples taken from the base of the excavation in Square 0N0E down to bedrock revealed no further major change in the sediment colour, texture or hardness. Based on extrapolation, occupation of Kria Cave may have begun around 8000 BP.

The stratigraphic boundaries in Square 0N0E are consistently 20 cm lower than in Square 1N1E at the front of the cave. This follows the estimated slope of the bedrock (at 190 cm sloping down to 235 cm depth) and suggests that the bulk of the deposit accumulated with a slight slope towards the rear of the cave.

The very close stratigraphic correspondence between the two squares indicates that the deposit has accumulated without any major disturbance. This is further supported by the excellent series of radiocarbon dates from Square 0N0E. Levelling of the surface has taken place during the accumulation of Unit I, which is considerably thicker in Square 0N0E towards the rear of the cave.

High proportions of both burnt bone and heated stone in Square 0N0E suggests that the general location of hearths was close to the back wall. The cave was probably used on a regular basis, considering the fairly large quantities of bone and stone throughout the deposit; however, the generally lower quantities of material in Unit III and IV suggests that the cave was visited less frequently during these periods.

Based on a model of interrupted deposition, with a significant hiatus between Units I and II, the following approximate age ranges can be suggested for each unit:

Table 3.3. Kria Cave, interpolated radiocarbon (uncalibrated) age ranges for the identified units; all ages are based on the equation given in Figure 3.24 (bottom), except for Unit I.

Unit	depth in cm		radiocarbon years BP
	Square 1N1E	Square 0N0E	
I	0-20	0-40	0 - >1840
II	20-60	40-80	4367-5106
III	60-85	80-105	5106-5567
IV	85-105	105-125	5567-5936
V	105-170	125-195	5936-7136
unexcavated deposit to bedrock	170-190	195-235	7136-7875

4. Occupation History of Toé Cave

4.1. DESCRIPTION OF THE CAVE

Toé Cave is located inside a low conical hill, situated on a small peninsula that juts into the southern margin of the western basin of the Ayamaru Lakes (see Fig. 2.1, Chapter 2, and Fig. 4.1). It lies near the villages of Men (or Semoe) and Kartapura and has the global coordinates of 01° 17' 5.6" S and 132° 13' 6.1" E. The cave has five entrances, two of which form narrow corridors into the cave; the remainder are small rockshelters or chambers leading into the cave. Two entrances are on the northern side, two on the southeastern side and one on the eastern side of the hill. The floors of the larger, inner chambers of the cave are scattered with large blocks of roof fall. A few stone artefacts were found on the surface of these chambers. As in Kria Cave, the atmosphere is humid and the inner walls are damp and mossy. No rock paintings or carvings were found.

The southeastern entrance (see Fig. 4.2), which was selected for excavation, is about 3 m wide and 3 m high and lies approximately 12 m above the present lake level. The floor is more or less horizontal for about 2.5 m, and then falls away back into the cave. This corridor merges with another corridor that runs to the two northern entrances on the opposite side of the hill. It also leads to an entrance (back to the southeast) immediately below the excavation site; this entrance is situated at lake level and has its floor under water. Where possible, the other entrances of the cave system were cored but these proved to have no archaeological material in their shallow deposits.

4.2. EXCAVATION IN TOÉ CAVE

The surface of the deposit in Toé Cave is essentially horizontal at the level of the test pits but slopes rapidly away from the entrance into the cave. The excavation grid, directed at 145° from north, was set up over the entire corridor surface. Two pits of 1 m^2 each were excavated at the entrance in a central position (labelled 0N0E and 1N1E), with two additional flanking pits, each of approximately 0.5 m^2 (labelled 1S0E and 2N1E), to complete a transect between the opposing walls of the cave (see Fig. 4.3). The horizontal zero point was established at the southwestern corner of Square 0N0E, and the vertical datum on a small ledge c. 61 cm above the surface in the southern wall near the entrance. All depth measurements have been converted into depth below the surface of each square. Bedrock was reached at depths between 110 and 140 cm.

The remaining length of the corridor was contoured and augured at the positions 1.1W0.5N, 2.8W0.5S, 4W1.4S and 5.2W2.5S. This demonstrated that the bedrock slopes between 15 and 25°; the thickness of the deposit reduces from 1.4 - 0.1 m (Jelsma 1998).

Figure 4.1. The karst hill containing Toé Cave; view from Lake Ayamaru (facing west). Courtesy J.Jelsma.

Figure 4.2. Toé Cave. Left: the entrance where the excavation took place. Right: the corridor leading into the cave, viewed from the more horizontal excavation area. Compare Figure 4.3.

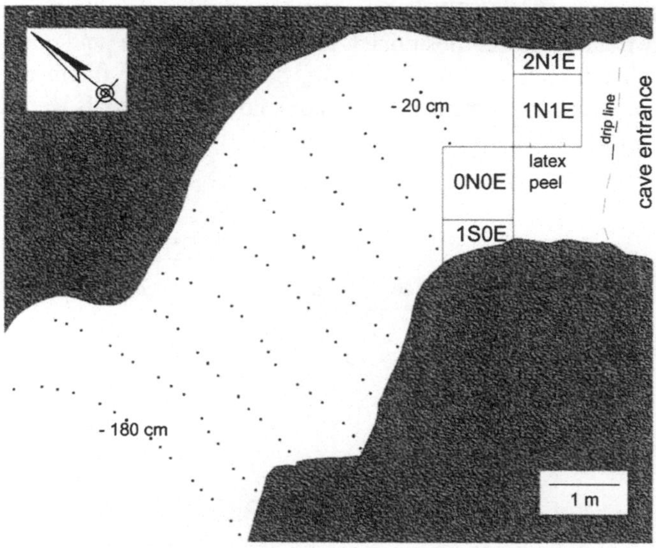

Figure 4.3. Plan of Toé Cave with location of the excavated squares.

4.3. OCCUPATION HISTORY OF TOÉ CAVE

Methods used to reconstruct the occupation history of Toé Cave are described in Chapter 2.

4.3.1. *Stratigraphy*

Excavation circumstances and light conditions in Toé Cave were similar to those in Kria Cave. Natural strata were similarly difficult to recognise during excavation or in the *in situ* profile, hence a latex peel was made of one of the profiles in order to study the stratigraphy. The peel was made of the central section of the southern wall of Square 1N1E (see Fig. 4.3), c. 50 cm wide and covering the total length of the profile down to bedrock.

The stratigraphy in Toé Cave was described earlier by Jelsma (1998). The latex peel shows a gradual change in colour from dark grey in the top layers, a lighter grey in the mid-section to a mottled orange-brown colour close to the bedrock (Jelsma 1998: 46; see Fig. 4.4). Closer inspection of the peel revealed a grey (10YR 6/1) layer in the top 7 cm followed by a light brownish grey (10YR 6/2) band until c. 10-15 cm below the surface. Below this level the sediment gradually changes to a greyish brown (10YR 5/2). Around 70 cm depth, the texture changes from a granular sediment in the upper unit to a smooth but 'blocky' sediment in the lower unit of. The colour change to brown (10YR 4/3) is more obvious and occurs at 80-85 cm depth.

No obvious pattern was visible in the presence of cultural material in the latex peel. Apart from two large rocks positioned at c. 55 cm depth and various randomly distributed smaller objects (presumably rocks, flakes and bone fragments – these were dislodged and removed when the peel was made), no other materials were attached to the peel.

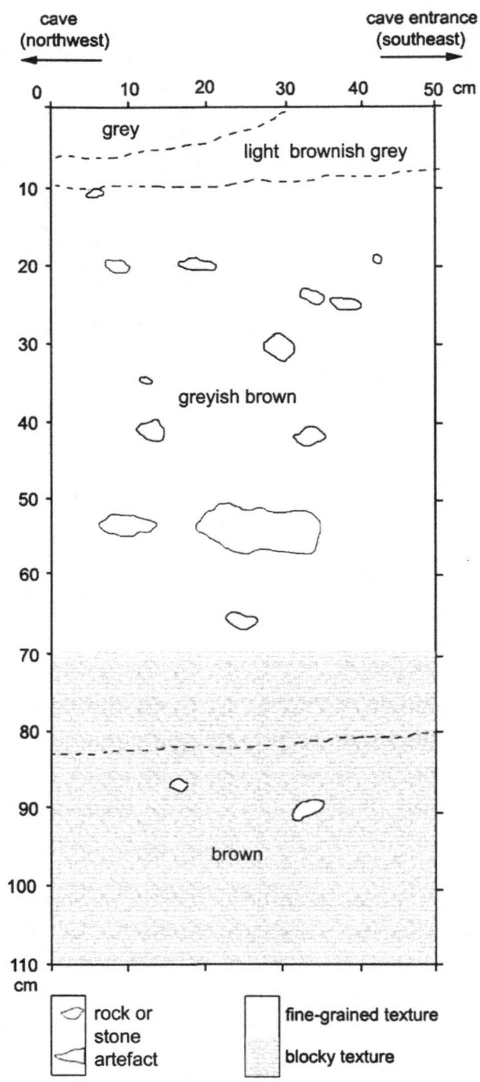

Figure 4.4. Schematic image of the latex peel from Square 1N1E in Toé Cave. Dotted lines indicate change of colour.

4.3.2. *Vertical distribution of cultural material*

As to be expected in sites located so close together, the cultural materials from Toé Cave are very similar in character to those from Kria [see Jelsma (1998) for an earlier analysis of the material]. Again, the bulk of the material consists of stone artefacts and animal bone, with lesser quantities of human bone fragments, ochre and molluscan shell, and small numbers of pottery sherds and bone points. More detailed analyses of various materials are presented in the following chapters and appendices. Figures of the distributions can be found at the end of this chapter. Note that the weights of all categories have been corrected for excavated area; see also Chapter 2. Values for 10-cm spits have been evenly split and presented as two 5-cm spits; see note 3 on page 21.

Faunal remains

Almost 5.7 kg (5695 g) of animal bone was recovered from Toé Cave, the highest densities being found in Square 1S0E. In all squares the lower half of the deposit produced the largest quantities of bone material (see Fig. 4.5 at the end of this chapter).

Two main peaks in the quantity of bone are observed in Square 2N1E: a major peak from bedrock to 80 cm below the surface, and a smaller peak between 45 cm and 10 cm below the surface. The zone from 60 to 80 cm below the surface produced much lower quantities of bone, with the lowest values seen at 50-60 cm. In Square 1N1E bone densities are uniformly low until c. 95-100 cm depth, below which level the amount of animal bone increases dramatically. The lowest quantities are observed around 50-60 cm, corresponding to the lowest levels in Square 2N1E. In Square 0N0E the quantity of bone is greatest just above bedrock and it declines steadily from there to the surface. Square 1S0E shows a different pattern again, with two strong peaks in the lower levels, reaching maxima at depths of 120-125 cm and 75-90 cm, and a gradual decline above those levels. The lowest quantities are observed at 15-35 cm with a slight recovery in the surficial layers.

The faunal collection includes essentially the same suite of species as found in Kria Cave, with an overall dominance of the forest wallaby *Dorcopsis muelleri.* Other species present include tree-kangaroos (*Dendrolagus* spp.), cuscuses (*Spilocuscus* spp. and *Phalanger* spp.), echidna (*Zaglossus bruijnii*), striped possums (*Dactylopsila* spp.), a sugar glider (*Petaurus* sp.), ringtail possums (*Pseudochirulus* spp.), bandicoots (*Echymipera* and *Microperoryctes* spp.), rodents, birds (including *Casuarius* spp.), snakes, lizards and bats. The lower levels produced a particularly diverse fauna, including a range of species not recorded previously from the Bird's Head region (Aplin et al. 1999). A few bone fragments from the surface are identified as pig; some fragments from within the deposit may originate from pig, but this could not be confirmed with certainty. Changes in the fauna through the deposit are thought to reflect significant changes in the local environment (Jelsma 1998; Pasveer & Aplin 1998; and see Chapter 9).

Human bone

A total of 132 g of human bone was recovered. The remains originate from five individuals and were scattered through the upper half of the deposit, from just below the surface until a depth of c. 75 cm. The majority of the fragments were found in Square 1N1E (see Fig. 4.6, p. 82). Four isolated teeth were used for DNA analysis (Capelli et al. 2001). Further details of these remains are discussed by Bulbeck in Appendix 1.

Molluscan shell

Molluscan shells were found mostly in the top section of the deposit in Toé Cave. They have a total weight of c. 286 g. Square 1S0E produced somewhat higher quantities of shell than the other squares, although there are isolated and seemingly uncorrelated peaks in each of Squares 2N1E and 0N0E (see Fig. 4.7, p. 83).

A selection of the more complete specimens was identified by Rob Moolenbeek (Zoological Museum, Amsterdam) and Katherine Szabó (Dept. of Archaeology and Natural History, Australian National University), either to genus or species level. *Melanoides* sp., *Bellamya* sp. (including *Bellamya* sp. cf. *B. kowiayiensis*), *Hyridella* sp. (including *Hyridella* sp. cf. *H. misoolensis*) and *Batissa* sp. are fresh water species; the genus *Polymesoda* sp. (including *Polymesoda viridescens*) occurs in fresh to brackish water. Some fragments could be no further identified than to *Batissa* or *Polymesoda* sp. The single fragment of a volute shell (*Voluta* sp.), found between 15-25 cm depth, is the only specimen that derives from a fully marine environment. Various land snails are represented, including *Cyclotus* sp., *Cyclotus hebraicus* (Lesson, 1831), *Leptopoma* sp. and an unidentified species of Helicarionidae. A more detailed analysis of molluscan shell is given in Chapter 7.

Apart from some very small fragments of *Hyridella* sp. and the land snail *Cyclotus* sp., molluscan remains were absent below 90-95 cm in all squares.

Avian eggshell

A total of 73 grams of eggshell was recovered, derived from eggs of *Casuarius* sp. and one or more megapode species. Fragments were found in most levels of all squares, but with different patterns of vertical distribution in each case (see Fig. 4.8, p.84). In Square 2N1E there are two peaks, one at 15-35 cm and another at 60-75 cm below the surface. The upper eggshell peak corresponds to a minor peak in the quantity of bone, but the lower one falls at a low point in the abundance of bone. Moderately large quantities of eggshell were found in the top 35 cm in Square 1N1E, below which the quantity remains low apart from a slight increase at 80-85 cm. In Square 0N0E there is less eggshell overall, with a slight concentration in the top 50 cm. Interestingly, Square 1S0E yielded very little eggshell in the upper levels, but large quantities in the lower half of the deposit, essentially below c. 110 cm depth and corresponding with the lowermost peak in bone quantities.

Figure 4.9. Toé Cave. Left: seed of *Celtis* sp.; specimen 133. Right: seed of family Musaceae (banana); specimen 49.

Botanic remains

As in Kria Cave, botanic remains are scarce in Toé. Fragments of *Celtis* seeds (see Fig. 4.9, left) were found mostly in Square 0N0E, three fragments between 25 and 60 cm depth, and one fragment in Square 2N1E at 30-35 cm depth. One fragment was found in a coring sample taken on the slope a bit further into the cave. The total weight of the four *in situ* fragments is 1.22 g.

Two black angular but rounded seeds were found in the top 15 cm, one in each of the two central Squares 1N1E and 0N0E (see Fig. 4.9, right). Each weighs 0.29 g. They were identified by Lyn Craven (Australian National Herbarium) as belonging to a member of the family Musaceae, or the bananas.

Carbonised fruits of the family Elaeocarpaceae were found in Kria Cave, but not in Toé.

Stone artefacts

Toé Cave yielded about 26.3 kg of stone artefacts, the majority of which (22.5 kg, or 1698 of the total number of 1775 fragments) consists of chert (see Chapter 5 for a detailed analysis of stone artefacts). The vertical and horizontal distribution of chert within the deposit is shown in Figures 4.10 and 4.11 (p. 85-86), expressed in terms of weight and the Minimum Number of Flakes (MNF).

The largest quantity of chert, as measured by weight, was found against the northern wall in Square 2N1E, and especially in the upper half of the deposit. Two major peaks are present, with maxima at 60-75 cm and 40-55 cm. The lowest values are found above 25 cm and at 75-85 cm, with two smaller peaks present below the lower depression. A similar pattern is shown by the MNF values, indicating a relatively even flake size through the sequence in this square.

All of the other squares produced much lower quantities of chert, with a general increase in weights between 25 and 80 cm, and lower quantities above and below those levels. Squares 1N1E and 1S0E show minor peaks in chert weight at 65-75 cm and 40-45 cm, corresponding reasonably well to the major peaks in Square 2N1E. However, the MNF values for these squares show a general increase to a single peak at 40-50 cm. Square 1S0E shows lower values overall but with minor peaks in MNF at 15-20 cm, 25-30 cm and 75-80 cm. The peak weights for chert in this square fall at 45-50 cm and 70-75 cm.

Bone points

Three bone artefacts were found in Toé Cave. One of these is a slightly modified fragment, which may have been unfinished at the time of loss or discard. The other two are broken bipoints of c. 2.5 to 3 cm long. The bipoints were found between 50 and 60 cm depth in Squares 0N0E and 1S0E, the 'unfinished' fragment between 30-35 cm in Square 0N0E. The great contrast in numbers of bone artefacts between Toé and Kria Caves is discussed in Chapter 6.

Pottery

Three pottery sherds were excavated from the top levels (5-15 cm) of the deposit. These have a combined weight of 13.3 g and are, apart from one groove, undecorated (see Fig. 4.12). Their temper is very similar to that found in the sherds from Kria Cave (see Chapter 3), including quartz grains and a feldspar-like mineral. The source of this material is likely to be found north and east of the lakes in the Tamrau and Arfak mountain ranges and the islands west of the Bird's Head (Waigeo, Batanta and Salawati; Ratman 1998: figure 7). No shell fragments were found in the temper, arguing against a coastal source of the pottery.

Figure 4.12. The three pottery sherds from Toé Cave. Specimen number for all sherds: 49.

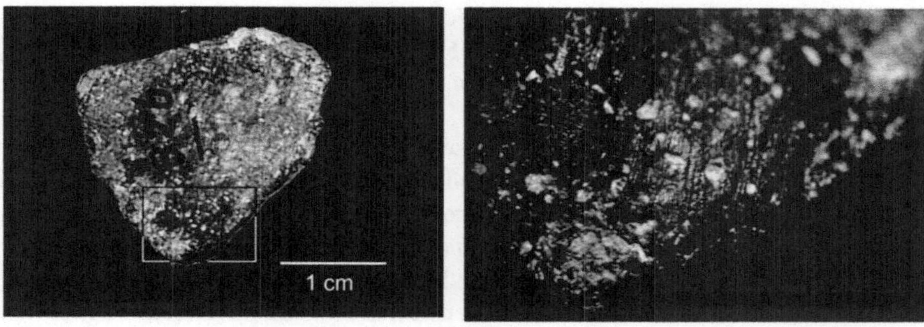

Figure 4.14. Toé Cave, red ochre fragment (with close-up on the right) showing signs of use. Specimen: 326.

Ochre

Red and yellow ochre of identical character to that described from Kria was found in Toé Cave. It occurs in virtually all levels and it is present in relatively large quantities, with a combined weight of c. 325 grams. Most of the ochre was found in Square 1N1E and in the adjacent Square 2N1E (see Fig. 4.13, p. 87). Square 1S0E has lower quantity but does show a peak in weight at 65-85 cm. This may correspond with a more focussed peak in Square 2N1E at 65-70 cm, and another in Square 1N1E between 65 and 85 cm.

One fragment from 30-35 cm depth in Square 2N1E shows clear signs of grinding on one surface (see Fig. 4.14). This demonstrates that the ochre was indeed used for some unknown purpose. However, it also highlights the rarity of evidence for use, and suggests that the ochre was either used elsewhere, or else normally processed in a way that leaves no archaeological trace, for example, crushed into water to make a paste.

4.3.3. *Weathering, corrosion and fragmentation of bone and stone*

Post-depositional damage of bone

Damage done by plant roots is minor in the bone material from Toé Cave, and if present at all, it is most pronounced in Square 1S0E (see Fig. 4.15, p. 88). Burnt bone is generally less affected than unburnt or lightly burnt material. As to be expected, Squares 2N1E and 1N1E both show slightly more root damage than Square 0N0E, situated further into the cave. Like the latter, Square 1S0E shows very little damage in the first 75 cm, but below that level, until c. 100-105 cm depth, there is an increase in root marks.

Tooth marks caused by animal scavenging are present on most of the bone material down to the lower levels. The highest level of tooth mark damage is observed in Square 2N1E (see Fig. 4.16, p. 89). A zone of little damage between 50 and 60 cm in this square is probably related to very small sample sizes. In Square 1N1E little tooth mark damage is found below 80 cm depth, and no tooth marks are seen on the bones below 105 cm. This is probably related to the high amount of burnt bone found in the bottom levels of this square (see below). Bones in Square 0N0E are the least damaged by scavenging, with very few tooth marks between 25 and 55 cm and below 105 cm. In Square 1S0E very little tooth mark damage is found below c. 110 cm. Again, the bone samples show a high proportion of burnt material.

Bones are most corroded in Square 2N1E (see Fig. 4.17, p. 90); this is probably related to the high number of tooth marks found on bone in this square, which penetrate the dense outer cortex and increase the surface area of the bone. Again a decrease in the proportion of corroded bone material is seen between 50 and 60 cm, perhaps due to small sample size. Square 1N1E shows a high proportion of corroded bone from the top of the deposit down to c. 60 cm. Below that level, the amount of bone corrosion falls, but increases again with depth, with another fall between 80 and 95 cm. The unburnt material in these levels is highly corroded, and even the burnt bone (but not calcined), of which there are high quantities in these levels, shows some degree of microbial breakdown. Degree of corrosion is generally lower in Squares 0N0E and 1S0E, with particularly low levels between c. 30 and 50 cm, and again below c. 105 cm in Square 0N0E.

Proportion of burnt vs. unburnt bone

The proportions of burnt and unburnt bone in Square 2N1E fluctuate through the deposit but without any overall trend or obvious relation to the degree of bone corrosion (see Fig. 4.18, p. 91). In Square 1N1E the fluctuations in proportion of heated bone partly seem to follow the pattern found in bone corrosion; especially in the bottom levels, where corrosion is high, the proportion of unburnt bone is reduced. However, the preservational circumstances in the bottom levels of this square are in general very bad as is shown by the fact that even the burnt material is to some extent corroded. Square 0N0E shows a sharp increase in the proportion of unburnt bone below c. 70 cm, possibly suggesting better preservational state. However, bone from these levels is highly corroded, which seems to contradict this notion. The patterning of burnt vs. unburnt bone and bone corrosion in Square 1S0E again appears to be unrelated. It must be concluded, therefore, that while the patterning in proportions of burnt and unburnt bone is clearly not random, it cannot be explained solely in terms of bone preservation, but must also reflect variation in original burning composition and thus some aspect of human behaviour.

Bone density

Bone densities of the Toé Cave material were determined by Jelsma (1998). He concluded from the densities as shown in his Figure 6 (1998: 52; reproduced here in a modified format on p. 92, Fig. 4.19, Square 1S0E) that there were two different gradients of fossilisation within the sequence. This conclusion was based on an apparent inversion of density trends between 65 and 80 cm depth, forming a 'layer' separating the overlying and underlying material. While Jelsma saw this 'layer' as evidence for two separate sequences in the Toé Cave deposit, this pattern is only visible in one of the four squares (1S0E). Fig. 4.19 also shows bone densities for the other three squares in Toé, which do not show any trends such as found for Square 1S0E. The pattern in this square may, in fact, have a simpler explanation. Square 1S0E lies against the southern cave wall, which slopes backwards at c. 70 cm below the surface to form a low overhang at this location (see Fig. 4.25, p. 76). This overhang may have served to protect any material that accumulated under it. Indications for better preservation below these levels are also found in a lower fragmentation both in bone and stone (see below).

Jelsma also argued that the bone density fluctuations, varying between 1.62 and 2.11 kg/litre, were so great 'that weight is not a good variable for the quantification of faunal remains' (1998: 52). However, the fluctuations shown for Square 1S0E (or any other squares, see Fig. 4.19) are no greater than those calculated for Kria Cave. Higher densities in the latter are argued to be the result of a number of possible variables, including the carbonate encrustation of many bones especially in the lower levels, burning state, and the presence of different species and different body parts in the sample. These variables must indeed be accounted for when bone weights are used in quantitative analyses.

Heating of chert

The proportion of heated chert material in Toé Cave shows strong patterning through the deposit, but with distinct contrasts between the squares (see Fig. 4.20, p. 93). Square 2N1E shows three distinct peaks in chert heating at 15-25 cm, at 30-65 cm, and from 80-110 cm. In Square 1N1E, the proportion of heated chert is high in almost all levels but with apparent peaks between 65 and 95 cm and above 35 cm. Square 0N0E shows little heating of chert below 75 cm, contrasted with uniformly high levels above that level, save for an absence between 20 and 30 cm. Square 1S0E shows less heating of chert overall, except for a consistent block between 45 and 70 cm depth. Although no hearth structures were detected during the excavation, the shifts in concentration are best explained as the result of shifting placement of fireplaces through time, from a location against the northern wall, then across the cave floor to close to the southern wall and back again.

This distribution does not clearly follow the distribution of burnt animal bone in the deposit. Larger quantities of burnt bone have accumulated in all squares in the lower levels, and against the walls (in Squares 2N1E and 1S0E) also in the uppermost part of the sequence. Shifting concentrations are much less obvious, although they can be seen in Squares 1N1E and 0N0E in the large amounts of burnt bone decreasing from the bottom to the top in Square 1N1E and the opposite occurring in Square 0N0E.

Fragmentation of chert

With a total number of flakes of 1698, and a Minimum Number of Flakes of 1165, Toé Cave has an overall fragmentation index of 1.46. The fragmentation index fluctuates markedly and shows no general pattern but increases in the lower levels of the deposit in Squares 2N1E and 0N0E (Fig. 4.21b, p. 95). The fragmentation as expressed by the weight divided by the total number of fragments shows considerable fluctuations (Fig. 4.21a, p. 94), due to the fact that large chert artefacts such as cores were included in this calculation.

Fragmentation decreases below c. 70 cm in Square 1S0E (Fig. 4.21b), presumably as a result of the protection provided by the overhang protruding from this southern wall. The size of the bone fragments in this square is also significantly larger below 70 cm depth than above that level (Jelsma 1998: 53).

Heat causes chert to fracture. However, comparing the fragmentation with the proportion of heated chert by weight per spit and per square showed only a very faint correlation between the two. Any trend in fragmentation is probably more likely to be related to other factors such as weathering, rather than to the amount of burning at the site. For example, the proportion of heated chert in the lower levels of Square 0N0E is low, yet fragmentation is high in these levels. Because the majority of the chert in Toé Cave is heavily weathered, and a large proportion has even turned into very powdery and fragile fragments, one would expect the fragmentation to be relatively high. However, although the correlation is clearly visible in very heavily weathered chert from Square 2N1E, the other squares show a much less obvious relation.

Weathering of chert

Toé Cave's unheated chert material is overall much more weathered than the Kria Cave assemblage. The Toé collection is dominated by heavily weathered (stage 3) material. The proportion of patinated material (stage 2) is considerably smaller than in Kria, but very heavily weathered specimens (stage 4) are relatively common especially in the lower sequence. As in Kria Cave, completely unweathered unheated fragments do not occur. The overall pattern of increasing proportions of stage 4 material in the bottom levels, and less weathered (stage 3 and 2) fragments higher up in the sequence is present in all squares (see Fig. 4.22, p. 96). Small numbers of stage 4 material are found in three of the four squares almost immediately from the surface downwards. However, at some deeper levels some patinated (stage 2) fragments occur. They are an obvious minority and may not be very significant. As will be discussed below, it is unlikely that the deposit has been heavily disturbed. Perfect sequences of A/I ratios of *Casuarius* eggshell amino acid racemisations (see Table 4.2, p. 77), provided by the University of Colorado, confirm this.

The generally higher degree of weathering of the chert in Toé is presumably related to the greater age of much of the Toé assemblage, together with a much lower sedimentation rate that would have left material at or near the surface for relatively longer periods of time. Otherwise, the conditions in both caves are broadly similar (as both excavated areas are close to the entrance of the cave/rockshelter, and both caves are humid to such an extent that water was dripping from the ceilings at all times), the upper sequence of Toé is of comparable age as Kria's deposit (see below), and the chert itself is probably of comparable quality, considering the short distance between the two sites, and the similar appearance of the chert in both collections.

As in Kria, the heated chert in Toé Cave is, in general, less weathered than unheated chert. Of the most heavily weathered fragments (stage 4), more than 90% is unheated, while all fragments with weathering stage 0 (no patination) are heated ones (Table 4.1).

Table 4.1. Weathering stages of heated and unheated chert in Toé Cave.

	no patination stage 0	lightly patinated stage 1	patinated stage 2	heavily weathered stage 3	very heavily weathered stage 4
heated chert	100% N = 3	88.4% N = 61	44.1% N = 164	8.2% N = 81	9.3% N = 24
unheated chert	0% N = 0	11.6% N = 8	55.9% N = 208	91.8% N = 909	90.7% N = 235

The proportion of heated chert in Toé seems to increase again very slightly in the very heavily weathered category. This may be related to the generally very heavily weathered state of all stone (and even burnt bone) material in the bottom levels. Obviously, heating of chert affects its preservation and hence the distribution of weathered material through the deposit.

Density of chert

The density of the chert artefacts from Toé Cave was calculated by Jelsma (1998) and the pattern shown for Square 1S0E in Figure 10 in his paper (1998: 58; reproduced here in a modified format on p. 97, Fig. 4.23). Artefacts below some 90 cm depth in this square clearly show a lower density than those from levels higher up in the sequence. Although this is consistent with the idea that chert loses some of its soluble constituents over time (Loughnan 1969), it is regrettable, however, that this pattern is only shown for Square 1S0E, as this square is the least representative of the four excavated squares, being located against the southern wall where the bedrock shows the steepest slope, and the wall forms a protective overhang at 70 cm below the surface. Based on the values for the density shown in this figure, Jelsma suggested that weight is not a reliable variable for the quantification of stone artefacts coming from limestone deposits (1998: 55). Indeed, stone weights should only be compared within levels, and the effects of burning on stone, which is likely to increase the density, as well as the proportion of cortex on the artefacts, should be accounted for. Despite these influencing factors, all squares show a relatively constant chert density with stronger fluctuations in the lower levels of the deposit and a faint overall trend declining towards the bedrock.

4.3.4. *Identification of stratigraphical units*

The distribution of cultural materials as described above clearly shows some recurrent patterns across the excavation squares. Animal bone and stone artefacts, which have the biggest sample size, give the best representation of this pattern. In Square 2N1E there is a clear break in quantities around 80-85 cm. Molluscan shell and eggshell closely follow this pattern. A similar but less obvious break appears around 60-65 cm depth, also seen in the distribution of eggshell. Square 1N1E shows breaks for all categories around 85-90 cm but with a lesser low at 75 cm again for stone artefacts and eggshell. Animal bone shows an extra low at 60 cm, also seen in the distribution of ochre at 65 cm. A similar pattern is seen in Square 0N0E where quantities of animal bone and molluscan shell show low values around 85 cm and stone artefacts and ochre around 95 cm. Only eggshell shows a break at 50 cm in this square. Square 1S0E has generally a deviating pattern, with breaks for animal bone and stone artefacts at 35 cm, and furthermore a clear break between 100 and 105 cm in bone, stone, eggshell and molluscan shell.

In summary, quantities of most of the cultural material reached particular low points in all squares when about 1/3 of the sediment was deposited: this boundary slopes down across the squares, from c. 80-85 cm in Square 2N1E, between 75 and 90 cm in Square 1N1E, between c. 85 and 95 cm in Square 0N0E and around 100-105 cm in Square 1S0E. This slope follows the slope of the bedrock in the cave. A less obvious low is seen in Square 2N1E and 1N1E, especially in the distribution of animal bone and to some extent in stone artefacts as well around 60-65 cm. Vague dips in quantities of ochre and eggshell in the other squares at similar depths may or may not correspond with this. The typical peak values for bone and stone in the first 35 cm in Square 1S0E may be an infill of the deposit, which effectively levelled the cave floor.

Based on the combination of sedimentological and cultural distributions given above, the stratigraphy in Toé Cave thus shows two major 'sequences': an upper unit, characterised by grey-brown, granular sediment, and large amounts of stone artefacts with smaller quantities of animal bone; and a brown lower unit of 'blocky', mottled sediment, containing large amounts of animal bone, and smaller numbers of stone artefacts. This basic division in two 'units' is also visible in the preservational state of bone and stone. The presence of pottery, restricted to the top 15 cm, may also indicate a separate unit; the faint difference in colour in the top 10-15 cm may be associated with this.

Whether the boundary between the upper and lower units is a sharp one or a gradual transition cannot yet be established. There are indications, though, that the deposit in Square 2N1E has been truncated at about 80 cm depth, while the other squares obviously show a more gradual change in quantities of both bone and stone.

The identification of animal species revealed the presence of a significant proportion of montane animals in the lower half of the deposit (confined to comparable boundaries), species that are virtually absent in the upper levels. In the latter, as to be expected, predominantly lowland species were found (Jelsma 1998; Pasveer & Aplin 1998). This forms important support for the presence of two separate units. The implications of the presence of these montane species will be discussed in Chapter 9.

4.3.5. *Correlation between squares*

Most of the peaks and lows in quantities of animal bone and stone correlate well in the four squares. However, the slope of the bedrock from Square 2N1E towards Square 1S0E and back into the cave, which is also clearly seen in the slope in the boundary between the top and bottom sequence in the four squares, obviously shifted some of the layers in the stratigraphy downwards. On the other hand, the slope of the bedrock and the virtually horizontal present cave floor demonstrate also that a fair bit of infill has been produced. Pulling the entire bone distribution in Square 1N1E 15 cm upwards, and the ones in Square 0N0E and 1S0E each 20 cm upwards, aligns the major peak in the lower section (Fig. 4.24), and shows the much higher amounts of bone deposited in the lower levels of all squares as well as in the upper section of Square 1S0E. In Squares 1N1E, 0N0E and 1S0E a regular and continuous infill was produced that gradually decreased the slope of the surface and eventually levelled the present floor. The recurring sharp break in the distributions of especially bone and stone at 80 cm depth in Square 2N1E may be the result of some removal of deposit in this square.

In the distribution of stone, the sequence with its peaks and lows is also essentially the same in Square 2N1E and 1N1E, especially in the upper sequence of the deposit. Square 1S0E shows a slightly different pattern, again due to the most severe slope in the bedrock at this side. Here, as well as in Square 0N0E, the matching peaks and lows are clearly visible, but various infill layers have formed through time.

The correlation between the squares in Toé Cave is obviously not straightforward. Aligning the major peaks in the lower half of the deposit does not simultaneously line up all minor peaks in the upper half. Although the sequence appears to be essentially intact, the sloping bedrock in the cave is likely to have caused local slumping and infill within the deposit.

4.3.6. *Horizontal distribution*

Horizontal patterning is seen in the distributions of animal bones and stone artefacts (see Figures 4.5 and 4.10). For most of the time during the formation of the upper unit, Squares 2N1E and 1N1E have accumulated the highest numbers and largest specimens of chert artefacts, with lesser but still substantial quantities in the other squares. In earlier times much less stone material was deposited in any square, but relatively large quantities are still to be found in Square 2N1E.

Animal bone, on the other hand, was deposited in the largest quantities during the formation of the lower unit and has particularly accumulated in Square 1S0E, with increasing amounts from c. 60 cm depth towards the bedrock. All squares produced relatively small amounts of animal bone in the upper unit, although Square 2N1E, as well as 1S0E (each against opposite walls), have accumulated the most.

Interestingly, the distribution of animal bone follows an almost inverse pattern to the distribution of stone artefacts (compare Figures 4.5 and 4.10 or 4.11). Whereas the weight of stone increases in the upper levels, bone weight increases in the lower half of the deposit; whereas large amounts of stone accumulated on the northern side of the cave floor (Square 2N1E and 1N1E), bone was discarded mainly at the southern end.

The distribution of heated chert also varies horizontally through the deposit. These concentrations presumably represent the shifting location of the fireplace in the cave. During the earliest periods (below c. 80 cm), hearths were evidently located mainly in Square 2N1E against the northern wall, extending into adjacent Square 1N1E. In later periods (the upper unit) the highest proportion of heated chert is found in Square 0N0E, extending somewhat into Square 1S0E against the southern wall. Most recently the hearth shifted again towards the northern wall, with the highest proportions of heated chert found in Square 1N1E.

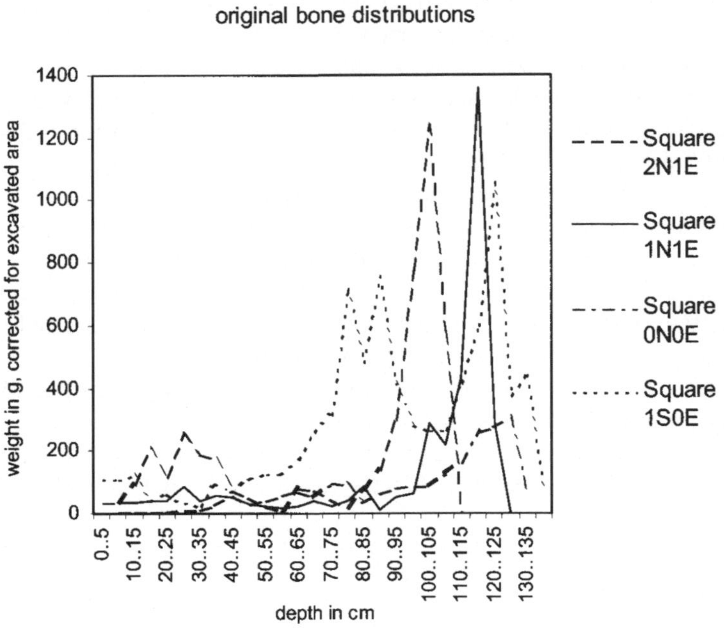

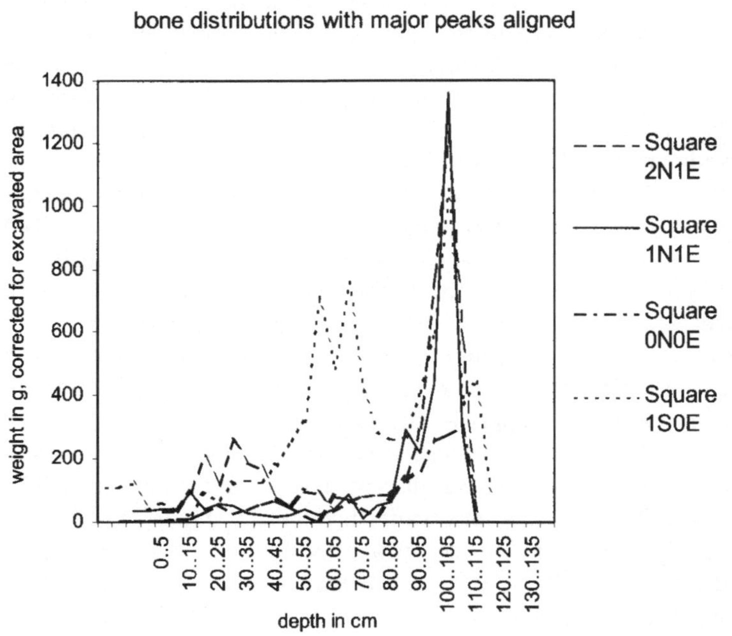

Figure 4.24. Toé Cave, correlation between squares of distributions of animal bone through the deposit.

Although the distributions of heated bone and stone follow each other to some extent, the proportion of burnt bone material is substantially higher compared with heated chert. Against the southern wall (Square 1S0E), for example, large amounts of burnt bone were recovered from the lower unit, but very little heated chert. This pattern is repeated in the lower unit in all squares except Square 2N1E, where larger amounts of heated chert than bone were found.

4.3.7. *Dating*

The stratigraphy of the Toé Cave deposit and the position of all radiocarbon-dated samples are shown in Figure 4.25. Attempts to date the sequence in Toé Cave initially met with limited success. Jelsma (1998) published the first four dates obtained for the site: a radiocarbon date on bone collagen of 2930 ± 65 BP from a depth of 35-40 cm below the surface; and three other dates on isolated charcoal fragments from below this level (between 45 and 70 cm) which produced ages of 220 ± 70 BP (45-50 cm), 111 ± 55 BP (60-65 cm), and 240 ± 40 BP (65-70 cm). Bone material from within the lower unit contained insufficient collagen for dating. An ESR date was performed on tooth enamel from a single specimen of an exclusively montane mammal species (*Mallomys* sp. cf. *M. aroaensis*) from 115-120 cm depth in Square 0N0E. Contrary to expectations, this specimen yielded a date of 3000 ± 300 BP (R. Grün, pers. comm.). Given the evidence from Kria Cave of an essentially lowland environment in the region since 6000 BP, as well as more general information of Quaternary environmental change in New Guinea (e.g. Hope 1989, 1996; see Chapter 9), it is unlikely that a montane animal such as *Mallomys* was present in the Ayamaru region around 3000 BP. This ESR result was therefore regarded as being almost certainly too young (Pasveer & Aplin 1998).

Recently, AMS dates were obtained from *Casuarius* eggshell fragments from Square 2N1E; the same samples were also analysed using amino acid racemisation (AAR). Three of these dates (5290 ± 75; 8860 ± 70; and 12,870 ± 110), and their associated A/I ratios form a perfect sequence in relation to their depth (Pasveer et al. 2002; see Table 4.2). The date of 12,870 BP was confirmed on a duplicated sample, giving 12,885 ± 90 BP, but its higher A/I ratio suggests that the latter eggshell fragment was burnt. Considering the securely dated stratigraphy in Kria Cave, and the comparable finds and faunal composition in the upper sequence of Toé Cave, these dates are considered reliable.

The fact that three eggshell fragments with mean ages of 13,230 BP, 20,430 BP and 23,140 BP were found in the same spit in the lower half of the deposit in Square 2N1E can be explained in several ways: it may mean that very little sediment accumulated in the cave over the period 23,000 to 13,000BP, presumably reflecting only occasional visitation; or it may point to some disturbance in this area of the site, be it a mixing of sediment or a truncation or erosion of the deposit. Either option is equally plausible. While there is no stratigraphic indication of major disturbance, some vertical mixing of archaeological remains might be expected given the shallowness of the deposit and the likelihood of trampling, and the possibility of small-scale cut-and-fill erosion related to the natural dip into the rear of the cave.

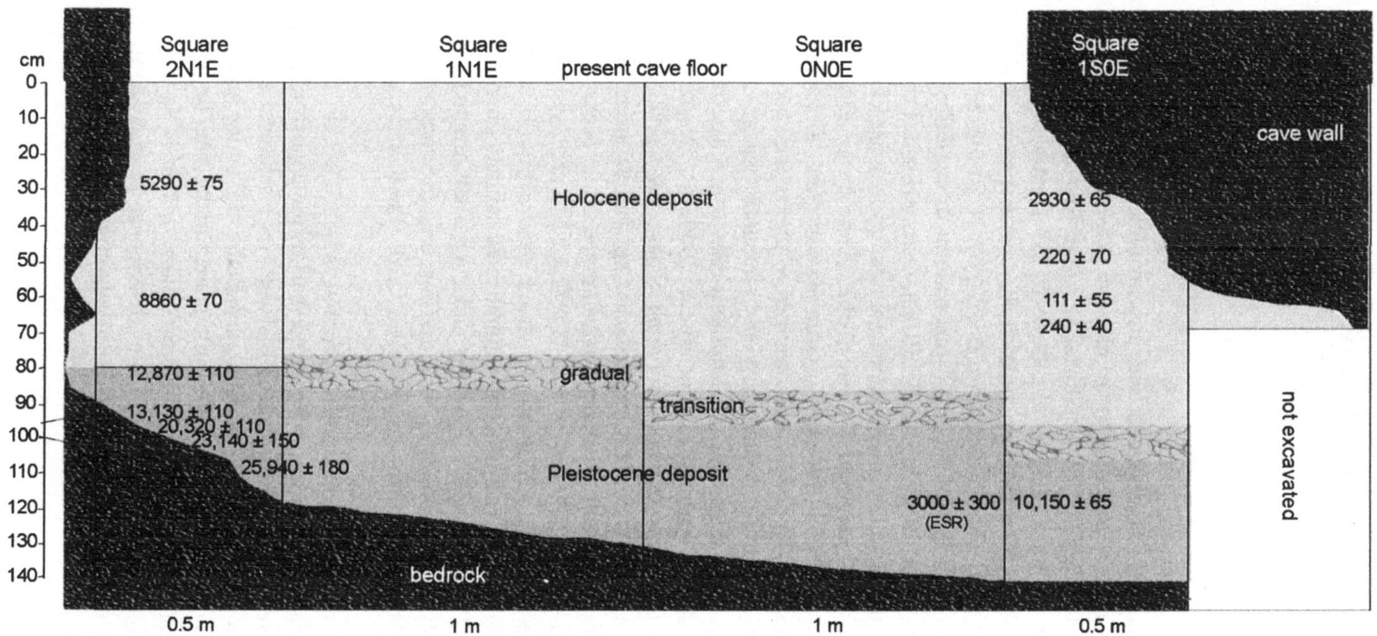

Figure 4.25. Toé Cave, schematic image of the stratigraphy, with position of uncalibrated radiocarbon dates (in years BP).

Table 4.2. Radiocarbon dates from Toé Cave.

laboratory ref. no.	depth below surface (cm)[1]	Square	radiocarbon years BP	$\delta^{13}C$ (‰, PDB)	calibrated age (2σ) in years BP[2]	A/I ratios	dated material
OxA 6201	35-40	1S0E	2930 ± 65	-20.1	3320-2870		animal bone[3]
OxA 6223	45-50	1S0E	220 ± 70	-25.3	460-0		charcoal
OxA 6224	60-65	1S0E	111 ± 55	-28.7	290-0		charcoal
OxA 6046	65-70	1S0E	240 ± 40	-25.7	425-3		charcoal
AA 32920	115-120	1S0E	10,150 ± 65	-12.4	12,320-11,340	0.361 ± 0.008	*Casuarius* eggshell
AA 32921	25-30	2N1E	5290 ± 75	-14.6	6280-5910	0.236 ± 0.001	*Casuarius* eggshell
AA 33404	60-65	2N1E	8860 ± 70	-15.1	10,210-9630	0.382 ± 0.001	*Casuarius* eggshell
AA 33405	80-85	2N1E	12,870 ±110	-12.4	15,990-14,410	0.499 ± 0.001	*Casuarius* eggshell
AA 33406	80-85	2N1E	12,885 ± 90	-12.8	15,990-14,440	0.552 ± 0.002[4]	*Casuarius* eggshell
OZF 251	95-100	2N1E	13,130 ± 110	-15.9	16,270-14,660	0.420 ± 0.011	*Casuarius* eggshell
OZF 516	95-100	2N1E	20,310 ± 110	-14.0	24,730-23,210[6]	0.513 ± 0.013	*Casuarius* eggshell
OZF 847	95-100	2N1E	23,140 ± 150	-15.0	-- [6]	0.665 ± 0.005	*Casuarius* eggshell
OZG 063	105-110	2N1E	25,940 ± 180	-13.0[5]	-- [6]	0.647 ± 0.000	*Casuarius* eggshell

Notes: The data in this table supersede those given in Pasveer (2003). [1]The deposit in Toé Cave is subject to a slope of some 20-30 cm, dropping from Square 2N1E to Square 1S0E over a horizontal distance of 3 m. [2]See note at Table 3.2. [3]Species not identified. [4]This eggshell sample is burnt, resulting in a relatively high A/I ratio. [5]Value estimated; radiocarbon age not corrected for isotope fractionation. [6]These uncalibrated radiocarbon dates fall beyond the limit of the dataset for atmospheric samples. The calibrated age range of OZF 516 is based on the maximum date for this dataset: 20,265 BP.

An eggshell fragment found just above the bedrock in Square 2N1E was dated to 25,940 ± 180 BP. Given the lack of coherent age-depth structure in the lower deposit of Toé Cave this is considered a minimum age for initial occupation of the site.

One fragment of *Casuarius* eggshell from the lower unit in Square 1S0E gave an AMS age of 10,150 ± 65 BP. This is somewhat younger than might be expected from its position in the deposit, and it too might be viewed as further evidence of local disturbance. As will be further discussed in Chapter 7, the detailed composition of the vertebrate fauna from the lower levels of Square 1S0E also differs from that of the basal levels of the other squares, and is perhaps more comparable to the fauna from levels dated to between 13,230 BP and 8860 BP in Square 2N1E. One possible interpretation is that the basal layers in Square 1S0E are indeed thoroughly disturbed, with a mixed fauna of Late Pleistocene and Holocene age. While this scenario cannot be dismissed out of hand, there is another possibility that warrants consideration – that this area of the cave was subject to a significant scouring event some time prior to 10,150 BP, after which the eroded area was filled in under more stable conditions. The fact that the bedrock in Square 1S0E slopes quite steeply, suggesting a natural channelling of water flow in this direction within the cave, and the location of the lower half of the deposit in Square 1S0E beneath a low overhang of the wall, forming a protected niche, supports this interpretation. The occurrence of a large quantity of relatively unfragmented (Jelsma 1998) and uncorroded bone with a relatively high degree of root damage in the lower levels of this square also seems to indicate more stable and protected conditions with a relatively rapid infill, following a potential scour event. In this scenario, the deposit built up during the early stages of occupation was lost in this square, but the sediment and material deposited after that event is essentially undisturbed.

The AMS date of 2930 ± 65 BP from the upper unit in Square 1S0E also seems to be slightly out of sequence, relative to the date of 5290 BP in Square 2N1E. Once again, the evidence points to possible erosional activity along the cave wall. Most importantly, the vertical distribution of cultural materials in the upper 30-35 cm of Square 1S0E shows major fluctuations in the quantity of archaeological remains, unlike the more even vertical distribution of remains in the upper levels of the other squares. This suggests a more varied sedimentation rate along the cave wall, and may reflect an ongoing process of small scale scouring and infilling that eventually levelled the cave floor.

Accepting the date of 2930 BP in Square 1S0E as reliable does mean that the cave floor must have shown a considerable slope until c. 35 cm below the surface. It also means that the sequence in Square 2N1E may have been truncated somewhere in the top levels, as well as at 80 cm depth, unless the sedimentation rate in these levels was considerably lower than in Square 1S0E.

Based on the securely dated sequence from Kria Cave, and the A/I ratios associated with the radiocarbon ages, it is assumed that the *Casuarius* eggshell from Square 2N1E have produced reliable dates for the deposit in Toé Cave. The following sedimentation rates have been established on the basis of these dates and their location in the deposit:

- above 30 cm (0-5290 BP): 0.006 cm/year;
- between 30 and 65 cm (5290-8860 BP): 0.01 cm/year;
- between 65 and 85 cm (8860-12,870 BP): 0.005 cm/year;

- between 85 and 110 cm (12,870-25,940 BP): 0.002 cm/year.

Since the lower unit is likely to have been truncated at 80 cm depth in Square 2N1E the calculated sedimentation rate between 65 and 85 cm may be an underestimate. The same may be true for the calculated sedimentation rate of the last 5000 years. The lower unit, the Late Pleistocene deposit, clearly accumulated more slowly than the overlying Holocene unit and therefore may have been more prone to trampling and local disturbance.

4.4. CONCLUSION

The Toé Cave deposit can be divided basically into two units: an upper unit (about two-thirds of the deposit) and a lower unit (c. one-third). The boundary between the two units, which slopes down following the slope of the bedrock, is not sharply defined and may indicate a gradual transition. The upper unit is characterised by a dark grey, granular deposit with relatively large amounts of chert artefacts, especially in Square 2N1E against the northern wall. Animal remains are abundant but in relatively low quantities compared to the lower unit. The lower unit consists of an orange-brown, more 'blocky', mottled sediment and has accumulated large quantities of animal remains, especially in Square 1S0E against the southern wall. Stone artefacts are present in relatively small numbers in this part of the deposit.

The top 10-15 cm of the deposit show a faint change in colour from grey to light grey-beige, containing a few undecorated pottery sherds. Apart from the appearance of pottery, the change in colour is not accompanied by changes in the kind or quantity of cultural materials.

The animal remains in the lower unit derive from a wide suite of species including both montane and lowland species, while the upper unit is dominated by the remains of *Dorcopsis muelleri* and other lowland species. This hints at significant changes in the local natural environment; this will be further discussed in Chapter 9.

Dating of the deposit in Toé Cave has been problematic. However, ten dates, nine based on cassowary eggshell and six with associated A/I ratios, are assumed to provide a realistic age of the cultural materials and events. The eggshell fragments with AMS dates falling in the time range 10-20,000 yr BP have associated A/I ratios that are consistently 2-3 times higher than those obtained from any Kria Cave samples; this is strongly supportive of a Late Pleistocene age. The fact that three pieces of eggshell from one spit at 95-100 cm differ in age by 10,000 radiocarbon years is somewhat anomalous, although it is perhaps not entirely unexpected given the relatively slow rate of build-up of the deposit and the fact that it has a natural dip into the rear of the cave, with clear scope for local erosion and conflation within the deposit (Jelsma 1998; Pasveer & Aplin 1998). Because the cassowary eggshell was brought to the site by people, it can be stated that occupation of the site and surrounding area had commenced by at least 26,000 BP.

Because the lower unit may have been truncated in Square 2N1E, the date of the transition zone between the upper and lower unit is difficult to establish. However, Unit V in Kria Cave, which shows remnants of the same montane species, provides indirectly a date for the termination of the transition zone of c. 6000 BP. The end of the occupation cannot be established based on the currently available radiocarbon dates, but the presence of pottery (which is similar in material and temper to the sherds from Kria Cave) suggests that the cave was still visited, at least infrequently, during the last 3000 BP.

Jelsma (1998) identified two 'occupation sequences' in Toé Cave's deposit, with a possible disturbed layer or stratigraphic inversion in between. The idea of a possible stratigraphic inversion was based on the first four dates obtained for the site, including a date of 2930 BP and three underlying sub-recent dates in Square 1S0E. However, the more detailed analyses conducted subsequently failed to find any evidence for a major disturbance in the deposit, and the sequence is considered essentially intact. Based on the correspondences in distributions between the squares, the similarities with the securely dated deposit in nearby Kria Cave (which also provided anomalous, sub-modern charcoal-based dates), and the available time-frame for environmental change in the region (Pasveer & Aplin 1998), four dates (ESR: 3000 BP; AMS: 220 BP; 111 BP; and 240 BP) are considered too young, possibly a result of sample contamination or intrusion during excavation; seven are considered reliable indications of their stratigraphic context (2930; 5290; 8860; 10,150; 12,870; 12,885; and 25,940 BP); and three (13,130; 20,320; and 23,140 BP) are argued to be reflective of minor local mixing or slumping of the deposit. Square 1S0E, because of its downslope position in the cave entrance, had the highest potential for sedimentary erosions and subsequent infill.

There are marked differences as well as similarities between Toé and Kria. The occupation in Toé Cave started almost 18,000 years earlier than that in Kria Cave, yet the site accumulated only a maximum of 140 cm deposit, as opposed to over 2 m in Kria. Comparing the contemporary (Holocene) portion of each sequence, the most remarkable similarity is in faunal composition, with the remains of the forest wallaby *Dorcopsis muelleri* dominant in each case (see Chapter 7 and 8). Pottery was rare in both sites. Rock art was not found in either of the caves despite the presence of large amounts of red and yellow ochre. A major cultural difference is found in relation to the bone artefacts; these are abundant in Kria but rare in Toé Cave.

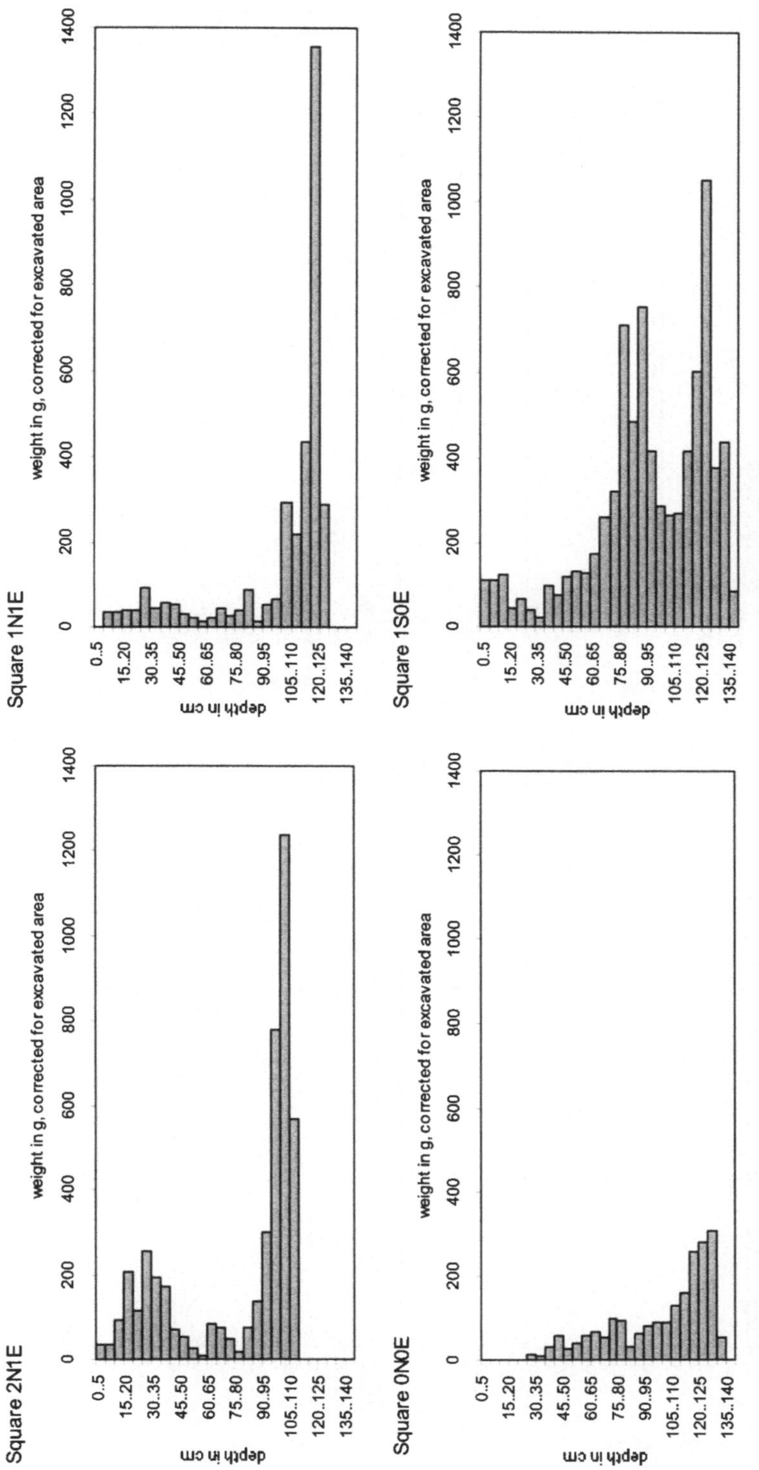

Figure 4.5. Toé Cave, distribution of animal bone through the deposit.

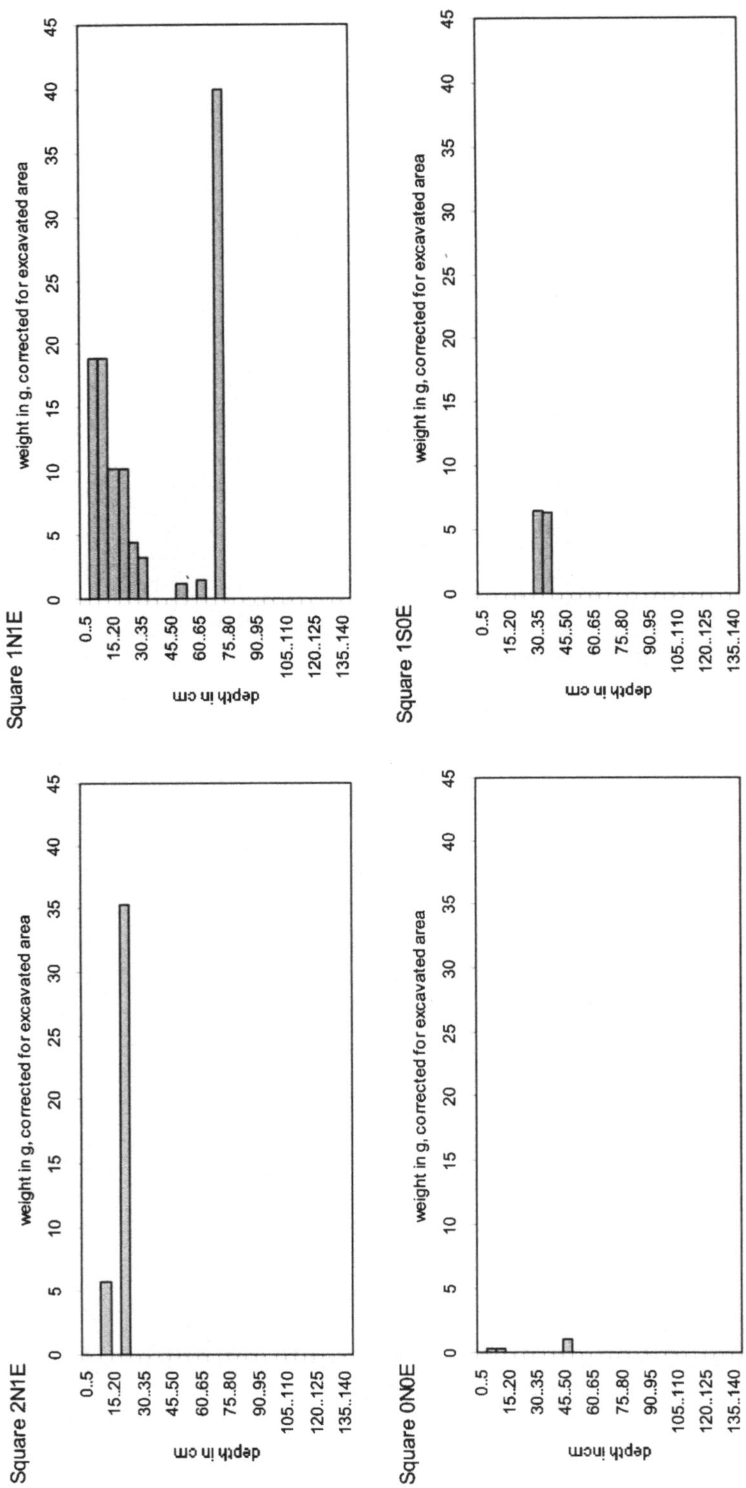

Figure 4.6. Toé Cave, distribution of human remains through the deposit.

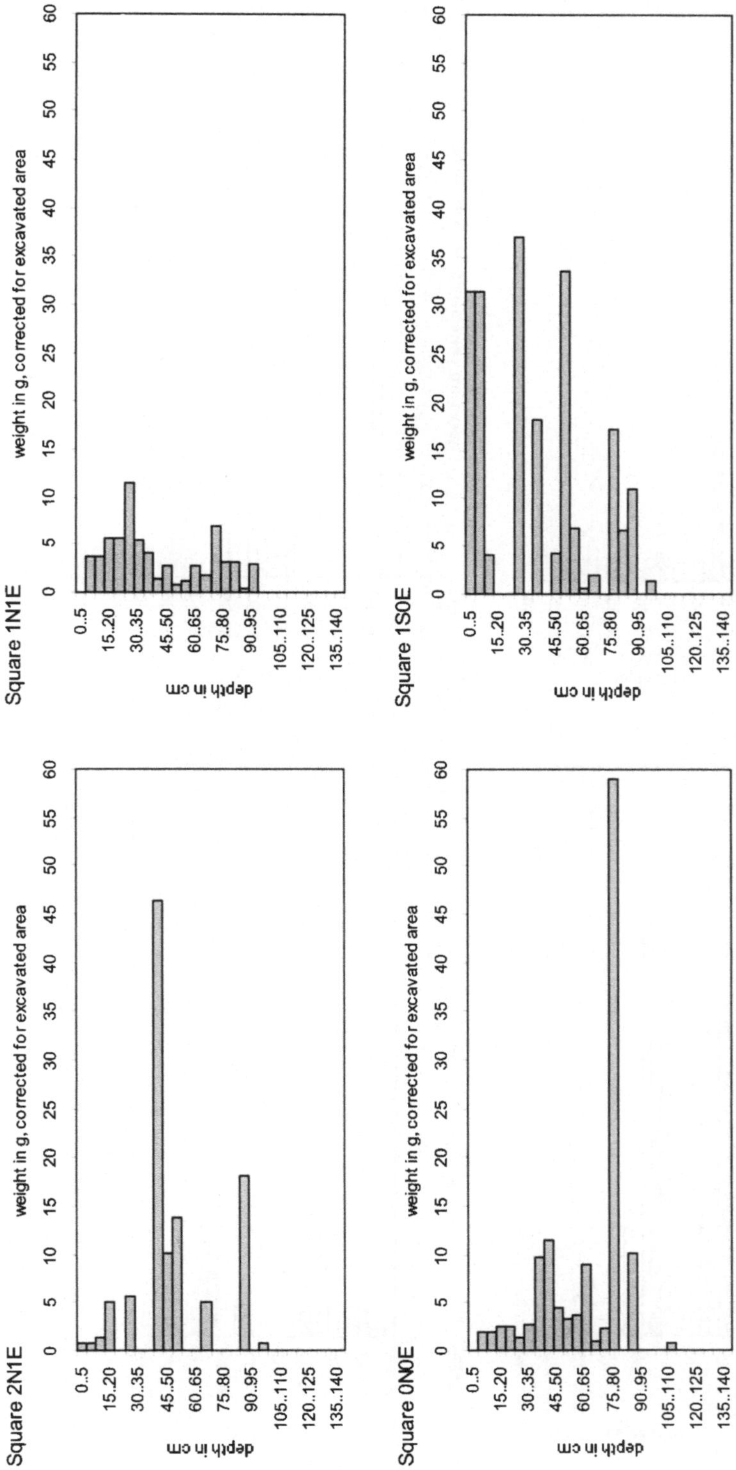

Figure 4.7. Toé Cave, distribution of molluscan shell through the deposit. See Chapter 7 for a more detailed analysis.

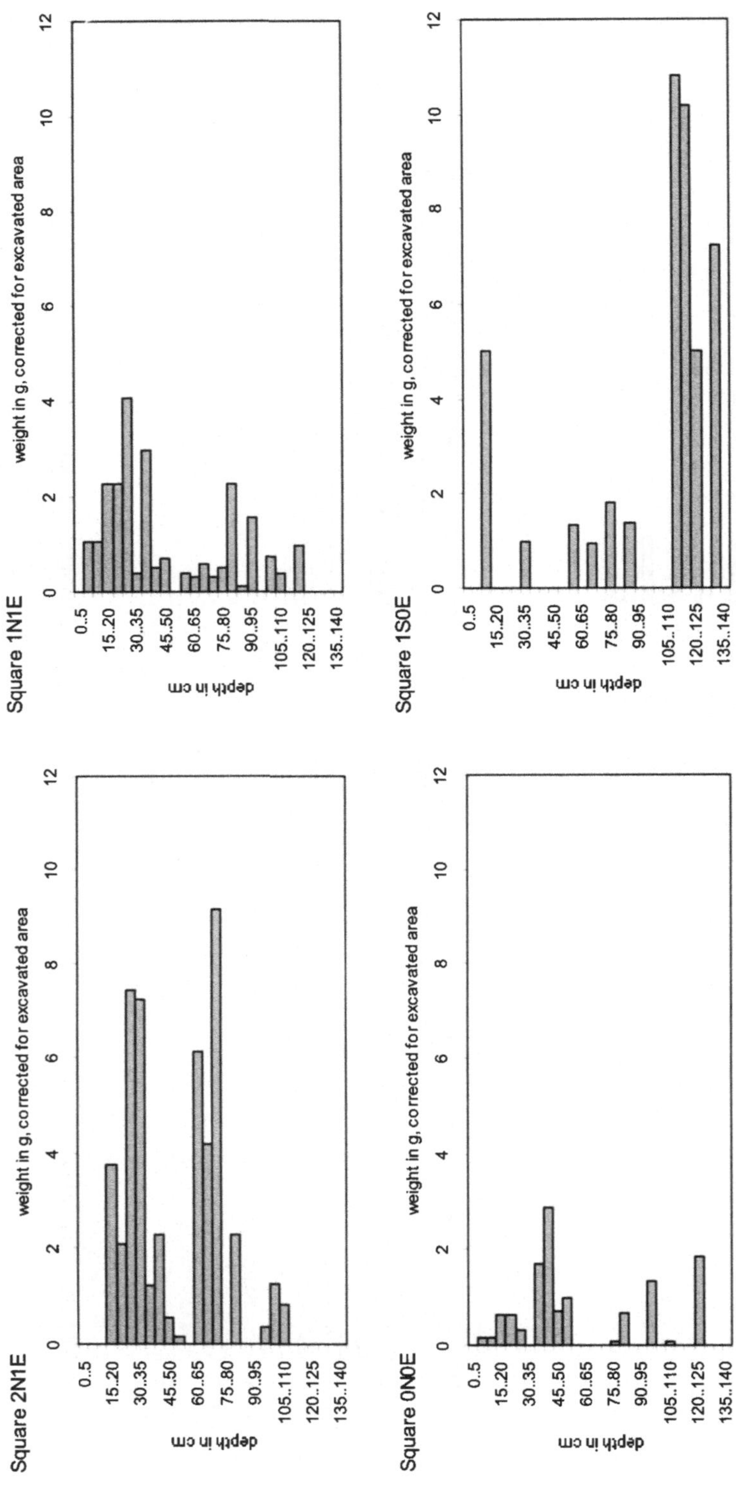

Figure 4.8. Toé Cave, distribution of avian eggshell through the deposit.

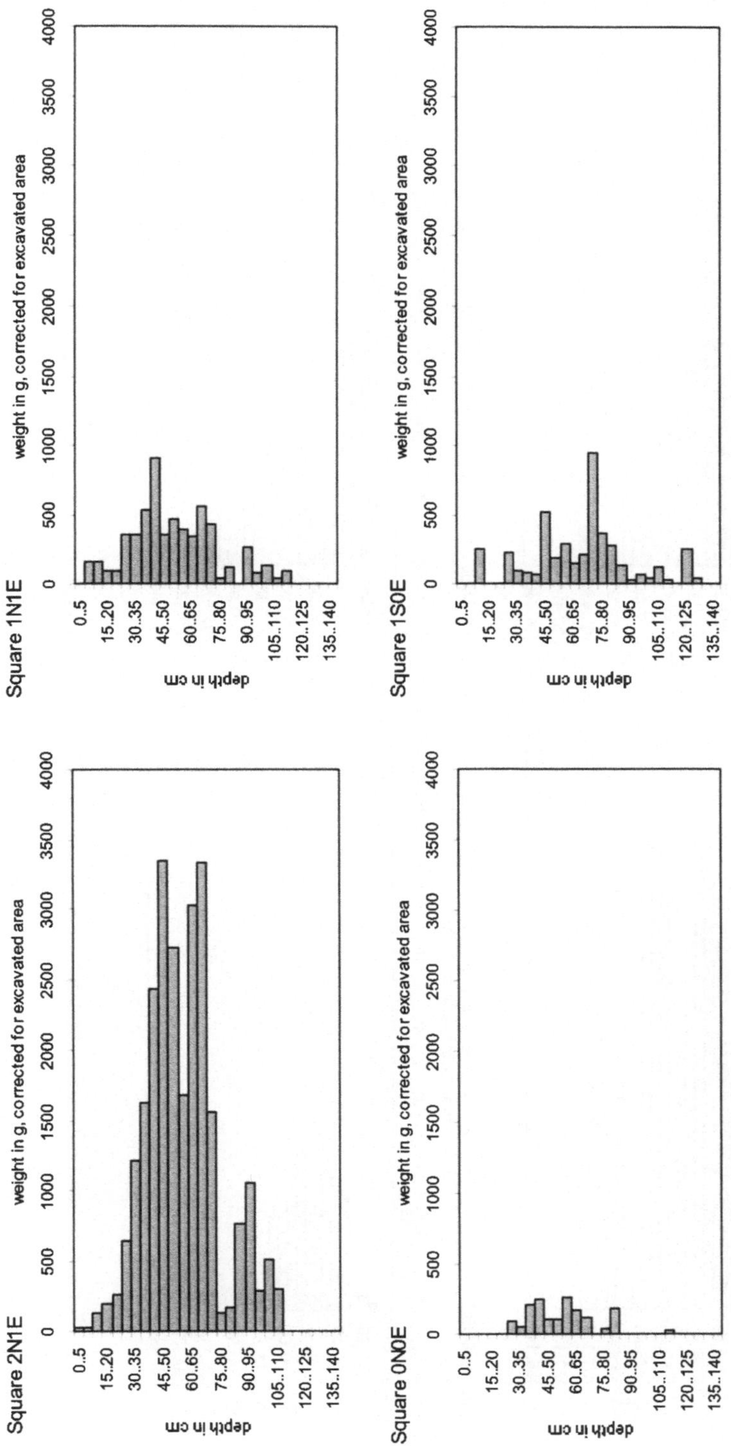

Figure 4.10. Toé Cave, distribution of chert artefacts through the deposit.

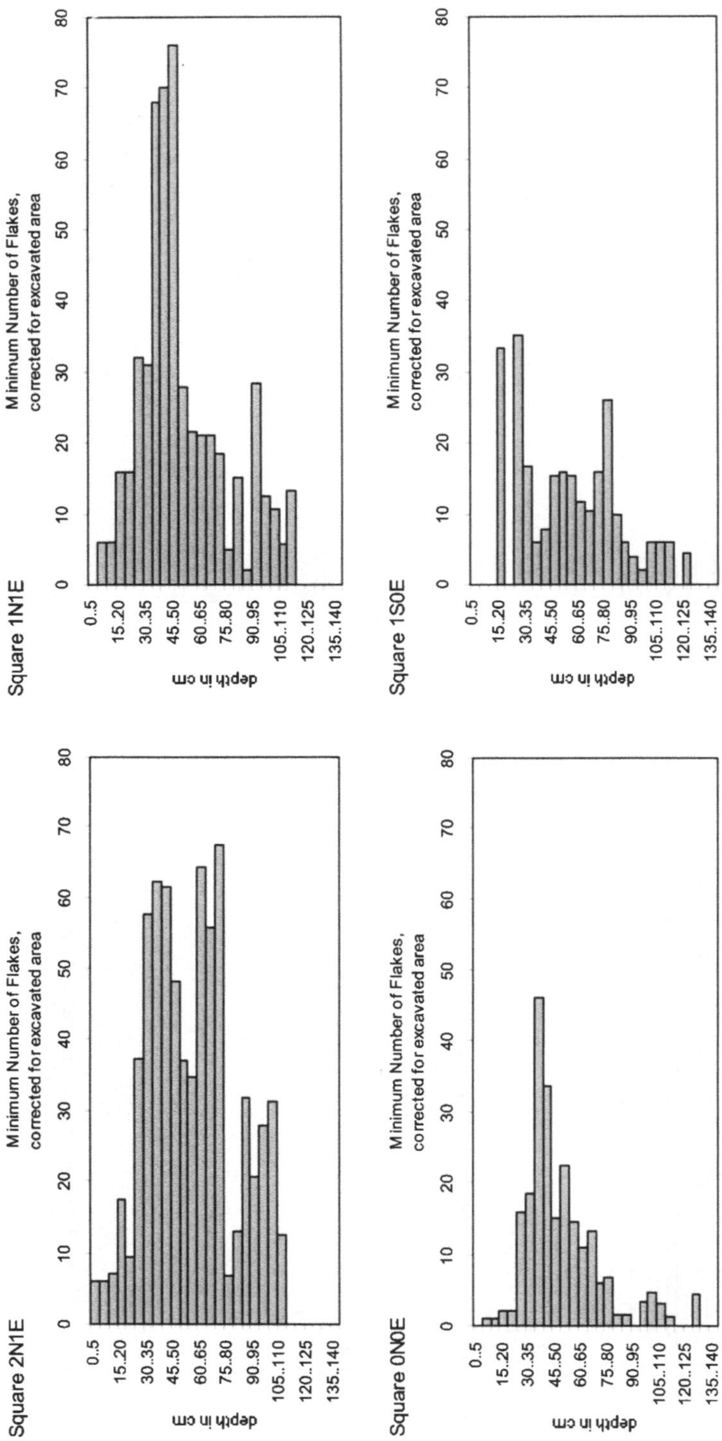

Figure 4.11. Toé Cave, distribution of Minimum Number of chert Flakes through the deposit.

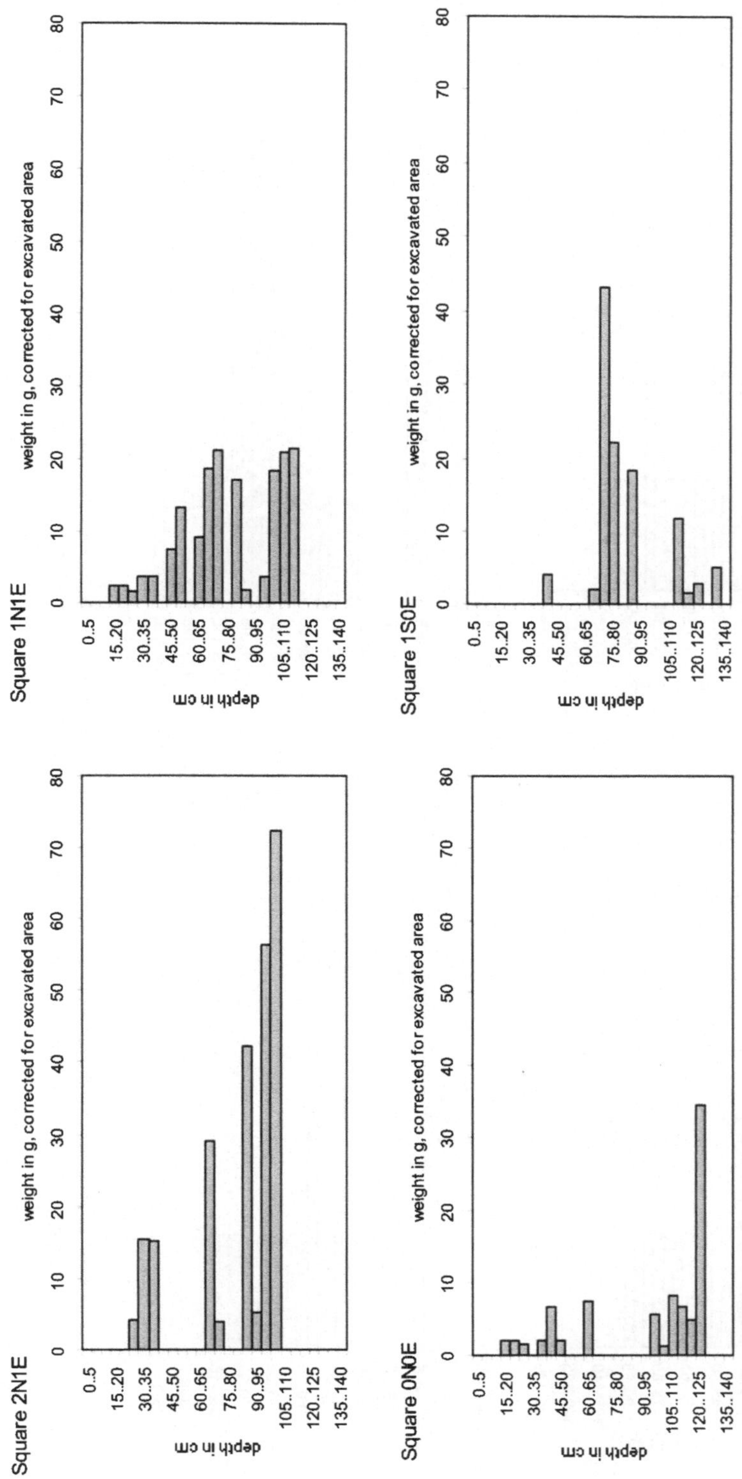

Figure 4.13. Toé Cave, distribution of ochre through the deposit.

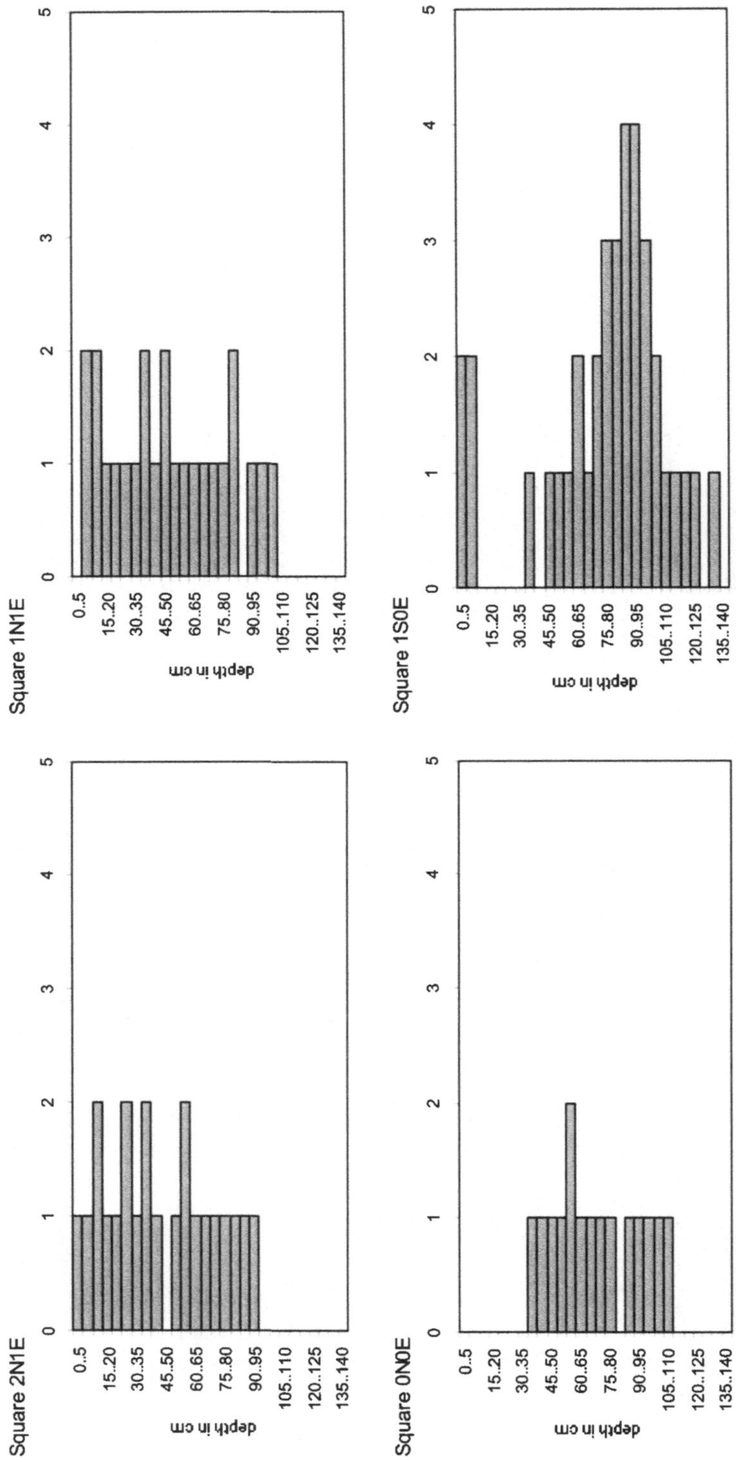

Figure 4.15. Toé Cave, degree of root damage on unburnt bone through the deposit. 0 = no unburnt bone present; 1 = 0% root damage; 2 = <10% root damage; 3 = 10-30% root damage; 4 = 30-50% root damage; 5 = >50% root damage. See Chapter 2 for more in formation on methods used.

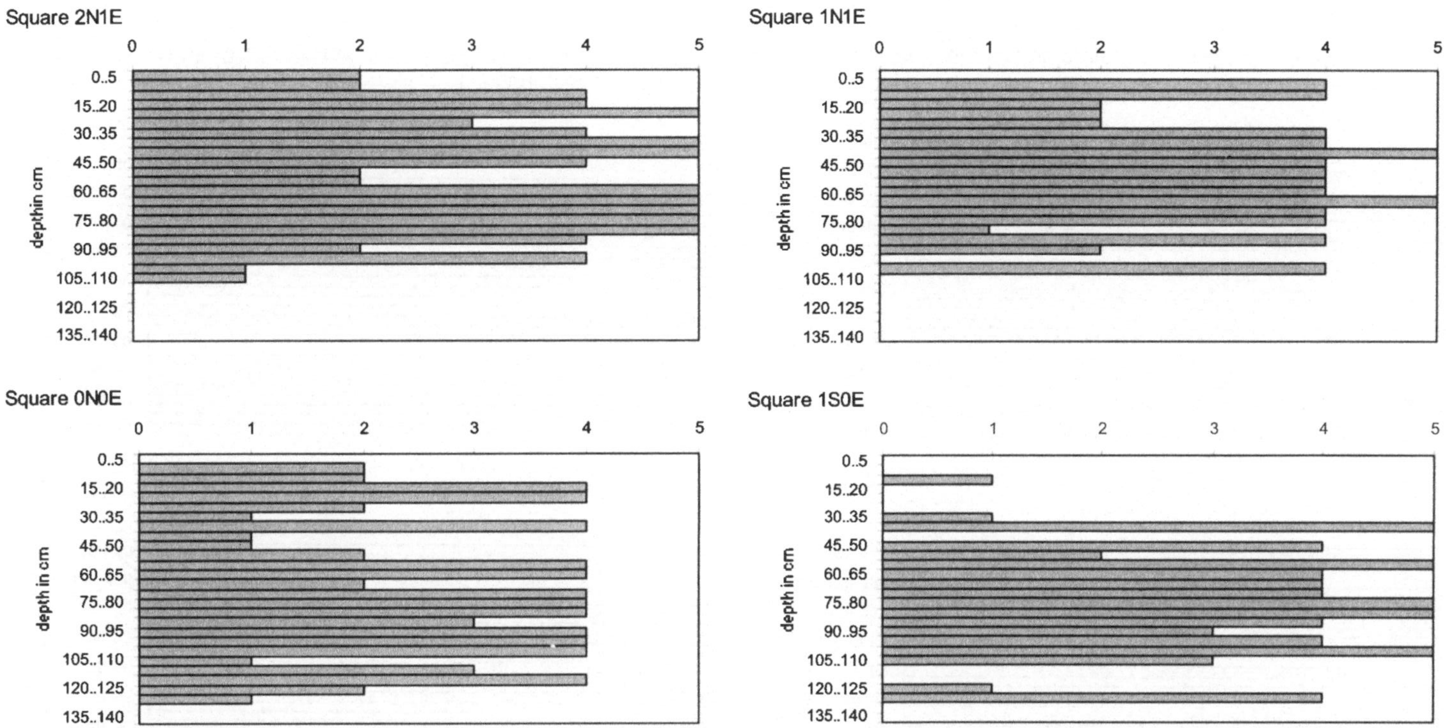

Figure 4.16. Toé Cave, abundance of tooth marks on unburnt bone through the deposit. 0 = no unburnt bone present; 1 = 0% tooth marks; 2 = <10% tooth marks; 3 = 10-30% tooth marks; 4 = 30-50% tooth marks; 5 = >50% tooth marks. See Chapter 2 for more in formation on methods used.

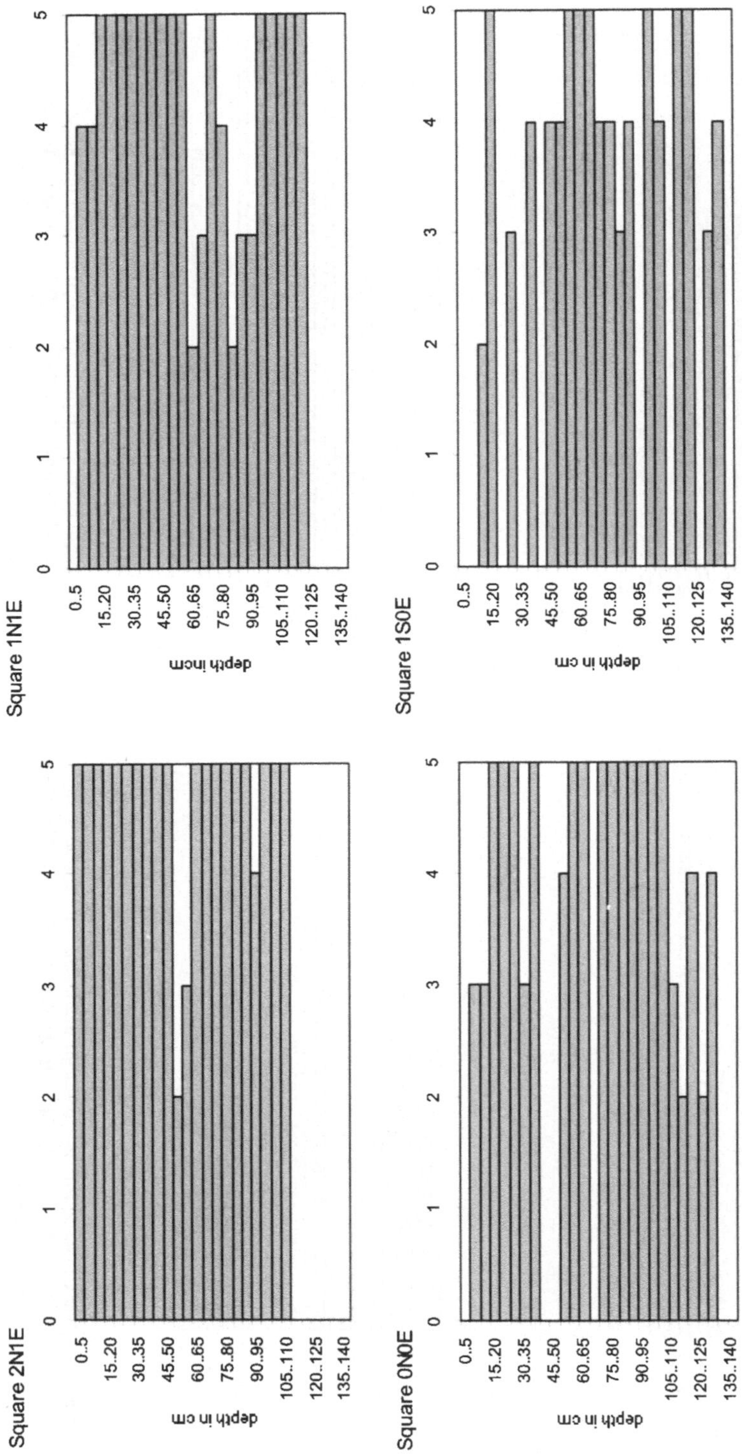

Figure 4.17. Toé Cave, degree of corrosion of unburnt bone through the deposit. 0 = no unburnt bone present; 1 = 0% corrosion; 2 = <10% corrosion; 3 = 10-30% corrosion; 4 = 30-50% corrosion; 5 = >50% corrosion. See Chapter 2 for more in formation on methods used.

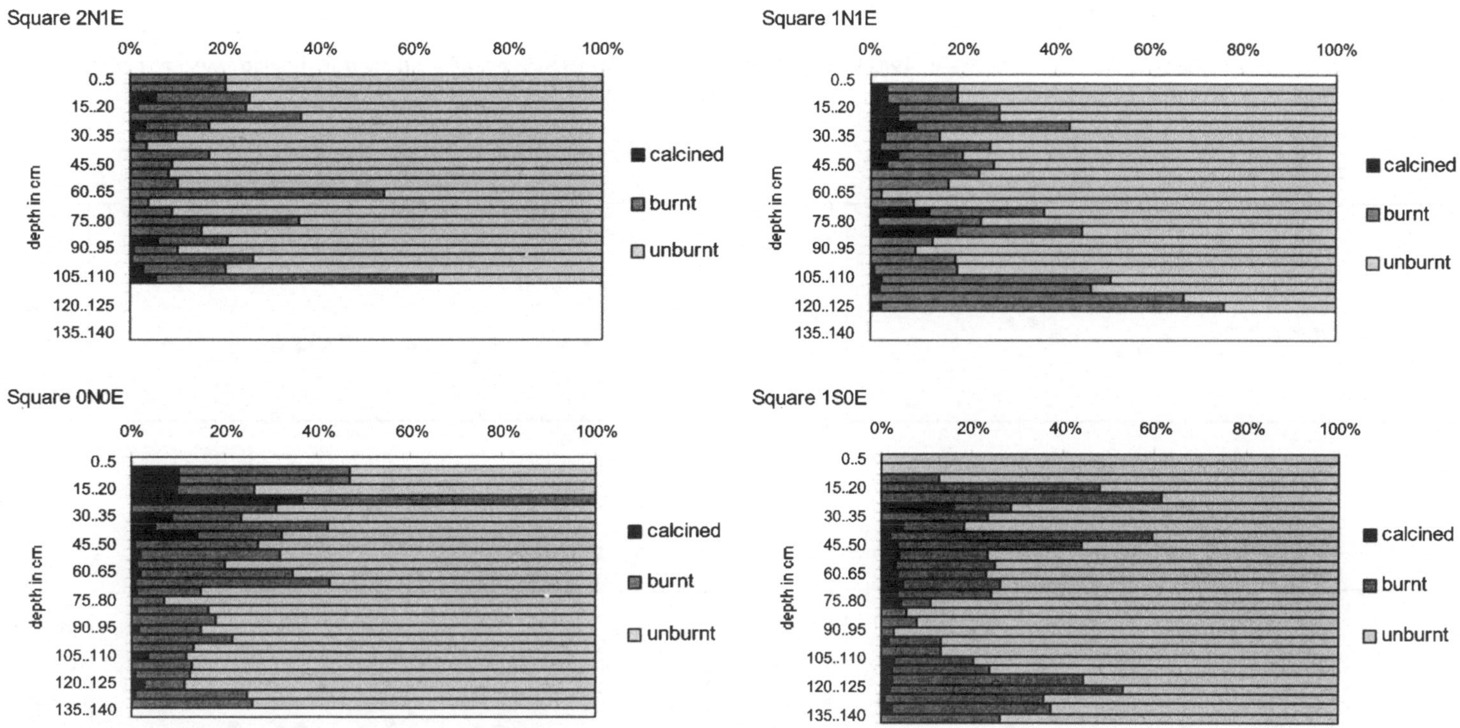

Figure 4.18. Toé Cave, proportions of burning categories of animal bone through the deposit.

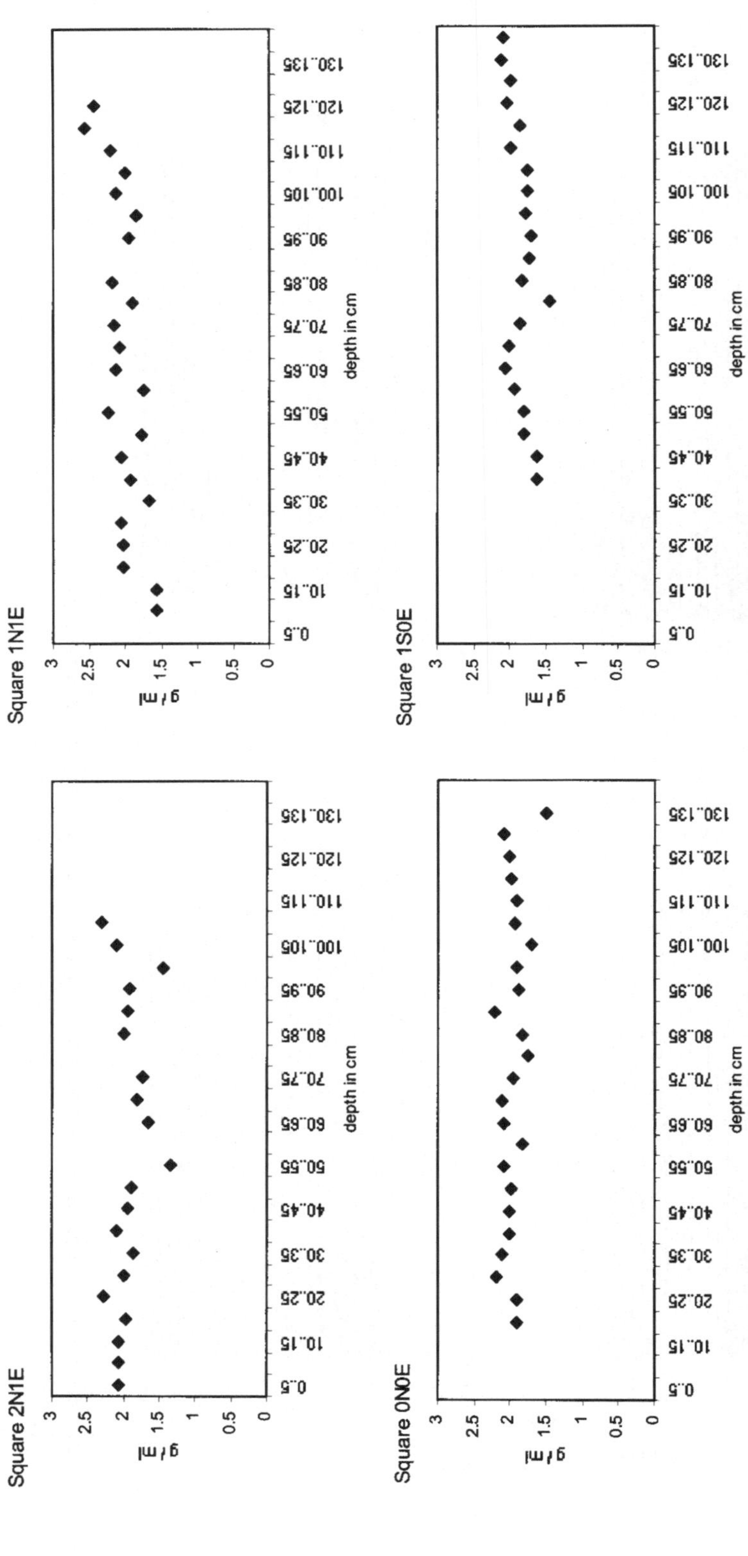

Figure 4.19. Toé Cave, density of bone through the deposit (Square 1S0E after Jelsma 1998).

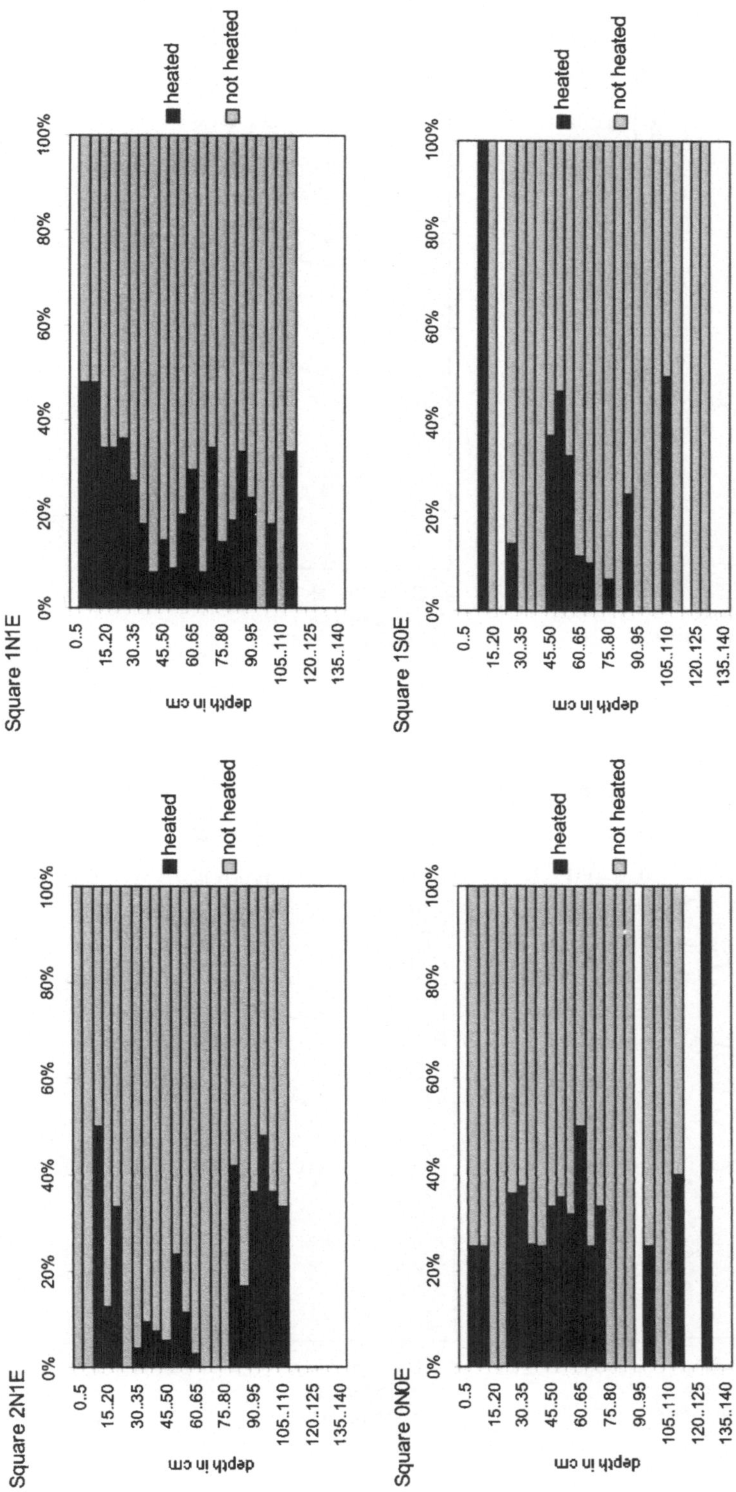

Figure 4.20. Toé Cave, proportions of unheated vs. heated chert through the deposit.

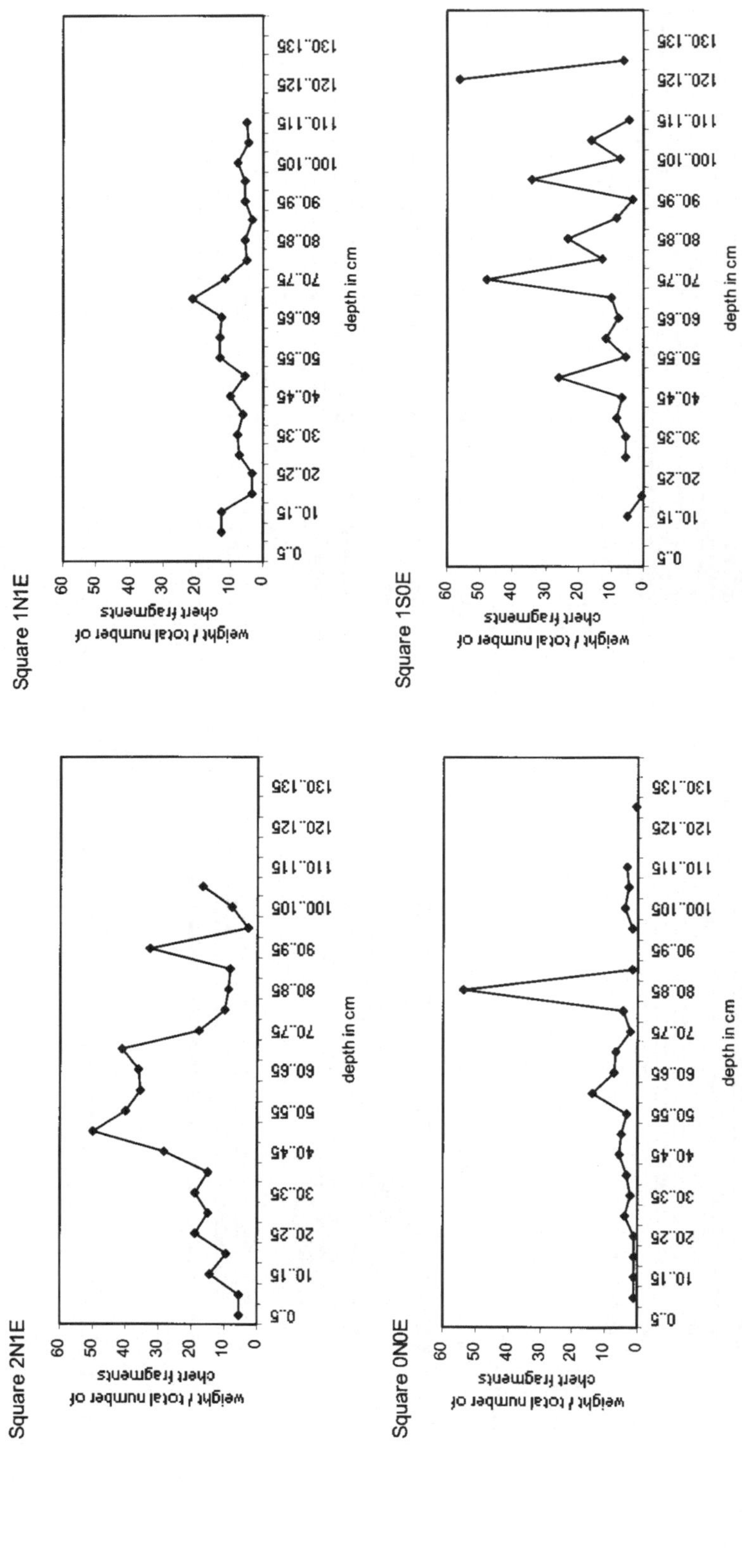

Figure 4.21a. Toé Cave, fragmentation of chert through the deposit, based on weight / total number of fragments.

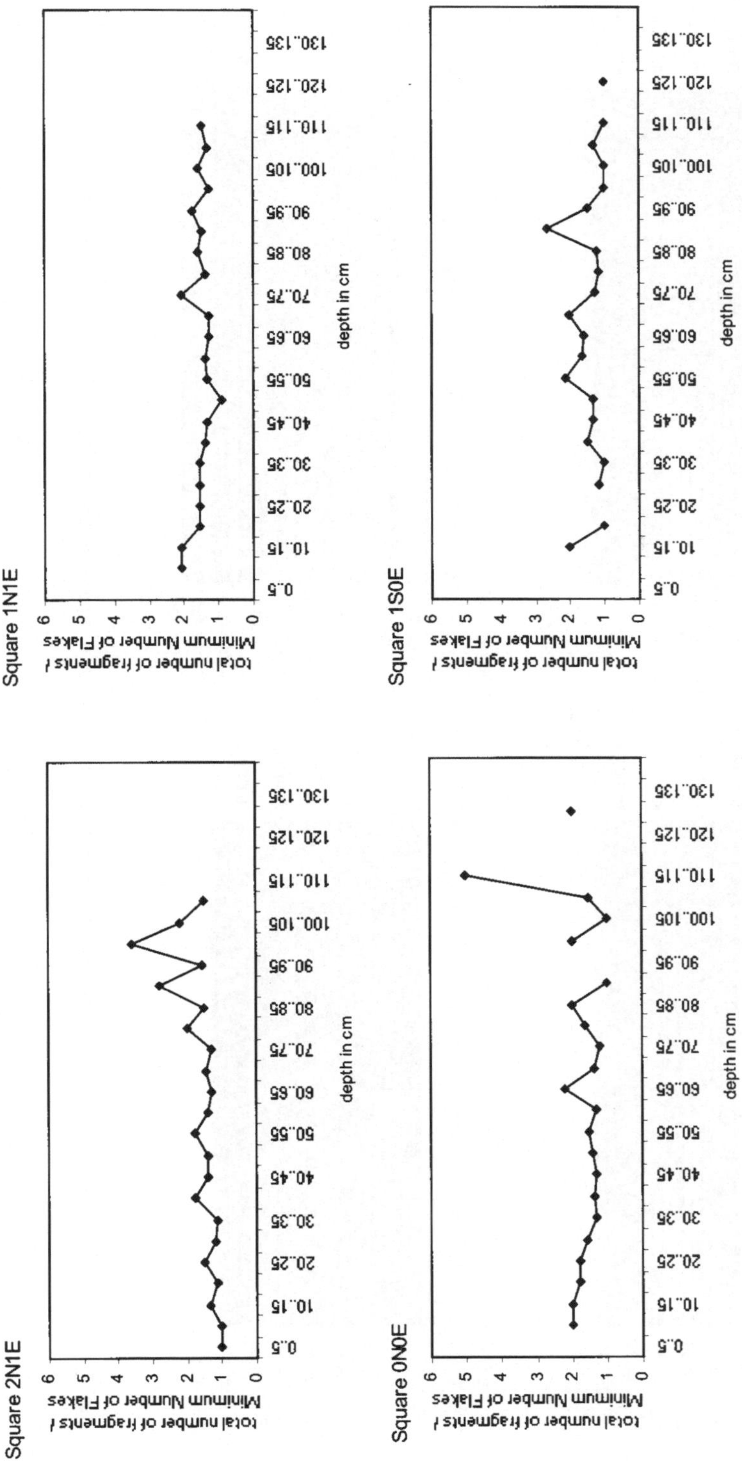

Figure 4.21b. Toé Cave, fragmentation of chert through the deposit, based on total number of fragments / Minimum Number of Flakes.

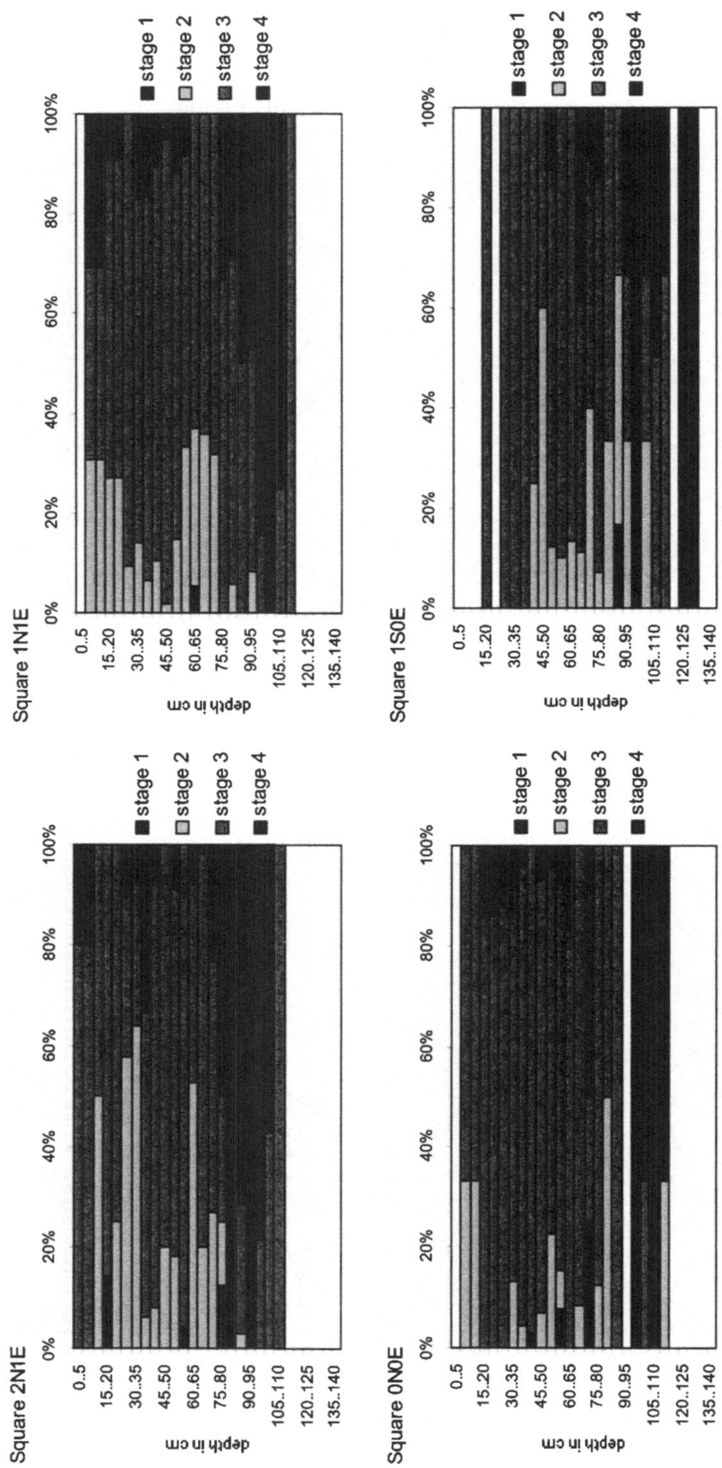

Figure 4.22. Toé Cave, degree of weathering of unheated chert through the deposit. No artefacts with stage 0 (no patination) are present.

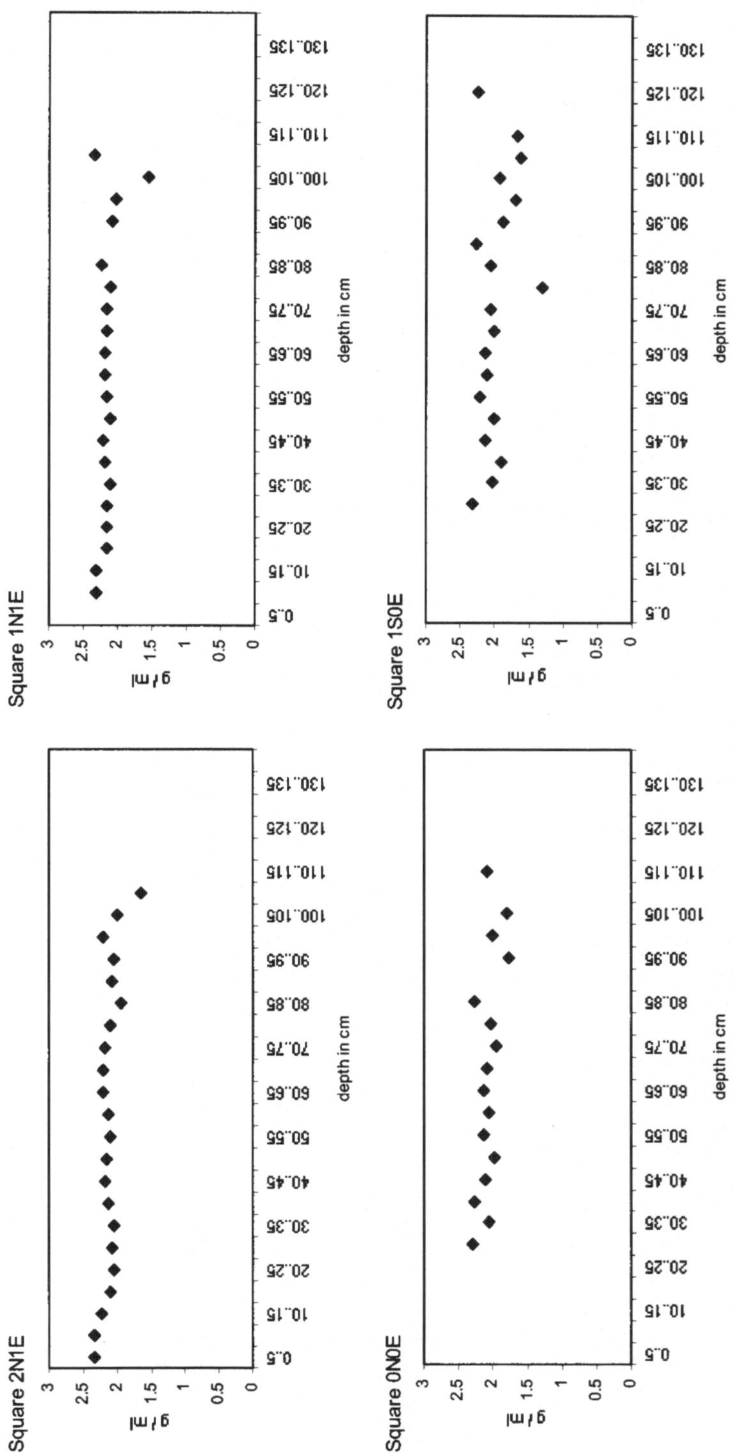

Figure 4.23. Toé Cave, density of chert through the deposit (Square 1S0E after Jelsma 1998).

5. Stone Artefacts

5.1. INTRODUCTION

New Guinean stone artefact assemblages are somewhat infamous for their amorphous quality (White 1972, 1977). A typical New Guinean lithic assemblage contains large quantities of debitage, a very low incidence of secondary retouch, and few if any recognisable, repeated forms or 'types'. Where stratigraphic sequences have been analysed, these typically reveal little or no change over time, even over long periods (White 1972). Naturally, there are exceptions to this generalisation – the major ones being the ground and polished stone adzes and axes, known from a number of Highlands sites in Papua New Guinea from 'about 4,000-3,000 BC' (Bulmer 1991: 475); the famous waisted blades known from various Late Pleistocene localities (Bulmer 1964, 1966; White et al. 1970; Groube 1984, 1989; Groube et al. 1986); and the distinctive sago-adzes described by Rhoads (1980) from the Gulf Province of Papua New Guinea. However, these rather specific exceptions merely highlight the fact that, for the most part, New Guinean lithic assemblages really are 'crude, colourless and unenterprising', as White (1977) suggested in a parody of early views of Australasian lithic assemblages in general.

The lithic assemblages from Kria and Toé Caves do little to change this general picture of New Guinean assemblages. Both assemblages consist predominantly of large numbers of fairly 'crudely' flaked pieces of chert, only a small proportion of which show evidence of subsequent modification and use. However, despite the many features in common between the two assemblages, they differ substantially in the average size of fragments and in the proportions of different kinds of debitage. This chapter will focus on the differences between the two assemblages in the hope that we might come to better understand some of the activities that took place at these small rockshelter sites over many thousands of years.

5.2. METHODS

Today, the study of lithic assemblages has a much wider focus than in the past. Studies involving technological analyses (Flenniken & White 1985; see also Odell 2000), spatial analyses (Koetje 1994; Montet-White & Williams 1994), sourcing analyses (Church 1994; Summerhayes et al. 1998) microwear and residues (Fullagar 1984; Fullagar & Field 1997) and properties of lithic materials themselves (Purdy 1974; Flenniken & White 1983; Griffiths et al. 1987), as well as the effects of taphonomy (Hiscock 1985,

1990) are directed to describe the human behaviour that underpins the production of lithic assemblages. Four of the major areas of stone artefact analysis are relevant to the study of the Bird's Head stone assemblages: the effects of taphonomy; stone-working technology; artefact function; and typology. The effects of taphonomy have already been discussed in Chapters 3 and 4, but certain aspects will be addressed again where relevant. In this chapter most emphasis will lie on technology and function, and the assemblages will furthermore be discussed in a comparison to other assemblages from the region.

5.2.1. *Technology*

Technological analyses of lithic assemblages are directed at understanding the various mechanistic processes that are involved in the physical reduction of the assemblage. Analyses of this kind can provide valuable cultural information, especially where highly standardised processes are involved, as for example in the production of elaborate 'points' or in the more highly developed blade technologies (Hiscock 1986, 1993; Mulvaney & Kamminga 1999: 230-231).

For crystalline raw materials such as chert, technological analyses generally use a fairly standard set of categories to describe stone-working debitage. The system used here is similar to that used by White (1972) for his major study of Papua New Guinean highland lithic assemblages.

The primary technological division is drawn between a core and a flake. In the simplest terms a 'flake' is the fragment that breaks away as a result of percussion or the application of pressure. Ideally it will preserve the classic signs of percussive origin – a striking platform, bulb of percussion and a positive bulbar surface (Crabtree 1972). The piece that was struck and that remains behind is a 'core'; in theory it will display at least one flake scar. On some occasions, percussion leads to general fracturing of a piece rather than the production of a neat flake and core. In such an event, the resulting amorphous debitage may include pieces that are clearly derived from stone-working activities but cannot be validly classified as either a flake or a core. Such pieces are here classified as 'flaked pieces' [Hiscock 1984: 139; called 'modified fragments' by White (1972)].

The boundary between flake and core is of course not absolute because a large flake can itself be used subsequently as a core. This is particularly prevalent in the Ayamaru assemblages where much of the raw material was probably brought to the site in the form of large primary flakes, which were subsequently used as a source for smaller flakes. These are distinguished as 'block-cores' and 'flake-cores', the latter retaining at least some evidence of a striking platform and/or positive bulbar surface. The Ayamaru assemblages contain relatively few block-cores and many potential flake-cores.

In some cases percussion or pressure is applied, not for the purpose of producing a new flake, but instead to modify either the shape or edge characteristics (e.g. angle, evenness or sharpness) of a stone artefact that has been selected for use. This special class of activity is often called 'retouching' but in some cases it can be difficult to decide what the purpose of the flake removal was. For this study, secondary removal of flakes from the ventral surface of a primary flake was in general scored as 'retouch'. Where retouching has resulted in the removal of very small flakes only, it was assumed that the stone-workers' purpose was mainly to improve the general shape of the artefact or its working-edge, hence the flake was simply scored as a 'retouched flake'. However, where the flakes being removed were of a suitable size for independent use (i.e. more than 20-30 mm), the large original flake was considered as a 'flake-core', the edges of which may again have been retouched or used.

Flakes are sometimes divided into various subcategories, mainly reflecting variations in the ventral and dorsal surfaces. For this study only two special categories of flakes were distinguished. The first are flakes that have remnants of the original weathered cortex of the nodule on their dorsal surface or on their striking platform, collectively termed 'cortex removal flakes'. These are typically flakes that have been struck at a relatively early stage in the reduction process.

The second category can be collectively termed 'edge-removal flakes' [White's (1972: 9) 'trimming flakes'], but actually includes two slightly different subgroups:

- *working-edge removal flakes*: these flakes are characterised by the presence of obvious, heavy usewear along the dorsal margin of the striking platform, indicating that the flake had removed all or part of a utilised and/or retouched working-edge. In many case, the percussion bulb lies exactly opposite the zone of intense usewear, producing a potential new working-edge of near-identical form. In other cases the flake has been struck behind but a little to the side of the zone of intense usewear, producing a potential new working-edge of somewhat different morphology (see for example Fig. 5.1).
- *platform removal flakes*: in some cases a flake was struck in such away that it removed a former striking platform, with or without associated usewear. This platform will be observed on the dorsal surface of the new flake; it is generally oriented either perpendicular or oblique to the striking platform of the flake.

Apart from the flaked chert, the excavated sites produced small quantities of stone that are exotic to the local limestone area and must therefore have been brought to the

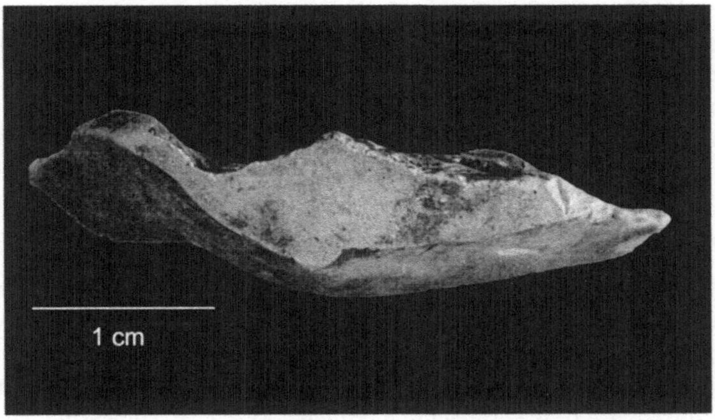

Figure 5.1. Working-edge removal flake. The percussion bulb lies exactly opposite the zone of intense usewear. Specimen from Kria Cave, number 710af.

cave. These 'manuports' include examples with obvious signs of use but no prior modification, such as hammerstones. However, other specimens show no obvious signs of use and the reason for their transport is obscure.

One final category, although not strictly a technological one, is 'potlid flake'. These small to very small, sub-circular fragments are the by-products of excessive heating of the stone (Purdy 1975). In many parts of the world siliceous stone was subject to deliberate heating as a way of improving its flaking properties (Crabtree & Butler 1964; Purdy 1974; Flenniken & White 1983). Typically, heating drives out the interstitial water and increases the number of internal micro-fractures, thereby increasing the brittleness of the material and making it easier to flake. As noted in Chapter 2, excessive heating or rapid cooling can result in spontaneous spalling to produce potlids, crazing of the surface, or 'crenation' and irregular, non-conchoidal fractures.

Rowney & White (1997) correctly emphasise the need to distinguish simple 'heating' that can occur through accidental burning of stone within or beneath a hearth, and 'heat treatment', which implies the deliberate heating of stone prior to flaking. They indicate that methods such as palaeomagnetism can be used to identify past heating, but that the sequence of events is better determined from analysis of flaking surfaces. In particular, they suggest that conchoidal surfaces showing a high lustre can be taken as evidence of flaking after the heating process, due to the fact that the lustre is produced at the time of fracture (see also Purdy 1974). However, the fact that an accidentally heated piece could be subsequently selected for flaking means that such evidence alone cannot be taken as evidence of intent. Griffiths et al. (1987) suggested that better evidence of intent could be found in an association of heated stone with particular artefact types (e.g. pressure flaked points). However, such patterns of association are ultimately subject to the same criticism, i.e. that they do not distinguish deliberate heating from *post hoc* selection of heated material. Flenniken & White (1983) claimed that the method of heat-treating stone was widely known and applied across prehistoric Australia, from Late Pleistocene times to the recent past. Accordingly, the method may have been known in the Bird's Head and elsewhere in New Guinea.

Where the heating of stone occurs accidentally, the proportion of potlids in an assemblage can be used as an index of the frequency of hearth construction within a particular site (Hiscock 1985, 1990). If broken, potlids can be confused with broken flakes, but they typically lack the diagnostic features of a percussion flake (e.g. a ring-crack and conchoidal waves).

5.2.2. *Function*

The analysis of stone artefact function has become a highly specialised area of study. The two major areas of study are concerned with what can be loosely termed 'usewear' and 'residues'. Usewear studies involve the microscopic examination of the edges or surfaces of stone artefacts to identify damage or other modification such as polish that has resulted from use of the artefact (e.g. Fullagar 1984). By comparing these patterns to examples produced under known circumstances of use, it is sometimes possible to infer the mode of use of prehistoric artefacts.

Residue studies, often undertaken in parallel with usewear analysis, seek to identify actual traces of organic or inorganic material adherent to the prehistoric artefacts. The 'residues' may include blood cells or even animal tissue where tools have been used to butcher animals or treat hides (Loy 1983, 1987), starch cells or other plant residues related to vegetable preparation (Fullagar 1988; Fullagar et al. 1992; Loy 1990; Loy et al. 1992; Barton & White 1993).

Both kinds of studies are technically demanding; moreover, they are best undertaken on an assemblage that has been excavated and processed with these goals specifically in mind. In the present case, the artefacts were thoroughly washed prior to transport to the University of Groningen, an act that might have removed microscopic residues. The transport itself has also undoubtedly caused some degree of damage to the delicate edges. While any loss of opportunity for future study is regrettable, under the circumstances alternative treatments of the collection were impractical on account of limitations of time and materials, and because of the need to satisfy quarantine requirements.

For the present purposes, the potential functional edges of all artefacts were examined for signs of obvious usewear or micro-retouch (generally characterised by fine step-fracturing). This was generally done with the naked eye but ambiguous specimens were examined with a light microscope to 40X magnification. All specimens were classified as either 'possibly used' or 'no usewear observed'. As a rule, caution was exercised in identifying an artefact as 'possibly used', on the grounds that some minor damage to edges might have occurred through trampling or during excavation, transport and handling. Many artefacts are also very weathered. The percentage of artefacts scored as having use-related damage thus is presumed to represent an underestimate of the true value in the assemblages.

Many specimens possessed a distinctly 'notched' form of usewear, with localised, heavy step-fracturing within a concave flake scar. These artefacts were noted specifically as bearing one or more 'notches', but no further attempt was made to investigate the microscopic morphology of these interesting wear features; their possible significance will be discussed in general terms in a later section.

No examples of use polish or 'silica gloss' were noted in either assemblage. This is significant in the light of Rhoads' (1980) discovery of high silica gloss on both ethnographic and prehistoric sago-adzes and on re-sharpening (or 'edge-removal') flakes derived from these adzes from the Papua New Guinean Gulf Province. Silica gloss on flakes is also reported from archaeological sites in East Timor (Glover 1986), Sulawesi (Glover 1976) and the Philippines (Scheans et al. 1970).

5.2.3. *Typology*

In recent years typological analysis – the process of classifying stone artefacts into certain formal types – has fallen somewhat into disrepute, especially among Australian and Pacific archaeologists. To some extent, this reflects the somewhat amorphous nature of many of these regional assemblages, and the discovery that several traditional stone tool-using groups in both Australia and New Guinea possess only very simple systems of ethnoclassification of stone artefacts, often with a strong emphasis on edges rather than overall shapes (White 1972; Gould 1980). However, it also reflects the changing views of archaeology as a wider science, with a greater focus on past human behaviour, rather than cultural affinities.

Despite this overall move away from typology, a strong case can be made to not completely abandon a typological framework where it has been applied in previous regional research and appeared to reveal changes in the composition of lithic assemblages through time and space. In Australia, few archaeologists would deny that some kind of 'cultural' change must have occurred in association with the emergence of the 'Small Tool Tradition' around 5000 BP, with its distinctive microlithic implements (Mulvaney & Kamminga 1999); or that the widespread stone point industries of the late Holocene in Northern Australia are not somehow indicative of shared cultural patterns across this area.

In the New Guinean context, the few widely recognised formal stone artefact types have been introduced already: ground edge axes and adzes; waisted blades; and sago-adzes. Several of these were defined by Bulmer (1964) and later discussed by White (1972: 5-7). White (1972) classified the majority of retouched artefacts from archaeological sites in the Eastern Highlands of Papua New Guinea as 'scrapers' and divided these into various sub-categories based on the location of the working-edges relative to the striking platform. However, in his opinion it was the properties of the edge that were significant, rather than its location. Jelsma (1998) presented a formal European typological classification for the collection from Toé Cave, as a means of objective description of the retouched artefacts, but without implying any cultural affiliations. At an early stage of this study, Marcel Niekus (Groningen Institute of Archaeology) applied a similar classification to the lithic assemblage from Kria Cave, as he did for Jelsma. His classifications are not presented here because of a lack of comparative data for prehistoric New Guinean assemblages. Furthermore, all of the retouched artefacts from Toé and Kria Caves would be classified as 'scrapers' and there seems little value in attempting any formal sub-division of this group along European typology. However, some metric analysis will be presented, based on a limited set of measurements.

5.2.4. *Recorded characteristics and measurements*

All artefacts and manuports were scored individually according to the following system:

1. Major category: core; flake; flaked piece; potlid; unmodified manuport; modified manuport.
2. Raw material: chert; red silicious tuff; limestone; quartzite; metamorphic rock.
3. Breakage: undamaged; proximal fragment; distal fragment; medial fragment (both distal+proximal missing); at least one margin damaged; longitudinal cone split (see also Chapter 2).
4. Cortex: present on core or dorsal surface of flake; absent from all surfaces.
5. Retouch: present; absent.
6. Usewear: present on one or more margins of a flake or core; not observed.
7. Edge removal: working-edge removal flake; platform removal flake.

8. Platform surface of flakes: conchoidal (one smooth surface); cortical (platform consists of cortex); facetted (on the platform negative scars of formerly removed flakes are visible).
9. Indications of heating (potlidding, crenation, irregular breaks, lustre): present; absent.
10. Weathering: no patination; lightly patinated; patinated; heavily weathered; very heavily weathered. For full definitions see Chapter 2, page 25.
11. Length (flakes only) = maximum length from the ringcrack on the striking platform to termination along the axis that runs along the direction of the bulb.
12. Width (flakes only) = maximum width perpendicular to the length axis.
13. Thickness (flakes only) = maximum distance from ventral to dorsal side.

5.3. THE LITHIC ASSEMBLAGE FROM TOÉ CAVE

For this analysis the Toé Cave assemblage is divided into Holocene and Late Pleistocene components, using the combination of radiocarbon dates, distribution of cultural materials, and the faunal composition to identify this boundary (Chapter 4). The very small sample of stone artefacts from the postulated 'transition' zone are included with the large Holocene sample in order to maintain the integrity of the smaller Pleistocene assemblage. About 40% of the habitable floor space of the Toé entrance was excavated, with a transect running across the full width of the entrance; hence the excavated lithic sample is considered representative for this site.

The Toé Cave deposit produced a total of 1775 fragments of worked or clearly exotic stone material, with a combined weight of 26,341 g. The great bulk of this material derives from the Holocene levels of the deposit, with only 10.8% by weight and 14.8% by number coming from the Late Pleistocene levels. The flaked stone consists primarily of a poorly silicified chert, much of it a pale beige colour but with an occasional blue fragment. This material accounts for a slightly higher proportion of the total assemblage in the Holocene than the Late Pleistocene levels (Table 5.1), especially as measured by weight. Other raw materials that show signs of flaking are a yellow to red siliceous stone (tentatively identified as a silicified tuff) and a moderately dense, crystalline limestone; together these account for less than 4% of the total assemblages by weight or number.

Table 5.1. Toé Cave, proportional representation of different raw materials of stone artefacts in the Holocene and Late Pleistocene units, calculated by number and by weight. Raw data are given in Table 5.2a and b, at the end of this chapter.

Unit	chert	silicified tuff	limestone	quartzite	chert?	limestone?	quartzite?	metamorphic	unknown
Holocene: count	95.6%	0.1%	2.0%	0.3%	0.2%	0.6%	0.1%	0.8%	0.3%
Pleistocene: count	95.8%	0.4%	1.5%		1.1%	0.8%		0.4%	
Holocene: weight	87.8%	0.2%	1.2%	0.6%	0.0%	1.9%	0.0%	8.2%	0.0%
Pleistocene: weight	67.0%	1.9%	0.2%		0.2%	0.1%		30.7%	

Other exotic stone includes rounded pebbles of quartzite and fragments of metamorphic rock. Some quartzite pebbles recovered from the Holocene unit show clear signs of use as hammerstones. In contrast, none of the metamorphic fragments show any definite sign of modification. One large metamorphic fragment from the Late Pleistocene unit accounts for 30.7% of this assemblage. If this piece is omitted, the percentage of chert by weight in this unit rises to 96.6%.

The sources of the various raw materials are unknown. The chert and crystalline limestone are potentially available from sources within the limestone anywhere on the Ayamaru Plateau. The silicified tuff and metamorphic rock are presumably derived from mountainous areas to the east or north of the Ayamaru Lakes, but the fact that they typically occur as river cobbles suggest that they may be locally available.

Overall the chert material is more weathered in the lower levels of the deposit, presumably reflecting its older age. More details on weathering and fragmentation of this material are given in Chapter 4.

The overall composition of the Holocene and Late Pleistocene assemblages is shown in Tables 5.2a and b (at the end of this chapter), divided into the basic technological categories as described under Section 5.2.1 (Technology). The two differ in the absence of any cores and of quartzite hammerstones in the Late Pleistocene assemblage, and the higher proportion of flaked pieces relative to flakes. The latter difference probably reflects the more weathered condition of the Late Pleistocene assemblage, with a greater degree of breakage leading to a higher proportion of non-diagnostic fragments (see below).

5.3.1. *Stone-working technology*

The dominant stone-working raw material in the Ayamaru Lakes area is an originally blue to beige-brown chert. In both sites, but especially in Toé, this has weathered into a predominantly beige to pale cream material. Occasional pieces of chert are mottled with iron-rich pigments, ranging in colour from bright yellow to red or purple, usually present as 'bands' against a cream background.

Throughout the collection, artefacts are observed with traces of a soft, chalky cortex. This points to a primary nodular source for the chert, from within the limestone or a weathered soil, rather than a secondary source from river cobbles. Many pieces also show internal air pockets ('vugs', see Fig. 5.2) or fossil inclusions, indicating that the original raw material was of very irregular quality.

Although a significant proportion of the Toé Cave chert artefacts show definite signs of heating (Chapter 4), there is no compelling evidence to suggest that this occurred as a result of deliberate 'heat treatment'. In particular, there are no specimens in which flake scars show a significantly increased 'lustre' in comparison to surrounding surfaces. Nor indeed is there any great variation in lustre within the assemblage, and it is possible that the rather poorly silicified chert of the Ayamaru region does not respond to heating in ways that are favourable for stone-working.

The Holocene assemblage produced a total of ten cores in the restricted sense, all but one of which are made of chert. The chert specimens range in size from 39-394 g (mean

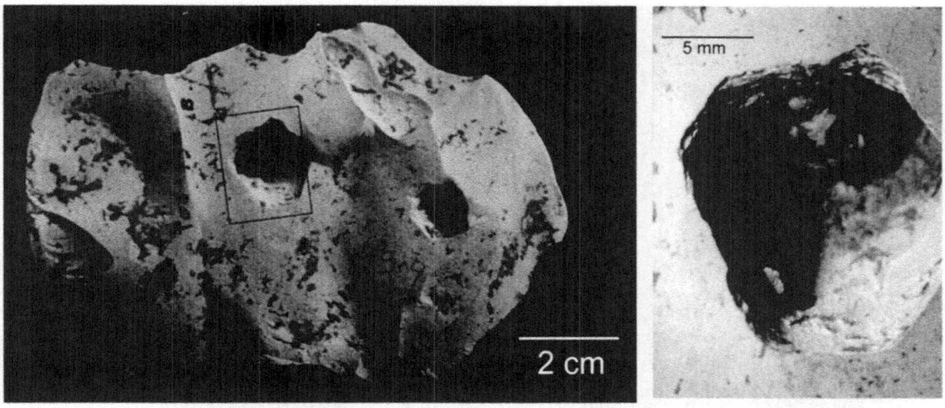

Figure 5.2. Chert artefact with air pockets or 'vugs' (left) with close-up of framed area (right). Note the unusual crushing along the edge of the air pocket (see close-up), perhaps comparable to that found in the distinctive notches (see below) and interpreted as functional usewear. Specimen from Toé Cave, number 492a.

± sd = 128.2 ± 124.3 g) and all but one retain patches of cortex. The majority of the cores display evidence of reduction from two or more striking platforms. The largest core is rounded in form and has cortex extending over approximately one-third of its surface area; it is clearly derived from a nodule rather than a flake (see Fig. 5.3). However, all of the other 'cores' could be derived from large flakes.

Nine specimens are tentatively identified as 'flake-cores' because all of the 'retouch' is very coarse. These range in weight from 45-208 g; all but the smallest one retain areas of cortex on the dorsal surface. Measurements are summarised in Table 5.3. An example of a potential 'blank' flake-core (without any retouch or larger flakes removed) is shown in Figure 5.4.

The flake-cores are reduced almost exclusively from the ventral surface, with flakes struck from both the distal and the lateral margins. In no case is there any indication of any special preparation of the striking platform. As noted above, some of the specimens classified as 'block-cores' could well represent later stages in reduction in which the original striking platform and associated bulbar region have been removed. However, only 14 platform removal flakes were recovered from the Holocene levels, suggesting that core rotation as a technique was practiced relatively infrequently.

The 'retouched flakes' from the Holocene levels include a number of large specimens with thicknesses comparable to those of the flake-cores. These specimens may include some functional tools; however, it is also likely that some at least represent flake-cores at earlier stages of reduction. Figure 5.5 illustrates a selection of the large retouched flakes, showing the apparent continuity in form between these classes. If the more heavily flaked specimens are used as guide, then flakes with a thickness of more than 25 mm were suitable for sustained flake removal. A total of 44 retouched flakes from the Holocene levels of Toé Cave conform to this criterion; measurements of these specimens are shown in Table 5.3.

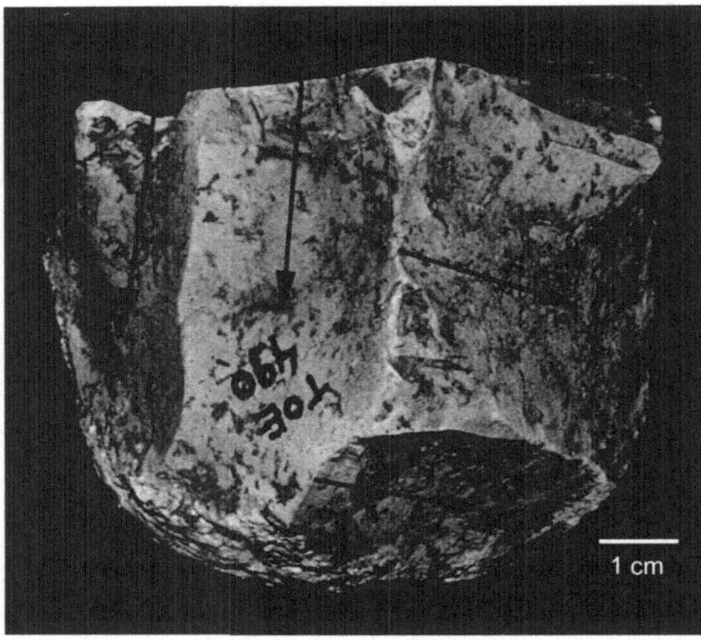

Figure 5.3. Block-core from Toé Cave, struck from multiple directions (indicated by arrows). Specimen 490a.

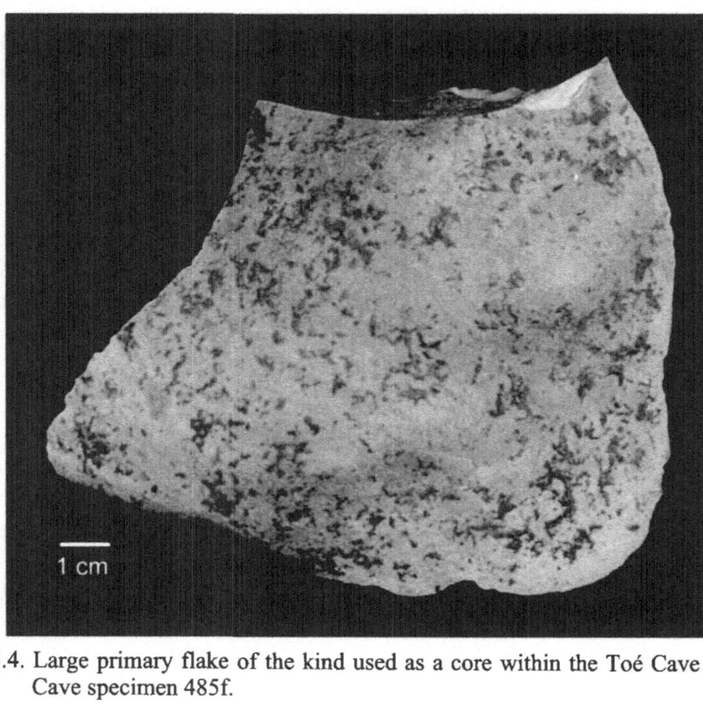

Figure 5.4. Large primary flake of the kind used as a core within the Toé Cave assemblage. Toé Cave specimen 485f.

Table 5.3. Dimensions of the nine chert 'flake-cores' and similar-sized 'retouched flakes' from the Holocene levels of Toé Cave.

	weight	length	width	thickness
flake-cores (N)	9	3	5	9
- range	45.0-208.0	58.4-79.7	61.0-87.2	24.9-52.3
- mean ± sd	116.2 ± 55.6	69.7 ± 10.7	69.6 ± 10.3	36.6 ± 8.4
large retouched flakes (N)	44	29	29	44
- range	46.0-322.0	28.4-101.5	40.9-93.2	25.0-46.5
- mean ± sd	119.1 ± 66.6	58.2 ± 19.2	62.8 ± 14.8	29.9 ± 4.3

The Late Pleistocene levels of Toé Cave did not produce any true cores. However, there are several retouched flakes that could potentially have served as cores. Two specimens exceed 25 mm thickness and one is only a mm below this lower limit, with weights ranging from 40-126 g. No platform removal flakes were recovered from the Late Pleistocene levels and none of the larger retouched flakes shows any indication of platform preparation or core rotation.

A total of 1167 chert flakes or flake fragments were recovered from the Holocene levels of Toé Cave, compared with 189 from the Late Pleistocene levels (Table 5.4). The character of the flakes, together with the recovery of quartzite hammerstones, suggests that the majority of the flakes were struck using the hard percussion technique (Jelsma 1998: 55). The Minimum Number of Flakes (MNF) in each of the Holocene and Late Pleistocene units is 1027 and 139, respectively.

The ratio of MNF to cores in the Holocene sample is 57.1 to 1 if only the 18 cores and flake-cores are taken into account, or 16.6 to 1 if the 44 larger retouched flakes are added in (this is in a later section compared to Kria Cave in Table 5.15, p.123). These values can be thought of as setting the limits to the MNF to core ratio for the Holocene assemblage. The MNF to core ratio for the Late Pleistocene sample, with a MNF of 139 and a maximum of three possible flake-cores, is 46.3 to 1 or higher. Although this value falls between the two extremes for the Holocene sample, it suggests that the Late Pleistocene assemblage has a deficit of cores relative to the younger assemblage.

The size of the complete flakes from the Holocene levels is compatible with the dimensions of the associated cores and larger retouched flakes. Most critically, the average length of the complete flakes (36.1 mm for cortex-bearing flakes; 25.5 mm for cortex-free flakes) is less than the thickness of the sample of flake-cores (mean = 36.6 mm; maximum = 52.3 mm). Indeed, with the exception of the largest cortex-bearing flakes, all of the recovered flakes could have been struck from cores of the kind recovered from these levels.

The complete flakes from the Holocene sample are on average slightly smaller in length and width than those from the Late Pleistocene sample (Table 5.4), but they are thicker and heavier. However, there is a higher proportion of flakes with remnant cortex in the Holocene sample compared to the Late Pleistocene sample (47.3% vs. 20.1% of all flakes and flake fragments). In both samples, flakes with remnant cortex are larger on average than those that lack cortex, which may partly account for the difference in average flake size between the two samples. The proportion of complete flakes that retain cortex is higher in the Holocene sample (273 with : 187 without cortex) than in the Late Pleistocene sample (14 : 32). A Chi-square test shows that this difference is highly significant ($X^2 = 14.24058$, d.f. 1, $p = 0.00016$).

The shape of unretouched flakes, as expressed by the ratio of length/width, is summarised in Table 5.5. Flakes with cortex are almost equal in length and width, while those without cortex are on average slightly more elongate. There is no difference in flake shape between the Late Pleistocene and Holocene units.

The difference in size between flakes with and without cortex is consistent with the notion that the entire sample of flakes is derived from a single reduction sequence. Typically, in any such sequence, the earlier struck flakes (those with cortex) will be larger on average than later struck, cortex-free flakes, especially where the original nodules are of modest size. The higher proportion of cortex-bearing flakes in the Holocene levels may indicate that more of the overall reduction process was performed

Table 5.4. Toé Cave, measurements and basic morphological characteristics of the Holocene and Late Pleistocene samples of retouched and unretouched chert flakes and flake fragments.

	Holocene	Late Pleistocene
total number of flakes/ fragments	1167	189
- with cortex remaining	552	38
- without cortex	615	144
- too weathered to identify cortex	-	7
number of complete flakes	460	49
- average length ± sd (range)	31.3 ± 14.8 (5.4-101.5)	33.0 ± 13.9 (13.9-68.2)
- average width ± sd (range)	32.7 ± 15.4 (1.1-124.4)	32.9 ± 15.0 (12.1-82.7)
- average thickness ± sd (range)	10.7 ± 6.8 (1.3-37.1)	10.5 ± 5.7 (2.4-26.0)
- average weight ± sd (range)	17.3 ± 35.2 (0.1-331.0)	13.1 ± 21.3 (0.2-126.0)
number of flakes with cortex		
- cortical striking platform	18	-
- conchoidal striking platform	425	30
- facetted striking platform	4	-
number of flakes without cortex		
- conchoidal striking platform	471	72
- facetted striking platform	4	-
number of complete flakes with cortex	273	14
- average length ± sd (range)	35.2 ± 16.3 (9.6-101.5)	40.1 ± 14.6 (14.1-62.4)
- average width ± sd (range)	38.0 ± 16.3 (3.7-124.4)	37.8 ± 10.4 (25.5-58.5)
- average thickness ± sd (range)	12.8 ± 7.3 (1.9-37.1)	12.5 ± 5.7 (4.0-26.0)
- average weight ± sd (range)	25.5 ± 43.1 (0.2-331.0)	16.3 ± 15.3 (0.5-45.0)
number of complete flakes without cortex	187	32
- average length ± sd (range)	25.6 ± 9.9 (5.4-56.7)	30.0 ± 13.2 (13.9-68.2)
- average width ± sd (range)	25.0 ± 10.0 (1.1-61.3)	30.5 ± 16.3 (12.1-82.7)
- average thickness ± sd (range)	7.7 ± 4.6 (1.3-31.9)	9.8 ± 5.8 (2.4-24.2)
- average weight ± sd (range)	5.3 ± 10.3 (0.1-102.0)	12.1 ± 24.3 (0.2-126.0)

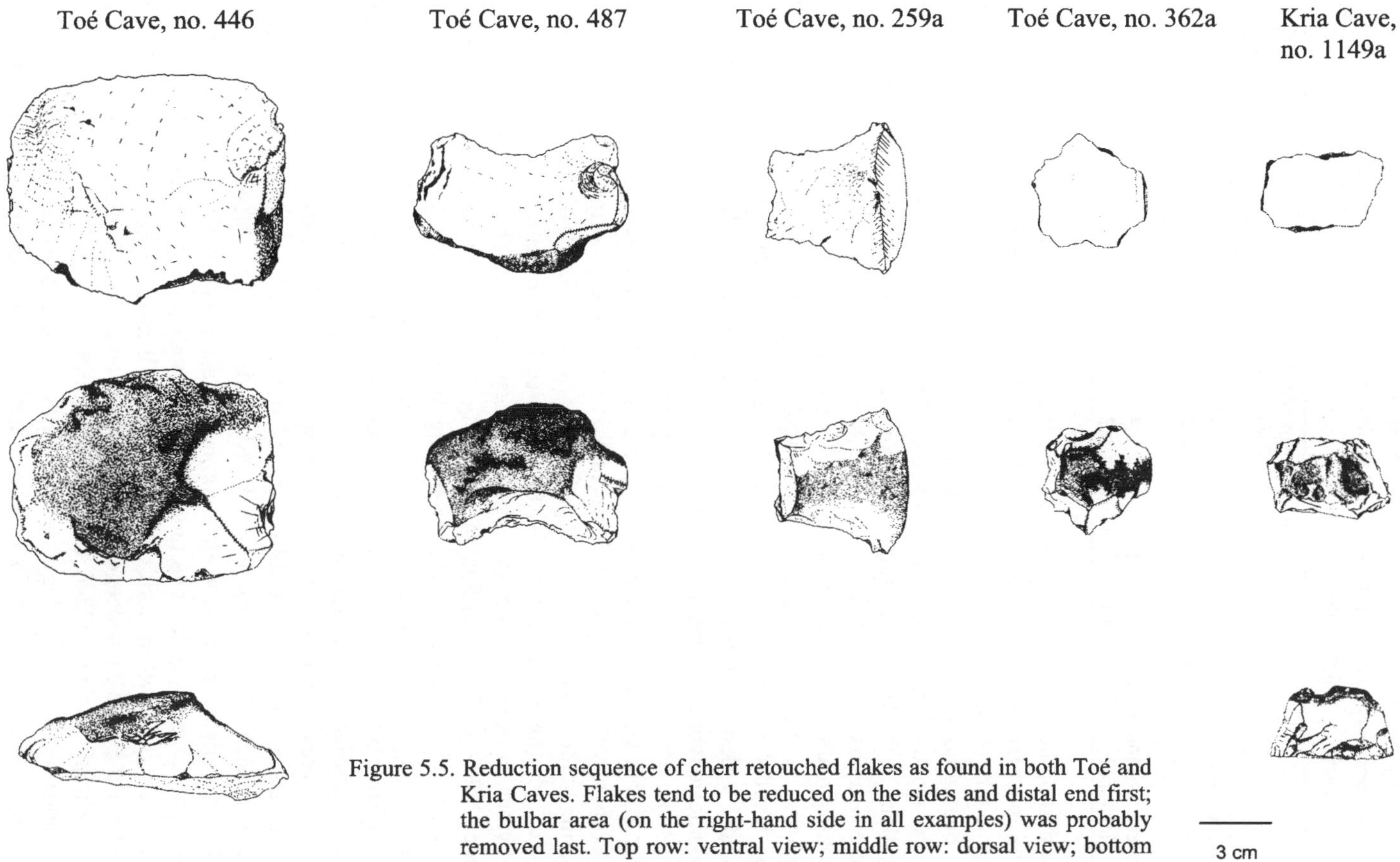

Figure 5.5. Reduction sequence of chert retouched flakes as found in both Toé and Kria Caves. Flakes tend to be reduced on the sides and distal end first; the bulbar area (on the right-hand side in all examples) was probably removed last. Top row: ventral view; middle row: dorsal view; bottom row: lateral view. Drawings: Nadine Guthrie.

Table 5.5. Comparison of shape of unretouched chert flakes and flake fragments, with and without cortex, between the Holocene and Late Pleistocene units of Toé Cave, as expressed by the ratio of length/width.

	mean	standard deviation	min.	max.	sample size (N)
Holocene, with cortex	1.0	0.4	0.4	2.5	221
Late Pleistocene, with cortex	0.9	0.3	0.6	1.6	16
Holocene, without cortex	1.1	0.4	0.3	2.6	206
Late Pleistocene, without cortex	1.1	0.3	0.3	1.7	34

on site during the Holocene than during the Pleistocene. This interpretation is consistent with the apparent lack of cores in the Late Pleistocene assemblage.

The nature of the striking platforms further supports the notion that the cores and flakes belong to a single reduction sequence. As noted above, the flake-cores are primarily worked from the ventral surface alone, with little evidence of platform preparation. In the Holocene sample of flakes the striking platform is almost always a simple conchoidal surface (97.2% of cases). A small percentage of flakes have cortex on the striking platform (2.4%), while a facetted platform is rarer still (0.4%). The flakes with cortex on the striking platform are mostly very large, with 14 of 18 examples having lengths in excess of 40 mm. The Pleistocene sample shows exclusively conchoidal striking platforms; however the total sample is smaller and the absence of other platform types is probably not significant.

The flake fragmentation index fluctuates markedly through the Toé Cave deposit (see Chapter 4), but with the highest values in the lower levels of Squares 2N1E and 0N0E (Fig. 4.21b, p. 95). The degree of fragmentation is not clearly correlated to the extent of weathering of the assemblage, except perhaps for the lower levels of Square 2N1E where there is a high proportion of very heavily weathered chert. Fragmentation does not show any obvious correlation with the extent of heat damage. The fragmentation index decreases below c. 70 cm in Square 1S0E (Fig. 4.21b), presumably as a result of the protection provided by the overhang protruding from this southern wall.

The breakage pattern of flakes is very similar between the Holocene and Pleistocene samples (Table 5.6). In both samples flakes that retain cortex are more often complete than those that lack any cortex, presumably reflecting the greater thickness and hence resistance to breakage of the cortex-bearing specimens. The relative frequency of other breakage categories does not differ greatly between the samples. The Pleistocene sample, consistent with its more weathered condition, shows fewer complete flakes overall and a higher proportion of amorphous flaked pieces. Chi-square tests of the breakage pattern between the two time periods do not show significant differences. The breakage pattern between cortex-free and cortex-bearing flakes is not significantly different for the Late Pleistocene sample, but the contrast is highly significant for the Holocene sample ($X^2 = 68.27240$, d.f. 5, $p < 0.00000$).

The stone-working technology does not appear to differ between the Late Pleistocene and Holocene levels in Toé Cave. The larger Holocene sample provides the most complete picture of the reduction sequence. This probably began at or near the unknown place or places of outcrop of the dominant chert raw material. Larger nodules were flaked or, in some cases, probably just shattered to expose the better quality chert within. However, much of the available material was apparently of quite poor quality, such that many of the pieces selected for transport either retained sizable patches of cortex or they contained vugs and other imperfections.

Large primary flakes ('blanks') and portions of shattered nodules weighing up to 500 g were carried to the site for subsequent reduction. For the primary flakes this was carried out primarily by unifacial flaking along the lateral and distal margins (see Fig. 5.5), with the flakes struck from the ventral surface (thus removing flakes from the dorsal side) with little or no platform preparation. The purpose of this exercise appears to have been to produce smaller flakes for use, but also to shape or resharpen the larger flakes themselves, many of which also show signs of use (see below). After the bulk of the original piece was removed in this fashion, or perhaps when step-fracturing made it difficult to remove further flakes along the margins, the flake-core was sometimes rotated to try new striking platforms, most often around the original bulbar region.

More rounded fragments of nodules were also worked primarily from one striking platform, although there is evidence of core rotation in some specimens. Further evidence of the relative infrequency of core rotation is seen in the small number of platform removal flakes and the very high proportion of flakes with simple conchoidal rather than facetted or cortex-bearing striking platforms. The large quantities of flaked chert in the Holocene levels suggests that flaking activities were carried out on a fairly

Table 5.6. Toé Cave, pattern of breakage of chert flakes in each of the Holocene and Late Pleistocene sample, shown separately for cortex-bearing and cortex-free flakes.

	Holocene		Late Pleistocene	
flakes with cortex				
	%	N	%	N
- complete flakes	50.9	273	38.9	14
- proximal fragment	7.7	41	11.1	4
- distal fragment	8.2	44	8.3	3
- medial fragment	1.1	6	-	-
- margins damaged	30.8	165	41.7	15
- conesplit	1.3	7	-	-
total number		536		36
flakes without cortex				
	%	N	%	N
- complete flakes	31.1	187	26.0	32
- proximal fragment	10.1	61	12.2	15
- distal fragment	4.2	25	14.6	18
- medial fragment	1.7	10	0.8	1
- margins damaged	51.7	311	45.5	56
- conesplit	1.3	8	0.8	1
total number		602		123
flaked pieces				
total number		213		48

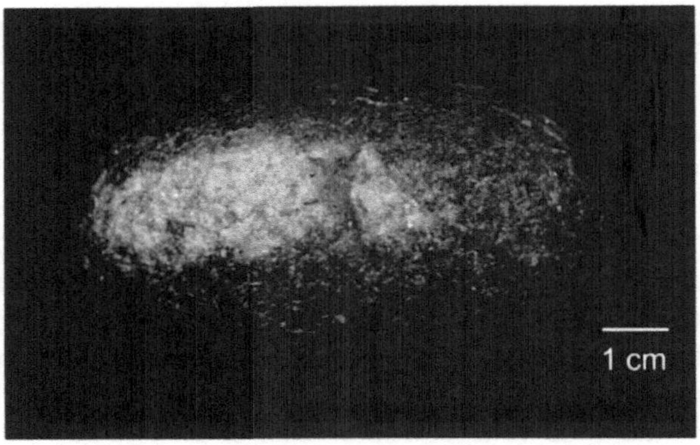

Figure 5.6. Hammerstone with bruised surface from use. Specimen from Kria Cave, number 1190a.

regular basis at the site; this is further indicated by the recovery of a number of quartzite hammerstones with heavily bruised margins from these levels (see Fig. 5.6).

The Late Pleistocene sample is very similar to the Holocene sample in the average size of flakes and the proportions of different flake fragments. However, there is an apparent lack of both very small and very large flakes, and of undoubted cores. Unfortunately, the overall small number of artefacts from the Late Pleistocene levels makes it impossible to rule out sample size as the explanation of these differences. The proportion of flakes with remnant cortex is lower in the Late Pleistocene than in the Holocene, presumably due to a greater initial reduction of cores away from site, or the preferential transport of small to medium-sized, cortex-free flakes to the site. The lower frequency of stone-working at the site during the Late Pleistocene suggests that hunters may have carried smaller pieces of chert with them for convenient use, with some limited further reduction at the site.

5.3.2. *Functional aspects*

A total of 172 Holocene specimens show signs of secondary retouch; 165 of these are themselves primary flakes, one is a block-core, and the remaining six are non-descript flaked pieces (Table 5.9). The Late Pleistocene levels produced a total of 16 retouched flakes, 15 of them made of the typical chert (Table 5.11) and one of silicified tuff.

The Holocene sample of retouched flakes contains a higher proportion of specimens with remnant cortex (86.1%) compared with the Late Pleistocene sample (40.0%).

The basic dimensions of the Holocene and Late Pleistocene retouched flakes are contrasted in Table 5.7. The mean lengths, widths and thicknesses are all remarkably similar between the two samples, although the Holocene flakes tend to average slightly narrower and the Pleistocene flakes slightly thinner. However the range is larger for each parameter in the Holocene sample. The presence of larger retouched flakes in the Holocene sample presumably accounts in part for the higher proportion of cortex-bearing flakes in this sample.

The shape of the retouched flakes can be expressed simply through the ratio of lengths to widths (Table 5.8). This varies in the Holocene sample from 0.5 (i.e. twice as wide as long) to 1.8 (i.e. almost twice as long as wide). The mean length/width ratio for the Holocene retouched flakes is 1.0 ± 0.3 (i.e. almost equal in length and width). The retouched flakes from the Late Pleistocene levels cover almost the same range of shapes and also have a mean length/width ratio of 1.0 ± 0.3.

The variation in shape of retouched flakes within and between the two age samples can also be examined graphically (Fig. 5.7, p. 118). One feature that is not so apparent from the summary statistics is the fact that the longest retouched flakes in the Holocene sample are all relatively narrow. This is partly a consequence of flake removal along the lateral margins, but may also reflect the original size and shape of the chert nodules. The largest retouched flakes from the Late Pleistocene levels are considerably broader, but the small sample rules out any meaningful statistical comparison.

Another interesting difference between the Holocene and Late Pleistocene samples is seen in a bivariate plot of thickness against width (Fig. 5.8, p.118). Compared with the sample from the Holocene levels, the Late Pleistocene retouched flakes show only a slight increase in thickness with width (or length). This difference might be put down to the fact that a higher proportion of the Pleistocene retouched flakes lack remnant cortex and thus are likely to be thinner for their length or width. However, the Holocene retouched flake sample does not show any comparable distinction between specimens with and without remnant cortex (Fig. 5.9). One possible interpretation is that the difference is related to the likelihood that the retouched flakes were transported during their functional life. As noted earlier, considerable stone-working activity was probably taking place within the site itself during the Holocene period, but much less so during the Late Pleistocene. Selection of flakes for use and subsequent retouch under these circumstances might be quite different. Flakes selected from local stone-working debitage and for use at the site most likely would be selected for edge and shape characteristics, but without too much regard for weight. However, if the Late Pleistocene sample represents flakes that were produced elsewhere and then selected specifically for transport, then they might be expected to be thinner (to cut down on weight) if it is the

Table 5.7. Comparison of chert retouched flake dimensions between the Holocene and Late Pleistocene levels from Toé Cave.

	length in mm	width in mm	thickness in mm	weight in g
Holocene				
mean	51.2 ± 16.3	53.4 ± 14.6	21.7 ± 8.1	63.9 ± 56.1
range	23.1-101.5	25.7-93.2	6.8-52.3	4.0-322.0
N	107	111	164	165
Late Pleistocene				
mean	52.5 ± 9.1	56.6 ± 19.3	18.8 ± 4.5	39.8 ± 26.7
range	39.7-68.2	32.7-88.4	11.7-27.7	13.0-126.0
N	11	10	14	15

Table 5.8. Comparison of shape between the Holocene and Late Pleistocene chert retouched flakes from Toé Cave, as expressed by the ratio length/width.

	Holocene	Late Pleistocene
mean	1.0	1.0
standard dev.	0.3	0.3
minimum	0.5	0.5
maximum	1.8	1.5
N	88	10

length and qualities of the cutting edge that were functionally important. The Holocene sample might then be expected to show a lack of large, relatively thin retouched or used artefacts as a consequence of preferential transport of these artefacts away from the site.

The great majority (91.3%) of retouched specimens also show distinct usewear (Table 5.9). Usewear is also clearly present in 47 of 1274 (3.7%) specimens that lack obvious retouch, including six of the block-cores. However, because a total of 626 unretouched chert artefacts are too weathered to confidently score usewear as either present or absent, the proportion of unretouched pieces with usewear may be more realistically put at 47 out of 648, or 7.3%. Even this estimate may be conservative given the cautious approach taken to identifying usewear.

White (1969) suggested that usewear on New Guinean stone artefacts usually can be distinguished from retouch. The most common type of usewear in the Toé Cave assemblage is an irregular, macroscopic chipping along the working-edge. This is generally unifacial, but if bifacial, then usually one face bears the majority of the flake scars. White described this type as 'bifacial chipping' or simply as 'utilization' (1969: 27). In the Toé Cave Holocene assemblage this type of usewear is often found on straight and relatively low-angled edges (Fig. 5.10). However, it also occurs on both concave and convex edges. The flake scars are rarely larger than a millimetre, and although visible with the naked eye, they are more easily detected using a magnifying glass or microscope. This kind of damage is probably caused by a wide variety of cutting activities.

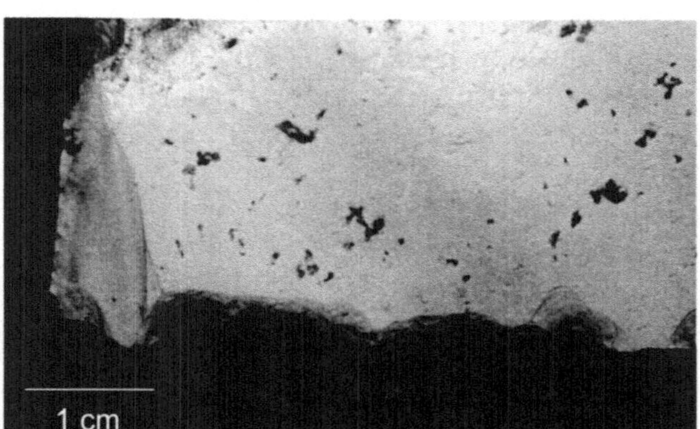

Figure 5.10. Usewear on a straight and low-angled edge. Kria Cave specimen 724a.

Table 5.9. Toé Cave, distribution of retouch and usewear within the Holocene chert assemblage.

	with retouch			no retouch			retouch uncertain			
	with usewear	no usewear	usewear uncertain	with usewear	no usewear	usewear uncertain	with usewear	no usewear	usewear uncertain	total
block-cores	1	-	-	6	2	-	-	-	-	9
flake-cores	7	1	1	-	-	-	-	-	-	9
flakes	143	6	7	35	380	568	1	1	26	1167
flaked pieces	6	-	-	5	170	23	-	-	9	213
potlids	-	-	-	-	21	-	-	-	-	21
manuport	-	-	-	-	27	-	-	-	-	27

Table 5.11. Toé Cave, distribution of retouch and usewear within the Late Pleistocene chert assemblage.

	with retouch			no retouch			retouch uncertain			
	with usewear	no usewear	usewear uncertain	with usewear	no usewear	usewear uncertain	with usewear	no usewear	usewear uncertain	total
block-cores	-	-	-	-	-	-	-	-	-	-
flake-cores	-	-	-	-	-	-	-	-	-	-
flakes	9	4	2	1	40	119	-	-	14	189
flaked pieces	-	-	-	1	30	3	-	-	14	48
potlids	-	-	-	-	5	-	-	-	-	5
manuport	-	-	-	-	10	-	-	-	-	10

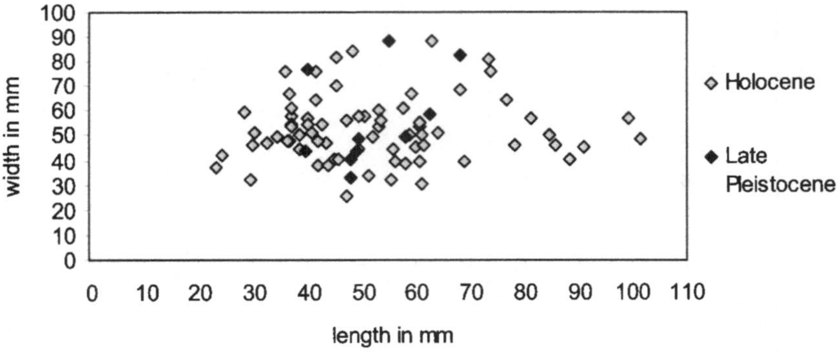

Figure 5.7. Toé Cave, relation between width and length of chert retouched flakes for the Holocene and Late Pleistocene samples.

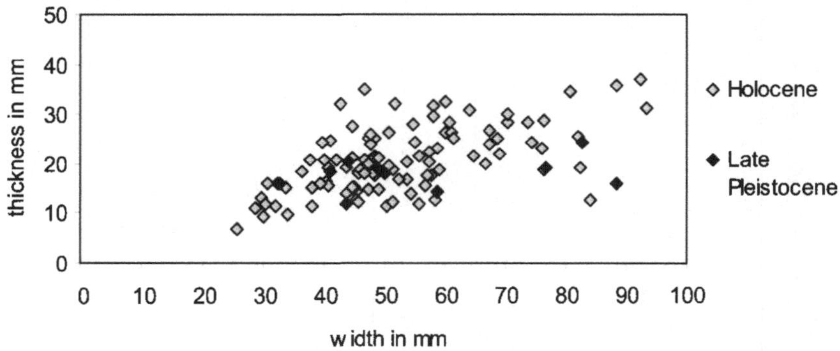

Figure 5.8. Toé Cave, relation between thickness and width of chert retouched flakes for the Holocene and Late Pleistocene samples.

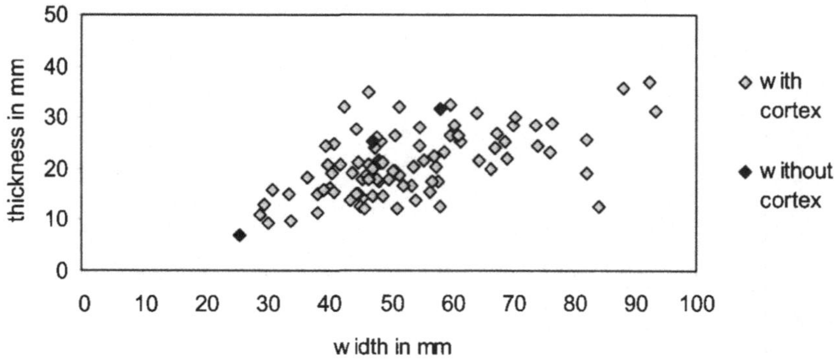

Figure 5.9. Toé Cave, relation between thickness and width of cortex-bearing and cortex-free Holocene chert retouched flakes.

A second type of usewear, very similar to that described by White as 'chattering' (White 1969: 27), is characterised by fine but intensive step-fracturing. This is almost always unifacial and is usually on steep angled edges. It is often concentrated in a small concavity, such as formed by a flake negative. In extreme forms the crushing is so intense over a small area that a small 'notch' is produced (Fig. 5.11). However, the same kind of crushing damage can be found extending along long straight edges.

Intensive step-fracturing of this kind is probably associated with some kind of shaving or planing activity. The more regular notch form suggests shaving of a resistant material such a hardwood, for example, for a spear or bow, or perhaps an arrow tip, or alternatively, the shaving of bone. This type of usewear was noted on 37 artefacts in the Holocene assemblage, with two notches on cores, 34 on flakes, and one on a flaked piece. Among the notch-bearing flakes, 21 are non-retouched, 12 are retouched, and one is too weathered to be certain. One flaked piece from the Late Pleistocene assemblage also bears a typical notch.

The varied kinds of usewear point to a range of functional activities at the site. However, without considerably more work including some experimental study, there is little more that can be said on this topic.

A total of 55 chert artefacts were classified as 'working-edge removal flakes', 53 from the Holocene deposit, and two from the Late Pleistocene. Twenty-two of these (all from the Holocene) have removed a prior 'notch'. Remnant cortex is present on 37 of the 53 Holocene specimens. At least some of these flakes were presumably struck to resharpen a working-edge; however, others may have been produced by general flaking of a previously utilised artefact. Their dimensions are presented in Table 5.10.

The Late Pleistocene sample produced 11 specimens with definite usewear (Table 5.11, p. 117). Nine of these are retouched flakes, while the others are a primary flake and a flaked piece. None of the Late Pleistocene retouched flakes shows the notch type of usewear. Two specimens were classified as working-edge removal flakes; these did not remove a notch but rather just general usewear. The presence of these specimens suggests that some flaking did occur at the site, although whether it was done to resharpen a working-edge or to produce flakes for subsequent use is unknown.

The frequency of usewear in the Late Pleistocene chert assemblage is 4.4% overall, or 11.0% if the heavily weathered specimens are excluded from the sum. These values are lower than those found in the Holocene levels (14.1% and 25.1% respectively).

Table 5.10. Toé Cave, dimensions of working-edge removal flakes from the Holocene chert assemblage.

	length in mm	width in mm	thickness in mm	weight in g
mean ± sd	31.5 ± 9.6	40.2 ± 11.5	11.9 ± 4.6	15.0 ± 20.4
range	15.5-57.6	23.0-76.4	4.0-29.0	1.8-131.0
N	44	33	51	53

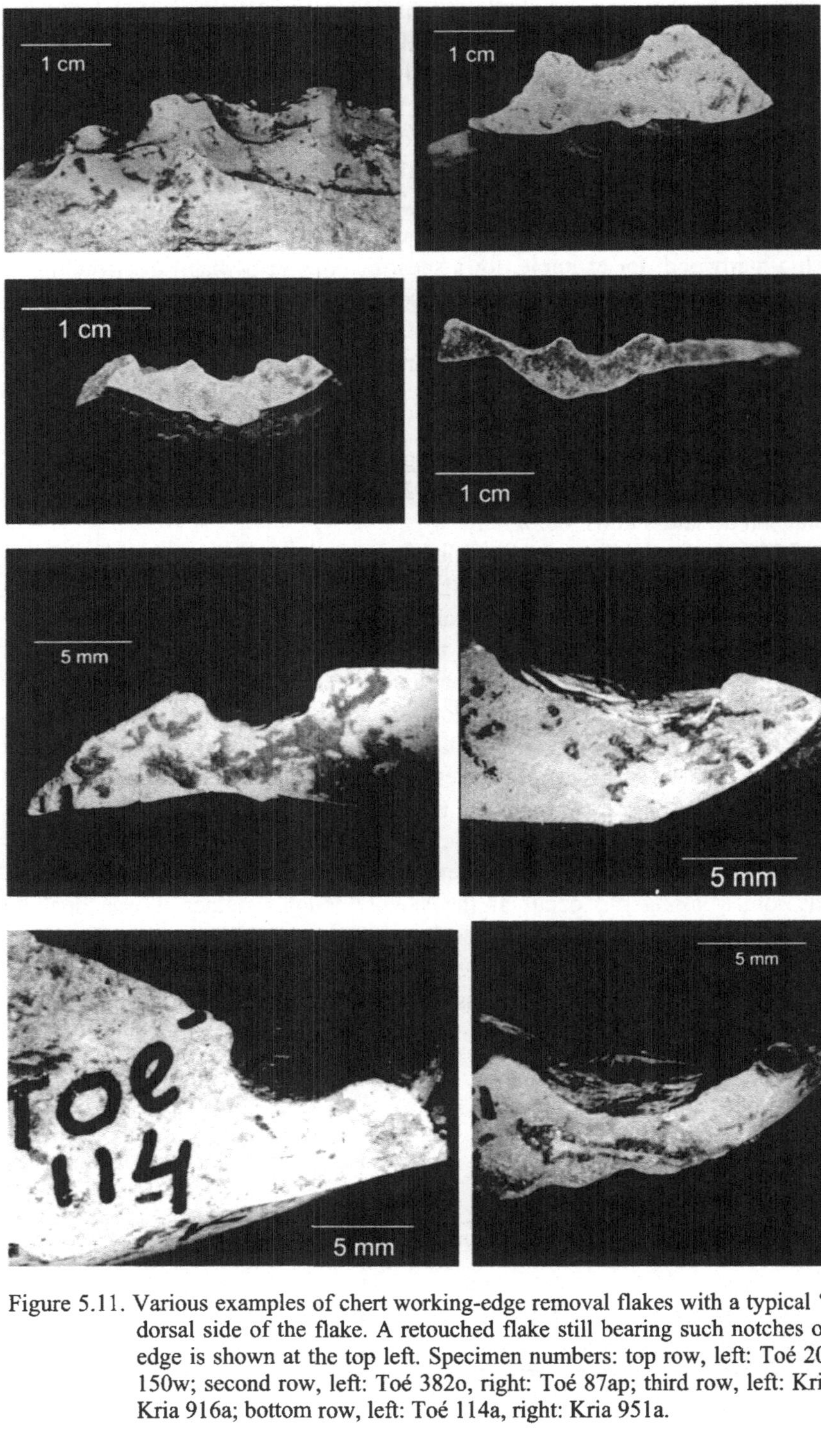

Figure 5.11. Various examples of chert working-edge removal flakes with a typical 'notch' on the dorsal side of the flake. A retouched flake still bearing such notches on its working edge is shown at the top left. Specimen numbers: top row, left: Toé 200a, right Toé 150w; second row, left: Toé 382o, right: Toé 87ap; third row, left: Kria 976a, right: Kria 916a; bottom row, left: Toé 114a, right: Kria 951a.

5.4. THE LITHIC ASSEMBLAGE FROM KRIA CAVE

For this analysis the Kria Cave assemblage is divided according to the five stratigraphic units as defined in Chapter 3. The total assemblage comprises 1528 artefacts and manuports, with a combined weight of 7386 g. The bulk of the collection was recovered from Units II-V, with only a trivial quantity of material coming from Unit I.

Chert dominates in all units except Unit I where quartzite and limestone assume greater importance (Table 5.12). The chert in Kria Cave is similar, if not identical, in properties to that found in Toé Cave. Other raw materials also duplicate those found in Toé, namely the reddish silicified tuff, a pale quartzite and several kinds of metamorphic rocks. Indeed, the overall composition is remarkably similar between the two sites, especially considering the distance between them and the difference in ages of the various assemblages.

The quartzite specimens include two sizeable cobbles, one with bruising from use as a hammerstone; these weigh 343 g and 396 g.

The overall composition of the assemblage in each of the five units is shown in Tables 5.13a-5.13e (at the end of this chapter), divided into the basic technological categories as described under section 5.2.1 (Technology). An unusual feature of the assemblage is the complete absence of chert cores; the solitary example recovered from Kria Cave is a highly reduced (probably exhausted) core made of silicified tuff.

Table 5.12. Kria Cave, proportional representation of different raw materials in the five units, calculated by number and by weight. Raw data are given in Tables 5.13a-e, at the end of this chapter.

	chert	silicified tuff	limestone	quartzite	chert?	limestone?	quartzite?	metamorphic
count								
Unit I	33.3%			50.0%		16.7%		
Unit II	95.8%	1.3%	0.5%	1.1%	0.2%	0.7%		0.5%
Unit III	96.4%	2.0%	1.2%	0.4%				
Unit IV	90.8%	2.2%			1.1%	4.3%	1.1%	0.5%
Unit V	87.9%	7.9%	0.5%	1.1%	1.3%	0.8%	0.5%	0.2%
weight								
Unit I	35.7%			57.1%		7.1%		
Unit II	86.7%	2.2%	0.2%	2.1%	0.02%	0.7%		8.1%
Unit III	91.4%	1.7%	0.6%	6.3%				
Unit IV	95.4%	0.4%			0.3%	2.9%	0.8%	0.2%
Unit V	67.3%	2.2%	0.4%	17.7%	2.1%	0.3%	0.7%	9.3%

5.4.1. *Technological aspects*

The Kria Cave assemblage was also scrutinised carefully for any indications of deliberate heating of the chert raw material. As with the Toé assemblage, no specimens could be said to show improved flaking qualities as a consequence of heating. To the contrary, heating in most cases seems to have caused internal fracturing that resulted in irregular shattering, without any obvious increase in lustre on flaked surfaces.

While the Kria chert assemblage lacks any primary cores, three specimens from Unit V were classified as flake-cores because all of the 'retouch' is very coarse. These range in weight from 66-400 g; two of the three retain areas of cortex on the dorsal surface. Measurements are summarised in Table 5.14. As in the Toé Cave assemblage, the Kria flake-cores are primarily reduced from the ventral surface, with flakes struck from both the distal and the lateral margins and there is no indication of preparation of the striking platform. A total of 13 platform removal flakes were identified from Kria Cave, distributed across all units; as in Toé Cave, core rotation was performed relatively infrequently.

All of the flake-cores in the Kria assemblage are thicker than 25 mm, the value identified from the larger Toé Cave sample as the probable lower limit for suitability as a flake-core. If the same value is applied to the Kria assemblage, a total of nine retouched flakes are identified as 'potential flake-cores'; measurements of these specimens are shown in Table 5.14. Six of these specimens came from Unit V, two from Unit IV and one from Unit II; all retain patches of cortex on the dorsal surface. Not surprisingly, given the basic similarity in raw materials and reduction processes between the two sites, the dimensions of the flake-cores and potential flake-cores are very similar between the two sites.

The number of flakes or flake fragments, the Minimum Number of Flakes (MNF) and the number of flake-cores and potential flake-cores from each unit in Kria Cave are shown in Table 5.15 (equivalent values are shown for the Toé Cave assemblages for comparison). The ratio of MNF to cores is lowest in Kria Unit V, with 116 flakes per core if only the three flake-cores are taken into account, or 39 flakes per core if the six larger retouched flakes from this unit are added in. This latter value is slightly lower than that obtained for the Late Pleistocene levels of Toé Cave (46.3:1; see Table 5.15). The other units in Kria produced even fewer cores or potential cores, hence the ratios are higher still, with indicative values of hundreds of flakes per core.

Table 5.14. Dimensions of the three flake-cores and similar-sized retouched flakes (potential flake cores) from the deposit in Kria Cave.

	weight (g)	length (mm)	width (mm)	thickness (mm)
flake-cores				
- N	3	2	2	3
- range	66.0-400.0	33.5-101.2	51.3-80.0	33.1-59.4
- mean ± sd	186.0 ± 185.78	67.35 ± 47.87	65.65 ± 20.29	42.03 ± 59.4
large retouched flakes (thickness >25 mm)				
- N	9	5	5	9
- range	43.0-166.0	29.8-69.9	42.3-70.3	26.5-43.2
- mean ± sd	81.22 ± 44.21	42.20 ± 17.52	59.32 ± 12.99	31.14 ± 5.19

Table 5.15. Ratios of Minimum Number of chert Flakes to cores and potential cores for each unit in Kria Cave, compared with equivalent values for Toé Cave.

	Unit II	Unit III	Unit IV	Unit V	Toé Holocene	Toé Late Pleistocene
number of flakes or flake fragments	361	195	141	435	1167	189
number of (flake-) cores	-	-	-	3	18	-
number of potential flake-cores	1	-	2	6	44	3
Minimum Number of Flakes	284	161	119	349	1027	139
ratio MNF/cores	284:1	>161:1	>119:1	116.3:1	57.1:1	>139:1
ratio MNF/cores + potential cores	284:1	>161:1	59.5:1	38.8:1	16.6:1	46.3:1

The approximate number of cores needed to account for the recovered flakes can be estimated from the relative weights of flakes to cores. For example, in Unit V, the average weight of all complete flakes is 7.2 g, and the flake-cores range in size up to 400 g (this being the largest core recovered from either site). Even with complete reduction, a 400 g core can only produce 56 flakes with an average weight of 7.2 g. Assuming incomplete reduction and a smaller initial size, a production ratio of around 30-40 flakes per core is perhaps not unreasonable. This figure is close to that derived from the numbers of flakes to cores for the Holocene levels of Toé Cave, but is very much lower than the estimates for all units at Kria and for the Late Pleistocene levels of Toé Cave, indicating that cores are very much underrepresented in these assemblages. This makes it perhaps more likely that some of the ‘potential flake-cores’ have indeed functioned as such.

There are four scenarios that might account for the almost complete lack of cores in Kria Cave. The first is that the original core size was very much larger than in Toé Cave, such that all of the reduction was done from a small number of originally massive blocks. However, this explanation seems very unlikely given the scarcity of very large flakes in the Kria assemblage and the fact that the proportion of cortex-bearing flakes is only slightly lower in Kria (47.3% in Holocene of Toé; 34.9-46.8% in Kria). The second scenario is that people simply took partly used and even depleted cores away from the site for further reduction. Although this hypothesis is difficult to test directly, the underlying implication that good flaking stone was in short local supply is supported by the smaller average size of the flakes in the Kria assemblage compared with that from Toé Cave (Tables 5.4 and 5.16). The third possible explanation is that cores were somehow destroyed on site, leaving no recognisable remnants. One way in which core reduction can continue to such extremes is to switch from hand-held to anvil-supported or bipolar reduction. This method of flaking is recorded ethnographically from New Guinea (e.g. Sillitoe 1988: Plates 16-17) and from archaeological localities in both

Table 5.16. Kria Cave, measurements and basic morphological characteristics of the samples of retouched and unretouched chert flakes and flake fragments, presented by unit.

	Unit I	Unit II	Unit III	Unit IV	Unit V
total number of flakes/ fragments	2	361	195	141	435
- with cortex remaining	0	146	68	66	160
- without cortex	2	215	127	75	275
number of complete flakes	0	119	79	53	145
- average length ± sd		22.47±8.62	21.26±8.23	24.90±8.64	23.18±9.66
(range)		(4.6-47.4)	(6.7-48.6)	(7.5-43.7)	(2.9-68.9)
- average width ± sd		21.55±8.26	20.69±8.60	25.78±11.16	24.53±13.01
(range)		(7.3-48.7)	(2.1-55.6)	(9.3-54.3)	(5.0-70.2)
- average thickness ± sd		6.01±4.46	5.92±4.05	8.44±6.04	7.51±5.73
(range)		(1.4-30.0)	(1.3-24.6)	(1.9-31.2)	(0.9-43.2)
- average weight ± sd		3.24±5.69	2.37±2.87	6.26±9.58	7.20±20.10
(range)		(0.1-49.0)	(0.1-15.0)	(0.2-47.0)	(0.1-166.0)
number of flakes with cortex					
- cortical striking platform	-	4	-	2	3
- conchoidal striking platform	-	95	42	50	116
- facetted striking platform	-	-	1	-	-
number of flakes without cortex					
- conchoidal striking platform	-	154	86	48	183
- facetted striking platform	-	-	-	-	-
number of complete flakes with cortex	0	58	32	30	81
- average length ± sd		24.68±8.60	25.14±9.58	28.18±7.58	24.77±11.10
(range)		(4.6-45.7)	(7.6-48.6)	(11.3-43.7)	(2.9-68.9)
- average width ± sd		23.12±8.97	22.47±7.66	30.56±10.85	27.90±14.23
(range)		(9.5-48.7)	(2.1-40.5)	(10.2-54.3)	(7.9-70.2)
- average thickness ± sd		7.11±5.45	7.51±5.05	11.05±6.53	8.95±6.88
(range)		(1.4-30.0)	(1.5-24.6)	(2.7-31.2)	(2.3-43.2)
- average weight ± sd		4.64±7.51	3.09±3.24	9.29±11.40	10.98±26.12
(range)		(0.1-49.0)	(0.1-15.0)	(0.4-47.0)	(0.2-166.0)
number of complete flakes without cortex	0	61	47	23	64
- average length ± sd		20.38±8.17	18.63±5.96	20.76±8.24	21.15±7.01
(range)		(6.6-47.4)	(6.7-37.1)	(7.5-40.7)	(5.1-47.2)
- average width ± sd		20.08±7.30	19.47±9.06	19.74±8.41	20.25±9.81
(range)		(7.3-36.2)	(7.5-55.6)	(9.3-42.5)	(5.0-51.7)
- average thickness ± sd		4.98±2.96	4.84±2.78	4.88±2.59	5.70±2.99
(range)		(1.5-14.4)	(1.3-12.8)	(1.9-11.2)	(0.9-16.7)
- average weight ± sd		1.91±2.52	1.87±2.51	2.14±3.55	2.42±3.89
(range)		(0.1-13.0)	(0.1-10.0)	(0.2-15.0)	(0.1-23.0)

Australia and New Guinea (Mulvaney & Kamminga 1999: 214-215) including sites of Late Pleistocene age. The Kria and Toé Cave assemblages do not include any cores or flaked pieces with the characteristic 'bipolar' shattering that is produced through use of this technique. On the evidence of the assemblage itself this mechanism of further reduction can be excluded. There is however another way in which a depleted core could be further reduced – through accidental or deliberate discard into a fireplace. With repeated firing, a chert core would be reduced into nondescript debitage or 'flaked pieces'. The proportion of flaked pieces in the total debitage (i.e. flakes + flaked pieces) in the Kria assemblage ranges from 10.1 to 19.0% (with no consistent trend through the deposit). This compares well with the Holocene assemblage (15.3%) from Toé Cave but is slightly lower than the Late Pleistocene Toé assemblage (20.3%), perhaps reflecting the more weathered condition of the latter sample. If chert cores were destroyed by heating in Kria Cave, then presumably they were already highly reduced and thus made only a minor contribution to the total assemblage. The fourth scenario is that the cores were placed or discarded elsewhere within the cave, in areas that were not excavated, perhaps in more peripheral areas to avoid standing or sitting on large pieces of stone. The evidence from Toé Cave provides some supporting evidence, as the majority of the cores and larger pieces of stone were recovered from Square 2N1E, positioned against the cave wall. However, this argument is weakened by the fact that Square 0N0E in Kria is also located close to the wall of the cave, in an area that might have attracted refuse of all kinds.

Of the various scenarios set out above to explain the lack of cores in Kria, it is possible to rule out only one of the four. The remaining explanations are each reasonable and some are actually supported by features of the assemblage; for example, the notion that stone was locally scarce is supported by other features relating to the degree of overall reduction. However, in the end it is perhaps most realistic to suggest that all of these factors might have contributed to the observed scarcity of cores in the deposit.

The character of the chert flakes in the Kria assemblage suggests that the majority if not all of the flakes were struck using the hard percussion technique. The recovery of one quartzite hammerstone and a number of small quartzite flakes that could have spalled off of similar hammerstones, further suggests that some stone-working was carried out at the site.

The size of the complete flakes from Kria (Table 5.16) is compatible with the dimensions of the few associated larger retouched flakes; all but a very few of the recovered flakes could have been struck from flake-cores of the kind recovered. In all units the average length of the complete flakes (<29 mm for cortex-bearing flakes; <22 mm for cortex-free flakes) is considerably less than the thickness of the recovered flake-cores (33.1-59.4 mm).

The complete flakes from Units V and IV average slightly larger in all measurements and are 2-3 times greater in average weight than those from Units II and III (Table 5.16). In Toé Cave, a similar difference was explained in terms of the differing proportion of cortex-bearing and cortex-free flakes, the latter being smaller on average than the former. As shown in Table 5.16, flakes with remnant cortex in each unit within the Kria Cave assemblage are again consistently larger on average than the associated cortex-free flakes. However, the proportion of cortex-bearing flakes fluctuates randomly through the Kria Cave deposit, ranging from 34.9-46.8% and with no overall trend.

When the dimensions of cortex-bearing and cortex-free flakes are calculated separately (Table 5.16), it appears that the cortex-bearing flakes are larger on average in the two lower units (IV and V) than in the two upper units (II and III). The differences in mean values are particularly marked for flake width, thickness and weight, and less so for flake length. Somewhat surprisingly, few of the contrasts are statistically significant, presumably because of the high standard deviations of most dimensions. In contrast, the cortex-free flakes differ only slightly in average size between the lower and upper units (Table 5.16). As with the cortex-bearing flakes, there is a hint of greater variation in size and shape in Units IV and V, especially in flake length and width.

The shape of unretouched flakes in Kria Cave, as expressed by the ratio length/width, is shown in Table 5.17. Cortex-free flakes are remarkably consistent in shape through the deposit. In contrast, the shape of cortex-bearing flakes changes slightly between the lower Units IV-V and the upper Units II-III, with a slight increase in length relative to width and an increase in variability (as reflected by higher standard deviations).

What these subtle differences in flake size and shape might mean in terms of stone-working technology is not immediately apparent. In general, a change in average flake proportions could be produced by a difference in basic stone-working technology or alternatively, by a change in core shape. In the present case, the fact that the cortex-free flakes remain fairly constant in size through the deposit suggests that the basic stone-working techniques did not change greatly through the period of occupation of Kria Cave. Minor changes in core shape thus seems likely, but without larger numbers of surviving cores it is difficult to say what such a change might have involved.

As in the Toé assemblage, the majority of flakes in all units have a simple conchoidal striking platform. Cortex striking platforms are uncommon, while a facetted platform is present in a single specimen.

As already reported in Chapter 3, the flake fragmentation index shows apparently random fluctuations through the Kria Cave deposit (Fig. 3.19). The pattern of breakage of flakes is very similar in all units (Table 5.18). As in Toé, flakes with remnant cortex are more often complete than those without cortex, presumably reflecting the greater thickness and hence resistance to breakage of the cortex-bearing specimens. The pattern

Table 5.17. Comparison of shape of unretouched flakes and flake fragments, with and without cortex, between the units from Kria Cave, as expressed by the ratio length/width.

	mean	standard deviation	min.	max.	sample size (N)
Unit II, with cortex	1.2	0.6	0.2	3.3	63
Unit III, with cortex	1.2	0.6	0.6	3.6	33
Unit IV, with cortex	1.0	0.4	0.4	2.5	35
Unit V, with cortex	1.0	0.4	0.1	2.5	80
Unit II, without cortex	1.1	0.5	0.4	2.3	72
Unit III, without cortex	1.1	0.5	0.4	2.2	53
Unit IV, without cortex	1.1	0.5	0.5	2.4	28
Unit V, without cortex	1.2	0.5	0.4	3.4	76

of breakage in flakes with and without cortex (see Table 5.17) is not significantly different between any of the units. In both groups of flakes, Units II-III show higher percentages of proximal fragments and smaller percentages of flakes with marginal damage than Units IV-V. The consistency of this difference suggests that it is somehow related to the technique of flake production. However, the significance of this contrast is unclear.

The basic stone-working technology in Kria Cave was probably very similar to that described above for Toé Cave. The major difference between the two sites is that the dominant chert material was consistently more reduced at Kria, resulting in smaller flakes and a notable lack of cores and large retouched flakes. This may indicate that sources of good quality stone were located considerably further away from Kria than from Toé, resulting in a higher energetic cost in acquisition and transport of the raw material. The relatively large numbers of small to medium-sized flakes in all levels suggests that flaking activities were carried out on site at Kria; this is further indicated by the presence of flakes derived from quartzite hammerstones and of one complete hammerstone with heavily bruised margins.

The subtle changes in flake size and shape, and in the pattern of breakage, through the Kria Cave deposit suggest that some change in stone-working technique occurred around the time of the Unit IV to Unit III transition, i.e. around 5500 BP. After that date, the flakes are on average narrower, thinner and lighter, with the cortex-bearing flakes showing a stronger decrease in size than the cortex-free flakes. Exactly what these

Table 5.18. Kria Cave, pattern of breakage of chert flakes in each of the units, shown separately for cortex-bearing and cortex-free flakes.

	Unit I		Unit II		Unit III		Unit IV		Unit V	
flakes with cortex	%	N	%	N	%	N	%	N	%	N
- complete flakes			41.7	58	50.8	32	46.2	30	52.6	81
- proximal fragment			9.4	13	9.5	6	3.1	2	4.6	7
- distal fragment			14.4	20	12.7	8	6.2	4	8.4	13
- medial fragment			1.4	2	-	-	6.2	4	0.7	1
- margins damaged			32.4	45	27.0	17	38.5	25	33.8	52
- conesplit			0.7	1	-	-	-	-	-	-
total number		0		139		63		65		154
flakes without cortex	%	N	%	N	%	N	%	N	%	N
- complete flakes			30.5	61	41.2	47	31.5	23	25.3	64
- proximal fragment			15.5	31	10.5	12	6.9	5	8.3	21
- distal fragment			8.5	17	2.6	3	13.7	10	7.1	18
- medial fragment			1.5	3	3.5	4	1.4	1	5.1	13
- margins damaged	100	2	43.5	87	40.4	46	45.2	33	53.8	136
- conesplit			0.5	1	1.8	2	1.4	1	0.4	1
total number		2		200		114		73		253
flaked pieces										
total number		0		49		22		22		103

changes might mean in terms of lithic technology is unknown, but the fact that some change occurred shows that the technological traditions of the Ayamaru Plateau were not entirely static through its period of prehistory.

5.4.2. *Functional aspects*

A total of 42 chert flakes from Kria Cave show signs of secondary retouch. The majority of these (92.9%) retain patches of remnant cortex, a higher proportion even than the Holocene sample from Toé Cave (85.5%).

The basic dimensions of the retouched flakes from each unit in Kria Cave are shown in Table 5.19. Retouched flakes from Unit V are larger in mean length, width, thickness and weight than those from the overlying units, but the differences are not statistically significant except for the contrast in weight between Units II and V.

The shape of the retouched flakes, as expressed by the ratio of length/width (Table 5.20), is essentially unchanging through the Kria sequence, with a total range from 0.6-1.5. This range is slightly more restricted than that observed in the Holocene assemblage from Toé Cave. Plots of retouched flake dimensions graphically illustrate the greater mean size of the sample from Unit V (Fig. 5.12 and 5.13). Interestingly, they also suggest that the retouched flakes from Unit II are relatively narrower for their thickness compared with those from Units IV and V. This is consistent with the overall narrower condition of the cortex-bearing flakes from Units II and III, described above.

If the Kria Cave and Toé Cave Holocene retouched flakes are plotted together, the two samples show almost complete overlap (Fig. 5.14). However, the Toé Holocene sample contains a significant number of retouched flakes that exceed in size any of those from Kria Cave. Many of these specimens are the potential flake-cores that, in Kria Cave, were presumably reduced more often to smaller size but retaining a comparable

Table 5.19. Kria Cave, comparison of chert retouched flake dimensions between the units.

		Unit I	Unit II	Unit III	Unit IV	Unit V
length in mm	mean ± sd	-	31.48 ± 10.56	21.60 ± 14.85	33.76 ± 5.53	49.27 ± 20.59
	range	-	17.3-45.7	11.1-32.1	27.4-38.4	29.8-101.2
	N	0	5	2	5	12
width in mm	mean ± sd	-	36.53 ± 8.77	29.70	44.46 ± 8.04	55.65 ± 13.15
	range	-	29.1-48.7	-	34.4-54.3	31.1-80.0
	N	0	4	1	7	14
thickness in mm	mean ± sd	-	16.45 ± 7.62	21.20 ± 4.81	19.19 ± 9.15	27.71 ± 11.17
	range	-	5.0-30.0	17.8-24.6	8.8-31.2	14.4-59.4
	N	0	11	2	7	17
weight in g	mean ± sd	-	20.36 ± 15.17	11.50 ± 6.36	25.22 ± 18.52	79.50 ± 84.24
	range	-	2.0-49.0	7.0-16.0	5.0-53.0	12.0-400.0
	N	0	11	2	9	20

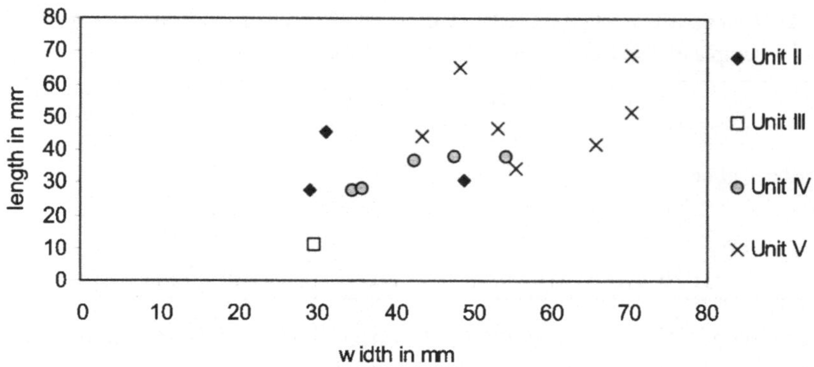

Figure 5.12. Kria Cave, relation between length and width of chert retouched flakes.

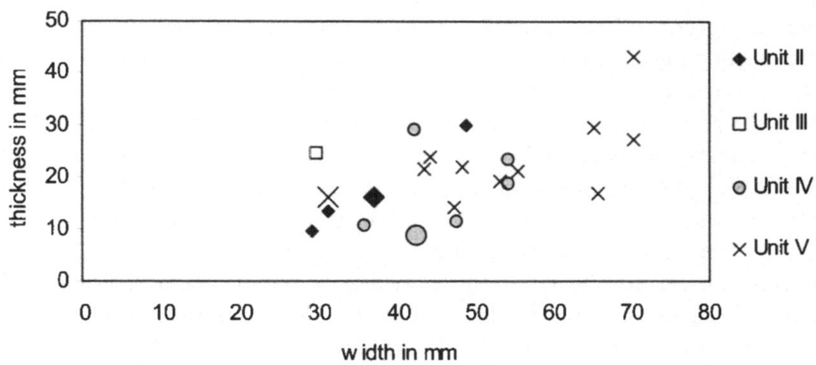

Figure 5.13. Kria Cave, relation between thickness and width of chert retouched flakes. Enlarged symbols (one in each of Unit II, IV and V) indicate specimens without cortex.

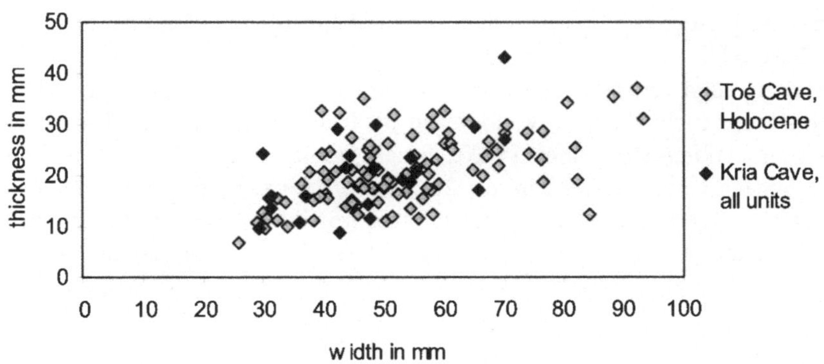

Figure 5.14. Comparison of the Holocene samples of chert retouched flakes from Toé and Kria Caves: relation between thickness and width.

Table 5.20. Comparison of chert retouched flakes between the units in Kria Cave and the Holocene sample of Toé Cave, as expressed by the ratio length/width.

	Unit II	Unit III	Unit IV	Unit V	Toé Holocene
mean	1.0	0.4	0.8	0.9	1.0
standard deviation	0.4	-	0.1	0.3	0.3
min	0.6	-	0.7	0.6	0.5
max	1.5	-	0.9	1.4	1.8
N	3	1	5	9	88

thickness. Indeed, the plot of width vs. thickness shows no discrimination between the two samples, unlike the earlier noted contrast between the Late Pleistocene and Holocene samples from Toé. This may indicate that transport costs were not important in the selection of flakes for use and subsequent retouch at Kria, even though the finer reduction of the flaking debris itself points to a more efficient use of the raw material at the site. Perhaps the difference here is that the majority of the retouched flakes found in the Late Pleistocene of Toé Cave were carried to the site as part of a transported toolkit, while those found in Kria Cave were selected from flaking debitage at the site and used locally, without any consideration of transport costs.

All of the retouched flakes in Kria also show distinct usewear (Table 5.21). Among the non-retouched flakes and other kinds of debitage, usewear is also clearly present in 56 of 1329 (4.2%) of specimens, including nine flaked pieces. However, because a total of 670 chert artefacts are too weathered or incomplete (e.g. lacking margins) to confidently score usewear as either present or absent, the proportion of unretouched pieces with usewear may be better represented as 56 out of 659, or 8.5%. This figure is very close the level of usewear calculated for the Toé Cave Holocene assemblage. As noted earlier, these estimates are probably conservative in view of the cautious approach taken to identifying usewear.

The Kria assemblage displays the same variety of usewear that was described for the Toé Cave material. The distinctive 'notch' type of usewear was noted on a total of 23 artefacts in the Kria assemblage. Nineteen of these were classified as 'working-edge removal' flakes, while one is an unretouched primary flake and three are flaked pieces. Notches are present in all units in Kria: four in Unit II, one in Unit III, three in Unit IV and fifteen in Unit V.

Working-edge removal flakes are present in all units except Unit I. The largest numbers were found in Unit V, followed by Units II, IV and III. The frequency of notches on working-edge removal flakes is very similar in Kria (19 of 52) and Toé (22 of 53). Remnant cortex is present on 26 of the 52 examples from Kria Cave. The dimensions of the working-edge removal flakes from Kria Cave are summarised in Table 5.22. There are no statistically significant contrasts between the samples from different units.

Table 5.21. Kria Cave, number of chert artefacts with retouch and usewear within the units. *silicified tuff

	with retouch			no retouch			retouch uncertain			
	with usewear	no usewear	usewear uncertain	with usewear	no usewear	usewear uncertain	with usewear	no usewear	usewear uncertain	total
Unit I										
block-cores										
flake-cores										
flakes						2				2
flaked pieces										
non-utilised manuport										
Unit II										
block-cores	1*									1*
flake-cores										
flakes	11			13	119	206			12	361
flaked pieces					49					49
non-utilised manuport					15					15
Unit III										
block-cores										
flake-cores										
flakes	2			5	80	101			7	195
flaked pieces					22					22
non-utilised manuport					15					15
Unit IV										
block-cores										
flake-cores										
flakes	9			5	49	75			3	141
flaked pieces					19	1			2	22
non-utilised manuport					2					2
Unit V										
block-cores										
flake-cores	3									3
flakes	17			24	139	217			38	435
flaked pieces				9	88	2			4	103
non-utilised manuport					6					6

Table 5.22. Kria Cave, dimensions of chert working-edge removal flakes from all units.

		Unit I	Unit II	Unit III	Unit IV	Unit V
length in mm	mean ± sd	-	28.3 ± 10.1	20.0 ± 8.0	25.9 ± 8.0	26.3 ± 10.8
	range	-	14.8-47.9	13.6-31.5	16.5-38.3	2.9-44.5
	N	0	10	4	6	21
width in mm	mean ± sd	-	29.6 ± 9.9	21.1 ± 1.0	24.6 ± 2.8	37.2 ± 11.7
	range	-	20.3-48.8	20.5-22.0	21.3-28.8	22.6-63.4
	N	0	9	3	5	18
thickness in mm	mean ± sd	-	9.1 ± 3.5	6.0 ± 2.0	9.3 ± 5.0	9.9 ± 4.3
	range	-	4.9-15.1	4.2-8.9	6.0-20.2	4.0-21.6
	N	0	14	4	7	24
weight in g	mean ± sd	-	7.2 ± 6.6	1.6 ± 0.4	9.9 ± 17.1	8.6 ± 11.2
	range	-	0.9-25.0	1.0-2.0	1.9-52.0	0.3-47.0
	N	0	15	4	8	25

5.5. SUMMARY AND CONCLUSION

The two lithic assemblages from the Bird's Head are in some respect fairly typical by New Guinean standards. Both are dominated by fairly crude chert flaking debitage and an absence of any formal 'types' among the artefacts. However, various specialised classes of artefacts that occur widely in this region, most notably, ground edge axes and adzes, but also waisted blades and silica-polished sago-adzes, are absent. The absence of ground stone artefacts, or even of flakes derived from these artefacts suggests that activities such as tree felling may have been avoided or were carried out using other methods such as burning (Groube 1989). Elmberg (1955: 56) notes that the Meybrat, who had steel axes already by the 1950s, were not familiar with stone axes, although they were referred to in old tales (Elmberg 1955: 12). The fact that characteristic silica gloss of sago-choppers was not found on any of the artefacts may suggest that sago was not exploited as a food resource, or that some other material was used instead of chert for the choppers. Crosby (1973: 173), for example, mentions a 'sago pith chopping implement' with a cylindrical wooden beating part, recorded from the eastern margin of the Bird's Head. Wollaston (1912: 90) also reports the use of hardwood for this purpose in the Mimika River area, southern Papua.

Prehistoric New Guinean lithic assemblages are generally thought of as fairly amorphous, with little change over time. Indeed, after studying large numbers of prehistoric stone artefacts from a variety of sites from the Highlands in Papua New Guinea, and observing ethnographic tool use, White (1972) came to the conclusion that it was not the shape of an implement that counted but the suitability of its edge. Ethnographic sources have shown that New Guineans, like Australian Aborigines, would often take a core, 'pound it up', and select fragments that looked suitable to do the job (Hayden 1977: 179; Sillitoe 1988: 58). Although New Guineans clearly did not put much effort in producing stone tools, their stone technology was obviously not that simplistic. While perhaps 'crude and colourless' (White 1977) to Australasian, and certainly to

European standards, the prehistoric assemblages from Ayamaru show evidence of standardised sequences of reduction and subtle changes in size and shape through time. Neither of the assemblages produced specific recurring 'types', but in both caves an orderly process of shaping, sharpening and reshaping can be followed.

The three assemblages investigated here (the Late Pleistocene assemblage in Toé Cave, the Holocene assemblage in Toé, and the Holocene assemblage in Kria) form a cross-section both through time and space. To return to the main focus of this chapter – to account for the differences between the assemblages in terms of the use of the cave sites over time – it may be useful to briefly summarise the points in which the assemblages correspond and differ. First of all, there are major differences in quantities and weight between the samples. Looking at only the chert material (comprising the great bulk of the assemblages), the Toé Pleistocene assemblage contained 252 chert artefacts with an average weight of 7.6 g per artefact. The Toé Holocene assemblage produced 1446 artefacts with an average weight of 14.3 g per artefact. Kria Cave produced 1402 artefacts weighing 4 g on the average. The Toé Holocene assemblage has visibly larger material than the other two samples, and it differs also in composition. Original nodules used as cores were not found in the Late Pleistocene assemblage or in Kria Cave (except possibly for one exhausted specimen made of silicified tuff), but these are plentiful in the Toé Holocene assemblage. Large flakes used as cores are again abundant in the Holocene collection in Toé, but present only in limited numbers in the Late Pleistocene of Toé or the Holocene of Kria. The amount of cortex present on the chert artefacts is similar (though slightly lower in Kria) in the two Holocene samples, but much less prevalent in the Late Pleistocene. The analyses of flake dimensions shows a wide size range in the Holocene of Toé Cave, including many very large and very small flakes. The Pleistocene sample, on the other hand, seems to lack the 'very small' as well as the 'very large' flakes; in other words, it produced only medium-sized flakes with a fairly narrow range, especially in thickness. The Holocene sample in Kria Cave contains some large flakes but the bulk consists of relatively small material. Clearly, the chert was reduced more thoroughly than the assemblages from Toé.

How can these patterns be interpreted in terms of chronological and spatial use of the area? Two major conclusions can be drawn, one related to the activities carried out on site, the other to the availability of raw material.

The Toé Holocene sample produced all elements necessary to support the interpretation that stone-working as well as tool maintenance was carried out on site. The assemblage contains ample core material, debitage, and edge-removal flakes showing the flaked-off exhausted edges of utilised tools. The presence of hammerstones with severe bruising from prolonged use supports the idea that active stone reduction occurred at the site throughout this period. The huge amounts of stone left behind at the site (20.5 kg of chert alone), including very large (clearly not exhausted) artefacts and cores, hints at an abundant source of raw material at a short distance from the cave.

Kria Cave has a deficit of cores and very large flakes. Its debitage is relatively small and the single core of silicified tuff found in the deposit is very much used up. Various reasons were explored as to why no block-cores and only very few potential flake-cores were found in Kria; in summary, these large chert artefacts were most likely either taken away from the site, destroyed (for example through firing), or discarded elsewhere either inside or outside the cave. Probably all of these factors were in operation, but the option that cores were removed from the site is interesting in the context of availability of raw material. The fact that the debitage as well as retouched flakes in this sample are relatively small, and that the single core present in the cave is very much exhausted, suggests that the core material and tools were reduced and resharpened in a highly efficient manner as to obtain a maximum yield of usable stone. A quarry with suitable raw material may not have lain close to the cave; in fact, the location where the Toé Cave inhabitants obtained their raw material may have been the closest source for the Kria occupants. Although Toé and Kria Caves are located only c. 12 km apart, the quarry may have been located even further to the south.

During the Pleistocene, the source for raw material in Toé Cave would have been just as close as it was during the Holocene. However, in contrast to the Holocene assemblage from this site, the Pleistocene sample contains no block-cores, only few potential flake-cores, and has a deficit of small debitage, as well as relatively large flakes, with an overall low proportion of cortex. Hammerstones are also absent. In combination, these features suggest that only limited stone-working was carried out on site. Virtually no core material was brought in, and the sample seems to be biased towards artefacts that were prepared before transport and selected on the basis of size. In particular, the larger retouched flakes are thin in comparison to those found in the Holocene samples from either Toé or Kria. Tool maintenance was nonetheless one of the activities performed at the site, as some working-edge removal flakes are present.

If the differences between the two age samples from Toé Cave are not related to availability of raw material, then it becomes likely that the function of the cave has changed through time. During the Pleistocene, stone-working activities were probably limited to maintenance of tools that were manufactured elsewhere and brought to the site. During the Holocene period large quantities of raw material were carried in and were flaked on the spot. Tools were probably created, used and resharpened on site. The raw material was evidently abundant locally so that little attempt was made to use this material as efficiently as possible. Why this change in site function occurred through time is obscure.

Suitable raw material was probably less readily availably in the northeastern regions of the Ayamaru Lakes. Kria Cave produced large quantities of fairly small stone debitage but very few cores and relatively few large tools. It seems that the site was used for highly efficient tool production and maintenance. Tools were certainly produced and used at the site, but artefacts that were still useful for future tasks and further reduction were probably kept and taken away from the site.

Interesting changes are observed in the size and shape of the unretouched flakes through time in Kria Cave. Complete flakes tend to be fairly large and relatively heavy in the two lower units (IV and V), and gradually decrease in size in Units II and III. This change is especially visible in the cortex-bearing flakes, which tend to be more variable in size. Also an increase in length relative to width of the cortex-bearing flakes was noted during more recent times in Unit II and III. The cortex-free flakes, interestingly enough, do not show much variation through time. This presumably reflects subtle changes in stone-working technology, perhaps related to core size or shape. These changes took place around 5500 BP.

Despite the lack of formal artefact 'types', the Ayamaru stone assemblages show considerable variation in both time and space. This variation probably reflects differing proximity to stone sources, as well as some minor changes in stone-working technology through time. Stone-working in this region of New Guinea clearly was not static, but such changes as did occur were subtle and perhaps explicable in terms of the economics of resource use within the prehistoric landscape.

Table 5.2a. Toé Cave, distribution of various artefact categories by raw material, presented by count and by weight (g) for the Holocene units.

		chert	silicified tuff	limestone	quartzite	chert?	limestone?	quartzite?	meta-morphic	unknown	total
core	count	9							1		10
	weight	1154.0							468.0		1622.0
flake (incl. flake-core)	count	1176	2	26		2	4		2		1212
	weight	18,133.2	51.0	263.9		2.3	17.6		164.0		18,632.0
flaked piece	count	213		4			4			1	222
	weight	1160.6		11.4			101.5			0.4	1273.9
potlid	count	21									21
	weight	22.2									22.2
utilised manuport	count				1		1	1	4		7
	weight				39.0		335.0	4.0	1201.0		1579.0
non-utilised manuport	count	27			4	1			5	3	40
	weight	170.8			105.0	1.2			87.0	5.0	369.0
total	count	1446	2	30	5	3	9	1	12	4	1512
	weight	20,640.8	51.0	275.3	144.0	3.5	454.1	4.0	1920.0	5.4	23,498.1

Table 5.2b. Toé Cave, distribution of various artefact categories by raw material, presented by count and by weight (g) for the Late Pleistocene units.

		chert	silicified tuff	limestone	quartzite	chert?	limestone?	quartzite?	meta-morphic	total
flake	count	189	1	4		1				195
	weight	1664.2	54.0	5.0		2.0				1725.2
flaked piece	count	48					2			50
	weight	177.9					2.1			180.0
potlid	count	5								5
	weight	7.8								7.8
non-utilised manuport	count	10				2			1	13
	weight	54.5				3.7			872.0	930.2
total	count	252	1	4		3	2		1	263
	weight	1904.4	54.0	5.0		5.7	2.1		872.0	2843.2

Table 5.13a. Kria Cave, counts and weights (g) of the various artefact categories by raw material.

UNIT I		chert	silicified tuff	limestone	quartzite	chert?	limestone?	quartzite?	metamorphic	total
core	count									
	weight									
flake	count	2			2					4
	weight	0.5			0.6					1.1
flaked piece	count				1		1			2
	weight				0.2		0.1			0.3
potlid	count									
	weight									
utilised manuport	count									
	weight									
non-utilised manuport	count									
	weight									
total	count	2			3		1			6
	weight	0.5			0.8		0.1			1.4

Table 5.13b. Kria Cave, counts and weights (g) of the various artefact categories by raw material.

UNIT II		chert	silicified tuff	limestone	quartzite	chert?	limestone?	quartzite?	metamorphic	total
core	count		1							1
	weight		24.0							24.0
flake	count	361	3	2	1	1	1		1	370
	weight	1067.4	4.0	2.2	0.3	0.3	6.0		36.0	1116.2
flaked piece	count	49	2		1		2		1	55
	weight	50.9	1.1		6.0		3.7		71.0	132.7
potlid	count	4								4
	weight	0.9								0.9
utilised manuport	count									
	weight									
non-utilised manuport	count	15			3					18
	weight	20.3			21.0					41.3
total	count	429	6	2	5	1	3		2	448
	weight	1139.5	29.1	2.2	27.3	0.3	9.7		107.0	1315.1

Table 5.13c. Kria Cave, counts and weights (g) of the various artefact categories by raw material.

UNIT III		chert	silicified tuff	limestone	quartzite	chert?	limestone?	quartzite?	metamorphic	total
core	count									
	weight									
flake	count	195	2	3						200
	weight	409.7	0.7	3.0						413.4
flaked piece	count	22	3		1					26
	weight	35.6	8.1		32.0					75.7
potlid	count	11								11
	weight	8.4								8.4
utilised manuport	count									
	weight									
non-utilised manuport	count	15								15
	weight	12.2								12.2
total	count	243	5	3	1					252
	weight	465.9	8.8	3.0	32.0					509.7

Table 5.13d. Kria Cave, counts and weights (g) of the various artefact categories by raw material.

UNIT IV		chert	silicified tuff	limestone	quartzite	chert?	limestone?	quartzite?	metamorphic	total
core	count									
	weight									
flake	count	141	1				7	1		150
	weight	659.1	0.6				21.4	1.5		682.6
flaked piece	count	22	3				1	1		27
	weight	89.3	2.5				1.3	5.0		98.1
potlid	count	3								3
	weight	2.1								2.1
utilised manuport	count									
	weight									
non-utilised manuport	count	2				2			1	5
	weight	2.1				2.5			1.9	6.5
total	count	168	4			2	8	2	1	185
	weight	752.6	3.1			2.5	22.7	6.5	1.9	789.3

Table 5.13e. Kria Cave, counts and weights (g) of the various artefact categories by raw material.

UNIT V		chert	silicified tuff	limestone	quartzite	chert?	limestone?	quartzite?	metamorphic	total
core	count									
	weight									
flake (incl. flake-core)	count	438	41	2	1	3	5			490
	weight	2900.3	101.4	11.0	1.9	66.0	13.0			3093.6
flaked piece	count	103	7	1	2	3		2		118
	weight	300.0	4.2	9.0	50.0	30.0		34.0		427.2
potlid	count	13	1							14
	weight	10.5	0.1							10.6
utilised manuport	count				1					1
	weight				343.0					343.0
non-utilised manuport	count	6	1		3	2		1	1	14
	weight	1.7	0.4		448.0	4.0		0.1	442.0	896.2
total	count	560	50	3	7	8	5	3	1	637
	weight	3212.5	106.1	20.0	842.9	100.0	13.0	34.1	442.0	4770.6

6. Bone Artefacts

6.1. INTRODUCTION

Bone is a versatile and readily available raw material that has been used for artefact manufacture throughout human history and in almost all parts of the world. In the wider Pacific region there is widespread evidence of the casual manufacture and use of bone artefacts, with many archaeological sites producing small numbers of artefacts of irregular form. Less commonly, evidence of more systematic and intensive manufacture of bone artefacts has been found, with the best known assemblages coming from Late Pleistocene and Holocene sites in Java, Sulawesi and southeastern Australia (Lampert 1966, 1971; Pickering 1979; Olsen & Glover in press). On occasion, bone artefacts are found in such large numbers that they are regarded as characteristic of a cultural horizon. For example, Hale & Tindale (1930) distinguished the late Holocene Mudukian culture at Devon Downs in South Australia on the basis of the abundant fusiform *muduk* points; similar bipointed bone artefacts have since been recorded from many sites in this region, albeit under various names and with various interpretations of their function.

Despite obvious interest in the functional and cultural significance of these bone artefact assemblages, there have been relatively few detailed studies of individual assemblages, and even fewer attempts at comparative analysis. Among earlier studies, the contributions of Lampert (1966, 1971) and Pickering (1979) are exemplary in their attempts at metrical and typological analysis of individual and regional bone artefact assemblages respectively. In more recent years, the emphasis in bone artefact analysis has shifted to microscopical analysis of usewear combined with experimental, replicative studies to determine prehistoric function. Important case studies include those of Webb & Allen (Webb 1987; Webb & Allen 1990) with Tasmanian bone artefacts, and that of Olsen & Glover (Olsen 1984; Olsen & Glover in press) for collections from Sulawesi. Similar methods have been applied with some success to bone tool collections elsewhere in the world (e.g. Tyzzer 1935-36; Olsen 1984; Campana 1989; LeMoine 1997).

A large assemblage of bone artefacts was recently described from Golo Cave on Gebe Island in the northern Moluccas (Pasveer & Bellwood in press). The abundance of bone artefacts in this cave is given even greater emphasis by the virtual lack of similar artefacts at contemporary sites on Gebe and nearby islands. Holocene archaeological sites in Java likewise show considerable variation in abundance of bone artefacts, even within a single region (Prasetyo 2002; Simanjuntak in press).

In this chapter, bone artefacts and utilised teeth from Kria and Toé Caves are described and analysed in their archaeological context to obtain insights into patterns of manufacture, use and discard of these artefacts and to understand the kinds of activities performed at the sites. Kria Cave is characterised by large numbers of these artefacts, while nearby Toé Cave has surprisingly few.

6.2. METHODS

The methods employed in this study were developed in the context of a wider study of bone artefacts from prehistoric sites in the northern Moluccas (Pasveer & Bellwood in press) and the Aru Islands (Pasveer forthc.). A standard terminology and related set of measurements were devised for all three regional collections to achieve a high level of comparability. The methods are similar to those employed by many lithic analysts, and emphasise aspects of raw material selection, methods of modification, basic metric attributes, the morphology of functional edges or points, and patterns of usewear and damage.

The approach used here is a compromise between a strictly typological approach and one that favours analysis of individual attributes related to aspects of technology and function. The identification of distinct classes or 'types' within the assemblage is supported by the good correspondence between the morphological groups and differences in the intensity and distribution of usewear and damage, as detailed below.

6.2.1. *Descriptive terminology*

For this study a bone artefact is defined as a piece of bone showing evidence of human modification other than simple breakage. The basic descriptive terminology and measurements for bone artefacts is illustrated in Figure 6.1.

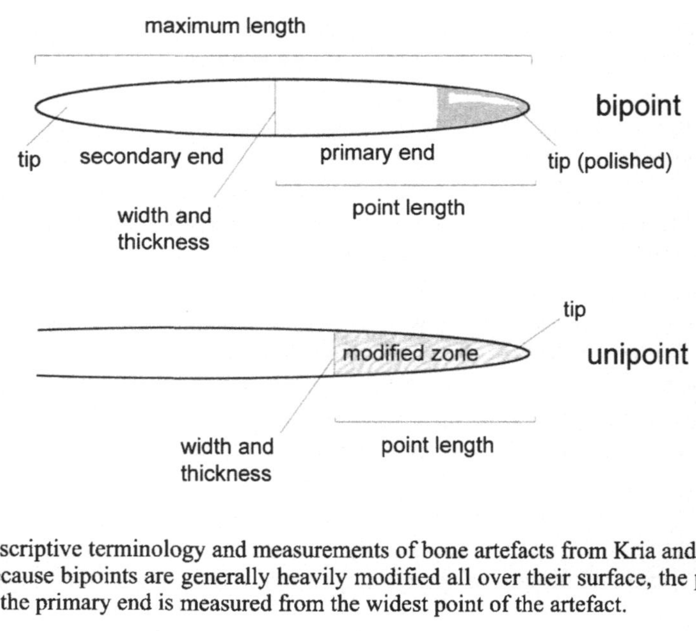

Figure 6.1. Descriptive terminology and measurements of bone artefacts from Kria and Toé Caves. Because bipoints are generally heavily modified all over their surface, the point length of the primary end is measured from the widest point of the artefact.

The following attributes were scored for each bone artefact:

1. Shape: two major 'classes' of bone artefacts were distinguished in the Bird's Head collection, based on the location and nature of modification. 'Unipoints' are relatively elongate, typically parallel-sided artefacts that are sharply pointed at one end only. 'Bipoints' are sharply pointed at both ends. A small number of morphologically irregular specimens are present; these show limited modification and little or no evidence of use and are identified as 'possibly unfinished' bone artefacts. However, this group might also include some 'expedient' artefacts that were produced, used and discarded on a casual basis.
2. Breakage: fragments of bone artefacts were classified into the following groups:
   - 'broken bipoints' that show obvious taper towards both ends;
   - 'point-bearing fragments' that retain one distinct point but do not obviously taper towards the other end; these might either be fragments of unipoints (or perhaps 'exhausted' unipoints if they have been subject to resharpening of the point) or bipoints that are broken at or before their maximum width; and
   - 'midsections' that show clear signs of modification but have lost one or both pointed ends; in the case of a unipoint the 'midsection' might also represent the original 'butt' of the artefact.

   All breaks were scored as either pre- or post-excavation in origin, based on the nature of the fracture surface (stained or encrusted vs. fresh).
3. Raw material: the bone artefacts are manufactured either from
   - splinters of cortical bone ('cortical') derived from the larger long bones (e.g. femur, tibia, humerus) of medium-sized to large mammals; these lack an intact medullary canal; or
   - sections of smaller, narrower long bones ('shaft') that contain an intact medullary canal; these are typically from smaller mammals (e.g. fibula, radius, wing elements of bats).
4. Burning state of the raw material: many of the artefacts are manufactured from bone that has been subjected to some degree of heating or burning. Three states are recognised, as also described in Chapter 2:
   - unburnt to lightly burnt (bone is pale, yellow to light brown and porous to moderately dense in texture). The lightly burnt condition is usually the result of cooking of bone inside a fleshed carcass rather than direct contact with flames. Staining often blurred the distinction between unburnt and lightly burnt bone, hence the two categories are grouped under the heading 'unburnt';
   - burnt (bone is very dark brown or dark grey to black and very dense). The burnt condition usually results from direct contact with flame or intense heat; and
   - calcined (bone is grey to white, very dense and sometimes crazed or warped). All organic material has been lost from calcined bone, resulting in shrinkage and increased brittleness.
5. Method of manufacture: three main techniques can be distinguished (see Fig. 6.2 for illustrations of each). Some artefacts show evidence of more than one method.
   - cutting (presumably with a stone flake; resulting in wide, flat surfaces with few striations);

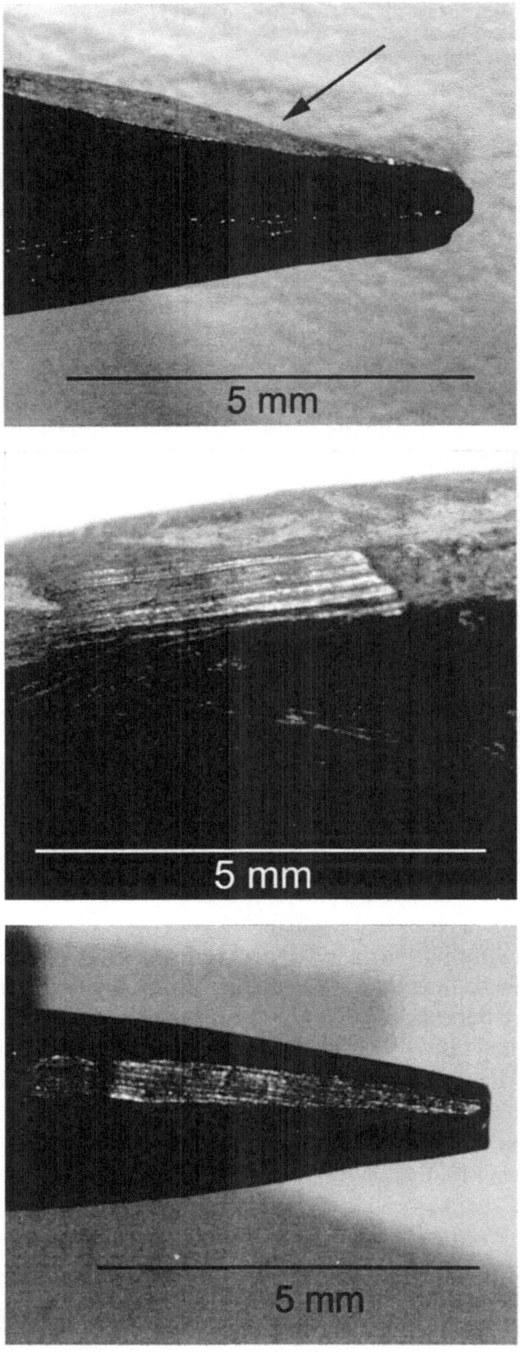

Figure 6.2. Bone artefacts from Kria Cave modified by cutting (top; specimen 1008); shaving (middle; specimen 1008); and grinding (bottom; specimen 854e).

Figure 6.3. Bone artefact from Kria Cave showing high polish at the tip. Specimen 804c.

- shaving (presumably with a stone flake or retouched flake; resulting in semi-regular, linear striations, generally flowing in one direction but with complex patterns of intersection and overlay); and
- grinding (presumably on a fine-grained sedimentary or metamorphic rock; resulting in flat facets with fine striations).

6. Primary vs. secondary ends of bipoints: for each bipoint one point was identified as the 'primary' functional end, based on the presence of localised polish and/or tip damage. The opposite or 'secondary' end generally showed less evidence of use. In the very few cases where the artefact lacked any localised polish or use-related damage, the more intensively modified end was designated as the primary end; in very few cases the decision was arbitrary.
7. Use polish: many specimens bear a localised polish on one or both points (see Fig. 6.3). In other cases the polish occurs as a light gloss over much of the bone surface. Four degrees of polish were scored: no polish; light polish; medium polish; and high polish. Apart from an occasional specimen that was completely covered with medium or high polish, unlocalised polish never scored higher than the 'light' category, while that localised on the tips was often of medium to high intensity. For bipoints, polish was scored separately for each of the primary and secondary ends.
8. Tip damage: this covers localised damage to the tip of the artefacts, as against more substantial 'breakage' of the artefact (see point 2). At least some of the tip damage presumably derives from use of the artefacts, although some might also have occurred following discard. Four kinds of damage were scored for each tip:
   - snap-fracture (usually transverse, see Fig. 6.4a);
   - step-fracture (directed along the length of the artefact and often hinged at its termination; see Fig. 6.4b);
   - crushing (see Fig. 6.4c); and
   - other (usually ragged, possibly the result of multiple step-fractures or combined step-/snap-fractures and crushing; see Fig. 6.4d).

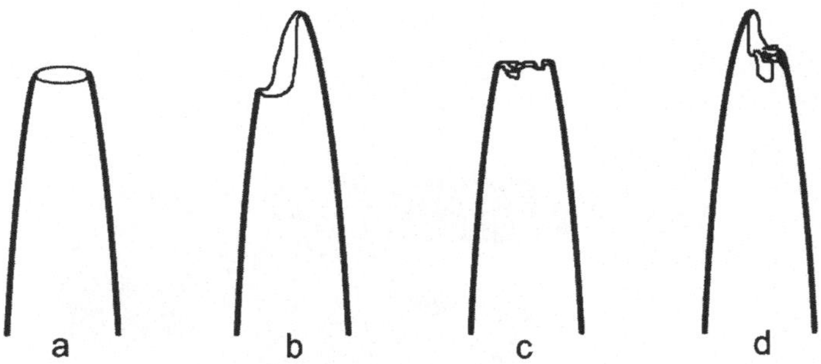

Figure 6.4. Types of damage found at the tip of bone artefacts: a: snap-fracture; b: step-fracture; c: crushing; d: multiple types of damage ('other').

Similar kinds of damage on the tips of bone and antler artefacts are reported by other researchers (e.g. Olsen 1984; Johnson 1985; Arndt & Newcomer 1986; Campana 1989). For bipoints, tip damage was scored for each of the primary and secondary ends.

9. Secondary burning: a small number of specimens were burnt after manufacture. This could be distinguished from the overall burning state of the raw material (primary burning, see point 4) by its highly localised occurrence and typically black discolouration. Obvious warping of an artefact was also taken as evidence of secondary burning. The location, severity and impact of this burning was recorded.

Incisors

All of the larger mammal incisors were examined for evidence of utilisation in the form of a non-random distribution of damage to the tip or margins. The incisors of Giant Rats (*Uromys* and *Mallomys* spp.), the larger possums and cuscuses (*Dactylopsila*, *Phalanger* and *Spilocuscus* spp.), and wallabies and tree-kangaroos (*Dorcopsis*, *Dorcopsulus* and *Dendrolagus* spp.) were considered as suitable for use. Evidence for utilisation was only observed on the lower incisors of the forest wallaby *Dorcopsis muelleri*.

For *Dorcopsis* incisors, each specimen was scored for the following attributes:

1. association: whether in alveolus or loose;
2. completeness of the tooth: scored as complete, root only or crown only;
3. burning state of the crown: scored as unburnt or burnt;
4. damage to the crown: scored as intact (no use-related damage), shattered, or with use-related damage.

For each incisor specimen showing evidence of utilisation, detailed notes were taken on the nature, intensity and distribution of damage.

6.2.2. *Measurements*

The following measurements (see Fig. 6.1) were taken on all bone artefacts with a vernier calliper, to a precision of 0.1 mm:

1. maximum length; no attempt was made to correct for minor damage that has affected the tip of many otherwise complete specimens. This damage has usually removed 1-2 mm or less, representing no more than 5% of the original artefact lengths, and will have little impact on the overall distribution of measurements.
2. width and thickness, taken perpendicular to each other and at the same position on the artefact. The width and the thickness of each artefact were defined in relation to the greatest width of the tip. For unipoints the width and thickness were measured at the base of the point/modified area. However, in some heavily modified unipoints the boundary between the point and the shaft of the artefact is unclear; in such cases the measurements were taken at the widest or thickest point of the artefact. For most specimens, the width exceeded the thickness, however the opposite was true for some nearly circular shaft bones. For bipoints the width and thickness were measured at the widest point.
3. point length, with the 'point' defined as the zone of modification or, in specimens showing general modification, as the distance from the tip to the widest point on the artefact. For intact bipoints, point length was measured for the primary end, with the length of the secondary end being obtained by subtraction from the maximum length. For broken bipoints the intact point of was measured. For 'point-bearing fragments', point length was in most cases the same as the total length of the artefact.

6.3. BONE ARTEFACTS AND UTILISED TEETH FROM KRIA CAVE

A total of 92 bone artefacts were recovered from Kria Cave, 51 from Square 1N1E at the front of the cave and 40 from Square 0N0E, close to the back wall. The provenance of one specimen is uncertain due to a labelling error. The site also yielded eight utilised teeth, five of these from Square 1N1E and three from Square 0N0E. Two other incisors from 1N1E were scored as possibly utilised.

All but one of the bone artefacts came from Units II to V. A single bone artefact was recovered from Unit I. Unlike the bulk of the bone material from this unit, this specimen is encrusted with calcium carbonate and this raises suspicions that it might have been derived from an underlying unit as a consequence of local disturbance. Disregarding this suspicious record from Unit I, the most recent bone artefact was found in spit 2 of Unit II in Square 0N0E, dating to just after 4370 BP. The oldest artefacts come from the deepest excavated spits of Square 0N0E and date to some 7000 BP (see Chapter 3). Still older specimens may be present in the unexcavated deposit below.

The utilised teeth are more abundant in the upper units. Unit II produced four definite and two possibly utilised teeth; Unit III produced three specimens; and Unit IV a single example. No utilised teeth were found in Unit V.

6.3.1. *Distribution of major classes of bone artefacts*

The majority of the more complete bone artefacts from Kria Cave can be clearly divided into two distinct classes based on a combination of their overall shape and raw material. The first class consists of relatively elongate, typically parallel-sided artefacts that are sharply pointed at one end only. These 'unipoints' are manufactured primarily from shaft bones that retain an intact medullary canal and typically have a near circular or triangular cross-section. The second class consists of specimens with points at both ends. These 'bipoints' are manufactured primarily from splinters of cortical bone and therefore lack an intact medullary canal; the cross-section is typically more ovate.

The collection includes 16 unipoints and 28 recognisable bipoints (Table 6.1). The remainder of the assemblage includes 43 fragments (of which 27 are 'point-bearing fragments' and 16 are 'midsections'), and a small number of morphologically irregular specimens. The majority of the fragments are of suitable size and shape to be derived from one or the other of the two primary classes. The irregular specimens show limited modification and little or no evidence of use. This group probably includes some unfinished bone artefacts but might also include 'expedient' artefacts that were produced, used and discarded on a very casual basis.

The distribution of the various classes of bone artefacts through the deposit is shown in Table 6.1. Both major classes of artefacts are represented throughout the deposit and while the proportion of unipoints to bipoints shows some variation between units (e.g. bipoints are more abundant in Units III and V), there is no clear directional trend. The bone artefact assemblage thus appears to be relatively stable in composition through time.

As noted above, utilised teeth are most abundant in Units II-III, rare in Unit IV and entirely absent from Unit V. This does not simply reflect the abundance of *Dorcopsis* incisors in each level; Unit III in fact shows the highest frequency of utilisation, with relatively few incisors present. Unit V, with no utilised specimens, actually produced almost two thirds of the total number of incisors from the site.

6.3.2. *Kria Cave bipoints*

A total of 28 bipoints were found in Kria Cave (some examples are illustrated in Figure 6.5). Eleven are complete with both ends more or less intact, while 17 have more significant breakage at one or both ends; all but one of the latter specimens was broken in antiquity.

Raw material and manufacturing

Twenty-six of the bipoints are made from cortical bone splinters, the remaining two from shaft bone fragments of smaller mammals (Table 6.2). Because of their high degree of modification the origin of the cortical splinters could not be identified. However, judging from the surface texture and general dimensions, it is likely that most of them are fragments of larger mammalian long bones. The largest taxa represented among the

Table 6.1. Kria Cave, distribution of bone artefacts and utilised incisors through the deposit.

	Unit I		Unit II		Unit III		Unit IV		Unit V		out of context	total	
	number of artefacts	%	number of artefacts	%	number of artefacts	%	number of artefacts	%	number of artefacts	%	number of artefacts	number of artefacts	%
unipoints			5	26.3	2	10.5	4	25.0	5	13.9		16	17.4
bipoints (incl. broken)			4	21.1	7	36.8	2	12.5	14	38.9	1	28	30.4
point-bearing fragments			6	31.6	5	26.3	7	43.8	9	25.0		27	29.3
midsections	1	100	3	15.8	4	21.1	2	12.5	6	16.7		16	17.4
unfinished			1	5.3	1	5.3	1	6.3	2	5.6		5	5.4
total bone artefacts	1		19		19		16		36		1	92	
utilised incisors	0	0	6	11.5	3	17.6	1	8.3	0	0		10	6.8
total number of incisors	2		52		17		12		64			147	

Table 6.2. Kria Cave, raw material used for the various bone artefact categories.

	shaft	cortical	total number
unipoints	15	1	16
complete bipoints		11	11
broken bipoints	2	15	17
point-bearing fragments	17	10	27
midsections	16		16
unfinished		5	5
total	50	42	92

Table 6.3. Kria Cave, number of artefacts manufactured from unburnt and burnt bone.

	cortical			shaft		total number
	unburnt	burnt	unknown*	unburnt	burnt	
unipoints	1			15		16
complete bipoints	11					11
broken bipoints	13	2		2		17
point-bearing fragments	8		2	16	1	27
midsections				14	2	16
unfinished	5					5
total	38	2	2	47	3	92

*calcined after discard

associated animal bones are the Brown Dorcopsis, *Dorcopsis muelleri*, various tree-kangaroos, and various species of cuscuses (*Phalanger* and *Spilocuscus* spp.). One possibly unfinished specimen from Unit II seems to have been manufactured of exceptionally thick cortex, which could be cassowary or human bone. The two specimens made from shaft fragments possibly originate from bat wing bones.

Twenty-four of the cortical bipoints and both of the shaft bipoints are manufactured on unburnt bone (Table 6.3). The remaining two cortical bipoints are manufactured on burnt bone. One otherwise unburnt cortical bipoint shows secondary burning at one tip. One of the two shaft bipoints also shows secondary burning on one side of the tip.

The bipoints show evidence of a variety of different manufacturing techniques (Table 6.4). Among the cortical bipoints, all but two specimens show the characteristic shaving marks produced by scraping with a stone artefact. Two specimens show one or more large cut marks, followed by shaving. In 15 specimens grinding occurred subsequent to shaving, presumably to finish the point or tip. Seven specimens were shaved exclusively, while two lack shaving marks but show evidence of grinding. In all cases it may be true, however, that cutmarks, shaving marks, or even grinding facets may have been obliterated by subsequent modification or use.

Table 6.4. Kria Cave, number of artefacts with evidence of particular manufacturing techniques.

	cortical					shaft			
	shaved	ground	shaved, then ground	cut, then shaved	cut, then ground	shaved	ground	shaved, then ground	total number
unipoints			1			3	2	10	16
complete bipoints	2		9						11
broken bipoints	5	2	6	2		1		1	17
point-bearing fragments		1	7	1	1	4	6	7	27
midsections						6	5	5	16
unfinished	2		3						5
total	9	3	26	3	1	14	13	23	92

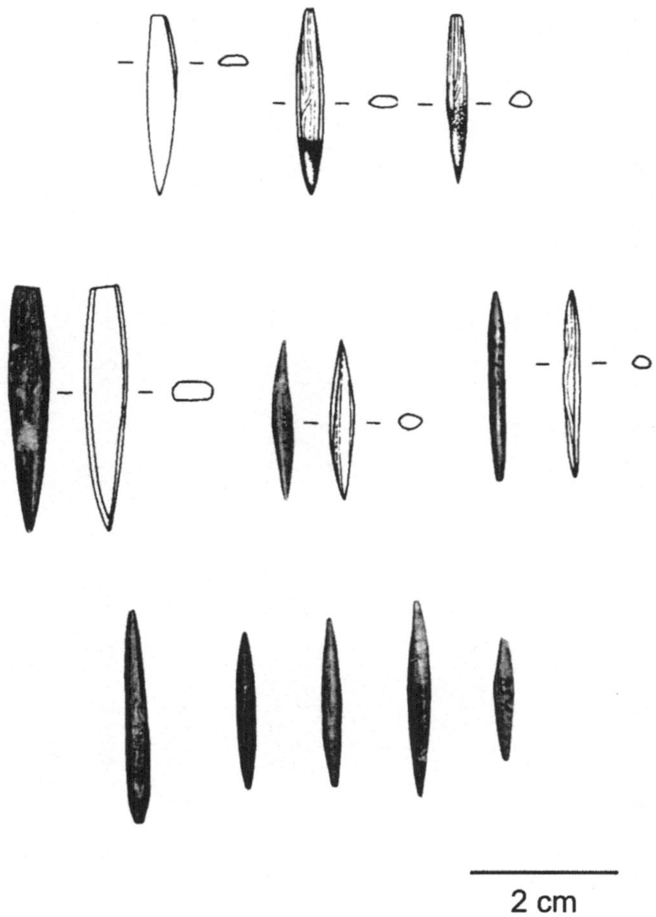

Figure 6.5. Some examples of the cortical bipointed artefacts from Kria Cave. Top row, left to right: broken bipoint (specimen 1044); bipoint with burning and high polish at one of the tips (1109a); broken bipoint with high polish at the intact tip (804c). Middle row: exceptionally wide broken bipoint, opposing sides (914); bipoint, opposing sides (854a); bipoint, opposing sides (1173). Bottom row, bipoints, left to right: specimens 854f, 896, 854e, 955a, 854c. Drawings: Mirjam Weijns.

Metrical analysis of bipoints

Metrical attributes of the Kria Cave bipoints are summarised in Table 6.5. Separate values are given for each of the cortical bipoints and shaft bipoints.

The 11 complete bipoints are all made of cortical bone. These are highly uniform in size and shape, ranging in length from 15.6 to 27.4 mm, with the majority falling between 20.1 and 27.4 mm. Although samples are very small, there is no obvious change in bipoint length between the various stratigraphic units (Table 6.5).

Table 6.5. Kria Cave, group statistics for each artefact category in each unit.

	N	length range in mm	mean	σ	N	width range in mm	mean	σ	N	thickness range in mm	mean	σ	N	point length range in mm	mean	σ
	Unit II				Unit II				Unit II				Unit II			
shaft unipoints	5	23.9-35.9	29.1	5.8	5	3.7-5.6	4.2	0.8	5	2.6-3.6	3.2	0.4	2	13.8-13.9	13.9	0.1
complete cortical bipoints	3	23.9-27.1	26.0	1.8	3	3.5-4.2	3.8	0.4	3	2.1-2.8	2.4	0.4	3	11.2-17.6	13.8	3.3
broken cortical bipoints	1	15.8			1	3.0			1	1.7			1	8.5		
shaft midsections	3	11.8-40.0	24.7	14.2	3	3.3-5.2	4.2	1.0	3	2.2-3.6	3.0	0.7				
	Unit III				Unit III				Unit III				Unit III			
shaft unipoints	1	31.9			1	3.1			1	2.6			1	10.0		
cortical unipoints	1	35.6			1	3.8			1	2.9			1	11.5		
complete cortical bipoints	4	15.6-27.4	21.5	4.8	4	2.6-3.2	2.8	0.3	4	1.8-2.3	2.1	0.2	4	4.1-11.4	8.7	3.2
broken cortical bipoints	3	13.6-24.7	19.5	5.6	3	2.8-3.8	3.1	0.6	3	1.5-2.2	1.9	0.4	3	7.5-12.7	10.9	3.0
shaft midsections	4	14.4-52.5	25.2	18.3	4	3.4-4.5	4.0	0.5	4	2.5-3.9	3.2	0.6				
	Unit IV				Unit IV				Unit IV				Unit IV			
shaft unipoints	4	27.1-51.2	36.9	10.5	4	4.0-4.4	4.2	0.2	4	2.7-3.5	3.2	0.3	3	14.0-23.6	17.6	5.3
complete cortical bipoints	1	20.1			1	2.5			1	2.2			1	10.3		
broken cortical bipoints	1	15.5			1	2.7			1	1.9			1	12.1		
shaft midsections	2	17.6-41.4	29.5	16.8	2	4.0-4.6	4.3	0.4	2	2.8-3.6	3.2	0.6				
	Unit V				Unit V				Unit V				Unit V			
shaft unipoints	5	26.9-50.9	41.3	10.9	5	3.6-4.7	4.2	0.5	5	2.7-3.6	3.3	0.4	4	7.2-26.9	17.4	8.1
complete cortical bipoints	3	25.1-27.3	26.2	1.1	3	2.3-3.5	3.0	0.6	3	1.7-2.0	1.9	0.2	3	5.9-15.5	12.0	5.3
broken cortical bipoints	9	11.9-25.7	19.0	5.4	9	2.3-4.0	3.2	0.7	9	1.4-2.0	1.8	0.2	9	7.9-20.0	12.0	3.7
broken shaft bipoints	2	40.2-40.6	40.4	0.3	2	3.5-3.9	3.7	0.3	2	3.0-3.1	3.1	0.1	2	10.1-12.4	11.3	1.6
shaft midsections	6	17.6-35.0	25.9	5.8	6	3.3-5.1	4.2	0.7	6	2.8-3.6	3.2	0.4				

The degree of asymmetry of the complete bipoints can be expressed by the ratio of primary point length to total length (ptl/l). A ptl/l value of 0.5 represents perfect symmetry; values below 0.5, specimens where the primary end is the shorter one; and values above 0.5, specimens where the primary end is the longer one. Among the sample of 11 complete bipoints, there are five specimens with a 'short' (ptl/l < 0.5) primary end and six specimens with a 'long' (ptl/l > 0.5) primary end. However, as shown in Figure 6.6, most specimens are only mildly asymmetric, with ptl/l values between 0.45 and 0.59. In contrast, two specimens have extremely short primary ends and in one specimen it is moderately long. Overall, the primary ends tend to make up slightly more than half the length of the artefacts, although slightly longer and much shorter primary ends are also represented.

Taken overall, there is no correlation between point length and maximum length of cortical bipoints ($R^2 = 0.036$). However, a plot of these variables (Fig. 6.7) shows that the smaller bipoints (up to about 25 mm length) maintain a high degree of symmetry. Above this size, point length of the primary end is highly variable and includes both rather long or extremely short examples. If the two artefacts with exceptionally short tips are excluded, the correlation between the total length and point length is moderately high ($R^2 = 0.68$; N = 9). If only specimens with long primary ends (i.e. ptl/l > 0.5) are considered the correlation between length and point length is very high ($R^2 = 0.895$; N = 6).

Bipoint thickness and width can be examined in the larger sample that includes both complete and broken cortical specimens (N = 26). Width of complete cortical bipoints ranges from 2.3 to 4.2 mm, and for broken cortical specimens, from 2.3 to 4.0 mm [one mislabelled specimen (and therefore without provenance) with an exceptional width of 5.6 mm is excluded from Table 6.5]; there is no significant difference in means. For thickness, the values are lower for each group, 1.7 to 2.8 mm and 1.4 to 3.1 mm respectively; these ranges overlap almost entirely and the difference in means is not significant. There are weakly significant differences in bipoint width and thickness between the various stratigraphic units (Table 6.5), however, these are caused by a single complete bipoint from Unit II that is relatively wide. When this specimen is omitted, the differences between the units are not significant.

For complete specimens, the correlation between bipoint width and length is moderately strong ($R^2 = 0.36$), while that between bipoint thickness and length is very weak ($R^2 = 0.16$). One possible explanation for this difference is that the thickness of a bipoint is more tightly constrained by the original parent bone than are either its width or length. However, measurements of the cortical bone thickness for each the two largest marsupials (*Dorcopsis* and *Spilocuscus* spp.) in the Kria Cave fauna suggests that the available raw material ranged between 1.3 and 5 mm, depending on the skeletal element and age of the animal. This suggests that original thickness of the cortical splinters was generally not a major constraint in the manufacture of bipoints, but rather that these artefacts were deliberately reduced, primarily by shaving, to a desired, relatively uniform thickness. This interpretation is supported by the heavily modified appearance of many of the bipoints. Width and thickness are moderately well correlated ($R^2 = 0.46$) in the combined sample of complete and broken bipoints.

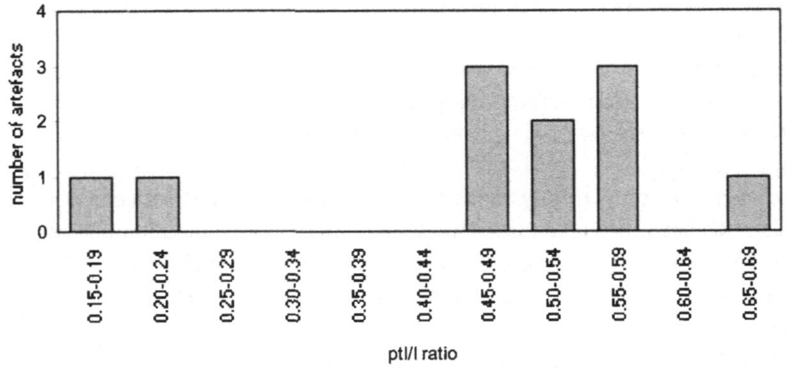

Figure 6.6. Kria Cave, asymmetry in complete cortical bipoints.

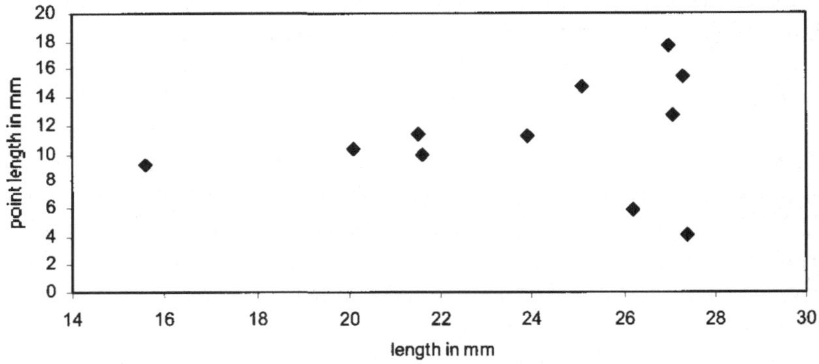

Figure 6.7. Kria Cave, relation between point length and total length of complete cortical bipoints.

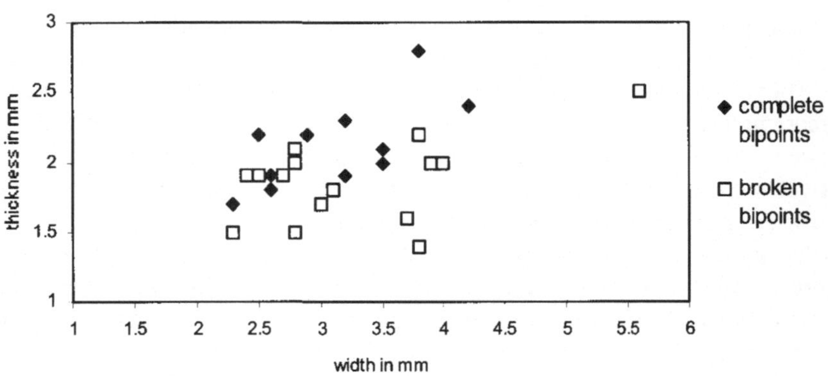

Figure 6.8. Kria Cave, comparison of thickness and width of complete and broken cortical bipoints.

Comparison of complete and broken bipoints

As noted above, broken bipoints average slightly thinner than complete bipoints. Interestingly, a plot of these variables shows that broken cortical bipoints tend to be thinner for their width than the complete specimens (Fig. 6.8). This is perhaps explicable in terms of the mechanical properties of the artefacts, with thinner specimens being more likely to break than thicker specimens of the same width. However, other explanations are possible, including a difference in functionality between the relatively thinner and thicker bipoints.

The length of the intact point of broken bipoints averages slightly longer (12.1 mm, range 7.5-20.0 mm) than the primary end of the complete bipoints (11.1 mm, range 4.1-17.6 mm), although the difference is not statistically significant. This suggests either that the broken bipoints were originally longer on average than the complete ones, or that the broken bipoints more often retain the long end of the bipoint, whether primary or secondary. If the latter were the case, then the average original length of the sample of broken cortical bipoints would perhaps be similar to, but not longer than, that of the complete bipoints.

These alternatives were further examined by estimating the original length of the broken bipoints, using the correlation between the long primary point length and maximum length of the complete bipoints. The equation based on the long primary ends was used for this calculation for two reasons; firstly because the intact end of the broken bipoint is longer on average than the primary ends of the complete bipoints; and secondly, because a regression formula based on the short primary ends could not be calculated because of the very small sample size. The result suggest that the broken bipoints were originally slightly shorter on average than the complete bipoints. However, because the estimations are based on the long primary end equation, the original length of any broken bipoints that actually retains the short end would be under- rather than over-estimated. Taking this factor into account, it seems more likely that the broken bipoints were similar in length rather than shorter than the complete bipoints.

The estimates of original length can also be used to estimate the likely proportion that was lost to breakage for each artefact. The resultant histogram shows a clear bimodality: six of the 10 broken bipoints evidently lost around 5 to 20% of their original length, while the remaining four lost 25-35% (Fig. 6.9).

Although there are no complete bipoints made of shaft fragments, two artefacts made out of shaft bone fragments were clearly pointed at both ends. Both specimens measure around 40 mm in length and are thus considerably longer than any of the complete cortical bipoints. They also exceed the estimated original length of the longest broken bipoints, none of which are likely to have been longer than 32 mm. The two shaft bipoints are similar in width to the cortical bipoints (3.5 and 3.9 mm), but both are relatively thicker (c. 3 mm) than most cortical bipoints, reflecting the more circular shape of the parent shaft bone.

In both shaft bipoints the medullary canal is exposed at one side of the tip. Except for its modified butt, one of these artefacts in particular bears a very close resemblance to the unipoints found at Kria Cave, to be discussed below.

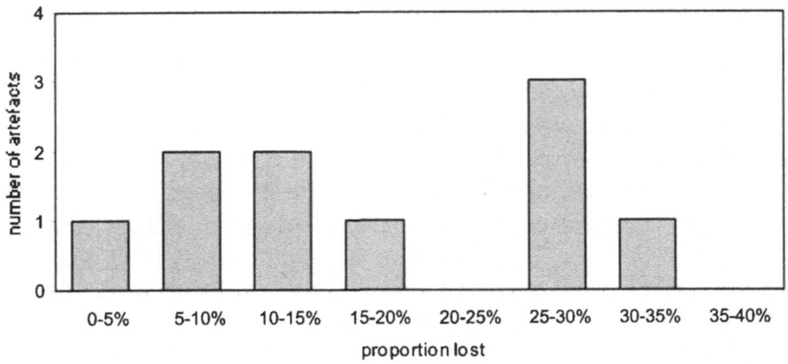

Figure 6.9. Kria Cave, proportion lost of broken cortical bipoints. The calculation was based on the equation for the long primary end.

Use-related polish of bipoints

Eight of the complete cortical bipoints show localised polish at the primary end (Table 6.6). This varies in intensity from light (N = 1) to medium (N = 6) and high (N = 1). Three complete cortical bipoints show little or no localised polish, but rather an overall light gloss. This might be due to repeated use of the artefact in a penetrative fashion; however, it might also be the product of frequent handling, either by the manufacturer or by the researcher. Unless non-localised polish of this kind attains a medium or high grade, it is not accepted as evidence of specific use.

By definition, it is the primary end that generally shows the most intense modification and highest degree of polish. Only one of the Kria cortical bipoints has polish at the secondary end, in this case of medium intensity. The occurrence of polish at neither, one, or both ends is not obviously correlated with overall bipoint length. Nor is it related to point length; medium and high polish was found on specimens with both very short and longer primary ends.

Ten of the 15 broken cortical bipoints show localised polish on the intact end (Table 6.6). Five show a light localised polish at the tip, three a medium polish, and two are highly polished. The remaining five specimens show virtually no localised polish at the tip, but some show a light gloss over their entire surface.

One of the shaft bipoints shows a localised medium polish at the intact tip; the other shaft bipoint is unpolished.

Use-related damage of bipoints

The primary ends of the complete cortical bipoints generally show some kind of damage (Table 6.7). Only one specimen is completely undamaged, while one other shows post-excavation damage that may obscure any original features. Six of the remaining nine artefacts show crushing of the tip and three show step-fractures. There is no clear

Table 6.6. Kria Cave, number of artefacts showing varying degrees of polish.

	no or unlocalised polish	light polish	medium polish	high polish	total number of no polish vs. polish
			cortical		
unipoints			1		
complete bipoints primary end	3	1	6	1	3 : 8
complete bipoints secondary end	10		1		10 : 1
broken bipoints intact end	5	5	3	2	5 : 10
point-bearing fragments	8			2	8 : 2
			shaft		
unipoints with 'open' tips	5		3		
unipoints with 'closed' tips	4		3		
broken bipoints intact end with 'open' tips	1		1		
point-bearing fragments with 'open' tips	6	1	4		
point-bearing fragments with 'closed' tips	5			1	

relationship to bipoint length, but it does appear that crushing tends to occur more often on longer primary ends.

The secondary end also frequently displays damage; only one specimen is undamaged and two show post-excavation damage. All other specimens show crushing (in four examples), minor snap-fractures (in three) or a step-fracture (in one). Snap-fractures are found on the longer secondary ends, while crushing tends to occur on the shorter secondary ends. Unlike polish, damage is thus commonly observed on both ends of the cortical bipoints.

Among the 15 broken cortical bipoints, the intact point is undamaged in two specimens and obscured by a recent break in one other (Table 6.7). In the remaining specimens the intact tip is most often crushed (seven examples), with fewer examples of step-fracturing (three cases), minor snap-fracturing (one case) and more complex fracturing (one case).

A cross-tabulation of damage against polish (Table 6.8) suggests that crushing could occur at more or less any stage in the lifetime of the artefact. In contrast, step-fracturing of the primary end is observed only on tips that display significant localised polish. This suggests that the likelihood of step-fracturing increased as a result of sustained use of the kind that is also likely to produce polish. Among the complete bipoints, minor snap-fractures are only observed at the secondary end.

Crushing of the tip was clearly the most common kind of damage to Kria cortical bipoints. Interestingly enough, this kind of damage equally affected both the primary and secondary ends of the bipoints and could occur at more or less any stage during the use of the artefacts. Step-fracturing occurred less frequently, and generally only at the

Table 6.7. Kria Cave, number of artefacts showing varying types of tip damage.

	no damage	snap-fractures	step-fractures	crushing	other	recent damage
			cortical			
unipoints						1
complete bipoints primary end	1		3	6		1
complete bipoints secondary end	1	3	1	4		2
broken bipoints intact end	2	1	3	7	1	1
point-bearing fragments	3	2	1	3		1
			shaft			
unipoints with 'open' tips	2	2	1	1	1	1
unipoints with 'closed' tips	1		5			1
broken bipoints intact end with 'open' tips	2					
point-bearing fragments with 'open' tips	3	5		3		1
point-bearing fragments with 'closed' tips	1	1	2	1		

primary end and as a consequence of sustained use. Snap-fractures were uncommon at the primary end but occurred at moderate frequency at the secondary end. The intact ends of the two broken shaft bipoints have undamaged tips.

Use-related breakage of bipoints

The broken end of incomplete cortical bipoints is generally truncated by a major snap-fracture (six cases) or by a more complex breakage involving combined snap- and step-fractures (six cases). In two specimens the major breakage was caused by a large step-fracture; in one case this fracture occurred to an end that otherwise displays a localised polish of medium intensity. In one specimen the major damage has occurred recently.

Major breakage may have occurred during manufacturing, during use or subsequent repair, or during or after discard. Three of these different scenarios might be expected to produce different patterns of association between polish, minor damage and major breakage. Artefacts broken during manufacture should show no localised polish or specific use-related damage such as step-fractures. The more heavily modified end (i.e. the primary end) might be expected to show the highest incidence of breaks. Artefacts broken as a consequence of use might show localised polish and/or tip damage, depending on the extent of wear, and the pattern of breakage might not be random if it is related to the specific modes of use. Artefacts broken after use and discard by trampling or other taphonomic processes are likely to show localised polish as well as damage, but the breakage of primary vs. secondary ends is likely to be random. Breakage during resharpening or other secondary modification might be difficult to distinguish from breakage during manufacturing or after discard, depending on the extent of prior usage and the extent of modifications prior to breakage.

Table 6.8. Kria Cave, association of polish and tip damage in cortical bone artefacts.

	primary end of cortical bipoints					
	no damage	snap-fractures	step-fractures	crushing	other damage	recent damage
no or unlocalised light polish				2		1
localised light polish				1		
localised medium polish	1		3	2		
localised high polish				1		
	secondary end of cortical bipoints					
	no damage	snap-fractures	step-fractures	crushing	other damage	recent damage
no or unlocalised light polish		3	1	4		2
localised light polish						
localised medium polish	1					
localised high polish						
	intact end of broken cortical bipoints					
	no damage	snap-fractures	step-fractures	crushing	other damage	recent damage
no or unlocalised light polish		1		2	1	1
localised light polish			1	4		
localised medium polish			2	1		
localised high polish	2					
type of breakage broken end		6	2*		6	1
	point-bearing cortical fragments					
	no damage	snap-fractures	step-fractures	crushing	other damage	recent damage
no or unlocalised light polish	2	2		3		1
localised light polish						
localised medium polish						
localised high polish	1		1			
type of breakage broken end		5	1		4	

Note: * one broken end shows localised medium polish

The presence of localised polish on many of the broken cortical bipoints can only mean that the majority were used prior to breakage and/or discard. This excludes the possibility that these particular specimens were broken during manufacturing. But did breakage occur mainly during use or after discard?

As indicated above, the key to distinguishing between breakage during use and breakage after discard might lie in random vs. patterned breakage of primary vs. secondary ends. To examine this possibility further we need to assume that the sample of complete and broken cortical bipoints are effectively drawn from the same original population, with approximately equal probabilities of breakage and discard for all specimens. The possible slight difference in original lengths and thickness of the two samples might caution against such an assumption, however, for sake of argument it is worth proceeding.

The 11 complete bipoints provide an estimate of the frequency of polish and damage within the population prior to breakage. As the bipoints can be polished at both tips, each end must be considered as an independent entity. The frequency of polish at the primary end is eight out of 11 tips; at the secondary end polish occurs in one of 11 tips. Based on the sample of complete bipoints, the ratio of unpolished to polished ends within the population is thus 13 : 9.

Three possible breakage scenarios can now be examined:

1. if it was always the primary end that broke off, we might expect the intact end of the broken bipoint to show a higher proportion of unpolished ends, approximating the ratio of 10 : 1 (unpolished : polished) as observed in the secondary ends of the complete bipoints (see Table 6.6).
2. if it was always the secondary end that broke off, we might expect the intact end of the broken bipoint to show a higher proportion of polished tips, approximating the ratio of 3 : 8 (unpolished : polished) as observed in the primary ends of the complete bipoints.
3. if breakage was effectively random, we might expect to find the ratio of unpolished vs. polished to approximate that of the entire population of ends among the complete bipoints, i.e. approximately 13 : 9.

For the sample of broken bipoints we do not know whether it was the primary end or the secondary end that has been lost. Among the sample of 15 broken bipoints, 10 ends show polish and 5 do not. The ratio of unpolished to polished tips in broken bipoints is 5 : 10. This is close to the ratio predicted by scenario 2 but is very different from the ratios predicted by each of scenarios 1 and 3. Given that some random breakage has almost certainly occurred after discard, it seems reasonable to conclude that most breakage did in fact occur during use, and further, that it was the secondary end that most commonly broke off, leaving the primary end intact.

Unfortunately, an equivalent test could not be applied to the distribution of damage at the tips. This is because the test would need to be done separately for each of the different kinds of damage, resulting in sample sizes too small for meaningful assessment.

Possible bipoint fragments

The category of 'point-bearing fragments' includes ten specimens made from cortical bone. Although some of these specimens closely resemble the population of complete and broken bipoints, the fact that they do not show any sign of tapering towards a second point leaves some doubt regarding their identity.

Five of these specimens, ranging in length from 9.1 to 17.8 mm, increase gradually in width and thickness back to the point of truncation. These specimens were all broken in antiquity and none of them shows any significant polish. All but the largest of these specimens fall within the metrical cluster of broken cortical bipoint in terms of their cross-sectional proportions (see Fig. 6.10). The simplest interpretation of these isolated tips is that they represent the secondary ends lost from cortical bipoints during use, as postulated above. The longest specimen may be an exception. Its considerable length and the fact that it has a more rounded cross-section invite comparison with the second major class of unipoints.

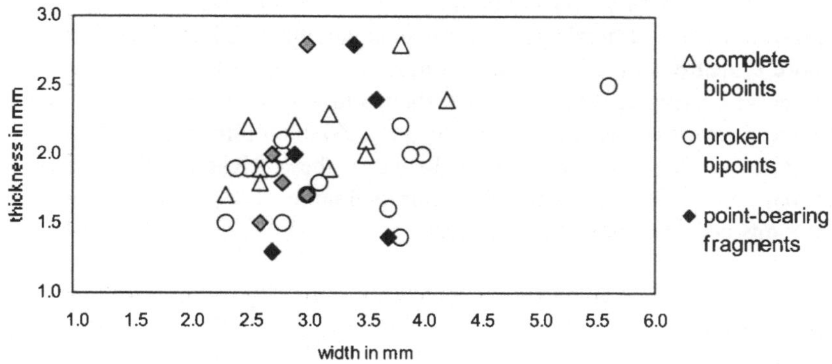

Figure 6.10. Kria Cave, comparison of thickness and width of point-bearing cortical artefacts. Grey diamonds are identified as isolated tips.

The remaining five specimens are even longer (18.9 to 24.8 mm) and they are also morphologically distinct, with relatively short points followed by a straight section that shows no indication of taper towards a second point. Importantly, this morphology in itself does not rule out the possibility that they represent broken bipoints, as there are indeed some complete cortical bipoints that have an elongate and straight midsection that tapers relatively abruptly to each of the two points. On the other hand, there is one very similar cortical specimen, to be described below, that is so long as to be classified with some certainty as a unipoint.

Among the ten point-bearing fragments, most have an ovate cross-section consistent with the population of cortical bipoints (Fig. 6.10). However, two specimens, including the longest one and the longest of the isolated tips, have more rounded cross-sections that further distinguish them from the bipoints.

Two of this group of ten artefacts have a high polish at their tips (Table 6.8); the remainder lack localised polish. One of these is undamaged, while the other has a small step-fracture. Among the unpolished specimens, two are undamaged, five have snap-fractures or crushing of the tip, and one tip was damaged after excavation.

The low incidence of localised polish (two out of ten) on the sample of point-bearing fragments further distinguishes them from the larger samples of complete and broken bipoints, both of which show a high frequency of polish of the primary or intact end. Considered in the context of the three scenarios for breakage of bipoints, as set out previously, the point-bearing fragments would have to be interpreted as primarily representing the secondary ends of bipoints. This interpretation poses some difficulties. If these are indeed secondary ends of cortical bipoints, then it is noteworthy that the larger specimens are already almost equal in length to the longest of the complete bipoints; this would seem to imply that this class of artefact includes specimens that were originally considerably longer than suggested by any of the available complete or undoubted broken specimens. On the other hand, if these specimens represent the major part of originally bipointed artefacts (or, in other words: are actually broken bipoints rather than isolated tips), then, based on the occurrence of polish, breakage must have involved loss of the primary end rather than the secondary end. This might imply a different pattern of use and breakage for these elongate bipointed artefacts as compared with the more typically formed cortical bipoints. Also noteworthy in this regard is the fact that the point-bearing cortical fragments show a low incidence of damage compared with either the primary or secondary ends of the undoubted bipoints. This observation further suggests a functional difference between these classes of artefacts. The possibility that some of the cortical point-bearing fragments are actually fragments of cortical unipoints is discussed in a later context.

Possible unfinished bipoints

One cortical specimen must either represent an unfinished bipoint or a crude parody of the typically finely-crafted item (see Fig. 6.11, specimen 1179a). The unburnt or lightly burnt splinter from which it was made is naturally pointed at both ends and only the midsection shows shaving marks; there is no evidence of grooving or cutting prior to splintering. The fragment measures 25.8 mm long, 4.4 mm wide and 3.1 mm thick and hence falls at the upper end of the variation for complete cortical bipoints. The specimen shows no damage or polish to either tip. With further modification, this specimen could very easily become a cortical bipoint of the kind represented in this collection. This specimen is also interesting insofar as it provides a further indication that some bipoints at least were produced at an originally quite small size, rather than necessarily being the product of breakage and resharpening.

Four other cortical artefacts are tentatively identified as unfinished specimens, although this would imply a very different process of manufacture for this class of implement. All four specimens are essentially crude splinters of unburnt or lightly burnt cortical bone but in each case they bear one well-formed point produced by shaving (two

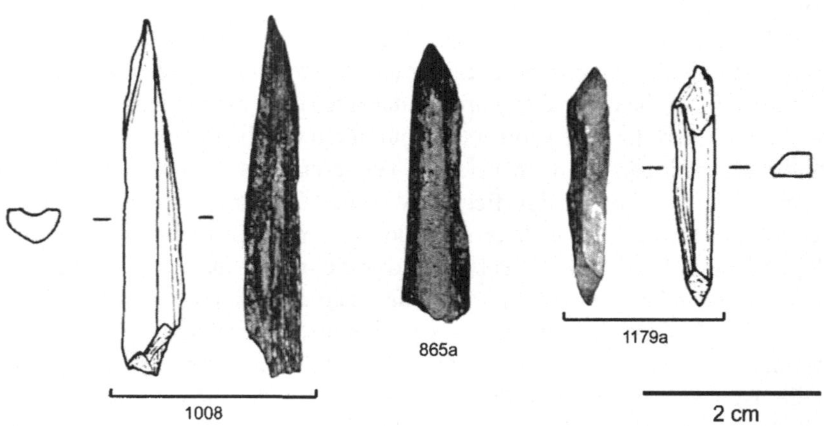

Figure 6.11. Some examples of the 'unfinished' artefacts from Kria Cave. Left specimen: photo and drawing of opposing sides (1008). Middle specimen: photo of one side only (865a). Right specimen: photo and drawing of the same side (1179a). Drawings: Mirjam Weijns.

examples are given in Fig. 6.11). In three of these artefacts the point is very finely produced, in two cases with grinding facets superimposed over the shaving marks; the fourth specimen is less well preserved. Apart from these well-formed points, the remainder of each splinter is unworked and they retain their crude original shapes. This shows little regularity; indeed, if anything, there is a hint of negative correlation between length and each of width and thickness: the longer specimens are actually narrower and thinner than the shorter ones. With a sample of four specimens, little significance can be attached to this observation. More important is the fact that none of the tips is damaged and three out of the four lack any trace of polish. However, one of the tips shows a light polish at the end, suggesting that it was utilised in some activity, although perhaps only briefly.

The general crude appearance of these artefacts, combined with the lack of damage and polish (except for one specimen) to the tips, suggests that they are unfinished specimens. Given the predominant use of cortical splinters in the production of bipoints, it seems most plausible to view them as partly formed cortical bipoints, although there is also an example of an undoubted cortical unipoint in the assemblage. However, unlike the clearly bipointed specimen described above, these specimens would imply a manufacturing process that involved initial production and even finishing of one complete point, followed by the creation of the second point from a crude butt. The reason for this difference in manufacture may simply lie in the size of the original bone splinter: the unfinished bipointed specimen (Fig. 6.11, specimen 1179a) would have been too small to completely finish one tip before the other. Since the majority of bipoints were first shaved and subsequently ground into shape, it is easy to imagine the difficulty of shaping the unworked end of such a small splinter without damaging or remodifying the finished end. This would be no problem if the bone splinter were large enough to maintain a full grip on the opposite end. Another reason may be that this manufacturing process enabled the manufacturer to utilise the functional tip before the artefact was fully formed. In fact, one of these four specimens bears evidence of utilisation.

6.3.3. *Kria Cave unipoints*

Sixteen specimens from Kria Cave were classified as unipoints (some examples are illustrated in Figure 6.12). These specimens are single-pointed and relatively long without any sign of modification at the butt. However, there are also 17 other point-bearing shaft fragments and 16 midsections that are tentatively associated with this class of artefact, based on similarities in raw material, cross-sectional dimensions and the pattern of use-related wear and damage. However, this association is made with considerable reservation, the reason being that the collection also contained two shaft bipoints, each with a minimum length of 40 mm, and each with a very similar overall morphology to the unipoints. Accordingly, it cannot be ruled out that some (or perhaps even many) of the 'unipoints' are actually fragments of original shaft bipoints. However, as shall be argued below, very long bipointed specimens of this kind are probably not functionally equivalent to the class of cortical bipoints described above.

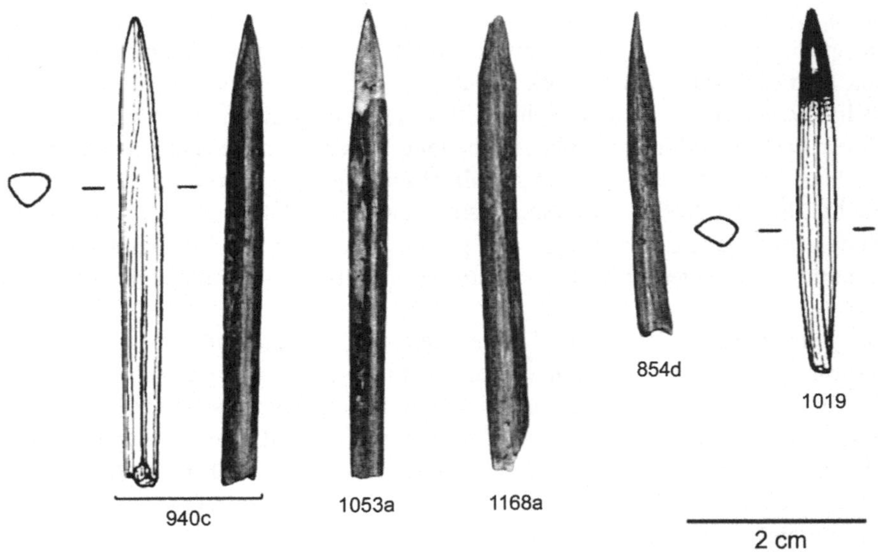

Figure 6.12. Some examples of the shaft unipoints from Kria Cave, in comparison to a cortical unipoint (854d) and a shaft broken bipoint (1019). From left to right: photo and drawing of shaft unipoint with 'open' tip, opposing sides (940c); photo of shaft unipoint with 'open' tip (1053a); photo of shaft unipoint with 'closed' tip (1168a); photo of cortical unipoint (854d); drawing of shaft bipoint with a burnt, polished, 'open' tip (1019). Drawings: Mirjam Weijns.

Raw material and manufacture

Fifteen of the longer unipoints are made from shaft bones. Although these are sometimes modified along the shaft section, the original cross section of the bone is generally still recognisable. The diameter, length and straightness of many of the parent bone strongly suggests that the majority of these artefacts were made from bones of megachiropterans or fruitbats, most likely the radius and the phalanges from the wings, but perhaps also the tibia. Many of these bones in fruitbats have a fairly large medullary cavity and this is consistent with the wide central canal seen in many of the artefacts. In some cases, the proximal part of the fibula of *Dorcopsis* is tentatively identified as the parent bone. *Dorcopsis* remains are abundant throughout the deposit in Kria Cave, but fruitbat bones are comparatively uncommon in all levels and are absent from the upper levels of Unit II and from Unit I. The common use of fruitbat bones for bone artefact manufacture might imply that supplies of these extremely useful bones were in fact carried around in prehistoric times, as indeed has been recorded ethnographically for the Wola of the Southern Highlands of Papua New Guinea (Sillitoe 1988).

Among shaft unipoints it is possible to distinguish between specimens made from bones with narrow cortex and a wide medullary canal, resulting in points with an 'open' or hollow tip (N = 8; see for example Fig. 6.12 specimens 940c and 1053a), and specimens made from bones with thicker cortex and a narrow canal, resulting in an effectively 'closed' or solid tip (N = 7; Fig. 6.12 specimen 1168a). Because these categories do not otherwise differ in dimensions (see below) it is possible that both 'types' were manufactured from the same parent bones, but presumably from different sections or ends. The 'open' vs. 'closed' categories are distinguished below in relation to patterns of polish and damage.

All 15 of the shaft unipoints are manufactured on unburnt to lightly burnt bone (Table 6.3). The same applies to the bulk of the fragmentary shaft specimens. One point-bearing fragment and two midsections are thoroughly burnt. Whether this occurred before or after manufacture is not clear; however three of the shaft unipoints and two of the point-bearing shaft fragments were clearly subjected to localised secondary burning, all but one of them around the tip. One midsection also shows localised secondary burning.

One artefact classified as a unipoint is manufactured from lightly or unburnt cortical bone. This specimen is very long and has a fine, well-made point at one end (Fig. 6.12 specimen 854d). The rounded shaft is also extensively modified. The butt end was broken in antiquity but shows some recent damage as well. This specimen closely resembles some of the shaft unipoints with a closed tip, described above, but it shows no trace of a medullary canal and shows heavier than usual modification along the shaft.

The method of manufacturing for the shaft bone artefacts is similar to that for cortical bipoints, except for a higher incidence of grinding as the exclusive method of modification (Table 6.4).

Metrical analysis of unipoints

The length of the shaft unipoints ranges from 23.9 mm to 51.2 mm. Because the butt is a simple fracture in all specimens, the lower cut-off for this group is quite arbitrary. This is illustrated by the fact that the maximum length for specimens classified as point-bearing shaft fragments is 24.4 mm, ranging down to 12.3 mm. Not surprisingly, unipoint length is not well correlated either with width or thickness. No significant difference was found between the dimensions of unipoints with open vs. closed tips and these are combined in the following analyses.

Width of the shaft unipoints ranges from 3.1 to 5.6 mm, their thickness from 2.6 to 3.6 mm. These measurements show virtually no correlation ($R^2 = 0.03$); however, this is improved considerably ($R^2 = 0.49$) by excluding one clear outlier (Fig. 6.13). This specimen is unusual in having a flattened, D-shaped cross-section in contrast to the more standard sub-circular form; it was most likely made from a fruitbat tibia.

Point length is highly variable within this group, ranging from 7.2 to 26.9 mm; this reflects the considerable variation in point morphology, some specimens having obvious points, others a more gradually tapering form where the distinction between point and shaft is not clear. Point length is not correlated either to width or thickness.

The one undoubted cortical unipoint measures 35.6 mm of which 11.5 mm is identified as the point. The shaft is nearly circular in cross-section, as in the majority of shaft unipoints.

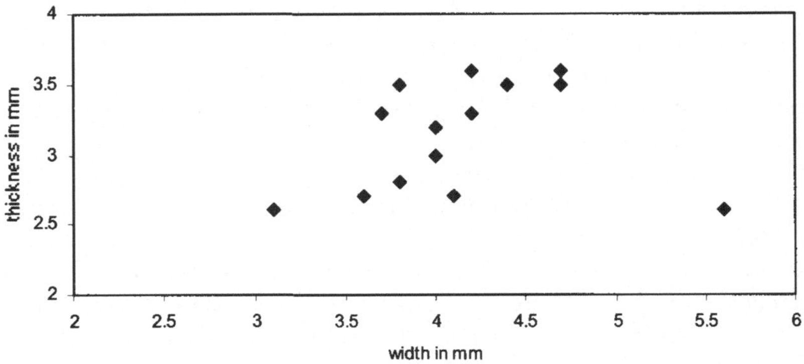

Figure 6.13. Kria Cave, relation between thickness and width of shaft unipoints.

Use-related polish of unipoints

Nine of the 15 shaft unipoints show little or no localised polish at the tip (Table 6.6). The remaining six specimens have a medium polish at the tip; this occurs in equal numbers of unipoints with open vs. closed tips. In one specimen with an open tip this polish extends a considerable distance back from the tip.

The cortical unipoint shows a medium polish at the tip, with a small, post-excavation step-fracture.

Use-related damage of unipoints

Two out of the 15 shaft unipoints showed modern rather than ancient damage; of the remaining thirteen, three have entirely undamaged tips, six have step-fractures, two have snap-fractures and one each have crushing or more complex fracturing. These categories of damage are not randomly distributed between the unipoints with open vs. closed tips (see Table 6.7). Specimens with closed tips actually account for five of the six cases of step-fracturing, with the remaining specimen showing an undamaged tip; in contrast the open tips show a variety of damage types. The high incidence of step-fractures in the shaft unipoints with closed tips is probably related to the fact that these tips are inherently stronger in comparison to the open tips, and thus less likely to snap or crush under use. However it is also possible that the two 'types' of unipoints were used in different ways.

Two specimens show post-excavation damage; one of these has medium polish on the tip.

Association of fragmentary shaft artefacts

The 17 point-bearing shaft fragments are consistent in all respects with the category of shaft unipoints. This is most clearly seen in a plot of width against thickness (Fig. 6.14) that shows complete overlap between these categories.

Eleven of the 17 point-bearing shaft fragments have open tips (Table 6.6), a slightly higher proportion than seen among the longer shaft unipoints. Among the specimens with open tips, four have medium polish (one extending back along the shaft) and one has light polish at the tip. In contrast only one of six specimens with closed ends has a localised high polish.

Snap-fractures are the most common form of damage amongst the point-bearing shaft fragments, with six examples (Table 6.7); this is followed by crushing, with four examples, and step-fracturing, with two. Four tips are entirely undamaged; three of these have localised polish. As described above for the more complete shaft unipoints, step-fractures are seen exclusively on specimens with closed tips, while snap-fractures and crushing are common on open tips.

Sixteen shaft midsections were found, ranging in length from 11.8 to 52.5 mm. The width and thickness of these fragments overlaps completely with the range of variation of the shaft unipoints (see Fig. 6.14). In many cases, the isolated midsections have a wide medullary canal, similar to those seen in the shaft unipoints and in the two shaft bipoints. Given this good overall correspondence in size and morphology, it is very likely that the midsections originate from breakage of similar artefacts.

The midsections generally show little or no polish. However, one specimen shows some light polish concentrated at one end, suggesting that it may have lost a relatively short section of the tip. In most cases the midsections terminate in simple snap-fractures at both ends. Some specimens show more complex fractures with multiple step-fractures and/or crushing; in contrast, simple step-fracture terminations are rare.

As mentioned earlier, five of the ten cortical point-bearing fragments are quite long (greater than 18 mm). These specimens also have a non-tapering shaft that would be unusual (although not unknown) for a cortical bipoint, and two of them have a sub-circular cross-section that further allies them with the broader group of unipoints.

Finally, the sample of cortical point-bearing fragments as a whole shows a relatively low frequency of use-related polish, a pattern that is both consistent with the unipoint category and the secondary ends of bipoints. Together, these characteristics thus make any allocation of the cortical point-bearing fragments to the bipoint or unipoint category inconclusive.

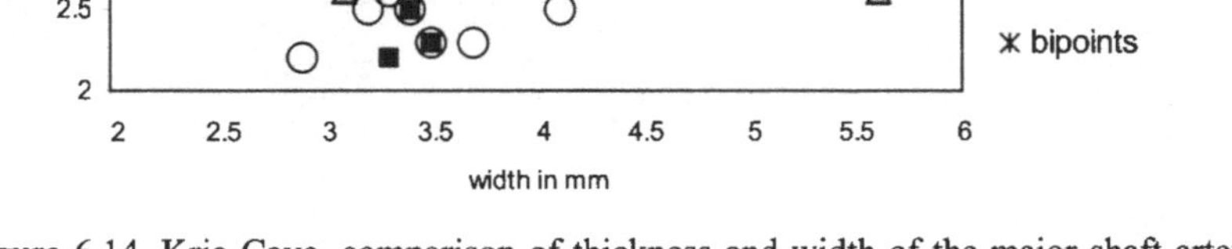

Figure 6.14. Kria Cave, comparison of thickness and width of the major shaft artefact categories and fragments thereof.

Use-related breakage of unipoints

The similarity in dimensions and pattern of polish and damage between the shaft unipoints and the majority of the shaft fragments (point-bearing fragments and midsections) constitutes strong grounds for associating these artefacts into a single category. However, as noted earlier, the evidence does not rule out the possibility that some, or even many, of these shaft artefacts may have originally carried two points. Rare bipointed shaft artefacts are indeed present in the Kria Cave collection and these are consistent in basic shaft dimensions to the shaft unipoint category.

An approach similar to that applied to the cortical bipoints to identify the circumstances of breakage, can be developed for the shaft artefacts. If the shaft uni- or bipoints broke during the final stages of manufacturing, one might expect the breakage to occur at more or less random positions on the artefacts; moreover, the tips would show no polish or damage. If these artefacts broke during use, then the tips would show polish and/or damage, and breakage would probably not be at random but related to the mode of use. If breakage occurred as a result of post-discard trampling or other taphonomic processes, the tips might show polish and/or damage but the artefacts should be broken at random positions. Random breakage patterns might be expected to reduce the shaft artefacts to fragments of a certain length (the 'breakage-resistant length') where they are more able to resist bending forces caused by trampling etc.

As noted above, the pattern of polish and damage at the tips corresponds closely between the longer shaft unipoints and the typically shorter, point-bearing fragments. Only very few of these artefacts show no polish or damage and it can be assumed that the majority were broken either during use or after discard. These different scenarios can be further explored by looking at the distribution of lengths of the artefacts.

Figure 6.15 gives the distribution by length for all of the point-bearing shaft artefacts. Interestingly enough, this plot shows a strong peak between 20 and 26 mm, with smaller numbers of longer artefacts, up to a maximum of 51.2 mm. Under a scenario of random breakage, the peak in point-bearing fragments at 20-26 mm might be identified as the 'breakage-resistant length' for these artefacts, and thus as a product of random breakage.

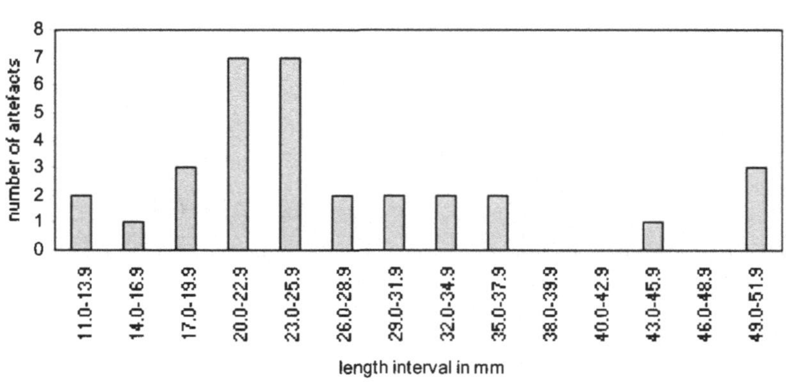

Figure 6.15. Kria Cave, distribution of all shaft unipoints and point-bearing fragments by length.

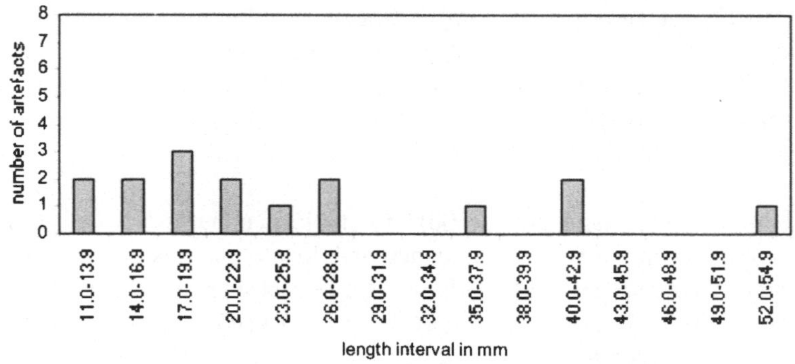

Figure 6.16. Kria Cave, distribution of all shaft midsections by length.

However, an alternative interpretation is that the large number of fragments measuring between 20 and 26 mm reflects the amount of the artefact that was subjected to bending forces during use, either set within a haft or held in the hand.

The distribution of midsection lengths might help to discriminate between these competing scenarios. Under the random breakage scenario, there is no obvious reason why the midsections, with identical width to thickness ratios, would not also show a comparable fragment length distribution, with reduction to an identical 'breakage-resistant length'. In contrast, under the scenario of breakage during use, the length of remaining midpoints would be related to the original artefact lengths, minus the lengths of the broken-off points. The broad and even spread of midsection lengths, as illustrated in Figure 6.16, is clearly more consistent with the latter scenario that implies breakage of the shaft unipoint (or bipoints) as a direct consequence of use rather than through random post-discard factors.

As noted above, the longest midsection measures 52.5 mm. Judging from the range of lengths of point-bearing fragments, this specimen probably exceeded 65 mm in original length if unipointed, and even longer if bipointed.

6.3.4. *Kria Cave utilised teeth*

A total of eight *Dorcopsis* lower incisors show definite evidence of utilisation (see Figure 6.17), with another two scored as possibly utilised. Because each specimen is slightly different in its pattern of damage and inferred mode of use, they are described individually below (from stratigraphically youngest to oldest).

Square 1N1E, Unit II, spit 3 (specimen 725): a complete, unburnt left incisor embedded in a jaw fragment. The crown is intact except for slight damage to the tip. The upper margin shows three small spalls from the buccal side located 5-6 mm from the tip and a zone of rounding for 3 mm anterior of these. The lower margin has a series of very small spalls from the lingual side, located from just behind the tip back to approximately 4 mm. The buccal enamel surface has a few longitudinal striations. This specimen has been lightly used, probably with a uni-directional drilling action.

Square 1N1E, Unit II, spit 3 (specimen 710a): a fragmentary but unburnt right incisor. The tip is missing and the root is freshly broken away. The upper and lower margins of the tooth are both heavily damaged, with a mixture of crushing and spalls from both directions. The enamel on the buccal face is glossy but lacks major striations. This specimen has been heavily used, probably with a bi-directional drilling action.

Square 1N1E, Unit II, spit 3 (specimen 710b): a complete, unburnt left incisor with intact root. The last 2-3 mm of both the upper and lower margins are intensively damaged with small step-fractures and crushing. The buccal enamel surface does not show any major striations. This specimen may have been used with a bi-directional drilling action, but the evidence is less compelling than for the other specimens.

Square 0N0E, Unit II, spit 7 (specimen 756): a complete, unburnt left incisor with intact root. The tip shows light crushing. The upper margin shows a zone of heavy step-fracturing, originating from the buccal side, in a zone 5-8 mm behind the tip. The lower margin shows an opposing zone of smaller step-fractures originating from the lingual side, located 6-7 mm behind the tip. The buccal enamel surface lacks major striations. This specimen has been heavily used, possibly in a drilling action to enlarge an existing hole, or in a shaving action against a narrow, rounded material (e.g. an arrow shaft), with alternate use of the upper and lower margins, these events separated by a 180^0 reversal of the tooth.

Square 1N1E, Unit II, spit 7 (specimen 774): a complete, unburnt left incisor with intact root. The last 4 mm of both the upper and lower margins are intensively damaged with small step-fractures, the upper series coming from the buccal side and the lower series from the lingual side. The buccal enamel surface shows transverse striations, the majority originating from the lower margin. This specimen has been subject to moderately heavy use, probably with a uni-directional drilling action.

Square 0N0E, Unit II, spit 8 (specimen 766; see Fig. 6.17): a complete, lightly burnt left incisor with intact root. The tip has been removed by a spall that emanated from the buccal side. The upper margin shows a 4-mm zone of crushing and spalling immediately behind the tip, mostly originating from the buccal side. The buccal enamel surface shows a distinct gloss located immediately below the zone of damage. The lower margin shows a light chattering near the tip. This specimen has been lightly used, probably in a shaving or cutting action.

Square 0N0E, Unit III, spit 3 (specimen 804): an unburnt left incisor crown, a fragment of jawbone adhering to the broken root. The tip has been removed by a longitudinal spall. The upper margin is crushed and rounded in a zone extending back 5 mm behind the tip, with occasional small spalls from the lingual side behind this. The lower margin bears a 5-mm zone spalled from the lingual side, followed by a second zone of equal length but spalled from the lingual side. The complex pattern of damage on this specimen indicates that it has been used as a drill, perhaps in several phases and from different directions.

Square 1N1E, Unit III, spit 3 (specimen 854): a complete, lightly burnt right incisor with intact root. The tip has been removed by a small step-fracture running back onto the buccal surface. The upper margin has a series of small spalls originating from the buccal side. The lower margin has similar spalls coming from the lingual side. On both margins the spalls occupy a zone from 3-6 mm behind the tip. The buccal enamel surface shows a network of fine longitudinal and transverse striations. This specimen has been lightly used, probably in a uni-directional drilling action.

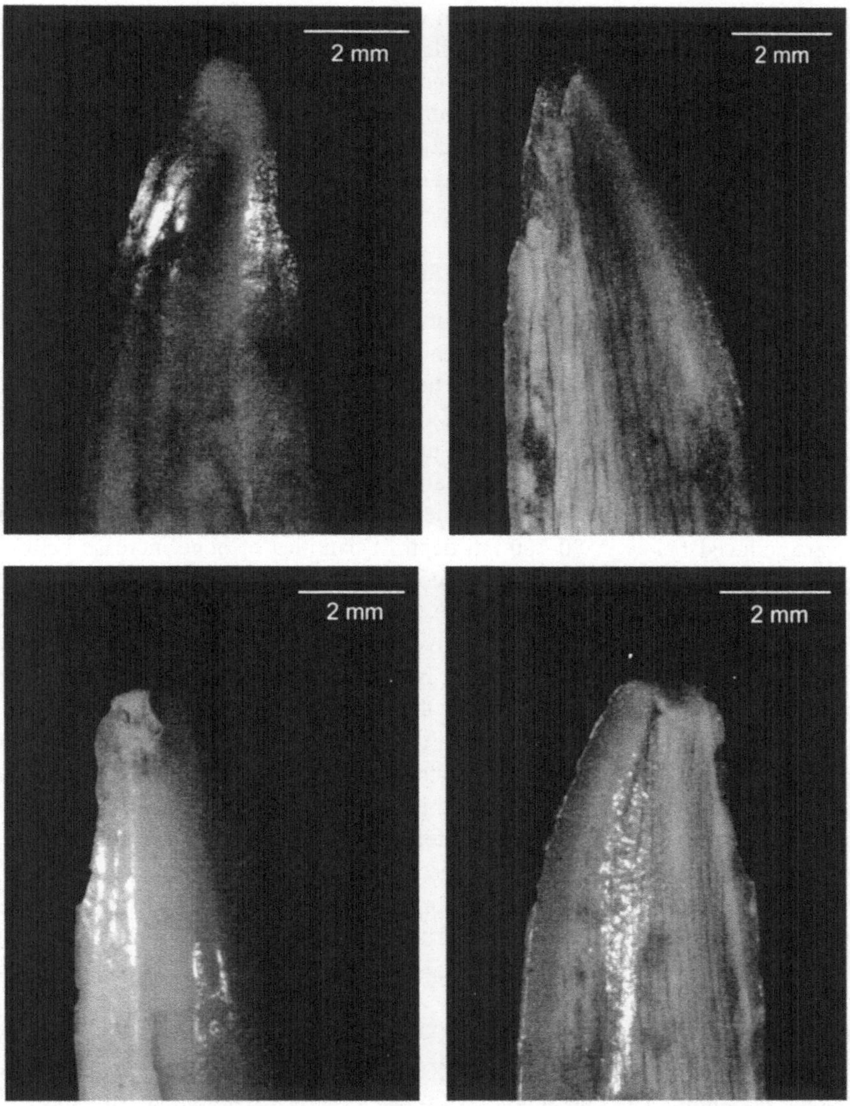

Figure 6.17. Utilised wallaby incisors from Kria Cave. Top: buccal (left) and lingual (right) view of specimen 766. Bottom: buccal (left) and lingual (right) view of specimen 922.

Square 1N1E, Unit III, spit 4 (specimen 871): a complete, unburnt right incisor with intact root. The upper and lower margins are damaged immediately behind an otherwise intact tip. The upper margin shows a 4-mm zone of rounding with associated transverse striations on the buccal enamel surface. The lower margin has step-fractures from the lingual side along a 2-3-mm zone. This specimen has been lightly used, probably in a uni-directional drilling action.

Square 1N1E, Unit IV, spit 2 (specimen 922; see Fig. 6.17): a complete, unburnt right incisor with intact root. The tip is broken off but the fractured edges are rounded. The upper margin shows numerous small spalls along an 8 mm zone running back from the tip; all of the spalls originate from the buccal side. The lower margin is step-fractured from the lingual side in a 4-mm zone behind the tip. The buccal enamel surface bears a network of transverse and oblique striations. This specimen has been subjected to moderately heavy use, probably with a uni-directional drilling action.

6.4. BONE ARTEFACTS AND UTILISED TEETH FROM TOÉ CAVE

Only three bone artefacts and a single utilised tooth were found in Toé Cave. These come from contexts that date to between c. 3000 BP and c. 9000 BP but their ages cannot be further refined at present. The bone artefacts were found at the southern side of the cave, in Squares 0N0E and 1S0E, two at a depth of 50 and 60 cm below the surface, and one at 30-35 cm depth. The utilised *Dorcopsis* incisor was recovered from 20-25 cm depth in Square 1N1E. Stone artefacts and faunal remains are abundant and well preserved virtually all through the deposit (see Chapter 4), although quantities of chert are reduced below c. 80-100 cm depth. Quantities of bone increase below these levels. Occupation of Toé Cave probably commenced around c. 26,000 BP, some 18,000 years earlier than in Kria Cave (Chapter 4).

The three bone artefacts from Toé Cave are all made of unburnt cortical fragments and show no sign of subsequent burning. In each case the extent of modification of the bone prohibited a precise identification of the skeletal element on which they were made and hence of the species. However, one of the specimens (the third one described below) is made of a thick cortical bone fragment that must be either from a large mammal, possibly human, or a very large bird such as a cassowary. All three artefacts have been shaved into their present shape. These artefacts are less well preserved than those from Kria Cave. They appear to show little or no polish.

One artefact may be a broken cortical bipoint. It shows some post-excavation damage at one end and pre-excavation crushing at the other end. It measures 24.3 mm long, 3.8 mm wide and 2.2 mm thick and thus falls within the range of variation for cortical bipoints from Kria Cave.

Considering its considerable length (27 mm), the second artefact is classified as a possible cortical unipoint, however it may also be a broken bipoint, albeit one that was originally very long. The one surviving tip has a post-excavation step-fracture, and the butt of the artefact was snapped off some time prior to excavation. The specimen is 4.4

mm wide and 2.9 mm thick; with these dimensions it falls outside the range of variation for cortical bipoints from Kria, but falls within the range of Kria's shaft unipoints.

The third artefact appears to be a badly eroded unipoint or unfinished artefact. However, since the tip has snapped off in antiquity, it is uncertain whether this part was casually or carefully formed, similarly to those described for Kria. The butt is unmodified. Original shaving marks on this cortical fragment are very faint. It measures 38 mm in length, 8.4 mm in width, and 4.6 mm in thickness.

The single utilised *Dorcopsis* incisor is an unburnt left incisor with intact root. The tip region has broken off in antiquity. The upper margin bears several small spalls immediately behind the fracture. The lower margin is more extensively spalled over a 4-mm-long zone, the majority arising from the lingual face. The buccal enamel surface bears numerous transverse to oblique striations, but these are largely confined to the lower half of the surface. The lower margin of the tooth was possibly used in a cutting or scraping action. However, because a large part of the tip is missing, it cannot be ruled out that the specimen was also used as a drill.

6.5. DISCUSSION

As described in the preceding sections, the Kria Cave bone artefact assemblage is highly structured, comprising essentially two distinct and almost exclusive, highly uniform 'classes' of bone artefacts. These are classified as bipoints, typically made on splinters of cortical bone of larger mammalian long bones, and unipoints, typically made on sections of small shaft bones of smaller mammals. The defining characteristics of each class of artefacts is summarised below as a prelude to addressing various issues related to their function and wider significance.

The bipoints found in the Kria Cave deposit can be characterised as follows:

1. they were made predominantly from unburnt or lightly burnt splinters of cortical bone, and only rarely from more intensively burnt bone;
2. they were generally shaved into shape with a stone tool, and often finished by grinding of the tips;
3. they are highly uniform in size and shape, and generally have a slightly flattened cross-section;
4. they are in most cases nearly symmetrical, although the primary end generally forms slightly more than half of the total length;
5. they are strongly asymmetrical only in cases where bipoint length exceeds 25 mm (the primary end can be either very short or moderately long);
6. they show a high incidence of medium to high polish which is generally confined to the primary end (72.7%) (the secondary end is rarely ever polished and few specimens show any polish extending along the length of the shaft);
7. they show a very high incidence of damage at both the primary and secondary end tips (> 90%);
8. they most often show crushing at both the primary and secondary end tips; at the primary end step-fractures are often found in association with polish;
9. they quite commonly broke during use, and this most often involved snap-fracturing of the secondary end.

The shaft unipoints exhibit the following characteristics:

1. they were made predominantly of unburnt or lightly burnt shaft bone fragments, and only rarely from more intensively burnt bone;
2. they were generally manufactured by a combination of shaving and grinding, but grinding was more often used exclusively than within the bipoints;
3. they show a greater degree of variation in size and tip form (perhaps reflecting use of different parent bones), but they usually have a more rounded cross-section than the bipoints;
4. in all cases they lack the articular end of the skeletal element; although this could be due to unintentional breakage, the ubiquity of this condition suggests intentional removal;
5. they show a lower incidence of polish of the tip (40%) than the bipoints, and this generally does not exceed medium intensity; only rarely does the polish extend back from the tip;
6. they show a lower incidence of damage at the tips (77%) than the bipoints;
7. they show a higher incidence of snap- and step-fractures than the bipoints, with step-fractures especially common on specimens with 'closed' tips;
8. they quite commonly broke during use, resulting in a fairly uniform length for point-bearing fragments.

A small number of specimens violate this basic dichotomy and serve to illustrate not only the fundamental versatility of bone as a raw material, but perhaps also the likely multifunctionality of these relatively simple implements. Among the bone artefacts made from shaft fragments, two specimens are undeniably bipointed. However, they show no other specific resemblance with the cortical bipoints, but rather conform closely to the metrical and morphological characterisation of unipoints as presented above. These artefacts are perhaps analogous to a pencil that has been sharpened at both ends. Nevertheless, their occurrence in the collection raises the possibility that this 'double-ended' form was actually quite common but is only recognisable in unbroken examples, quite possibly a rare occurrence for such an elongate and obviously fragile implement.

One well-preserved unipoint is made of cortical bone. This specimen differs from the majority of the cortical bone artefacts in terms of its general dimensions and cross-sectional shape, but sits comfortably with the remaining (shaft) unipoints. There may be more such artefacts amongst the broken cortical bone artefacts, several of which are indeed rather long to be broken bipoints.

Five other modified cortical bone splinters are identified as possible unfinished bipoints. In one case, the splinter is naturally bipointed and shows minimal modification on either end, but is shaved in the mid-part. In functional terms, it may be ready for use, however its crude manufacture contrasts markedly with the typically well-finished bipoints. In the remaining four specimens, the splinter is less regular in shape but one end has been carefully shaped to produce a fine point; one specimen shows signs of having been used in this condition, but this does not preclude an intention to perform further shaping of the butt of the artefact. As noted earlier, all five of these artefacts may be unfinished bipoints but, if so, it implies some variation in how this morphologically uniform class of artefacts was produced (crude shaping, then finishing vs. finishing of one end at a time). It is also possible (but perhaps less likely considering their frequency) that these single-pointed artefacts were half-products of cortical unipoints, rather than bipoints, in which case the difference in manufacture is associated with the production of different classes.

The presence of these two distinct groups, albeit with occasional exceptions, raises a number of important questions. First, how was each group used (e.g. were they hafted or hand-held) and what were they used for? Second, was each of the different kinds of artefacts manufactured and used on site? And last, were they resharpened after becoming worn or suffering minor breakage, or were they discarded under these circumstances? The presence of utilised teeth is significant in several respects. First, it reflects additional diversity in the selection of raw materials. And second, the very sensitive nature of the incisor margins to damage allows a more detailed assessment of their function. This in turn may inform on the mode of use of other components of the assemblage.

6.5.1. *Mode of use of bone artefacts and teeth*

Because organic material such as resin, string or cane was not preserved in either cave, there is no direct evidence for or against hafting of the bone artefacts as part of composite tools. Accordingly, whatever case is made in this regard must derive from the morphology and pattern of use-related wear and breakage of the artefacts themselves. Although this gives less chance of certainty, I believe that there is sufficient evidence to support a basic dichotomy within the bone artefact assemblage between bipoints that were most likely hafted and unipoints that were more likely hand-held (see below). The wallaby incisors show several distinct patterns of use-related wear that is consistent in some respects with the inferred mode of use of some of the bone artefacts.

Bipoints

The careful manufacture, degree of uniformity and abundance of the bipoints, coupled with their very specific pattern of use-related wear and breakage, suggests that they were produced for a specific purpose or narrow range of purposes by highly skilled manufacturers. The highly localised polish, almost always confined to one tip only, indicates that these artefacts were used in a precise and repetitive way that generated pressure and friction to the tip. This effectively rules out a purely decorative role for these artefacts, including suggestions that they were produced as nose plugs or hairpins.

It is difficult to imagine that these bipoints could have been hand held. Their small size, usually falling between 16 and 27 mm, almost implies that they must have been hafted as a part of a composite tool. The typically ovate cross-section of the bipoints would make them relatively easy to mount, either in the end a split stick or reed, with subsequent binding, or into a groove on the side of a projectile point as a barb.

Apart from their size and morphology, various other lines of evidence support the notion that the Kria Cave bipoints were hafted. One strong indication of hafting comes from the fact that they are so often polished at one end only. Intuitively, this would be unlikely to occur if they were hand held. Most of the bipoints are nearly symmetrical and bear well-produced tips at both ends, hence there is no obvious reason why they would always be oriented the same way in the hand during repeated episodes of use. In contrast, repeated use of a hafted bipoint will produce polish at one end only, unless of course the bipoint is reversed at some stage within the haft. Pasveer & Bellwood (in press) reported evidence for reversal of hafted bipoints in the Golo Cave collection from the northern Moluccas, associated in this instance with other morphological evidence for resharpening and reuse of broken bipoints (see below). Reversal of bipoints appears to have been much less common at Kria Cave.

Another indication that the Kria Cave bipoints were hafted derives from their regular pattern of breakage. As noted above, Kria Cave bipoints most often broke during use and this generally involved loss of the secondary end. If the artefacts were indeed hafted, this implies that breakage occurred more often within the hafted section than the exposed section. Although this might seem counterintuitive, the inferred pattern of breakage is supported by the results of various experimental studies. Arndt & Newcomer (1986: 167) described breakage at the base of bevelled bone points hafted as projectiles; they suggested that the somewhat asymmetric nature of the bevel 'encouraged this end to bend and finally snap'. Guthrie's experiments with bone and antler projectile points also showed frequent breakage of the artefacts within the hafted zone (Guthrie 1983: 289-290, Fig. 9). Apparently, bending forces produced during impact or under sustained pressure applied to the tip are translated back into the hafted region where they may be focussed by any irregularity of fit and resulting uneven support of the artefact. Breakage of the secondary end of a bipoint would be less likely to occur if the artefact was hand-held.

Ethnographic sources from the Australo-Melanesian region contain little specific information on the use and manufacture of bone points and this is especially true in the case of bipointed artefacts. However, ethnographic collections do contain significant numbers of both spears and arrows that incorporate bipointed bone artefacts either as tips or barbs (e.g. McCarthy 1940; Cranstone 1961: 63-64, 66; Schrire 1982: 36, Plate 4; Mulvaney & Kamminga 1999: 288). This fact has encouraged a succession of authors to identify prehistoric bipoints as components of projectiles (e.g. McCarthy 1940; Lampert 1966, 1971; Schrire 1982; Jones & Johnson 1985: 60); however, the suggestion has not been subject to critical assessment through examination of patterns of usewear and damage. Lampert (1966) and Jones & Johnson (1985: 60) describe remnants of hafting gum on the midsection of archaeological bipoints, strong support for their claims that these artefacts were hafted as spear barbs or tips.

Kria Cave produced abundant remains of several medium-sized to large animals and it is conceivable that the bipoints served as tips and/or barbs of projectiles used in hunting. However, the very specific pattern of polish and damage on the artefacts is difficult to reconcile with this particular mode of use. Experimental studies on prehistoric projectile points (e.g. Tyzzer 1935-36; Guthrie 1983; Olsen 1984; Arndt & Newcomer 1986; Webb 1987) have demonstrated various kinds of damage for end-mounted points including crushing, snap- and step-fracturing, and minor rounding of the tip. In general, points used in these experiments sustained little damage unless impact involved contact with cortical bone (in the carcass) or some other, equally dense material (hardwood or stone). Where such hard impact does occur, long step-fractures or longitudinal spalls running back from the tip are commonly observed (Tyzzer 1935-36:

267; Olsen 1984: 163; Arndt & Newcomer 1986: 166). Other experiments conducted by Tyzzer (1935-36: 267) demonstrated that penetration of a bone point into loamy soil or gravel produced only a superficial smoothing, even after fifteen repeats. Tyzzer (1935-36: 268) suggested that projectiles would accumulate high polish through extremely long periods of non-damaging use and favoured the view that high polish on bone points is most likely 'intentionally bestowed', i.e. as part of a finishing process to round off sharp edges and broken tips. Interestingly, Tyzzer's experiments also indicated that while bone arrow tips are intrinsically quite resilient, once damaged they will deteriorate quite rapidly unless repaired.

Much less attention has been given to the possible use of bone bipoints as barbs. Intuitively, the exposed portion of a barb might accumulate polish as it moves within or is removed from a carcass, and if it manages to do so many times without breaking. The occurrence of step-fractures, however, would be highly unlikely on the tips of such barbs. Campana (1989: 90-91) suggests that bone bipoints used as barbs might produce polish on the embedded end if they were twisted into wooden hafts, but it seems unlikely that specimens with very finely pointed tips could withstand adequate pressure without snapping.

Another common suggestion for the function of bone bipoint is that they represent a component of fish gorges. The ethnographic evidence for use of fish gorges in Australasia was last reviewed by Massola (1956). There is good evidence for the post-contact use of hard wood bipoints in this fashion among aboriginal groups living along the Victorian coast (Smyth 1878: 391) and along the Murray River (Massola 1956: 4-5; see also Mulvaney 1960: 77). Similar uses have been recorded for bone bipoints in the Solomon Islands (Codrington 1891: 316), and elsewhere in Oceania (Anell 1955: map 8). Although the latter author provides some description of attachment of the gorge to the line, details of binding methods are generally lacking. Hale & Tindale (1930: Fig. 242) suggested a 'double-hitch' method based on a Victorian example, but Massola (1956: 4) countered that 'in most places where fish gorges are used, they are attached by the middle and not double hitched'. There appears to be no detailed study of the morphological attributes of fish gorges or of patterns of use-related polish or damage.

The Kria Cave bipoints are similar in morphology to the somewhat smaller examples among the ethnographically documented fish gorges. However, it is difficult to imagine that bone bipoints used in this manner would acquire such intense polish or step-fractures at the tip, or why this wear and damage would tend to accumulate at only one end of most examples. Another factor that would seem to caution strongly against identifying the Kria Cave specimens as fish gorges is the fact that fish remains are extremely rare in both sites and the native fishes of the Ayamaru Lakes are all far too small to be captured with such 'large' implements.

Sillitoe (1988: 64) provides a rare but extremely detailed and relevant description of the use of bone as borers among the Wola of the Southern Highlands of Papua New Guinea. The Wola use a variety of raw materials as borers, including chert blades, large rodent and marsupial incisors, sharpened slivers of wood and 'pointed pieces of mammal bone'. The choice of borer reflects the hardness of the material worked on. Chert blades and mammal incisors are used for piercing harder materials, such as wood and bamboo, while mammal bone is used 'for boring less hard but tough materials, like some barks'.

Sharpened slivers of wood are used to pierce soft materials like bark-cloth and leaves. Sillitoe notes that 'Except for the latter, they fabricate these implements as required and usually discard them when they have finished with them, hence their working lives are brief'. He further notes that 'For bone implements they prefer the bones of pigs, *mapun* wallabies [*Thylogale* or *Dorcopsulus* spp.] and the wings of *tagem* black flying foxes [*Dobsonia magna*], as these are particularly strong' and 'Bones they simply break to produce jagged points, sometimes paring them to give sharp tips'.

Among the Wola, borers are sometimes hand held. However, hafting enhances efficiency or precision of work (Keeley 1982: 799). Sillitoe (1988: 64) recorded the mounting of mammal tooth and chert borers in wooden handles, as follows: 'After shaping a piece of wood to the size required, a man splits it lengthwise down the centre sufficiently to accommodate the bit, extracted from its mandible if an incisor, or knapped to a suitable 'nose' if chert, and binds it tightly in place by winding a strand around the split, pinching it firmly closed, securing the strand with a couple of hitches'. The Wola can take between ¾ to 1 hour to produce a borer with a mounted incisor, but much of this time is spent in extracting the incisor from the socket. Less than 15 minutes was required to fabricate the handle and binding material, and to insert and bind the tooth. Among the Wola 'Only men make these various boring implements. They are all capable, manufacturing them as they are required. They produce them in considerable numbers'.

The utilised *Dorcopsis* incisors from Kria Cave include at least seven specimens that show evidence of drilling action, with fine spalling of the upper and lower margins in directions consistent with rotation under pressure. In several cases the incisor tip also bears a longitudinal spall suggestive of apical pressure. In at least one case the incisor has probably been used while attached to a lower jaw or jaw fragment, but in many other cases the teeth were clearly dissociated from the lower jaw in antiquity, and may have been hafted prior to use, precisely as described by Sillitoe for the Wola. However, for the present, the more important inference is that some kind of drilling activity took place in Kria Cave, and that this may have involved hafted bone points as well as hafted wallaby incisors. For the bipoints, this interpretation represents a radical departure from previous suggestions that they were used either as tips or barbs of projectiles, or as fish gorges, but it is far more consistent with the observed pattern of use-related polish and damage.

Unipoints

The unipoints are more variable in their cross-sectional dimensions and tip morphology than the bipoints and they show a less specific pattern of wear and damage. Overall, they present a more 'casual' appearance than the bipoints, although in some cases their manufacture would have entailed a considerable amount of work.

Although the original length of the unipoints is unknown, there are several specimens that exceed 50 mm in length. Unlike the situation with the bipoints, length in itself does not provide any grounds to suspect that these artefacts were hafted. Indeed, their slender form would tend to suggest more delicate hand-held use. The light gloss present over the entire surface of many of these artefacts also suggests that they may indeed have been hand held. On the other hand, ethnographic sources and collections suggest that even quite long bone points could be used as barbs, or spear or arrow tips, although in general these are more robust than the Kria Cave specimens.

Fine-pointed unipoints, often referred to as 'needles' in the ethnographic literature, are mentioned in relation to diverse functions. For example, hollow bat bones were used amongst the Meybrat in the Ayamaru region in Papua for 'sewing' pandanus leaves (Elmberg 1955: 14). The local schoolteacher Elimas Kambuaya also mentioned the sewing of 'bags' made of bark, and 'umbrellas' (made of palm leaves), which were used for shelter, to sit or sleep on, and to carry foods or other things (pers. comm. 16.3.1998). Sillitoe describes needles made of hardwood or bone and used amongst the Wola; these are pointed at one end and rounded at the other, usually perforated at the blunt end, and rather long, perhaps up to 15 cm (Sillitoe 1988: 65). Among the Wola they are used for stitching up rain capes and for making string bags. The Wola also produce decorative hairpins made of wood, bamboo or bone. These are pointed at both ends and usually quite long, up to 15 cm; the most commonly used bones are flying fox wing bones, wallaby femurs and pig ribs. Hairpins are quickly made and just as quickly discarded after use (Sillitoe 1988: 67). Fine bone needles are also mentioned as being used for surgical procedures such as removing splinters or lancing a boil (Blackwood 1950: 32).

In the case of the Kria Cave unipoints, the variable and localised damage to the tip of many of these artefacts suggest that most were probably used in utilitarian ways rather than as decorative items. Most were probably hand held. The general lack of polish extending back along the shaft argues against any function that involved passage through an abrasive material, such as might be expected if they were used for sewing. However, with such generalised artefacts, it would be unwise to exclude any possible function. In all likelihood the Kria Cave unipoints were used for a variety of functions.

Utilised teeth

As noted above, the majority of the utilised *Dorcopsis* incisors were apparently used in some kind of drilling activity. This is shown very clearly by the opposing damage on the upper and lower margins, the consistent direction of spalling from these edges, and the presence of transverse or oblique striations of a kind not found on teeth which otherwise lack evidence of utilisation. One utilised specimen only remains inserted in a fragment of lower jaw, hence this incisor was presumably not hafted. Vanderwal (1978: 423) mentions wallaby incisors with usewear from the site Oposisi on Yule Island, off the south coast of Papua New Guinea. These are dated to c. 2000 years ago. Three out of four from this site retain the mandibular jawbone, presumably used as a handle, similar to the specimen from Kria. The fourth shows grooving on the root, suggesting that this specimen was hafted in some way. At least some of the remaining examples from Kria were presumably similarly hafted, and used as 'borers', perhaps in the very simple way documented ethnographically by Sillitoe (1988: 64) for the Wola.

A few specimens show patterns of wear that are more consistent with a simple scraping or cutting action. These are examples to remind us that all bone and tooth artefact types were most likely multifunctional and further, that individual artefacts may have been used in a variety of ways at different times during their functional life.

Almost 85% of all macropodid incisors have shattered crowns. For burnt incisors this figure is 100%, presumably as a direct consequence of firing. Among unburnt incisors 69.1% are shattered, and at least some of these were probably damaged as a consequence of deliberate use. The number of teeth showing evidence of utilisation is thus likely to be a significant underestimate of the actual utilisation rate.

6.5.2. *Raw material selection*

The great bulk of the Kria Cave bone artefacts are made of originally unburnt or lightly burnt bone fragments, with a lower usage of heavily burnt material. Artefacts made of shaft fragments show a slightly higher proportion of burnt bone compared to cortical bone artefacts.

The predominant use of unburnt and lightly burnt bone for artefact manufacture might be explained in several ways. The simplest explanation is that it represents a random selection from the assemblage, in which unburnt and lightly burnt bone are indeed the most abundant raw materials at the site (Chapter 3). Alternatively, the observed pattern might represent a process of deliberate prehistoric selection for specific kinds of bone. The slight difference in proportions between shaft and cortical bone would appear to favour an argument for selectivity over randomness. However, this difference might also exist within the wider assemblage; after all, delicate shaft bones originating from smaller animals may be more often cooked or scorched in a fire than large mammal long bones which tend to be protected from heat by thicker layers of flesh.

Opinions vary somewhat over what constitutes the best kind of bone for artefact manufacture and use. Nyungar Aborigines of southwest Australia reported to Bird and Beeck (1980) that very green, unburnt bone is hard to modify. They further specified that somewhat weathered and desiccated bone would be best for bone point manufacturing, and that bone from a cooked kangaroo was often used for this purpose (Bird & Beeck 1980: 169). Olsen (1984: 185-187) and Campana (1989: 28-29) present a contrary view based on their replicative studies. They claim that fresh or 'rehydrated' bone is easiest to modify using stone tools. On the other hand, Knecht (1997: 200) claims that soaking or even boiling of bone does not appear to soften unburnt bone. Evans (1973) and Guthrie (1983: 278) showed that fresh (or at least wet) bone is more resilient and less brittle than dry bone. Heating of bone will initially make the material more dense and harder than fresh bone. However, excessive heating and burning of bone will remove the organic part of the matrix and render the bone increasingly brittle (Campana 1989: 36). It is therefore not surprising that very few of the Kria bone artefacts are made from burnt bone.

The simplest explanation for the predominant use of unburnt and lightly burnt bone for bone artefact manufacture at Kria Cave is that these were the most readily available raw materials. However, this pattern of usage is also broadly consistent with the differential properties of the raw materials themselves. Since extreme burning of bone does seem to limit the usefulness of bone for tool manufacturing, the burnt and calcined state of some artefacts is probably a result of secondary burning. This could have occurred either in or under a fireplace following disposal.

6.5.3. *Manufacturing, use and discard*

The remarkable difference in numbers of bone and tooth artefacts between Kria and Toé Caves raises important questions regarding the circumstances under which these artefacts were manufactured, used and discarded. In the case of Kria Cave it is clear that significant numbers of both bipoints and unipoints, as well as utilised teeth, were discarded at the site. But were these artefacts actually manufactured and used on site, or were these activities performed elsewhere, with Kria Cave perhaps acting as a place where maintenance of tools was carried out?

If bone artefacts were manufactured on site we might expect to find specimens at all stages of manufacture, including specimens that broke during manufacture and were discarded on the spot, and perhaps also any materials used in their manufacture, such as scraping or grinding tools. As described above, the Kria Cave assemblage contains one cortical bipoint that is either unfinished or very crudely finished, and four cortical single-pointed artefacts that could conceivably represent an intermediate stage in bipoint (or unipoint) manufacture. These specimens may constitute evidence that cortical bipoints and/or unipoints were manufactured on site. Similar evidence for shaft unipoint manufacture is lacking. However, these artefacts are so simple that no intermediate stage of production is really feasible.

The Kria Cave assemblage contains a small number of cortical and shaft artefacts that lack any trace of polish or damage. Although not in abundance, these artefacts are consistent with the notion that bone artefacts were manufactured on site at Kria Cave. However, these 'fresh' artefacts might just have easily been produced elsewhere and transported to the site. Among the sample of broken bipoints, all specimens show either polish or damage of some kind on the surviving tip, hence the majority probably broke during or after use rather than during manufacture. Interestingly enough, a higher proportion of shaft artefacts lack evidence of use. This might be because these artefacts were used in a less intensive fashion that took longer to produce polish or damage, or it might be a corollary of the greater simplicity of these artefacts, compared with bipoints, perhaps making them more casually expendable.

No artefacts have been identified that can be definitely linked to bone artefact production. However, the shaving marks that occur on the majority of bone artefacts were clearly produced by scraping with a stone artefact as described by Sillitoe (1988: 98, Plate 24). As described in Chapter 5, many of the utilised chert tools from Kria Cave possess notches of an appropriate size and shape to have been involved in bone artefact manufacture. It is much less obvious how the grinding facets observed on many cortical and shaft artefacts were produced. A suitable grinding medium would be a fine to medium-grade sedimentary or metamorphic rock. No specimens of these rock types were recovered during the excavation of Kria Cave.

Evidence that many of the bipoints and unipoints actually broke during use was presented earlier. The fact that both the broken artefacts and the broken-off parts seem to be present at Kria Cave is strong evidence that the breakage itself actually occurred on site. If the bipoints were hafted, this would also imply that the haft itself was either discarded or that the broken part of the bipoint was removed and the bipoint either repaired or replaced. If the shaft artefacts were not hafted but hand held, then recovery of the broken tip sections would be an even stronger indication for on-site use of the artefacts.

One other possibility should be mentioned in relation to the pattern of use and discard. Secondary burn marks occur on a small proportion of the bone artefacts, and more commonly on fragments of shaft artefacts than on cortical bone artefacts. It is possible that these artefacts were brought to the site as fragments of projectile points or barbs inside shot game (see for example also Keeley 1982: 803). These fragments might then burn or scorch during cooking of the carcass.

6.5.4. *Evidence for resharpening?*

Both of the major categories of bone artefacts found in Kria Cave could potentially have their functional life extended through resharpening. In the case of bipoints this might involve actual resharpening of a blunt tip or the production of a new point in the event of minor breakage. Similarly, for unipoints the functional end might be resharpened or a new point produced after snapping. This process might occur multiple times, until the shaft became too short for the artefact to be effectively used.

Pasveer and Bellwood (in press) present evidence for resharpening of cortical bone bipoints from Golo Cave in the northern Moluccas. These artefacts are remarkably similar to those from Kria Cave, although they are slightly longer on average. One indication of resharpening among the Golo Cave bipoints is the fact that many of them show localised polish on both ends of seemingly intact artefacts, presumably as a result of reversal in a haft. Double-ended polish occurs most commonly in Golo Cave bipoints that are strongly asymmetric. A second indication of resharpening is the fact that the shorter bipoints within the Golo Cave assemblage tend to be more asymmetric than the longer specimens; this would be expected if broken points were simply given a new point rather than being completely remodelled.

Polish on both tips has been recorded on some Kria Cave bipoints, but it is relatively rare in comparison with the Golo Cave assemblage. Damage is commonly observed on both tips among the Kria bipoints, but this is likely to reflect damage to the secondary end within the haft, rather than any process of functional reversal. However, the fact that considerable effort was put into producing both ends of the bipoints, coupled with the occasional specimen with double-ended polish, suggests that optional reversal of bipoints might have been an integral part of the design of these artefacts.

The relationship between bipoint length and asymmetry in the Kria Cave bipoints is the exact opposite of the pattern observed in the Golo Cave sample. As described earlier, the shorter bipoints in the Kria Cave sample are in fact strongly symmetric, and while asymmetry is present within the sample, it is confined to bipoints that are longer than c.

25 mm. If asymmetry in these larger specimens is a product of resharpening, it implies that some of the bipoints were originally much longer than any of the complete examples recovered from the site. This possibility was raised earlier in relation to the recovery of some very long point-bearing cortical fragments. However, as explained there, these fragments might also be derived from cortical bone unipoints.

An alternative explanation for the unusual relationship between bipoint length and asymmetry in the Kria Cave assemblage is that bipoints required a certain minimum length for each of the primary end (to provide a functional tip) and the secondary end (to facilitate hafting). For the very small bipoints, both of these limits might have been reached simultaneously, resulting in the high degree of symmetry. However, for the larger specimens, where the limits were clearly exceeded, there was room for these artefacts to be variable in their original form. Overall, especially when compared with the Golo Cave assemblage from the Moluccas, there appears to be relatively little evidence for resharpening and repair of the Kria Cave bipoints.

Evidence for resharpening or repair of broken unipoints is even more difficult to obtain. A unipoint with a worn or broken tip is easily modified without altering any dimension except for its length; unfortunately, owing to the fact that the articular ends of these shaft bones was evidently removed at an early stage in manufacture, the original length cannot be determined for any individual unipoint.

6.5.5. *Comparison between Kria and Toé Cave: contrasting patterns of site use?*

The paucity of bone artefacts in Toé Cave stands in marked contrast with their abundance in Kria Cave. There are various possible explanations for this contrast, variably relating to sampling procedures, the age of the deposits, the length of occupation, the preservational environment and the functional or cultural characteristics of the sites. These different options will be examined in turn.

In Kria Cave most of the bone artefacts came from layers dating between c. 7000 BP and c. 4300 BP. Calculated as a rate of bone artefact deposition, Unit II has the slowest rate with one bone artefact per 50 years, and Unit IV has the highest rate, with one artefact per 19 years. The bone artefact deposition rate in Unit III is only slightly lower than in Unit IV, despite the fact that faunal remains and lithic artefacts are relatively less abundant in this unit compared with the underlying and overlying units. The virtual absence of bone artefacts from the upper unit of Kria Cave reflects a general paucity of all cultural material from these layers, but perhaps also the different nature of the occupation (see Chapter 3). The few bone artefacts from Toé Cave date between c. 9000 BP and c. 3000 BP. Even restricting the comparison to a comparable period of 7000 - 4300 BP, the discard rate in Toé Cave could not be higher than one bone artefact per 900 years, a dramatic contrast with Kria. The small number of artefacts from Toé Cave is even more remarkable given the much longer total duration of occupation, dating back to c. 26,000 BP.

Preservation is also unlikely to be responsible for the difference in bone artefact numbers between the two sites. Both Kria and Toé Caves produced abundant, well-preserved faunal material, with no obvious difference in preservation state (see Chapters

3 and 4). Unburnt and lightly burnt bone, the dominant material for manufacture of bone artefacts, is equally dominant in the overall faunal assemblages from both sites. Sampling can also be ruled out; both sites were excavated and sorted using identical procedures and even the same individuals.

If none of these sampling or taphonomic factors adequately explains the difference in bone artefact numbers, then this can only be explained in functional or cultural terms. Webb (1987: 80) suggested that differences in bone artefact density between sites might reflect variation in the nature of occupation. Cave sites used by small groups and for single purposes (e.g. transit camp between major sites, casual consumption of game etc.) might be less likely to accumulate bone artefacts than sites that have been occupied by more balanced groups and on a more permanent basis where we might expect to see evidence of a wider range of activities including maintenance of hunting gear, the repair or decoration of clothing and the production of other items of material culture.

As argued before, the bone artefacts from Kria Cave include a range of possible unfinished and unused artefacts suggestive of on site manufacture, as well as a series of complete and broken pieces indicative of on site breakage and perhaps repair. This evidence suggests that the site was used fairly intensively and that a variety of activities were carried out on site. The abundance and diversity of other cultural materials and animal food remains, together with the rapid deposition rate at this site point in the same general direction. Using the same logic, Toé Cave, with its relatively low number of bone artefacts – yet higher proportion of surface area excavated than Kria – would appear to have been used in a more restrictive way, with fewer activities, or at least with activities that were not related to manufacturing, maintenance or use of bone tools. However, Toé Cave contains abundant animal remains (so there was certainly no shortage of raw material) and even more abundant stone artefacts, whose size and degree of reduction in fact suggests on site manufacture (see Chapter 5).

Cultural differences as an explanation seem less likely in this case. Today the Meybrat do make a distinction between the area northwest of the lakes and that to the southeast (Miedema 1986: 3), and Kria Cave and Toé Cave indeed fall in these different areas where even different Meybrat dialects are spoken (see figure on p. 2 in Elmberg 1955). However, given that today the Ayamaru Lakes separate these northern and southern areas, the boundaries are unlikely to have remained absolutely stable over very long periods, especially because the lake system was probably not present to its full extent until the mid-Holocene. There is, however, ethnographic evidence to suggest that not all peoples in western New Guinea made extensive use of animal bone for tools and weapons. For example, Pospisil (1963: 17) reported that Kapauku Papuans in the Wissel Lake area had only a few surgical and carving tools made of rat teeth or hawk claws, and further, that the use of bone for arrow tips was exceptional.

Somewhat surprisingly, equivalent differences in bone artefact numbers between geographically and temporally related sites have been reported elsewhere in the wider region. The most immediate example involves Golo Cave and Um Kapat Papo on Gebe Island in the northern Moluccas (Pasveer & Bellwood in press); the numbers of bone artefacts found in each site was 130 and 1, respectively. Other examples of such differences include sites in the Gunung Sewu area of Central Java (Simanjuntak in press); sites in Victoria, southeastern Australia (Pickering 1979); and sites on New

Ireland, where the inland site of Balof Shelter contained many bone artefacts, contrasting with the lack of bone artefacts in contemporaneous coastal sites on the island (Downie & White 1978: 777-779). In my view, it is unlikely that all of these sites had stable and strict cultural boundaries through time. It is more plausible that they differed in each case in functional characteristics related to the local environment or the nature of the visiting groups.

6.5.6. *Wider comparisons*

As far as is possible to establish from the available site reports and other publications, it appears that bone artefacts such as described here occurred from Tasmania, through to southeastern, western and northern Australia, into Melanesia (including at least the Bismarck Archipelago, Papua New Guinea and Papua), and further west from the Aru Islands, the northern Moluccas, through to Sulawesi, Java, and other parts of Indonesia, the Philippines and mainland Southeast Asia (see for a review of sites with bone artefacts in Southeast Asia Olsen & Glover in press). Expedient blanks with usewear have unfortunately rarely been recognised, due to their minimal (or partial - Webb & Allen 1990: 77) modification and the unawareness of the researchers (Lampert 1966; Pickering 1979; Webb & Allen 1990).

Bone artefacts were not only used widely throughout the region but also through long periods of time. Bone points of this kind have been used from at least c. 30,000 BP [e.g. Bone Cave in south Tasmania (Webb & Allen 1990); Niah Cave in Sarawak (Majid 1982)], but were most common from Late Pleistocene - early Holocene times onwards. In sites where larger numbers of bone artefacts were found, reports mention that there was no change in morphology or parent bone through time (see for example Olsen & Glover in press). Within the c. 3000 years of their occurrence, the bone points identified in Kria Cave show also no clear relation to any particular period in time. All varieties appear to have been present throughout this time span.

Interestingly, not all varieties of bone points that are known from other sites in this region were found in the Bird's Head sites. Spatulae manufactured from wallaby fibulae are perhaps the best-known type of bone artefact throughout the wider Australasian and Melanesian region. Their absence in the Kria Cave assemblage is all the more remarkable given the fact that *Dorcopsis* wallaby makes up the great bulk of the faunal remains. Pointed artefacts including the joint of the parent bone (often called awls) are another type of bone artefact that has been found throughout the region, yet not in the Bird's Head sites. The same applies to expedient tools (although some examples may be present amongst the unfinished artefacts), which are clearly casual implements, used on the spot.

While the function of artefacts such as awls and spatulae is unknown (although many speculations have been offered), their different shape and size (especially the rounded tip of the spatula) suggest that they may have been involved in various activities. One aspect that seems to be characteristic of the spatula is the polish extending over much of the surface of the artefact. This implies that it was used for activities involving the movement or penetration of the artefact through a softer material, such as in sewing, fastening of a cloak, or perhaps the use of a nose plug. Artefacts with polish extending beyond the tip are rare in Kria and have not been found in Toé; it is possible that such activities were not carried out on site, or that artefacts used for such tasks were carefully kept and maintained, rather than discarded.

6.6. CONCLUSION

The abundant bone artefact assemblage from Kria Cave is characterised by the occurrence of two distinct classes of artefacts, unipoints and bipoints, found in roughly equal numbers. Toé Cave produced far fewer bone artefacts, but the small number of specimens is consistent in basic form with those from Kria. Other types of bone artefacts recorded from sites in adjacent regions (e.g. the northern Moluccas), such as spatulae or chisels, are not represented in either of the Ayamaru sites. Both sites also show a surprising lack of expedient tools suggesting that casual use of bone was avoided in favour of the use of the more specialised artefacts.

In general, the Ayamaru artefacts show evidence of careful manufacture and remarkably little variation in size or shape within each class. There is an almost complete division in raw material used to make each class of artefacts. Unipoints were almost always made from fragments of long bones derived from small animals including fruitbats and they retain an intact medullary canal. In contrast, bipoints were almost always made from cortical splinters derived from larger bones of medium-sized to large animals. The few exceptions include two bipoints made on shaft fragments; one well-made unipoint made from a cortical splinter; and four minimally modified artefacts made from crude cortical splinters but bearing one well-made point.

From the frequency and pattern of usewear on the Ayamaru bone artefacts it is clear that both the uni- and bipoints were functional tools rather than decorational items. In both classes, the tip (rather than the shaft) was clearly the main functional part of the tools; hence, neither class of artefacts is likely to have been used in any repetitive, penetrative task such as sewing. The analysis of usewear and breakage patterns suggests that the bipoints were probably hafted and used for drilling holes or carving work; a few specimens were clearly used on both ends, suggesting an occasional reversal in their hafts. The unipoints were probably hand held and might have been used in a variety of ways, some of which could well have overlapped the function of hafted bipoints. Other, more casual uses, such as removing splinters, cannot be ruled out.

The breakage pattern of the bi- and unipoints suggests that both classes were used as well as maintained on site at Kria Cave. This conclusion, together with the abundance of the artefacts, suggests that the site was used for a variety of activities that required the use, maintenance and repair of bone points, whether or not these artefacts were part of a composite tool. The lack of bone artefacts in Toé Cave is difficult to explain other than in relation to differences in site function, involving fewer activities and perhaps associated with less frequent occupation compared to Kria.

7. Prehistoric Exploitation of Food Resources

7.1. INTRODUCTION

Through time, the floors of Kria and Toé Caves were littered with human food refuse, much of it burnt and splintered as a consequence of cooking and consumption. Some of these remains found their way into the deposit, being those that survived the attention of small scavengers such as rats, and the processes of weathering. A sample of the preserved remains was recovered by our field team. Excavations in both caves produced large amounts of faunal remains, including bone, molluscan shells and bird eggshell. Much less commonly, botanical remains were recovered, including seeds and carbonised fruits.

Archaeological faunal remains can be used to reconstruct changes in subsistence patterns, or perhaps more accurately, changes in consumption patterns of the people who visited the cave through time. However, reconstruction of prehistoric diet is often fraught with difficulties relating to the inevitable bias against the more perishable items; questions as to how representative the site is for a wider region; and methodological problems associated with estimating the relative abundance of each food category. Another common problem that besets faunal analysis in many parts of the world is the separation of food remains taken into caves by human and non-human predators. This issue is less relevant in lowland New Guinea, where there were no large predators except possibly the dog-sized Tasmanian Tiger or Thylacine, a species hitherto only recorded from prehistoric sites in highland Papua New Guinea and one which is not known to have accumulated bone in cave environments.

The large quantity of well-preserved bone from Kria Cave (a larger, better dated and better stratified sample than that from Toé) thus provides an opportunity to explore some of these methodological problems; and this exercise in turn sheds light on some aspects of human behaviour at the sites.

The bulk of the faunal material from both sites comes from the Brown Dorcopsis, *Dorcopsis muelleri*[1], or *Djief* as the animal is called in Meybrat language (see Appendix 3). This is a medium-sized, ground-dwelling member of the kangaroo family, known from scattered zoogeographic records in lowland western New Guinea. *Dorcopsis* is a relatively shy, secretive animal, and while it is still occasionally hunted, local informants report that it is nowadays scarce in the Ayamaru area.

---

[1] The Brown Dorcopsis will henceforth be called *Dorcopsis muelleri*, or simply *Dorcopsis*, as it is the only species of this genus found in this part of New Guinea.

Three main questions need to be asked in relation to the abundant *Dorcopsis* remains in Kria and Toé Caves. Firstly, how important was *Dorcopsis* in the prehistoric diet, especially in comparison with other prey items such as snakes or smaller mammals such as cuscus? Secondly, to what extent does the importance of *Dorcopsis* in the sites reflect preferential hunting rather than a natural abundance of this species in the prehistoric environment? And lastly, what can we learn from the remains about the impact of sustained hunting on the *Dorcopsis* population, in particular noting its reported scarcity in the area today? This last question will be taken up in Chapter 8, based on a detailed analysis of the *Dorcopsis* remains.

Investigation of archaeological food remains from well-stratified, well-dated cave deposits can sometimes yield insights into environmental change through time. The archaeological deposits in Kria and Toé Caves, together spanning the period from around 26,000 BP to within the last few thousand years, provide such a record of changing environments; this subject will be explored in detail in Chapter 9.

7.2. METHODS OF RECOVERY AND IDENTIFICATION

7.2.1. *Field recovery, cleaning and treatment*

For reasons given in Chapter 2, it was not possible to sieve the excavated cave earth from either site. As an alternative, all samples were carefully screened by hand, with the same people performing this task through the two excavations. Bulk samples of every excavated spit were collected and later wet sieved to check the rate of recovery of very small fragments. The 'manual screening' method resulted in the retrieval of very small items down to the size of individual rodent teeth, small fish vertebrae, and other fragments down to 2-3 millimetre in size. The wet-sieved samples did not produce any additional categories of remains, nor did they suggest that large numbers of items were being missed. Clearly though, the rate of recovery of these very small items is not as good as it might have been had we been able to use wet sieving.

The organic material was provisionally cleaned in the field to allow transport to the Groningen Institute of Archaeology, University of Groningen. There, the material was more thoroughly washed, dried, and then individually numbered with black ink. Especially in the lower levels, many bone fragments were thickly encrusted with calcium carbonate and some required treatment with dilute acetic acid prior to examination. This treatment resulted in some specimens becoming brittle and these were hardened with an acetone-soluble polymer (Mowital). The treated specimens[2] are easily recognised as fairly light (yellow to light brown) in colour (the carbonate crust prohibited discolouration through the surrounding soil) and somewhat glossy as a result of the hardening material.

[2] The find numbers containing animal bone material that has been treated with acetic acid and Mowital are listed in my unpublished PhD thesis (Pasveer 2003: 184).

7.2.2. *Identification and quantification*

Macrobotanical remains

Only a small collection of botanical remains is available from Kria and Toé Caves. The carbonised fruits were examined by Kathleen McConnell (then Dept. of Archaeology and Natural History, the Australian National University; see her report in Appendix 2) and the *Celtis* and banana seeds by Lyn Craven (Australian National Herbarium, Canberra).

Soil samples for pollen analysis were taken from both caves. Samples from Toé Cave were investigated by Sytze Bottema (Groningen Institute of Archaeology, University of Groningen), but no pollen were found (Jelsma 1998: 60). No further attempts were made to isolate pollen from other samples of either cave.

Molluscan shell

Following an initial sort, a sample of the more complete molluscan shells from both caves were examined by Rob Moolenbeek (Zoological Museum, Amsterdam). Specimens identified by Moolenbeek were then used as a reference set to identify the remaining samples from Kria and Toé Caves. Katherine Szabó (Dept. of Archaeology and Natural History, the Australian National University) subsequently checked and refined the identification of some specimens.

For each sample, the number and combined weight of fragments were recorded for each taxon.

Crustacean exoskeleton

Small fragments of crustacean exoskeleton were only recovered from Kria Cave. No attempt was made to further identify these specimens. Lowland rainforest habitats in New Guinea support a variety of aquatic and terrestrial crustaceans; Brongersma (1958: 57) reported that freshwater crayfish was locally abundant in the Ayamaru Lakes.

For each sample, the number and combined weight of fragments were recorded.

Bird eggshell

The bird eggshells fall into two varieties, a thicker kind with a distinctive, coarse-pitted outer layer; and a thinner kind with a smooth surface. In the New Guinean context, the former morphology is diagnostic for members of the genus *Casuarius*, the cassowaries. The thinner, smooth eggshell is less diagnostic but is most likely derived from members of the family Megapodidae, the mound-building Brush-Turkeys or megapodes.

Bird eggshell is an ideal material for radiocarbon dating and for amino acid racemisation studies because the dense calcite structure protects indigenous inorganic carbon (Pasveer et al. 2002). Material from both sites was included in a pilot amino acid racemisation study by Gifford Miller of the University of Colorado, and in a subsequent, more detailed study by Simon Clarke of the University of Wollongong (Clarke forthc.). The resulting radiocarbon dates and A/I ratios are discussed in Chapter 3 and 4 in the context of the dating and stratigraphy of the sites.

For each sample, the combined weight of fragments was recorded.

Vertebrate bone

The large and complex task of identifying the abundant vertebrate remains was undertaken in Australia, where adequate reference specimens were readily available. Identifications were performed by comparison with modern New Guinean and Australian mammal specimens from the collections of the Western Australian Museum in Perth and the Australian Museum in Sydney, as well as subfossil material from several archaeological localities in Papua New Guinea. Attempts were made to identify both cranial and post-cranial body parts, although post-cranial material could seldom be determined below family or genus level. For mammals, species-level identifications are in most cases based on teeth or tooth-bearing elements (dentary, maxilla, premaxilla).

Species-level identifications were carried out for mammals by Ken Aplin (CSIRO, Australia); for birds by Walter Boles (Australian Museum, Sydney); and for snakes by John Scanlon (South Australian Museum, Adelaide). Aplin et al. (1999) provided full taxonomic details of the species identified from Kria and Toé Caves.

During the study of the collections, several previously undescribed species of mammals were identified. Two of these were named and published by Aplin et al. (1999): a very small striped possum, *Dactylopsila kambuayai* Aplin 1999, named after our 'right hand' in the field Mr Elimas Kambuaya (see his contributions in Appendix 3), and a small ringtail possum, *Petauroides ayamaruensis*[3] Aplin 1999. Another new but as yet undescribed species is a small rodent of the genus *Pogonomelomys*. Type specimens of these species and representative examples of the other identified species have been temporarily lodged in the vertebrate palaeontological collection of the Western Australian Museum (registration numbers WAM 98.7.9-98.7.57), pending their transfer to an Indonesian institution. At the moment of lodging this dissertation, the remainder of the bone (and other) material from the Bird's Head is currently stored in the Dept. of Archaeology and Natural History of the Australian National University in Canberra, pending its repatriation to Indonesia.

All bone material was initially allocated to one of four classes:

- Mammalia (mammals)
- Aves (birds)
- Reptilia (reptiles)
- Teleostomi (fishes)

[3] In Pasveer & Aplin 1998, this animal is referred to as '*Pseudocheirus*' *ayamaruensis*. At that stage the generic affinity of this ringtail possum was not fully understood. The paper by Aplin et al. 1999, which was written at the same time as Pasveer & Aplin 1998 but was published a few months later, contains the description of this new species, allocated to the genus *Petauroides*.

Within each class, specimens were then allocated to lower level taxonomic categories (family, genus, species), depending on the availability of reference material, the completeness of the remains and the utility of the available body parts for identification. Vertebrae and ribs of mammals are much less diagnostic than their cranial and limb elements, hence no attempt was made to identify these elements below the level of class. In contrast, reptile vertebrae are highly diagnostic down to family level, and even to genus or species in the case of snakes.

For otherwise unidentified remains within each class the combined weight of fragments was recorded. Mammalian vertebrae and ribs were similarly treated.

The following characteristics were recorded for all bone material identified to family level or below from Kria Cave (the stratigraphy in Toé Cave is not sufficiently controlled to warrant such detailed recording):

- taxon;
- skeletal part (e.g. cranial, humerus, pelvis, etc);
- part of the element (e.g. proximal, shaft, ilium, ischium, etc);
- age class (adult or juvenile, based on epiphyseal fusion and/or cortical bone texture, and on tooth eruption), and;
- burning condition (for definitions see Chapter 2).

Cutmarks were not observed on any bone.

The system of identification and recording allows for considerable flexibility in data analysis. Allocation of all bone fragments to at least class level means that the composition of the *entire* faunal assemblage can be quantified at the highest level. Similarly, by allocating virtually all mammal remains (except for vertebrae and ribs) to at least the family level, it is possible to provide a nearly comprehensive breakdown of the assemblage at this level. By recording both weights and numbers of fragments, it is further possible to assess the relative abundance of taxa either by calculating the relative weight of their remains, or by calculating the relative number of identified specimens (NISP). The detailed system of recording of skeletal parts for each taxon also allow the Minimum Number of Individuals (MNI) to be estimated, using a variety of different skeletal elements (e.g. cranial only vs. all elements; with or without information on age). Details of how these values were calculated and discussion of the relative merits of each of these approaches will be found in later sections of this chapter.

7.3. THE ORGANIC ASSEMBLAGES FROM BOTH SITES

Table 7.1 shows the contribution of the various categories of organic remains to the total assemblages from each of the two sites, not including the human remains (these are discussed in Appendix 1). In both sites vertebrate bone makes up more than 90% of the assemblage by weight, with molluscan shell being the next most abundant class of remains at around 5%. The total assemblage from Kria Cave weighs 12,691 g. Toé Cave produced less than half the total quantity of organic remains from Kria, despite its longer occupation history and larger excavated area. Toé yielded proportionally more bird egg-shell than Kria, but failed to produce any crustacean remains; the latter class of remains is consistently present but rare in Kria Cave. Macrobotanic remains are rare in both sites.

Table 7.1. Weights and proportional composition of the organic remains from both sites.

	Kria Cave		Toé Cave	
	weight in g	%	weight in g	%
vertebrate bone (excl. human)	12,691.0	94.27	5695.0	94.04
bird eggshell	57.0	0.42	73.0	1.21
molluscan shell	712.0	5.29	286.0	4.72
crustacean exoskeleton	0.3	<0.01	-	-
macrobotanical remains	1.5	0.01	1.8	0.03
total organic remains	13,461.8		6055.8	

7.3.1. *Taxonomic representation*

Botanic remains

Carbonised plant remains found in Kria were subjected to a detailed study (see report by Kathleen McConnell in Appendix 2), but this failed to secure any precise identifications; some specimens were tentatively identified as fruits belonging to one or more members of the family Elaeocarpaceae. These were found only in Unit V in Kria and date to the period 6100-6800 BP.

In Toé, fragments of *Celtis* seeds were present down to 60 cm in the deposit, but these were absent from the Late Pleistocene levels. Two rounded seeds found in the top 10 cm in Toé were identified by Lyn Craven as seeds of a member of the family Musaceae, the bananas. Their position so close to the surface suggests that they may be of fairly recent age, but they could be potentially up to c. 2-3000 years old.

Molluscs

A total of nine taxa of aquatic molluscs were identified in the combined assemblages, with eight recorded from Kria and seven from Toé Cave (see Table 7.2). Four of these taxa originate from a freshwater environment (*Melanoides* sp., *Bellamya* sp., *Hyridella* sp. and *Batissa* sp.), one from brackish water (*Polymesoda* sp.), three from a mangrove environment [*Ellobium aurismidae* (Linnaeus 1758), *Anadara granosa* and *Saccostrea cucullata*], and one from fully marine conditions (*Voluta* sp.). None of these taxa could be expected to occur naturally in the cave environment, hence all of the remains may be assumed to result from human harvesting. The distribution of each molluscan species through the Kria and Toé Cave deposits is summarised in Figures 7.1 and 7.2.

In Kria, molluscan shells are concentrated in the top unit, with two-third (63.4%) of the total assemblage found in the top 40 cm. The bulk of the material from these levels was identified as the freshwater species *Melanoides* sp., most of which specimens were concentrated in a small shell midden in Unit I.

Melanoides is a genus of freshwater gastropods that includes species typically found in lakes and rivers. The majority of these shells from both caves lack their tips. It was suggested by Sapri Hadiwisastra, who was on the excavation team, that the tips were

Table 7.2. Molluscan taxa recorded from each of Kria and Toé Caves.

taxon	habitat	Kria Cave	Toé Cave
Bellamya sp. (incl. *B*. sp. cf. *B. kowiayiensis*)	freshwater lakes and rivers	+	+
Hyridella sp. (incl. *H*. sp. cf. *H. misoolensis*)	freshwater lakes and rivers	+	+
Melanoides sp.	freshwater rivers	+	+
Batissa sp.	freshwater rivers		+
Polymesoda sp. (incl. *P. viridescens*)	brackish rivers		+
Polymesoda sp./*Batissa* sp.	brackish / freshwater	+	+
Ellobium aurismidae	mangrove	+	-
Anadara granosa	mangrove	+	-
Saccostrea cucullata	mangrove	+	+
Voluta sp.	marine	-	+
unidentified aquatic spp.	---	+	+
Leptopoma sp.	land snail	-	+
Helicarionidae indet.	land snail	+	+
Cyclotus sp. (incl. *C. hebraicus*)	land snail	+	+
Palaina sp.	land snail	+	-
Pupinella sp.	land snail	+	-
Papuina sp.	land snail	+	-
Clausiliidae indet.	land snail	+	-
Pila sp.?/*Camena* sp.?	land snail	+	-
unidentified terrestrial spp.	---	+	+

intentionally broken off to suck out the mollusc, a custom he had observed in rice fields on Java. However, *Melanoides* species in Fiji are reported to commonly lose their tips if they live in flowing rather than static water (Ryan 2000: 48). *Melanoides* and other mollusc species were also found below Unit I, but not in such large amounts. Interestingly, the remains of *Melanoides* sp. are scarce in Toé Cave.

Among the other freshwater molluscs, *Batissa* species typically come from a river environment (with running water), while *Hyridella* species are also found in more static water, such as in lakes. Not much is known of *Bellamya kowiayiensis*, but another species of this genus, *Bellamya javanica* (V.d. Busch 1844), is found in rivers, lakes or wet rice fields (Dharma 1988), suggesting that it is both tolerant of running and still water environments. *Bellamya* sp. occurs only in Unit I in Kria, but extends down to c. 80 cm in Toé. It is relatively scarce in both caves. *Hyridella* sp. is also scarce in Unit I of Kria Cave but it is consistently present throughout the lower units; it is also present throughout the Holocene part of the Toé deposit.

Polymesoda sp. lives in rivers with fresh to brackish water and must have been obtained, or traded in, from an area closer to the coast, which today lies 30-40 km to the southwest. Rob Moolenbeek identified examples of both *Batissa* sp. and *Polymesoda* sp. among the more complete specimens from Toé Cave, but the majority of fragments from both caves ascribed to either *Batissa* sp. or *Polymesoda* sp. could not be further identified due to their similar morphology and high degree of fragmentation. However, since *Batissa* would have been locally available, the bulk of the fragmented shell is most likely from this species.

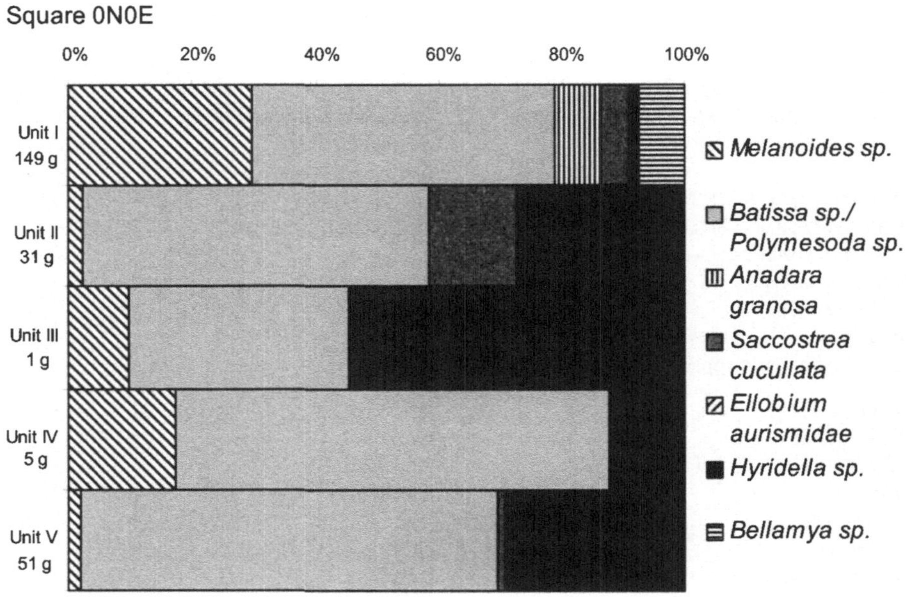

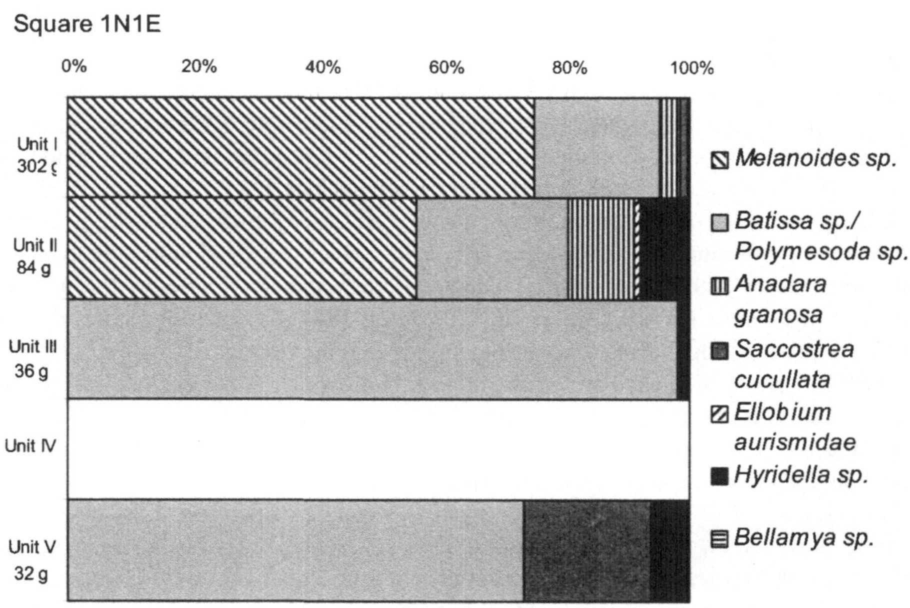

Figure 7.1. Kria Cave, proportional representation of aquatic molluscs, based on weight. No shell was found in Unit IV of Square 1N1E. The total sample weight is given on the Y-axis, below the Units.

Polymesoda/Batissa are the dominant taxa in the Holocene levels of Toé Cave. Some specimens were identified as *Polymesoda viridescens*, including one found in the upper part of the Late Pleistocene unit, dating to c. 13,000 BP. As noted above, this species favours brackish water and must have come from an area towards the coast. These molluscs may have been carried around as potential raw material for tools or as decoration; however, no sign of modification or use has been observed on any of the shell fragments.

Polymesoda/Batissa have both the highest NISP values and the highest shell weights in Toé Cave, but the latter is disproportionally high due to its heavy individual shells. Kria has very similar quantities of this species compared to Toé, but the shells are much more fragmented. The proportional representation of *Polymesoda/Batissa* in Kria is much lower than in Toé Cave because this value is suppressed by the abundance of *Melanoides* sp.

Ellobium aurismidae, *Anadara granosa*, and the oyster *Saccostrea cucullata* are found in brackish water and are typical mangrove species. In Kria Cave, these taxa were found mainly in Units I and II. *Saccostrea cucullata* was also recovered in very low numbers from the deepest levels of Kria Cave's deposit, therefore dating to around 6900 BP. *Anadara granosa* and *Ellobium aurismidae* were absent in the Toé assemblage, and *Saccostrea cucullata* is represented by one fragment, from a depth of 30-35 cm, dating to around 5290 BP. At that time, the nearest mangroves probably lay at least 40-50 km to the southwest, indicating a distant origin for these shells.

Toé Cave produced the only true marine species: a fragment of a *Voluta* shell, found at 15-20 cm depth probably dating to after c. 3000 BP. No typical marine species were found in Kria.

A variety of land snails were also present in both caves, but generally not in quantities that would suggest these to be the product of systematic human harvesting. The total weight of the land snail sample from Kria is 23.5 g and from Toé 5.3 g.

Land snails are most abundant and diverse in Unit I of Kria Cave. *Cyclotus* sp. and *Pila* sp./*Camena* sp. are present in all units except III and IV, which is probably due to overall small sample sizes in these units. Some land snails can be found living in or close to cave environments, and their presence may have little meaning from a cultural point of view. In Toé Cave land snails are most abundant in the upper half of the deposit.

Interestingly, molluscs of all kinds are decidedly scarce in the Late Pleistocene levels of Toé Cave. This may be the result of severe fragmentation and/or poor preservation of the remains. However, changes in the aquatic and terrestrial environment cannot be ruled out as a cause of this contrast.

One hint of significant changes in later times comes from the dating of basal lake sediments of the westernmost Lake Ayamaru to c. 4000 BP (R. Dam and S. van der Kaars pers. comm.; see Chapter 9), corresponding to the very top of Unit II in Kria Cave. However, since it seems likely that the western- and easternmost lakes are fed through different sources (see Chapter 2), it is equally possible that the other lake basins maintained standing water bodies from much earlier times. Unfortunately, none of the mollusc species from the sites is absolutely characteristic of still water conditions, although *Hyridella* sp. is abundant in such contexts.

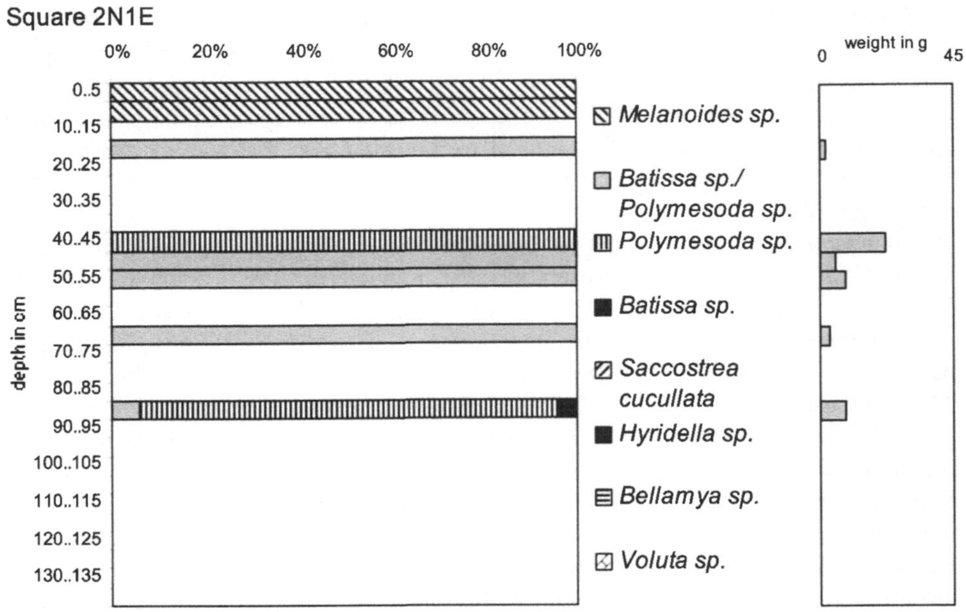

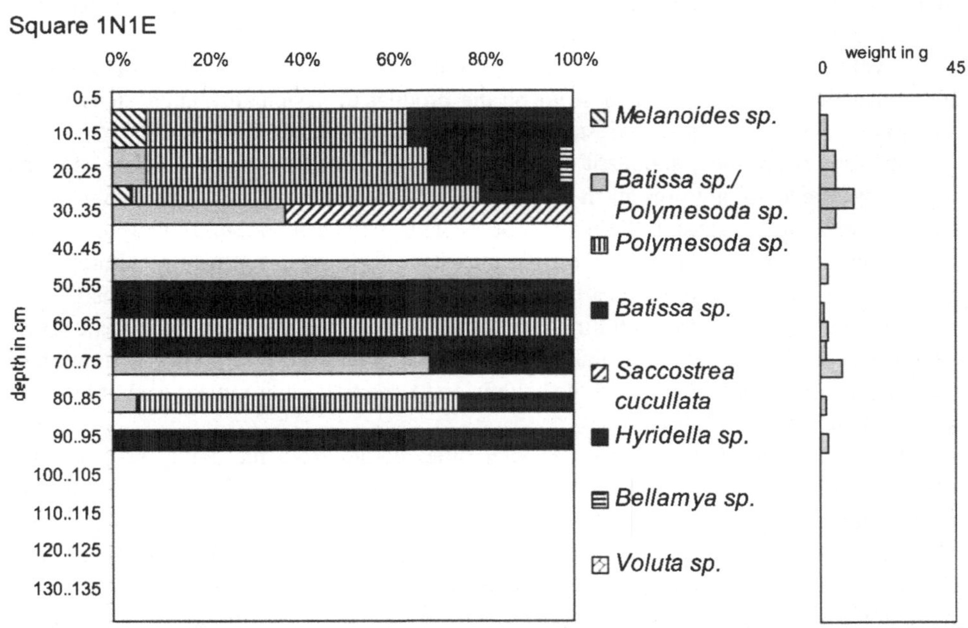

Figure 7.2. Toé Cave, proportional representation of aquatic molluscs, based on weight. The upper boundary of the Late Pleistocene deposit lies at c. 80 cm (Square 2N1E) and c. 85 cm (Square 1N1E). 10-cm spits are represented as two 5-cm spits (see note 3 on page 21).

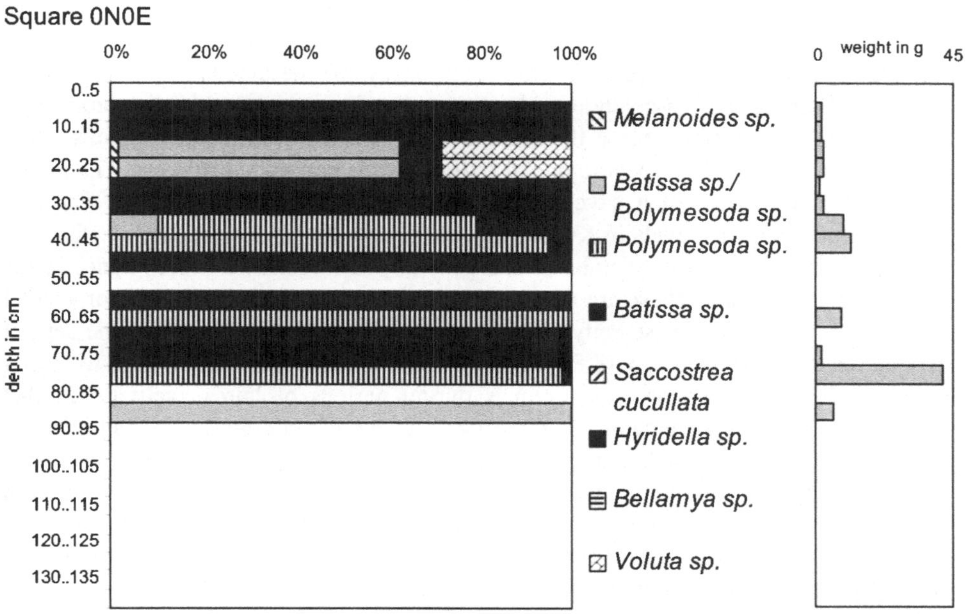

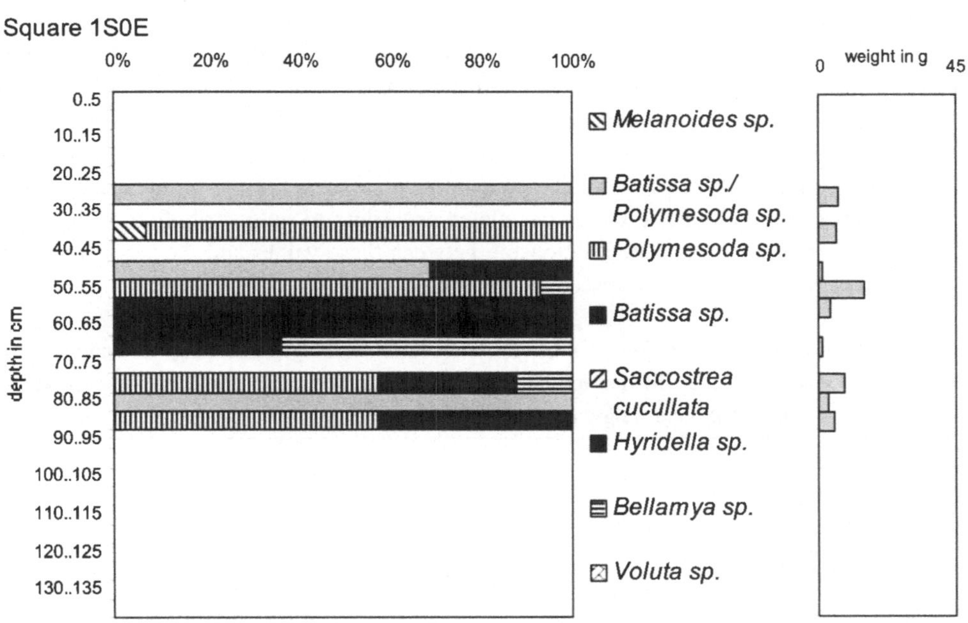

Figure 7.2 (continued). Toé Cave, proportional representation of aquatic molluscs, based on weight. The upper boundary of the Late Pleistocene deposit lies at c. 95 cm (Square 0N0E) and c. 105 cm (Square 1S0E).

Vertebrates

The bulk of the vertebrate remains from both caves are presumed, by virtue of their degree of burning and association with other cultural materials, to be the product of human activities in and around the sites. The Kria Cave assemblage contains small quantities of remains derived from one or more possible non-human sources. The first is the occasional remains of a cave-roosting, insectivorous bat, *Hipposideros diadema*, which occur at various levels in Kria Cave and may be the result of natural deaths. The second is a minor quantity of unburnt and generally unfragmented rodent bones in the upper levels of the site. These are likely to be prey remains of a raptorial bird or a small mammalian carnivore. The scarcity of such remains through the remainder of the deposit may be the result of one or more factors: 1) that the cave was less likely to be used as a roosting place for owls or other animals during periods of human activity; 2) that unburnt small mammal remains are more likely to be affected by decay than burnt or cooked bone, and thus less likely to survive; and 3) that during periods of more intensive human activity, any non-human prey remains present in the cave are likely to be burnt in and beneath fire places, and to become fragmented through trampling. As such they may become difficult to distinguish from human food remains. The negligible quantities of small vertebrate remains through the deposit suggest that one or all of these factors were in operation through the greater part of the occupation period.

Table 7.3 (at the end of this chapter) shows the full range of vertebrate taxa identified from the combined assemblages from Kria and Toé Caves, along with some basic ecological information on each taxon. The mammal remains are especially diverse and include a range of species that are not recorded as part of the modern fauna of the Bird's Head. The palaeoenvironmental and zoogeographic significance of these new distribution records are discussed at length in Chapter 9.

For the following analyses, taxa are grouped by major taxonomic groups (e.g. families of mammals, reptiles and birds; fish), presented in Figure 7.3 and Table 7.4a and b for Kria, and Figure 7.4a and b and Table 7.6 for Toé. For the initial analysis, the relative abundance of each taxon is calculated according to its contribution to the total weight of vertebrate bone. The relationship between these values and the various counts based on NISP and MNI is discussed in a later section.

Table 7.4a. Composition of the Kria Cave vertebrate assemblage, including the unidentified mammal bone, divided into stratigraphic units. The composition is summarised by proportional bone weight per taxonomic class.

class	whole deposit	Unit I	Unit II	Unit III	Unit IV	Unit V
total mammal	93.40 %	82.49 %	92.74 %	93.80 %	90.76 %	94.37 %
total reptile/frog	5.77 %	15.26 %	6.54 %	6.12 %	8.28 %	4.66 %
total bird	0.82 %	2.14 %	0.71 %	0.08 %	0.95 %	0.96 %
fish	0.01 %	0.11 %	<0.01 %	-	0.01 %	0.02 %

Kria Cave

About 60% by weight of the Kria bone assemblage consists of unidentified mammalian bone, including mammalian ribs and vertebrae (10% by weight). The remaining c. 40% was identified to various taxonomic levels. With the exception of the fish, almost all non-mammalian material could be identified at least to family level.

Macropodids make up the bulk of the vertebrate remains in Kria Cave (Table 7.4b). By far the most commonly identified macropodid in the total sample is *Dorcopsis muelleri*, the Brown Dorcopsis, but one or more species of tree-kangaroos (*Dendrolagus* spp.) and a smaller forest wallaby (*Dorcopsulus* sp.) are also represented.

Given the overwhelming predominance of *Dorcopsis* in the assemblage, it is reasonable to infer that the bulk of the unidentified macropodid remains, as well as the bulk of the mammal vertebrae and ribs, and indeed the bulk of unidentified mammal bone, are also derived predominantly from *Dorcopsis*. If only the remains identified to genus are considered, the proportion of *Dorcopsis* is 72.2%. However, if the other components (including the unidentified mammal bone) are included, the maximum contribution by weight of *Dorcopsis* to the assemblage rises to 90.6%. The true abundance of *Dorcopsis* probably lies somewhere between these values. Fluctuations through the deposit in the proportion of macropodid to 'unidentified mammal' probably reflect a variable degree of fragmentation of the remains (which will increase the 'unidentified mammal' category) as well as genuine changes in the proportional abundance of *Dorcopsis* compared with other major prey items such as snakes.

A wide variety of other mammalian families are represented in the Kria assemblage. With the exception of the Muridae (rodents), none of these contribute more than 1-2% overall, either in weight or in NISP. As noted above, the rodent remains probably include a non-human component concentrated in Unit I. Pig remains were found only in Unit I of Kria Cave, perhaps belonging to a single individual.

Snakes are relatively abundant in the Kria assemblage. The sample contains mainly vertebrae and ribs, with occasional cranial and dental material, all of which are highly distinctive. Snake remains are readily identified to at least family level. The bulk of the Kria snake remains are from members of the family Boidae (the pythons). John Scanlon (South Australian Museum) was able to identify examples of *Morelia* spp. including *M. amethistinus* and possibly *M. boeleni*. Other kinds of snakes were also identified, including a semi-aquatic file snake (*Acrochordus arafurae*), and unidentified members of the venomous snake families Colubridae and Elapidae. Most of the unidentified material originates from large snakes, predominantly pythons.

Bones of several lizard taxa were found in Kria Cave, including varanids and agamids, but all in very low quantities. Occasional frog remains were also identified.

Birds are represented by cassowaries (*Casuarius* spp.) and megapodes. Cassowaries are dominant amongst the bird taxa by weight (obviously on account of their much larger bones), as well as by NISP. At least one species of cassowary was identified from both caves, *Casuarius bennettii*, the Dwarf Cassowary. Amongst the megapodes, *Aepypodius arfakianus* (Wattled Brush Turkey) was identified in the Kria assemblage.

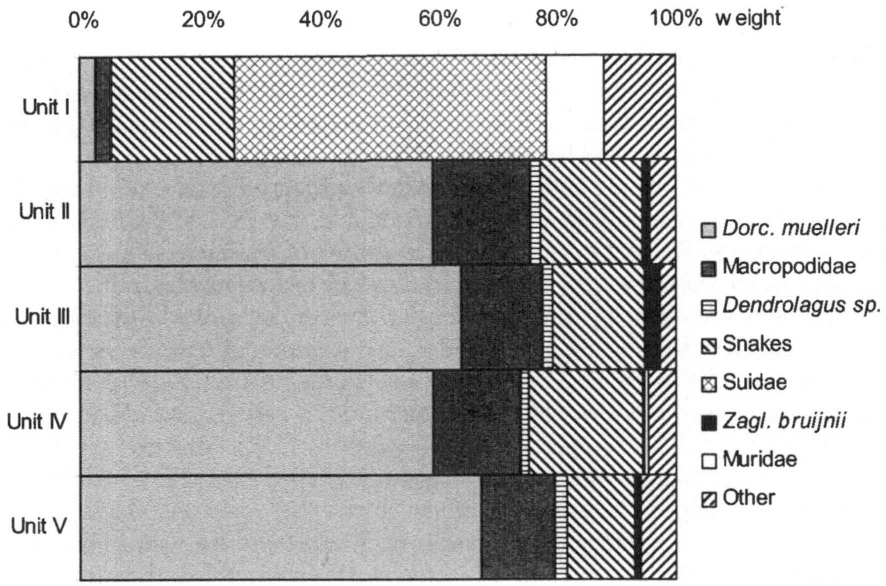

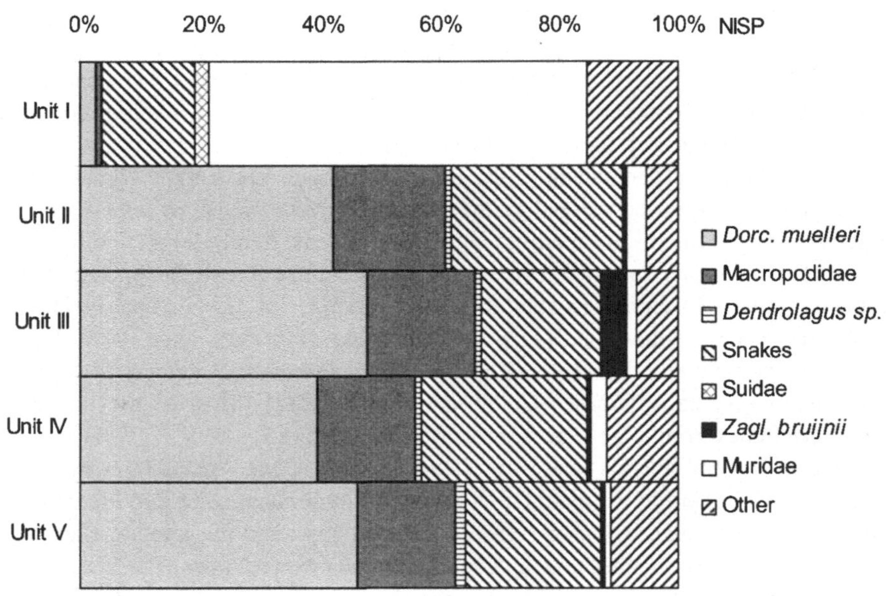

Figure 7.3. Kria Cave, simplified proportional representation of vertebrate taxa per unit; both squares combined. Top: based on weight; bottom: based on Number of Identified Specimens (NISP). For more details of identified families and groups, see Table 7.4b.

Table 7.4b. Taxonomic composition of the Kria Cave identified vertebrate assemblage, divided into stratigraphic units. The composition is summarised by proportional bone weight and by proportional Number of Identified Specimens (NISP).

taxon	all units		Unit I		Unit II		Unit III		Unit IV		Unit V	
	weight %	NISP %	weight %	NISP %	weight %	NISP %	weight %	NISP %	weight %	NISP %	weight %	NISP %
Tachyglossidae	0.88	1.06			0.85	0.73	2.66	4.54	0.39	0.56	0.69	0.64
Macropodidae	77.68	60.86	5.28	3.74	77.02	61.86	79.36	66.98	75.01	56.85	81.50	64.11
Phalangeridae	1.69	1.58	5.73	2.80	0.93	1.19	1.17	1.41	0.47	0.56	2.21	1.99
Pseudocheiridae	0.04	0.12	1.38	0.93	0.01	0.06			<0.01	0.19	0.02	0.11
Petauridae	0.08	0.58			0.09	0.17	0.05	0.31	0.14	2.44	0.07	0.60
Peroryctidae	0.53	1.13	1.67	2.80	0.69	1.02	0.25	0.63	1.15	2.44	0.34	0.94
Dasyuridae	0.02	0.17	0.16	0.93	<0.01	0.06					0.03	0.26
Suidae	1.25	0.09	52.56	2.34								
Muridae	0.46	4.26	9.60	63.08	0.41	3.22	0.30	1.72	0.43	2.63	0.10	1.16
Pteropodidae	0.42	1.58			0.16	0.45	0.75	2.03	0.66	3.56	0.48	1.95
Hipposideridae	0.02	0.46	0.04	<0.01			0.04	1.41			0.04	0.67
total mammal	83.09	71.89	76.42	76.62	80.16	68.76	84.58	79.03	78.25	69.23	85.49	72.42
snakes	14.42	24.50	20.48	15.42	17.38	28.87	15.12	19.87	19.32	27.77	11.62	22.78
Varanidae	0.25	0.50			0.49	0.85	0.05	0.16	0.10	0.19	0.20	0.45
other lizards	0.10	0.70	0.08	0.47	0.02	0.17	0.05	0.63	0.07	0.75	0.15	1.09
frogs	0.02	0.05									0.03	0.11
total reptile/frog	14.79	25.75	20.56	15.89	17.90	29.89	15.22	20.66	19.50	28.71	12.00	24.43
Casuariidae	1.79	0.67	2.37	0.47	1.92	1.02	0.15	0.16	2.12	1.13	1.94	0.49
other birds	0.31	0.41	0.51	0.93	0.02	0.11	0.05	0.16	0.12	0.19	0.53	0.67
total bird	2.09	1.08	2.88	1.40	1.94	1.13	0.20	0.31	2.24	1.31	2.47	1.16
total fish	0.03	1.27	0.14	6.07	0.01	0.23			0.02	0.75	0.04	1.99

Table 7.5. Fishes presently found in the Ayamaru Lakes.

species	common name	size	origin
Neosilurus brevidorsalis	Short-finned Catfish	to 17 cm	indigenous[1]
Melanotaenia ajamaruensis	Rainbow fish	to 7 cm	indigenous[1,2]
Melanotaenia boesemani	Rainbow fish	to 9 cm	indigenous[1,2]
Pseudomugil reticulatus	Blue-eye	very small	indigenous[1,2]
Glossogobius hoesei	Goby	very small	indigenous[1,2]
Oxyeleotris fimbriata[1]	Gudgeon	to 20 cm	indigenous[1]
Trichogaster pectoralis	Three Spot Gourani	to 10 cm	imported[1,2]
Helostoma temmincki	Labyrinth fish	-	imported[2]
Cyprinus carpio	Carp	-	imported[2]
Osphromenus goramy	Labyrinth fish	-	imported[2]

Note: [1] G. Allen, pers. comm. [2] Miedema 1986; Boeseman 1959; Reeskamp 1959.

Small fish vertebrae, many of them burnt or calcined, were found in low numbers throughout the deposit in Kria Cave (Table 7.4b). They have not been further identified, but the majority measure some 2-3 mm in diameter, which is consistent with the size of the native fish currently found in the Ayamaru Lakes and associated rivers (Table 7.5). Many early reports from the Ayamaru region consistently mention the low fish populations in the lakes (e.g. Brongersma 1958: 57), and Elmberg (1955: 59) reported that fishing had only recently become a popular occupation amongst the Meybrat. Various larger fish species were introduced by the Dutch in the 1930s (Reeskamp 1959).

The archaeological evidence suggests that fish were utilised throughout the prehistoric occupation of Kria Cave, but it is difficult to estimate their economic importance. Fish remains are scarce in the Kria Cave assemblage, but the fact that the majority were retrieved from the sieved sub-samples rather than from manual picking of the excavated soil, suggests that a large proportion may have been missed during this process. However, the majority of the sieved sub-samples of the 5-cm spits produced no more than one or two vertebrae, and the single 10-cm spit in Square 0N0E that was sieved in its entirety produced only 33. Other factors apart from recovery methods might also be important, including the likelihood that fish of such small size were consumed whole, resulting in the complete destruction of the bones, and the further consideration that such small bones are very prone to decay due to their large surface area in relation to their volume. In this light it is perhaps surprising that any fish remains were found at all.

The presence of a stingray spine fragment in the basal levels of Unit V (associated with ^{14}C age of 6080 BP) is noteworthy, as this object must have come originally from a marine or estuarine habitat. It further supports the evidence from the molluscan shell that people had contact with coastal areas, either direct or through trade, from at least 6-7000 BP.

The composition of the vertebrate fauna varies little through Units II to V (see Fig. 7.3 and Table 7.4b), with the only discernible trend being a slight decrease in the contribution of both *Dorcopsis muelleri* and macropodids as a whole between Unit V and Unit II, with a corresponding increase in the proportion of snake remains. The smaller possums (Pseudocheiridae and Petauridae) and bandicoots (Peroryctidae) are more abundant in Unit IV, and the Echidna, *Zaglossus bruijnii* (family Tachyglossidae), more so in Unit III; these fluctuations may be related to the small size of the samples from these units.

As described in Chapter 3, Unit I of the Kria deposit is radically different from the underlying units (Fig. 7.3). Besides the sudden and 'overwhelming' introduction of pig (represented by a few large bones, possibly from one individual), this unit is marked by a strong decline in macropodids relative to other prey items such as *Phalanger*. The weight and NISP data for this unit are very different from those for the other units. The bone shows a high proportion for the larger vertebrate taxa (pig, cassowary), which make up 53% of the identified bone by weight. In contrast, the NISP values emphasise the small vertebrates, among which rodents alone contribute 62% to the total NISP (but only 10% by weight). Remains of large vertebrates contribute only 4% to the total NISP. These contrasts further support the view that the pattern of occupation in Kria Cave changed dramatically during this period. However, the small sample size in Unit I is no doubt also partly responsible for these strong contrasts with the underlying units. The methodological issue of bone weights vs. NISP as measures of relative abundance will be taken up again below.

Toé Cave

A little less than 30% of the vertebrate assemblage from Toé Cave was classified as unidentified mammal, including 5% mammalian vertebrae and ribs. This value is considerably lower than in Kria, due to the slightly better preservation of the bone material in Toé Cave, to a lesser degree of fragmentation, especially in the Late Pleistocene levels, and to the higher proportion of both reptiles and small to medium-sized mammals, the latter with more diagnostic limb elements.

As discussed at length in Chapter 4, the Toé Cave deposit is stratigraphically complex and spans a longer time period than that in Kria Cave. Toé Cave also contains more dramatic changes in faunal composition through time, and there is little benefit to discussing the assemblage as a whole. The vertebrate fauna from each square will be discussed separately prior to attempting any general conclusions (see Figures 7.4a, b).

Square 2N1E: The abundant faunal remains from the lowermost 30 cm of deposit in this square contains a high diversity of taxa with no one group contributing much more than 25% by weight. The typical co-dominant taxa of these levels are *Spilocuscus* spp., *Zaglossus bruijnii*, *Dendrolagus* spp. and boid snakes, with cassowary and *Dorcopsulus* sp. important in some levels. *Dorcopsis muelleri* is present in most levels but is not especially common. As described in Chapter 4, this lower assemblage probably dates to the period c. 12,500 – 26,000 BP.

The zone above 80 cm and through to 50 cm produced much less bone (see Fig. 4.5), but there is a clear shift in composition with increased *Dorcopsis* (usually 50-75% by bone weight). Snakes decline at first, then increase in the levels above 65 cm to around 35-50%. *Zaglossus bruijnii* is moderately abundant only between 65-75 cm, while *Dendrolagus* spp. and *Spilocuscus* spp. are only occasionally represented. A radiocarbon date of 8860 ± 70 BP at 60-65 cm in this square suggests that the upper part of this zone might correspond to Unit V of Kria Cave. The faunal composition is certainly consistent with this interpretation.

Bone is moderately abundant again above 40 cm in Square 2N1E. Above the 50-cm level the vertebrate fauna is even more dominated by *Dorcopsis*, with values usually above 80% by weight. Snakes are less abundant than in the underlying zone, usually falling well below 10% by weight. *Zaglossus bruijnii* is moderately abundant in several levels and there are occasional minor peaks of *Dendrolagus* spp. and *Spilocuscus* spp. A date of 5290 ± 75 BP from 25-30 cm serves to link this zone to Units II to V of the Kria Cave deposit and the faunal composition is again quite consistent with this interpretation.

Square 1N1E: The lowermost 25 cm of deposit in this square produced faunal remains with essentially identical composition to the basal zone in Square 2N1E. The quantity of bone falls abruptly at the 100-cm level and shows only a very slight peak to correspond with the upper zone of Square 2N1E. In all levels above 90 cm, *Dorcopsis* forms at least 45% of the remains by weight, with this taxon exceeding 80% in most levels. Snakes and a variety of minor mammal taxa make up the remainder of the sparse remains. No radiocarbon dates were obtained from this square, but there is no reason to suspect any difference from the chronology of Square 2N1E.

Square 0N0E: This square produced much less bone in the lower levels, and only modest quantities through the upper part of the sequence. No radiocarbon dates are available. The lowermost levels between 105-135 cm produced a diverse fauna consistent with that in the other squares. Above the 105-cm level, the proportion of *Dorcopsis* increases sharply and then remains high through to near the surface. Snakes are moderately abundant (20-40%) in virtually all levels.

The uppermost 15 cm in this square produced only very small quantities of bone including the remains of *Zaglossus bruijnii.*

Square 1S0E: This square produced a large quantity of bone from the lowermost levels, at depths between 110 and 140 cm, and a strong secondary peak at 70-95 cm (see Fig. 4.5). Much less faunal material was present above the 60-cm level.

The assemblage from 105-140 cm is broadly similar to those from the lower levels of the other squares, but differs in having a somewhat higher proportion of *Dorcopsis* (10-42%, usually 10-15% by weight). *Spilocuscus* spp., *Dendrolagus* spp., *Zaglossus bruijnii* and snakes are otherwise co-dominant and there is a variety of other taxa represented including *Phalanger* spp. and several pseudocheirids. This difference in composition supports the suggestion made in Chapter 4 that the deposit in Square 1S0E may have been subject to a major scour event sometime prior to 10,000 BP, after which sediment and cultural materials accumulated rapidly and relatively undisturbed in this protected area below the overhanging cave wall. The generally good condition of the bone material also argues against major disturbance of this part of the deposit. This would explain the relatively low position of the radiocarbon date of 10,150 ± 65 BP from a depth of 115-120 cm in this square.

The zone between 80-105 cm, spanning the major secondary peak in bone abundance, is characterised by a marked increase in the abundance of *Dorcopsis* (55-89%) and a decline in the quantity of snake (below 10% in all but one level). Snakes increase once again above 80 cm and remain consistently co-dominant with *Dorcopsis* in all but a few levels through to the surface, with fluctuations probably due to small samples in some levels.

The initial decline and subsequent return in the relative abundance of snakes in the 1S0E sequence is mirrored to some extent in all squares and may also be related to the trend towards increased abundance of snakes through time in the Kria deposit.

The relatively small quantity of bone from the uppermost 60 cm of the deposit in Square 1S0E is dominated by the remains of *Dorcopsis* and snakes, with occasional fragments of cassowary, *Spilocuscus* spp., *Phalanger* spp. and fruitbat. A radiocarbon date of 2930 ± 65 BP was obtained from 35-40 cm in 1S0E. Interestingly, this falls within the potential time frame covered by Unit I of Kria Cave. However, the fauna from above this level in Square 1S0E in Toé is the same in all respects as that of the earlier part of the Holocene. Apart from the appearance of scarce remains of pottery, the upper levels of Toé bear no resemblance to that of Kria Cave Unit I.

General features of the Toé Cave vertebrate fauna

The Toé Cave vertebrate fauna shows a dramatic change in composition between the Late Pleistocene and the Holocene periods. Levels dated to before 12,000 BP contain a diverse and balanced fauna in which various mammals and reptiles are co-dominant. *Dorcopsis muelleri* is generally present in these levels but is a minor element only. Levels dated to around 10,000 BP and through to c. 8000 BP show a moderate increase in the abundance of *Dorcopsis*, but without losing all elements of the earlier fauna. More recent levels show a most complete dominance of *Dorcopsis*, although snakes usually also make a significant contribution.

Toé Cave produced remains of several taxa not encountered in Kria Cave. The majority of these were confined to the lower unit and thus date to the Late Pleistocene. One taxon of interest here is the Sooty Owl (*Tyto* sp. cf. *T. tenebricosa*) that is represented amongst the bird remains from Toé. This species probably does not roost in caves (Rand & Gilliard 1967). It is known to have been hunted by people in the New Guinea Highlands (Majnep & Bulmer 1977) and it may well have been consumed by the prehistoric inhabitants of the cave.

The vertebrate fauna from the Holocene levels in Toé Cave is similar in composition to that from the equivalent time period in Kria Cave. *Dorcopsis muelleri* is numerically dominant through the Holocene in both sites, with a secondary emphasis on snakes. Some minor differences are explored in the following section.

The Toé Cave deposit does not contain any horizon to compare with Unit I of Kria Cave. No pig remains were identified from the excavation in Toé Cave, although some pig fragments were collected from the surface. However, highly fragmented shaft bone or vertebrae of pig can be difficult to distinguish from those of cassowaries and it cannot be excluded that some fragments of pig may be present in the category 'cassowary'. The most recent specimens in this category came from levels potentially dating to 3000 BP.

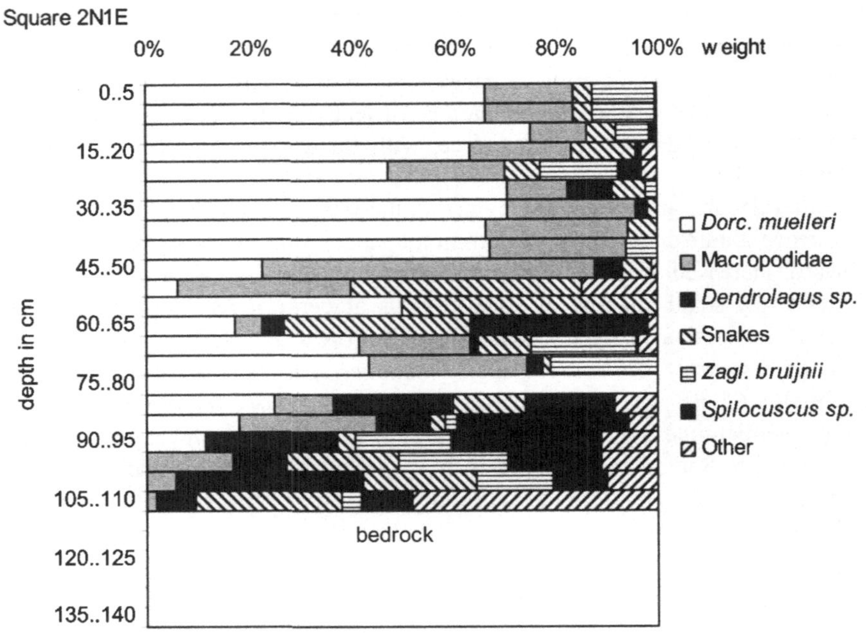

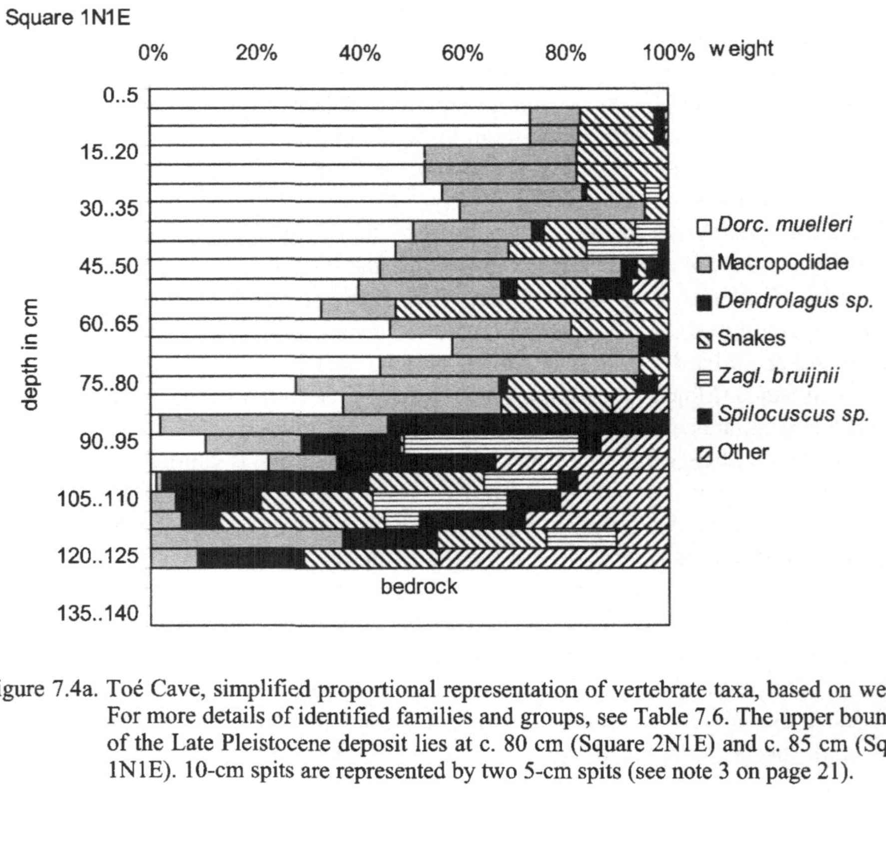

Figure 7.4a. Toé Cave, simplified proportional representation of vertebrate taxa, based on weight. For more details of identified families and groups, see Table 7.6. The upper boundary of the Late Pleistocene deposit lies at c. 80 cm (Square 2N1E) and c. 85 cm (Square 1N1E). 10-cm spits are represented by two 5-cm spits (see note 3 on page 21).

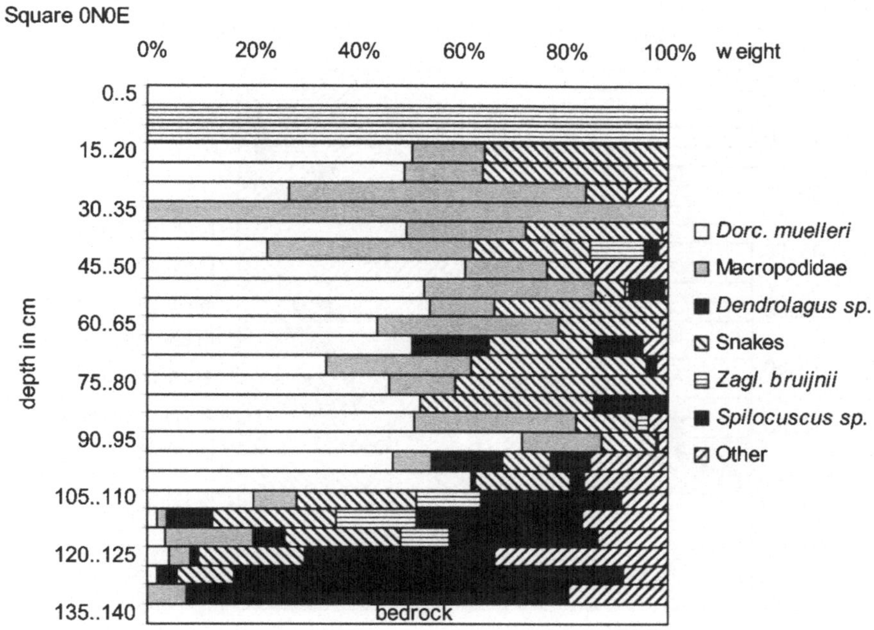

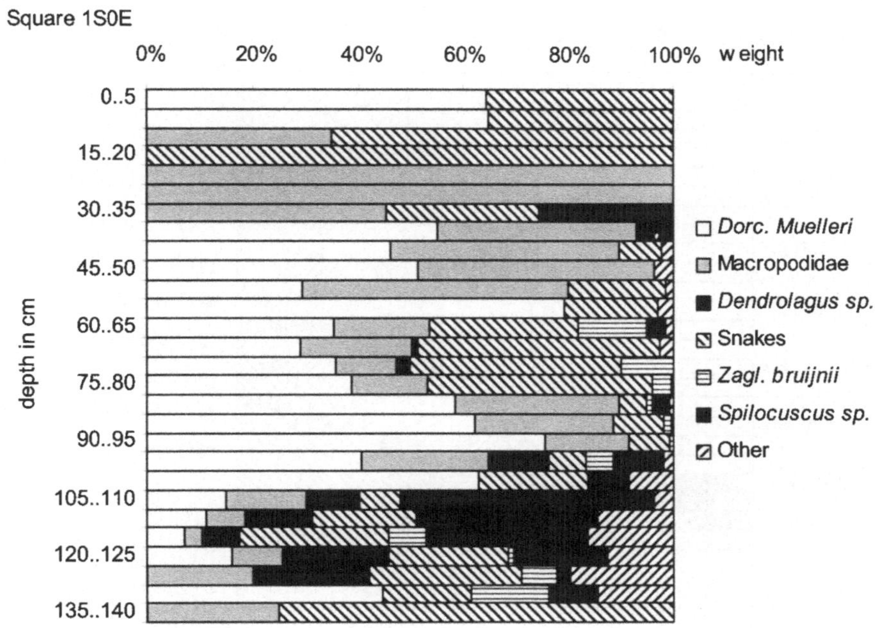

Figure 7.4a (continued). Toé Cave, proportional representation of vertebrate taxa, based on weight. The upper boundary of the Late Pleistocene deposit lies at c. 95 cm (Square 0N0E) and c. 105 cm (Square 1S0E). Bedrock in Square 1S0E at 140 cm.

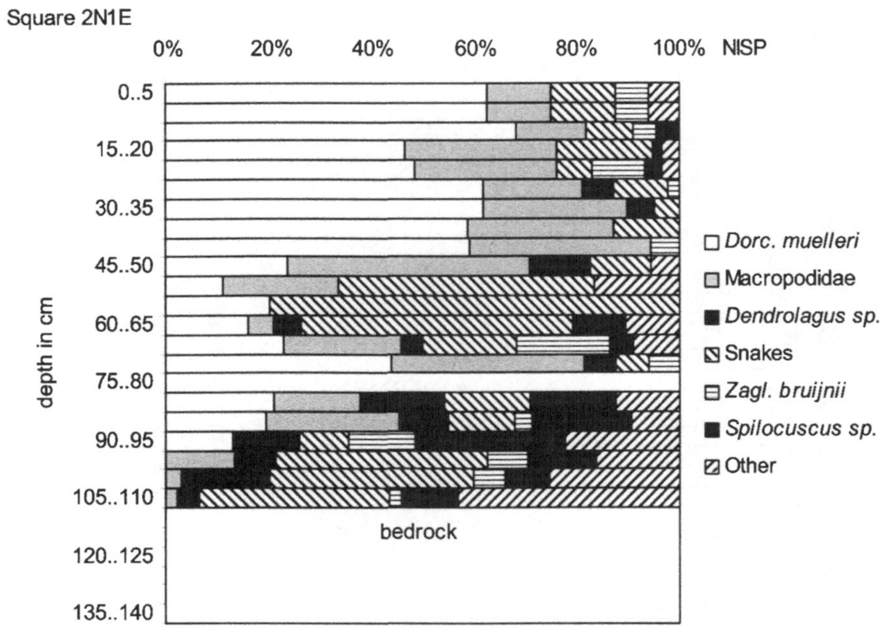

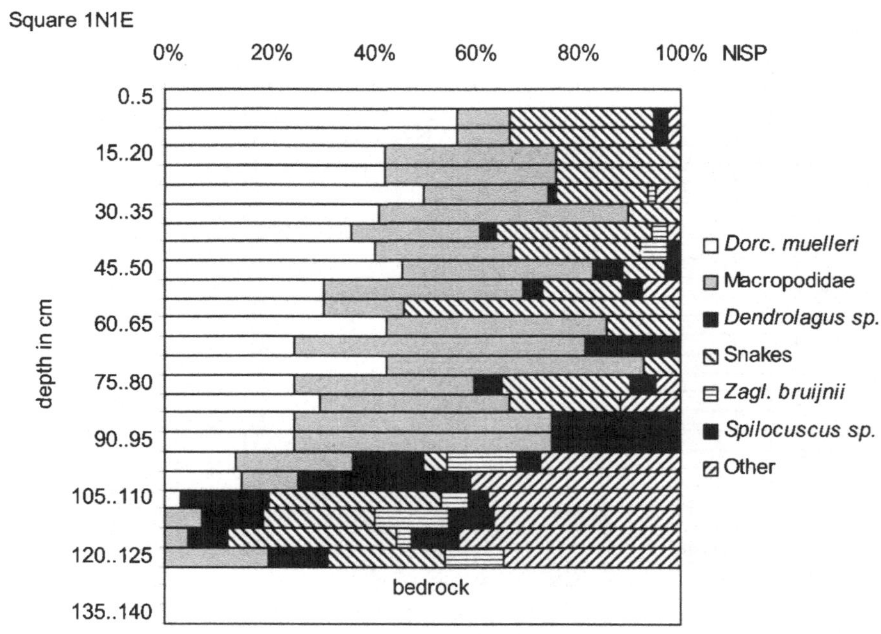

Figure 7.4b. Toé Cave, proportional representation of vertebrate taxa, based on NISP. For more details of identified families and groups, see Table 7.6. The upper boundary of the Late Pleistocene deposit lies at c. 80 cm (Square 2N1E) and c. 85 cm (Square 1N1E). 10-cm spits are represented by two 5-cm spits (see note 3 on page 21).

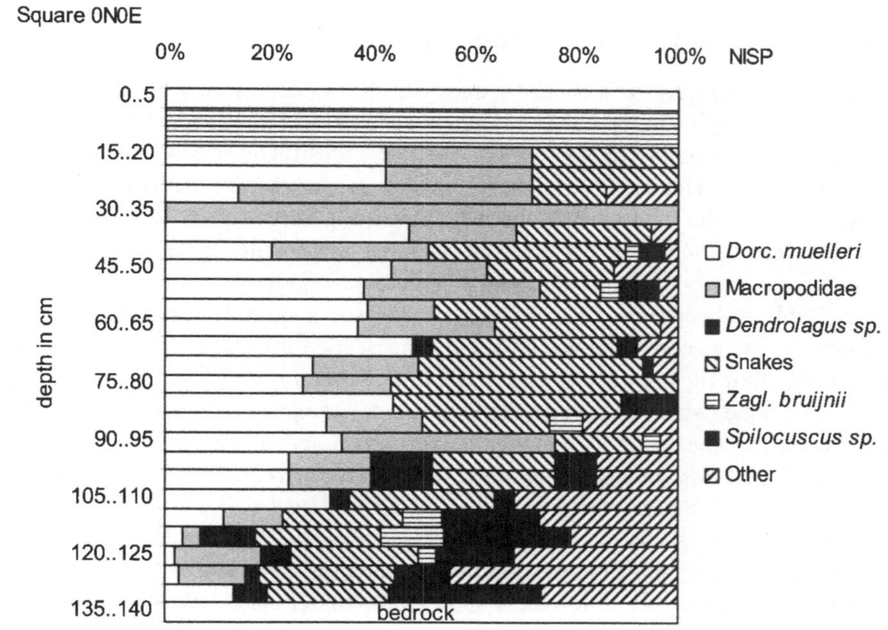

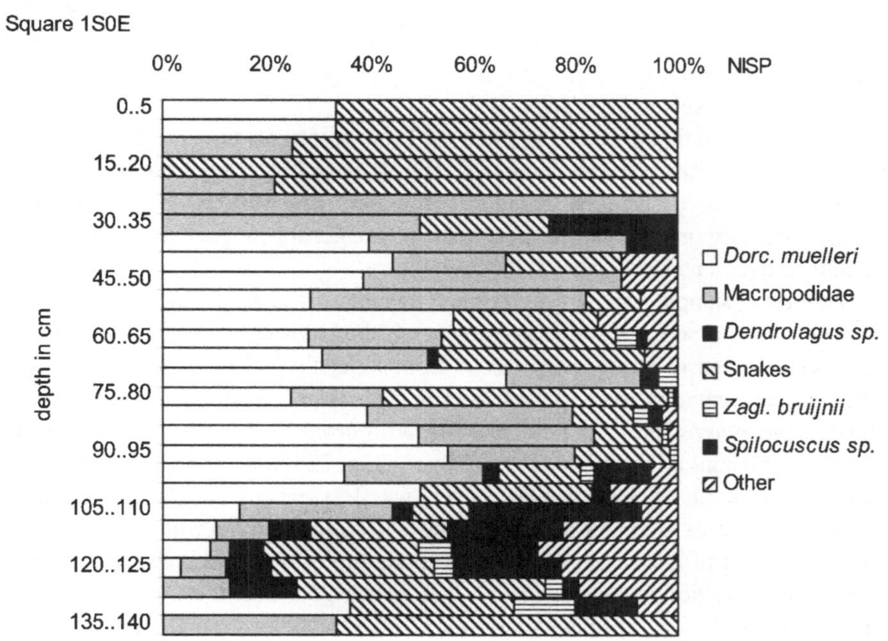

Figure 7.4b (continued). Toé Cave, proportional representation of vertebrate taxa, based on NISP. The upper boundary of the Late Pleistocene deposit lies at c. 95 cm (Square 0N0E) and c. 105 cm (Square 1S0E). Bedrock in Square 1S0E at 140 cm.

Comparison of vertebrate faunal assemblages across space and time

The complex stratigraphy of Toé Cave does not allow the faunal assemblage to be divided into stratigraphic units with the same precision as in Kria Cave. However, major faunal changes are certainly present within the Toé deposit, and it would be invalid to simply combine the remains from all levels within this cave. An alternative approach was taken in which the total assemblage was effectively 'sampled' by selecting narrow stratigraphic 'blocks' or time periods that satisfied three criteria: 1) they contain a relatively large quantity of bone (as shown in Figure 4.5); 2) they can be directly related to the available radiocarbon dates within the deposit; and/or 3) they show internal consistency in faunal composition. To satisfy the second criterion, the samples were drawn mainly from Squares 2N1E and 1S0E (see Table 7.6 and Fig. 7.5); spits with relatively little bone, which could bias the proportional composition, were excluded. The basal, Late Pleistocene levels of Squares 2N1E, 1N1E and 0N0E could be combined because the faunal assemblage from these levels is consistently different from the overlying material. The basal levels in Square 1S0E produced a slightly different faunal assemblage that is also accompanied by a slightly younger radiocarbon date.

The following stratigraphic samples are defined for Toé Cave: Square 2N1E: 15-40 cm and 60-80 cm; Square 1S0E: 0-40 cm, 40-80 cm, 80-105 cm and 105-140 cm; and one sample that combines Square 2N1E (80-110 cm), Square 1N1E (85-125 cm) and Square 0N0E (110-135 cm). For Kria Cave the analysis will focus on Units II and V; Units III and IV have been left out on account of the smaller samples. However, Unit I has been included, because of its radical different faunal composition and so that it can be compared with similarly dated levels in Toé Cave.

The faunal composition is shown in Figure 7.5 for the various time periods for each of the two sites. The contrast between the pre-12,000 BP assemblage and those from the mid- to late Holocene is particularly striking. However, the various intermediate-aged assemblages are also transitional in composition, suggesting a gradual change in the hunting returns over a period of several thousand years.

The Holocene samples show less marked temporal variation but there are some consistent differences between the sites. In general, the Holocene assemblages from Toé contain a wider range of species than those from Kria; they also show slightly higher proportional representation per spit of snakes and larger mammals including tree-kangaroos (*Dendrolagus* spp.) and Echidna (*Zaglossus bruijnii*). The mammals in question are relatively intolerant of disturbance (see Table 7.3, p. 229) and it seems likely that the area around Toé, south of the Ayamaru Lakes, was less heavily exploited through this period than was the area around Kria to the north. This suggestion fits with the over-all less intensive occupation of Toé Cave, and with the suggestion that it was not used as a venue for maintenance activities on composite tools that included bone artefacts.

The striking contrast in species composition between the post-3000 BP faunas from Kria and Toé Caves is further evidence of some functional differentiation between the two sites. At Toé similar activities were probably carried out as in previous times by the cave occupants, while at Kria there was a radical change in the nature of site use, with a virtual cessation of *Dorcopsis* consumption at the site and a change to other activities including human burial, and the consumption of shellfish, pig and cassowary.

Table 7.6. Taxonomic composition of the Toé Cave identified vertebrate assemblage, divided into 'time blocks' (compare Figure 7.5).

Taxon	whole deposit		Square 2N1E 15-40 cm		Square 2N1E 60-80 cm		Pleistocene levels		Square 1S0E 0-40 cm		Square 1S0E 40-80 cm		Square 1S0E 80-105 cm		Square 1S0E 105-140 cm	
	weight	NISP	weight	NISP	weight	NISP	weight	NISP	weight	NISP	weight	NISP	weight	NISP	weight	NISP
Tachyglossidae	7.10	3.86	2.43	1.81	10.72	7.66	11.17	5.56	0.00	0.00	4.34	0.97	1.69	1.65	4.23	4.38
Macropodidae	50.30	39.07	88.15	83.55	52.38	53.17	36.29	24.38	67.02	32.08	56.84	58.09	82.55	74.01	35.57	26.80
Phalangeridae	14.96	10.77	1.12	1.42	16.01	7.30	20.81	14.73	1.60	1.23	0.74	1.61	3.75	3.95	27.54	20.69
Pseudocheiridae	2.47	6.35	0.00	0.00	0.00	0.00	3.98	10.60	0.00	0.00	0.25	0.67	0.09	0.33	3.53	7.16
Petauridae	0.28	0.35	0.00	0.00	0.00	0.00	0.55	0.56	0.00	0.00	0.00	0.00	0.00	0.00	0.12	0.42
Peroryctidae	0.92	2.02	0.90	0.94	0.95	1.52	1.19	2.77	0.00	0.00	0.26	1.14	0.10	0.33	0.95	1.92
Dasyuridae	0.06	0.16	0.00	0.00	0.00	0.00	0.12	0.28	0.00	0.00	0.03	0.16	0.00	0.00	0.00	0.00
Suidae	0.00	0.00	0.00	0.00	0.00	0.00	0.00	0.00	0.00	0.00	0.00	0.00	0.00	0.00	0.00	0.00
Muridae	0.71	3.76	0.00	0.00	0.51	3.45	1.19	6.19	0.00	0.00	0.12	1.78	0.07	0.66	0.63	1.65
Pteropodidae	0.40	1.05	0.00	0.00	0.00	0.00	0.65	1.66	0.94	0.59	0.14	0.31	0.05	0.66	0.10	0.21
Hipposideridae	0.00	0.00	0.00	0.00	0.00	0.00	0.00	0.00	0.00	0.00	0.00	0.00	0.00	0.00	0.00	0.00
total mammal	77.20	67.38	92.60	87.71	80.56	73.09	75.94	66.74	69.57	33.90	62.72	64.74	88.30	81.58	72.67	63.24
snakes	18.87	28.66	6.94	11.84	19.44	26.91	19.16	27.96	30.43	66.10	34.41	32.90	8.87	15.79	21.88	31.69
Varanidae	1.13	0.74	0.00	0.00	0.00	0.00	1.89	1.13	0.00	0.00	0.06	0.16	0.00	0.00	1.44	1.23
other lizards	0.01	0.12	0.00	0.00	0.00	0.00	0.00	0.00	0.00	0.00	0.02	0.93	0.00	0.00	0.02	0.48
frogs	<0.01	0.01	0.00	0.00	0.00	0.00	0.00	0.00	0.00	0.00	0.00	0.00	0.00	0.00	0.00	0.00
total reptile/frog	20.00	29.54	6.94	11.84	19.44	26.91	21.05	29.09	30.43	66.10	34.49	34.00	8.87	15.79	23.34	33.39
Casuariidae	1.60	0.91	0.46	0.45	0.00	0.00	1.01	0.56	0.00	0.00	2.79	1.26	2.47	1.97	2.59	1.23
other birds	1.20	2.17	0.00	0.00	0.00	0.00	2.00	3.61	0.00	0.00	0.00	0.00	0.40	0.66	1.40	2.14
total bird	2.79	3.08	0.46	0.45	0.00	0.00	3.01	4.17	0.00	0.00	2.79	1.26	2.84	2.63	4.00	3.37

Note: Values are in percentages. The composition is summarised by proportional bone weight and by Number of Identified Specimens (NISP). The raw data were corrected for excavated area. The values for fish are unknown. 'Pleistocene levels' are Squares 2N1E (80-110 cm), 1N1E (85-125 cm) and 0N0E (110-135 cm) combined.

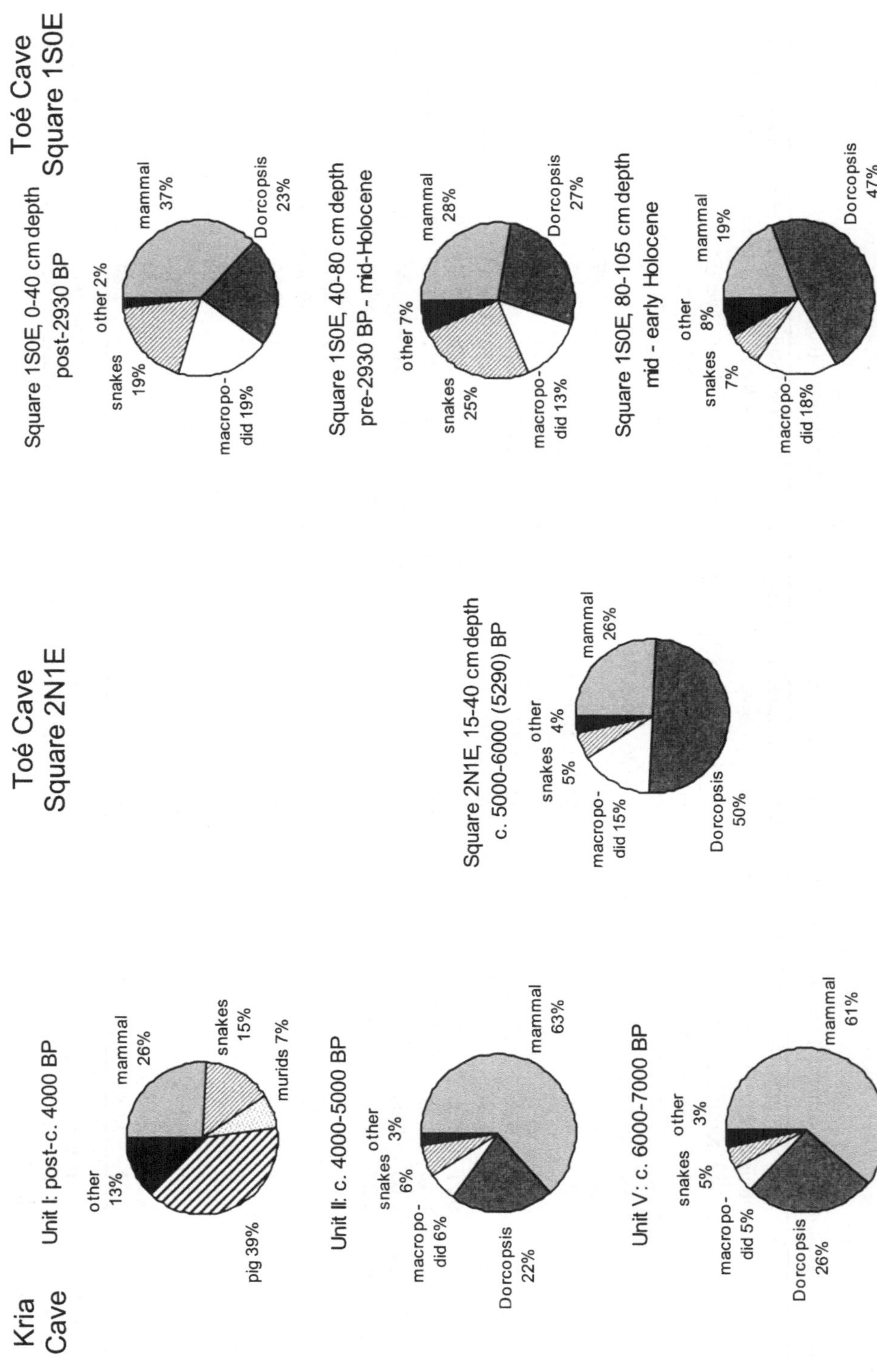
Kria Cave
Unit I: post-c. 4000 BP
other 13%
mammal 26%
snakes 15%
murids 7%
pig 39%
Unit II: c. 4000-5000 BP
macropo-did 6%
snakes 6%
other 3%
mammal 63%
Dorcopsis 22%
Unit V: c. 6000-7000 BP
macropo-did 5%
snakes 5%
other 3%
mammal 61%
Dorcopsis 26%
Toé Cave Square 2N1E
Square 2N1E, 15-40 cm depth
c. 5000-6000 (5290) BP
macropo-did 15%
snakes 5%
other 4%
mammal 26%
Dorcopsis 50%
Toé Cave Square 1S0E
Square 1S0E, 0-40 cm depth
post-2930 BP
snakes 19%
other 2%
mammal 37%
Dorcopsis 23%
macropo-did 19%
Square 1S0E, 40-80 cm depth
pre-2930 BP - mid-Holocene
other 7%
mammal 28%
snakes 25%
Dorcopsis 27%
macropo-did 13%
Square 1S0E, 80-105 cm depth
mid - early Holocene
other 8%
mammal 19%
snakes 7%
Dorcopsis 47%
macropo-did 18%

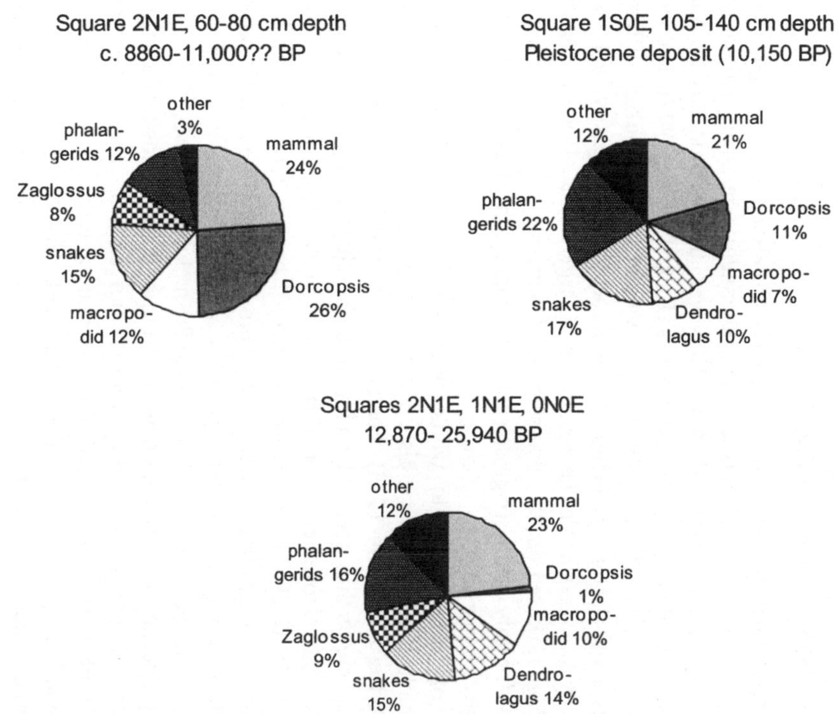

Figure 7.5. Proportional representation of the major vertebrate groups in both caves through time. The category 'other' contains all species represented by less than 5%. In the Late Pleistocene deposit from Squares 2N1E, 1N1E and 0N0E the category '*Dorcopsis*' (1%) has been presented separately for comparison.

7.4. PATTERNS OF PREHISTORIC FAUNAL EXPLOITATION

7.4.1. *Quantifying the pattern of faunal exploitation*

Determining the proportional contribution of each taxon based on the weight of its identified remains provides a straightforward and internally consistent estimate of relative abundance of major taxa through time. However, this method has attracted criticism as a means of quantifying faunal assemblages (Reitz & Wing 1999). The most important criticisms include the different bone to body weight ratios per taxon; different pre- and post-depositional factors (reduction or increase in bone density, e.g. with burning); the tendency to overestimate the abundance of large animals (see for example the pig remains in Fig. 7.3); and the fact that selective butchering can strongly alter the bone to body ratio for any taxon (e.g. selectively bringing the meaty portions of carcasses of some species to a site).

The proportion of bone weight to whole body weight varies greatly between vertebrate classes such as mammals vs. fish, but somewhat less so within one class (Prange et al. 1979; Reitz & Wing 1999). For an accurate estimate of dietary contribution, relative bone weights thus need to be adjusted before they can be used as a proxy for relative meat weights. However, once this correction has been made, it seems reasonable in most cases to take this step. For example, if rodents form 2% of a bone assemblage by weight and pigs 60% by weight, then provided the bone to meat ratio is comparable, the 1:30 ratio in bone weights is probably a fairly accurate reflection of the relative contribution of these two groups of animals to the prehistoric diet.

A considerable proportion of the bone material from both sites shows evidence of burning (see Chapters 3 and 4). This proportion is not constant across the major vertebrate taxa (see Table 7.7). In the case of bird/cassowary and fish the variations may simply reflect the small sample sizes. However, snake bone is clearly less often burnt (though somewhat more often calcined) than mammal bone. This presumably results in some bias in its relative contribution to the total assemblage as measured by bone weight, and may lead to an underestimation of its contribution in terms of total body weight. However, this would not be sufficient to alter the overall conclusion for Kria Cave that macropodid remains are dominant, with snakes forming a second major component. The differences in burning rates evident from Table 7.7 nevertheless do indicate the need for caution in the use of bone weight as a measure of taxonomic composition in archaeological assemblages where burning intensity may vary among major taxa.

As will be shown in the following chapter, *Dorcopsis muelleri*, the most abundant and also one of the largest prey species in the Kria Cave assemblage, shows little or no bias in the representation of major body parts (hindlimb, forelimb, head); it thus seems reasonable to conclude that both assemblages are unlikely to be heavily biased in this respect. Differential transport of body parts with unequal bone to meat weight ratios is more likely to be a problem where very large prey (e.g. large ungulates, elephants) are dismembered prior to transport.

Table 7.7. Kria Cave, relative proportions of burning classes for bone of each major group of vertebrates.

	unburnt	burnt	calcined
macropodid	79.36%	20.08%	0.56%
mammal other than macropodid	79.86%	19.00%	1.13%
snake	91.31%	4.49%	4.20%
bird/cassowary	92.06%	6.35%	1.59%
fish	62.16%	14.86%	23.00%

Wing & Brown (1979: Table 7.2) and Prange et al. (1979) provide allometric equations to calculate skeletal mass (column 2 in Table 7.8) from body mass (column 1) for various taxa, including mammal, bird, snakes and fish. Sources for average body weights for each taxon are given in the footnote to Table 7.8. For each of the vertebrate families identified in the Kria assemblage, the relevant allometric equation was used to estimate the proportion of skeletal weight to body weight (column 3). These values were then used to calculate the total body weight represented by the excavated bone weight of each taxon in the assemblage. No allometric equations were available for lizards and frogs, but these are very minor elements in both assemblages. Columns 4 and 5 of Table 7.8 show the proportional representation (as %) of each major taxon in the Kria Cave assemblage based on skeletal weight and estimated original body weights respectively. As expected, the potential dietary contribution (assuming that all parts are consumed) of larger animals weighing more than c. 5 kg is slightly underestimated from the excavated bone weights, while that of smaller animals is generally overestimated. However, the correction for different skeletal to body ratios does little to alter the overall conclusion that *Dorcopsis* and snakes made up the bulk of the animal foods consumed at Kria Cave.

Two other methods that are commonly used to quantify faunal assemblages involve calculating the Number of Identified Specimens (NISP) and the Minimum Number of Individuals (MNI). The level of detail recorded for the vertebrate fauna from Kria Cave allows each of NISP and MNI to be calculated for comparison with the results based on bone weight. For the Toé Cave assemblage only NISP and bone weight were recorded.

The NISP is widely used as a measure of relative abundance of taxa. However, this measure can be seriously affected by one or more of four factors (Klein & Cruz-Uribe 1984; Reitz & Wing 1999): 1) differential fragmentation; 2) differential preservation; 3) recovery techniques; and 4) different number of skeletal elements per taxon and different identifiability of these. In general, NISP values tend to underestimate the contribution of large animals to the diet.

Some skeletal parts are more robust and resistant to breakdown than others, even within a single taxon. Within a single animal class (e.g. Mammalia), however, decay rates for a particular skeletal element are more or less comparable, or may be related mainly to bone size. However, between classes (e.g. Mammalia and Teleostomi) decay rates may differ considerably; in particular unburnt fish bones tend to decay much more rapidly than unburnt mammal or reptile bones due to their lower density (Aplin 1980).

Table 7.8. Kria Cave, proportional representation of each taxon in the assemblage based on total bone weight, with and without corrections for differences in skeletal weight relative to body weight.

	average body weight (g) of one individual	calculated skeletal weight (g) of one individual	skeletal weight as percentage of body weight	proportional representation (%) of each taxon in the deposit based on total bone weight	proportional representation (%) of each taxon in the deposit corrected for differences in relative skeletal weight to body weight
Suidae	60,000	3147.96	5.25	1.45	1.77
Casuariidae	36,500	2334.79	6.40	2.07	2.07
Dendrolagus spp.	11,500	691.54	6.01	1.92	2.04
Tachyglossidae	7800	484.32	6.21	1.02	1.05
Dorcopsis muelleri	6000	380.71	6.35	72.24	72.79
Serpentes	5000	320.74	6.41	16.69	16.64
Spilocuscus spp.	4800	310.23	6.46	0.57	0.56
Phalanger spp.	2200	151.65	6.89	1.39	1.29
Dorcopsulus sp.	1800	126.15	7.01	<0.01	<0.01
Aves sp. indet.*	1500	118.56	7.90	0.36	0.29
Peroryctidae	750	56.50	7.53	0.61	0.52
Petauroidae	560	43.22	7.72	0.14	0.12
Dasyuridae	500	38.95	7.79	0.03	0.02
Pteropodidae	400	31.74	7.94	0.49	0.40
Muridae	200	16.81	8.40	0.53	0.41
Hipposideridae	40	3.84	9.60	0.03	0.02
Pisces sp. indet.	15	1.18	7.87	0.03	0.03

Note: Average body weight is taken from Flannery (1995a) unless stated otherwise. Skeletal weight is calculated from the allometric formulae given by Wing & Brown (1979: Table 7.2) and Prange et al. (1979), in which y = body weight, x = skeletal weight: mammals: $\log y = 1.09 (\log x) - 1.2147$; birds: $\log y = 1.071 (\log x) - 1.1871$; snakes: $\log y = 1.0107 (\log x) + 1.1660$; fish: $\log y = 0.8682 (\log x) + 1.1136$. The average body weight for fish is calculated from the equation given by Casteel (1974) and rounded up: live weight = $0.9138 (\text{vertebral width})^{2.4497}$, with an average vertebral width of 3 mm. The body weight for Suidae is estimated for a subadult pig, as found in Kria Cave. The body weight for *Dorcopsis muelleri* is taken slightly less than the average to account for any juveniles in the assemblage. Body weights for bird and *Casuarius* are rounded values from Prange et al. (1979). The body weight for snakes (Serpentes) is estimated. No equations were found for varanids, lizards and frogs, hence these taxa were excluded from the comparison. * includes predominantly megapodes.

'Identifiability' is also a critical factor in the interpretation of NISP values of each species. The bones of bandicoots, for example, are highly distinctive and even small fragments are readily distinguished from other mammal bones. Similarly, the vertebrae of snakes with their distinctive bulbar joint and complex articular processes, comprising the great bulk of their skeleton, are easily identified at least to family level, even when heavily fragmented, and can be expected to be overrepresented by NISP values. The relative abundances of the major vertebrate taxa, calculated by weight and by NISP, are shown in Tables 7.4b and 7.6 (see also Fig. 7.3 and 7.4a and b) for each of Kria and Toé. The relationship between these values is shown graphically in Figures 7.6 and 7.7.

Unit I in Kria is different from the underlying Units II to V and shows clearly that both pig and rodents are dominant in this unit, while this was not visible using only one or the other method. Units II-V show the clear dominance, both in weight and in NISP, of *Dorcopsis*, macropodid and snake, and show little change in this pattern over time. In Toé, the top 40 cm in Square 2N1E and the top 105 cm in Square 1S0E show a similar dominance of these three groups. The pattern changes gradually towards the bedrock in this cave, showing a decline in *Dorcopsis* and a rise in other species, especially phalangerids, *Dendrolagus*, *Zaglossus* and petauroids. This change is hinted at already from 60 cm downwards (dated to c. 9000 BP) in Square 2N1E and is consistent throughout the Late Pleistocene levels in all squares. The bottom 105 cm in Square 1S0E shows a similar pattern but with a relatively large proportion of *Dorcopsis*.

Estimates of faunal composition based on bone weights and NISP values are strongly correlated in Unit II to V in Kria, as well as in most of the upper part of the deposit in Toé. For these samples, bone weight and NISP thus provide valid estimates of the relative contribution of the various major taxa. However, the values for Unit I in Kria shows no such correlation; bone weight values emphasise the importance of larger taxa, while NISP values emphasises the smaller taxa. A similar lack of correlation between NISP and weight is visible in the first 40 cm of Square 1S0E and in the Pleistocene levels of Toé. For these samples, use of only one or the other method would give a potentially biased impression of faunal composition. The choice between the two methods must then be based on the research questions asked: bone weight is clearly the superior measure if it is the contribution of different groups of animals to diet that is of interest, while NISP is more informative if it is the taxonomic composition of the hunting returns. Since the subject of this chapter is the exploitation of food resources, I have continued to use the results based on the bone weights.

A third method is available that can provide estimates of both the relative abundance of individuals and their relative contribution to diet. This is the calculation of Minimum Number of Individuals, based on actual counts of various identified body parts, with the MNI values converted back into meat weights based on data from modern specimens. The method is generally applied as a means of comparison between taxa from one assemblage, but can also be employed between assemblages, provided the method is applied in a consistent manner between taxa and between sites. However, this issue of comparability between taxa and sites is rarely adhered to and there is no strict consensus as to exactly how MNI should be calculated (see for example the many references with criticism on MNI cited in Reitz & Wing 1999: 195).

One very serious problem with the calculation of MNI is that the values are strictly determined by the extent to which an assemblage is subdivided or aggregated (Grayson 1984). For example, in the present context, the MNI for *Dorcopsis* in the Kria Cave

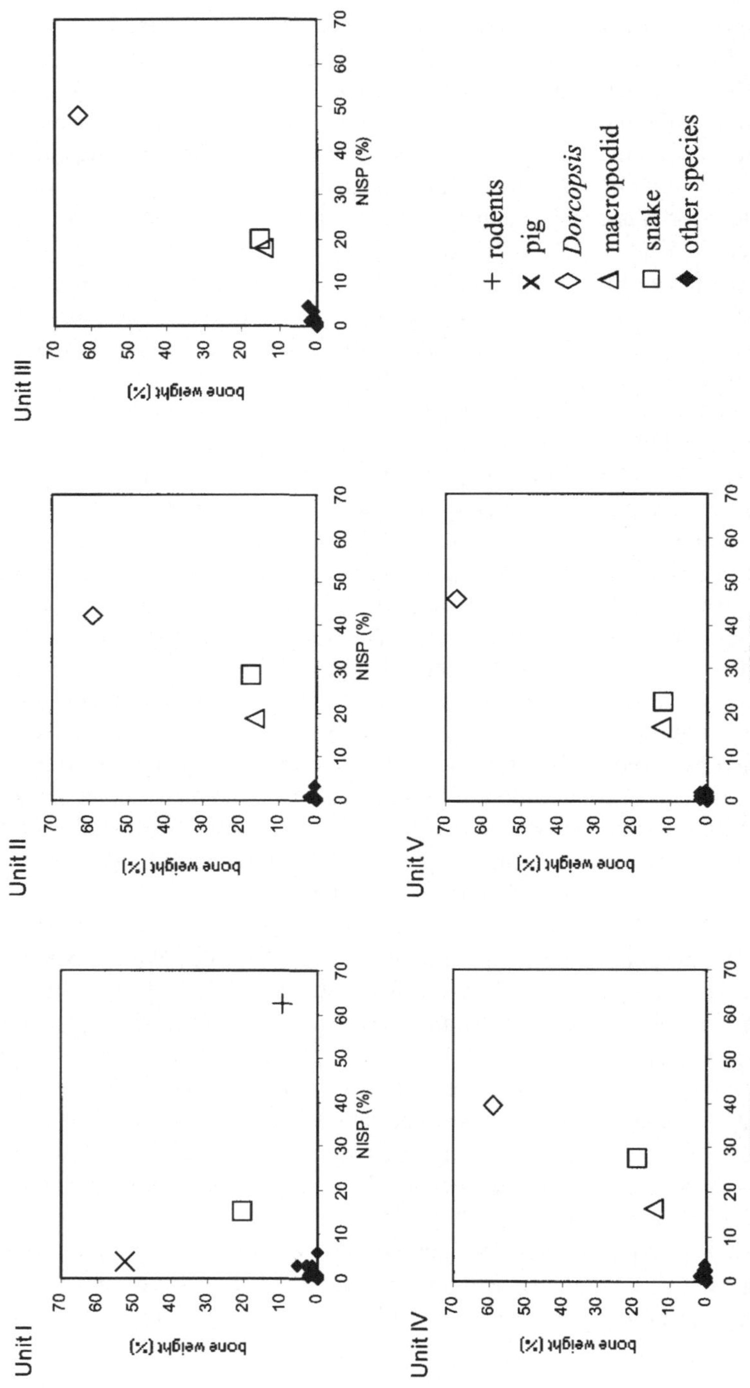

Figure 7.6. Kria Cave, proportional representation of various vertebrate groups for each of the stratigraphic units, as calculated from NISP values and bone weights.

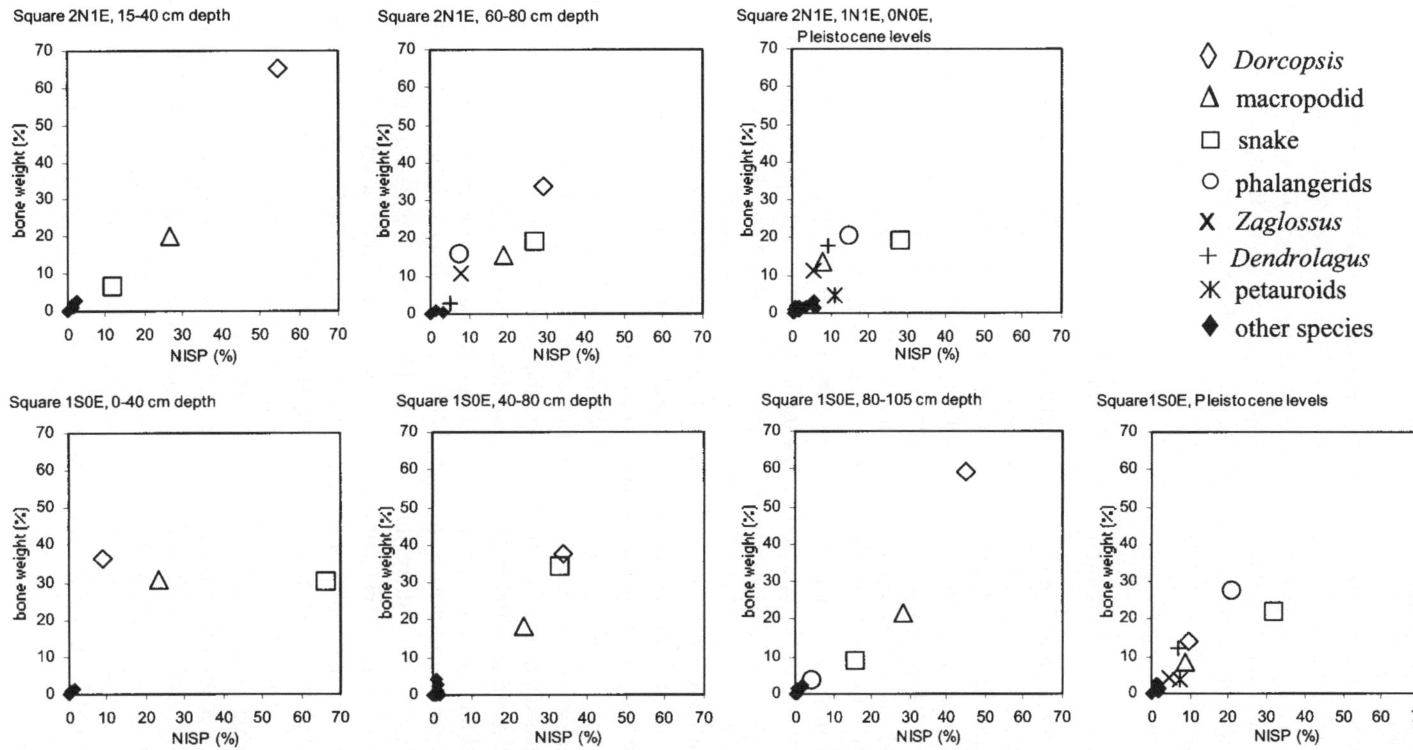

Figure 7.7. Toé Cave, proportional representation of various vertebrate groups for each of the stratigraphical 'time blocks', as calculated from NISP values and bone weights.

assemblage will vary enormously depending on whether each level is treated as an independent sample or whether the levels are grouped into Units or even pooled for the entire site (see Chapter 8).

Although MNI values were calculated by stratigraphic unit in Kria, the MNI for the majority of the species was 1 or a few at the most. Only for *Dorcopsis* were higher MNIs obtained; these values are not presented here as they provide little information beyond confirming the dominance of this wallaby in the Holocene assemblages. The MNI values for *Dorcopsis* will be discussed again in Chapter 8, where they will be used to provide insights into the intensity of site use.

7.4.2. *Interpreting the changing pattern of faunal exploitation*

The faunas from Kria and Toé Caves reveal that the pattern of faunal exploitation on the Ayamaru Plateau has changed radically through time. Assemblages dating from the Late Pleistocene period in Toé Cave contain a wide variety of vertebrate species, with no single taxon dominating the assemblage. This pattern changed through time, partly influenced by climatic changes (see Chapter 9), and by at least 9000 BP (7000 BP in Kria) the faunal assemblages in both caves are dominated by the remains of one species of forest wallaby, the Brown Dorcopsis (*Dorcopsis muelleri*).

Another change in faunal exploitation occurred during the late Holocene. The differences between the Kria and Toé assemblages during this period (post-3000 BP) might reflect specialisation within the landscape. In Kria Cave there was no longer emphasis on the consumption of wild game, but instead on an occasional meal of shellfish, and at least one pig was brought in. The cave was furthermore used for human burial. In Toé, no indications of such clear and dramatic changes in use were found. Perhaps the site was still used as a casual hunting camp.

Returning now to questions posed in the introduction to this chapter, just how important was *Dorcopsis* in the prehistoric diet, especially in comparison with other prey items such as snakes or smaller mammals like the cuscus? To answer this question it is necessary to determine whether the archaeological remains from the two sites can be taken as representative of the prehistoric subsistence pattern for each time period. The similarity between the two early to mid-Holocene assemblages indicates that they do represent a consumption pattern that was probably practised widely in the area. This seems especially true since this pattern is consistent through time in Units II to V in Kria Cave, as well as the upper part of the deposit in Toé, despite the fact that differences in stone and bone artefacts between the sites suggest that their functions were different.

But is this consumption pattern representative of the total diet? In particular, we might ask to what extent animals were consumed at locations other than rockshelters? There is little relevant information on this point, but it is interesting to note that Rhoads (1980) found differences in faunal composition between village and rockshelter sites during the ceramic period in the Kikori region in Papua New Guinea. Older pigs (12-24 months) were especially consumed in the villages, while the remains of pigs of all age classes were found in the rockshelters. He suggested that the people visiting the caves consumed feral pigs hunted from these sites, but that people in the village 'harvested' older pigs from domestic herds (Rhoads 1980: 230). While it may be unlikely that settled villages were present in the Ayamaru region during the early Holocene, it exemplifies the possibility that consumption patterns may vary between locations.

As argued earlier, the relative weight of bones of different taxa provides a good measure of the contribution of each animal group to the total sample of recovered remains. However, two further conditions must be met before an archaeologically recovered sample can be used to reconstruct human diet, namely 1) that all taxa caught by the prehistoric inhabitants were equally likely to be brought into the cave for consumption; and 2) that food preparation and consumption methods have no influence on the preservation and recovery of certain taxa.

Some evidence suggests that these conditions may not be met in the case of the Ayamaru assemblages. Ethnographic information on the Meybrat people from the Ayamaru region suggests that protein was mainly derived from small animals: 'Snakes, grubs, caterpillars, rats, fledglings, lizards – everything is caught with enthusiasm and put on the fire.' (Elmberg 1955: 62). Game such as pigs, possums and kangaroos appeared to be mainly hunted during the dryer season (October to December) and was considered especially important for ceremonial exchange during feasts (Elmberg 1968: 16). Why then, do both prehistoric faunal assemblages from the Ayamaru region show such a strikingly different pattern through the Holocene period from that documented by Elmberg?

The low proportion of small vertebrates in both assemblages may be the product of several factors: 1) especially in hot and humid tropical areas weathering, decay and breakage are more pronounced for smaller bones than for larger fragments, due to their lower density and higher surface area to mass ratio; 2) small bones are more easily destroyed or carried away by scavengers such as rats and marsupial carnivores; 3) small animals receive generally little preparation or dismemberment prior to cooking, and are often consumed whole by humans, including most of the bones (Bulmer 1976: 180); 4) small animals are often obtained casually and consumed at the time and place of capture rather than being carried back to a habitation site; and 5) in Kria Cave, some of the shaft bones of small animals may have been removed from the site for the manufacturing of bone points (see Chapter 6).

All of these factors ensure that small vertebrates generally leave less archaeological trace of consumption than larger game. On top of these factors, as has already been mentioned, are the necessarily crude recovery methods used in the field which also had some effect on the sample size of especially the smallest vertebrates. But while the proportion of small game in the diet may thus have been considerably higher than the Holocene remains from both sites might suggest, the contrast between this material and that from the Late Pleistocene levels points to a genuine shift in emphasis from broad-based hunting to the more specialised procurement of *Dorcopsis muelleri*.

How large was the relative contribution of plant food and invertebrates? Living on a diet of predominantly forest wallabies would cause a serious deficiency in fat and carbohydrates. Fat may have been obtained from eggs, the remains of which were clearly present at both sites, or from other animal sources, but carbohydrates would have had to originate from vegetables and fruits. Scarce remains of fruits were found in both Kria and Toé, but it must be assumed that far more vegetable matter was consumed to provide for the necessary carbohydrates.

As suggested by Dwyer (1986, 1988), for people living in rainforest, the collection of daily food is governed by the balance between obtaining essential nutritional elements and the time available to collect it. The hunting of large animals provides for large amounts of protein, and, depending on the animal, fat, but is time-consuming. Carbohydrates need to come from a different (plant-based) source, and, since, in this example, most time was spent on the hunting expedition, this source must be readily available and easy to harvest. Conversely, if much time was spent on the procurement of carbohydrates, for example, from sago, then the source of protein must come from a restricted area and be easily caught or collected. In many rainforest populations, such 'procurement systems' (Dwyer 1986: 348) either involve less time-consuming trapping or fishing, or form the basis for a clear division of labour.

In the 1950s, when Elmberg made his recordings about the Meybrat in the Ayamaru region, the staple food was taro, with occasional yams, sweet potato, sugarcane, banana and corn (Elmberg 1955: 56). Sago is a very recent introduction to the local diet and is mostly obtained through trade with coastal people (ibid: 58), although some sago palms were present when we visited the area in 1995. Although information from ethnographies cannot be directly extrapolated into the past, it seems reasonable to speculate that the prehistoric Ayamaru inhabitants complemented their wallaby-based diet with the consumption of root crops, rather than through the labour-intensive procurement of sago. This suggestion finds support in the absence of any polished adzes (sago-pounders) such as were identified by Rhoads (1980) amongst the material excavated from the Kikori area in Papua New Guinea.

Molluscs formed only a small part of the diet, although they were brought into Kria Cave from at least the early Holocene onwards. The possible increased rate of exploitation of gastropods during Unit I times coincides broadly in time with the development of the westernmost lake basin, but it is not known when the other basins were formed. The relatively high numbers of gastropods found in this unit of Kria may in fact be the result of only one or a few large meals. It is interesting that some mollusc species from a brackish or mangrove habitat were found in both caves throughout the Holocene deposit, albeit in very small numbers. One specimen from Toé Cave may date up to c. 13,000 BP. These brackish and mangrove-inhabiting species clearly were not brought into the sites for their food value but rather as valuable objects. Whether this occurred through a process of direct visits to the coast or through some kind of trade system linking inland with coastal peoples is unknown.

7.4.3. *Is there evidence for selective exploitation of* Dorcopsis*?*

Selectivity is difficult to demonstrate in a prehistoric faunal assemblage because it assumes that the structure of the natural animal community through time is known. The most direct evidence would come from comparison of an archaeological assemblage with a non-human assemblage of faunal remains that has accumulated under relatively unbiased circumstances, for example, in a pitfall cave where animals have fallen in more or less at random. Such assemblages are not available for the Ayamaru region (or anywhere else in New Guinea).

General biological principles may also provide some information about the natural structure of animal communities in lowland rainforests. It is generally believed that tropical rainforests are rich in vertebrate species. Fleming (1973: 556-557) estimated that such forests usually accommodate twice as many species on average as a forest in a temperate region. However, much of the species richness in these environments is contributed by small-bodied vertebrates (Stahl 1995: 167) and by arboreal species, the latter reflecting the fact that a large part of the available energy is concentrated in the canopy. Arboreal life places strong constraints on body size and the few large-bodied arboreal mammals, such as South American sloths, are generally solitary, relatively inactive and hence notably inconspicuous. The largest arboreal mammals in New Guinean lowland rainforests are cuscuses of the genus *Spilocuscus*, and tree-kangaroos (*Dendrolagus* spp.); the latter are partly terrestrial.

The largest mammal species in most rainforest communities are usually terrestrial. However, the fact that the understorey stratum is relatively poor in resources compared with the canopy generally constrains the population density of these species, and necessitates a high mobility (Stahl 1995: 168). Because of their relatively slow biological turnover, large mammals are also particularly susceptible to overpredation. For this reason, the larger terrestrial rainforest mammals are almost always cryptic and solitary, using camouflage rather than flight as a means of predator avoidance.

Another general observation on rainforest animal communities is that mammal densities can vary enormously, even within a single biogeographic region. Glanz (1990) compared mammal densities at a number of neotropical sites in Central and South America and found variations ranging from 10-200 animals per km^2 for small mammals (e.g. opossums of c. 250 g), and less than 1 to 633 animals per km^2 for larger mammals such as the sloth (c. 4-8 kg). This degree of variation is observed in all mammalian groups and across all major vegetation strata, but with the exception that the larger terrestrial mammals are uniformly scarce (Glanz 1990: tables 16.5-16.8). Human hunting and poaching may partly account for this variation, but it probably also reflects underlying environmental variation in soil type and climatic variability, size of the habitat, fluctuations in food availability, and the abundance of predators.

Given these general constraints, is it possible that *Dorcopsis muelleri*, a purely terrestrial mammal, really was the most abundant mammal in the Ayamaru rainforests? Perhaps so. The forests of the Ayamaru Plateau are relatively low in both stature and species diversity, presumably reflecting the poor nutritional status of the limestone soils. The canopy is therefore unlikely to support high densities of arboreal mammals, whereas the understorey is probably less impoverished compared to rainforests with a more closed forest cover. *Dorcopsis muelleri*, as a solitary, cryptic and moderately mobile mammal, has all of the key attributes to perform well under such circumstances. Under natural conditions and in the absence of any effective predation, it may well have made up a significant proportion of the total mammalian biomass. What this 'significant proportion' might have been is impossible to say. However, even under the most extreme conditions it is unlikely that *Dorcopsis* would have contributed 80% or more to the mammalian biomass, as the archaeological evidence for the Holocene would suggest.

The discussion of 'selectivity' thus far has focussed on the inference of bias in hunting returns in relation to natural animal communities. In this sense, the Ayamaru hunting returns were undoubtedly a selective 'catch', although perhaps only in degree, if *Dorcopsis* did in fact make up a 'significant proportion' of the local mammalian biomass. But were the hunters actually targeting the *Dorcopsis* over other species? Or was the method of hunting such that *Dorcopsis* was simply more likely to be intercepted than other species because of its terrestrial habits? Was *Dorcopsis* actually preferred over other animal foods to the point that other potential prey would be ignored in pursuit of *Dorcopsis*? Or was the selectivity due to technological factors that made *Dorcopsis* particularly vulnerable to whatever capture methods were being used (e.g. snares or nets)?

Any suggestion that the Ayamaru hunters were 'selective' to the extent that they would ignore other prey is at odds with the notion that the Ayamaru Plateau environment was a relatively harsh and unproductive environment through much of the Holocene. Moreover, a suggestion of selectivity based on cultural preference would also need to account for the consistent strong bias towards *Dorcopsis* in the Holocene archaeological record, not only within the Kria Cave sequence, but also between Kria and Toé Caves.

The very strong bias in the archaeological record towards *Dorcopsis* almost certainly implies some selectivity. However, this selectivity may well have been an unintentional outcome of the intersection between the biology of *Dorcopsis* and the kind of hunting methods used through the Holocene period. The fact that the emphasis on *Dorcopsis* remained unchanged over many millennia perhaps favours a non-cultural explanation, but the evidence does not rule out an additional cultural preference: after all, it was one of the largest mammals in the region, it was moderately abundant, at least in relation to other prey, and as a terrestrial species, it was readily accessible and relatively easy to catch. One final question on selectivity remains: how did the Ayamaru hunters manage to hunt this animal for thousands of years without causing its demise? This question will be taken up in detail in the following chapter.

7.5. DISCUSSION AND CONCLUSION

The faunal remains from Kria and Toé Cave together provide a long record of the changing pattern of animal exploitation on the Ayamaru Plateau over 26,000 years.

The Late Pleistocene record from Toé Cave spans the Last Glacial Maximum and the subsequent climatic amelioration. During this period, visitors to the area captured a wide range of animal prey including a range of small to medium-sized taxa such as small forest wallabies (*Dorcopsulus* spp.), cuscuses (*Phalanger* spp., *Spilocuscus* spp.) smaller possums (*Pseudochirops* spp.; *Dactylopsila* spp.), giant rats (*Mallomys* sp.), pythons and monitors, and megapodes (all weighing up to c. 5 kg), and occasional larger species such as tree-kangaroos (*Dendrolagus* spp.) and Echidnas (*Zaglossus bruijnii*), with body weights between 5 and 10 kg or more. Some of these species persist in remote areas of the Bird's Head today, but others are no longer found in this region. The Brown Dorcopsis, *Dorcopsis muelleri*, was caught less often during Late Pleistocene times than subsequently. It may have been less prolific in the local area, perhaps due to a combination of physiological constraints under different climatic conditions (see Chapter 9) and increased competition within a more diverse mammalian fauna.

The transition to the early Holocene saw a decline in the archaeological representation of many animal species, and a corresponding rise in the abundance of *Dorcopsis muelleri*. This change is recorded in both sites, although only the final stages of the transition are represented in the basal unit of Kria. The only taxon that does not register significant change in relative abundance across this transition is Serpentes (snakes).

In Kria the faunal assemblage remained stable from before 7000 BP through until c. 4000 BP. Some time after that, the faunal composition registers significant change, with associated differences in the mode of use of the cave. Freshwater gastropods were consumed in larger quantities, fish seemed to have been caught more frequently, and the Brown Dorcopsis was no longer caught, or was no longer brought to the cave for consumption. A pig was brought in some time before 1840 BP; this introduced species was apparently absent from the area before 4000 BP. The remains of smaller animals, especially rodents, became more abundant in the cave, although some of these were probably brought in by roosting owls and other predators. Human remains of an adult and a 12-year-old child were also deposited in the cave (see Appendix 1). It is uncertain whether these people were buried in the cave, but if they were, then their grave was shallow and their remains were scavenged. There is no evidence that they represent human food remains.

In Toé Cave the preferred consumption of *Dorcopsis* continued until at least 3000 BP, and probably later, and no change in the use of the cave appears to have taken place. Pig remains were recovered from the surface, but the amount of shellfish did not increase in the upper levels of the deposit, and no distinct human burials were found. The apparent differences between the two caves in the final periods of their occupation may be related to a difference in the nature of the human activities and environment between the north and south side of the lake basins – with perhaps more settled activity and some gardening on the north side, and continued hunting in forest habitats on the south side.

It needs to be stressed again that this dietary reconstruction and consumption pattern is almost certainly not representative of the total daily diet of the Ayamaru people. Paoletti et al. (2000) point out that invertebrates such as snails, insects, worms, grubs and larvae, form a significant food resource amongst Amerindian groups living in the Amazon basin, and not necessarily so only when game and fish are scarce. In fact, many invertebrates are valued for their flavour. Because of their abundance and ready availability they have a much higher 'success-rate' in their catch than vertebrates. As noted earlier, Elmberg (1955) also commented on the importance of invertebrates in the diet of Meybrat people and this may have been the case for Ayamaru people in the past. Thus, the actual proportion of vertebrate meat on the daily menu is likely to have been much lower than the faunal remains might suggest. The proportion of the daily protein and carbohydrates that has been obtained from sources that leave no archaeological trace whatsoever may have been considerable.

Table 7.3. List of vertebrates recorded from the combined Kria and Toé faunal assemblages, after Pasveer & Aplin (1998). See Chapter 9 for more information on species distribution through time.

scientific name	common name	present in Kria Cave	present in Toé Cave	modern Bird's Head fauna	body weight in g (males and females)	elevational range in metres	habitus	habitat and comments	activity period
Tachyglossidae									
1 : *Zaglossus bruijnii*	Long-beaked Echidna	x	x	x	5900-9800	sea-level/600 - 4150	terrestrial	disturbance intolerant	diurnal
Dasyuridae									
2 : *Myoictis melas*	Three-striped Dasyure	x	x	x	±200	sea-level - 1450	terrestrial	mainly lowland forests; disturbance tolerant	diurnal
3 : *Dasyurus albopunctatus*	New Guinea Quoll	x	x	x	580-710	sea-level - 3500	terrestrial	mainly montane	diurnal
Peroryctidae									
4 : *Echymipera rufescens*	Long-nosed Echymipera	x	x	x	no info	sea-level - 1200	terrestrial	mainly lowland forests	nocturnal
5 : *Echymipera kalubu*	Common Echymipera	x	x	x	450-1500	sea-level - 2000	terrestrial	disturbance tolerant	nocturnal
6 : *Microperoryctes longicauda*	Striped Bandicoot	x?	x	x	350-670	1000 - 3950	terrestrial	mainly montane forests; disturbance intolerant	nocturnal?
7 : *Echymipera* sp. indet.	Echymiperas	x			--		terrestrial		
Macropodidae									
8 : *Dorcopsis muelleri*	Brown Dorcopsis	x	x	x	5000-±10,000	sea-level - unknown	terrestrial	low altitudes; disturbance intolerant	crepuscular
9 : *Dendrolagus inustus*	Grizzled Tree-kangaroo	x	x	x	10,600-17,000	100 - 1400	arboreal, sleeps on ground	disturbance intolerant	crepuscular
10 : *Dendrolagus* sp. cf. *D. goodfellowi*	Goodfellow's Tree-kangaroo group	x	x	-	6700-9100	680 - 2865	arboreal, sleeps on ground	> 680 m; disturbance intolerant	crepuscular
11 : *Dendrolagus* sp. indet.	tree-kangaroos	x	x		--		arboreal, sleeps on ground		crepuscular

Table 7.3 (continued). List of vertebrates recorded from the combined Kria and Toé faunal assemblages, after Pasveer & Aplin (1998).

scientific name	common name	present in Kria Cave	present in Toé Cave	modern Bird's Head fauna	body weight in g (males and females)	elevational range in metres	habitus	habitat and comments	activity period
12 : *Dorcopsis*/ *Dendrolagus*	large wallabies or tree-kangaroos	x	x		--		--		
13 : *Dorcopsulus* spp. (incl. *D. vanheurni*)	Small Dorcopsis	x?	x	-	1500-2340	800 - 3100	terrestrial	montane forests only; disturbance intolerant	crepuscular
Phalangeridae									
14 : *Spilocuscus rufoniger*	Black-spotted Cuscus	x	x	-	5500-6600	sea-level - 1200	arboreal	< 1200 m; disturbance intolerant	nocturnal
15 : *Spilocuscus maculatus*	Common Spotted Cuscus	x	x	x	2300-6000	sea-level - 1200	arboreal	< 1200 m; disturbance tolerant	nocturnal
16 : *Spilocuscus* sp. indet.	Spotted Cuscuses	x	x		--		arboreal		nocturnal
17 : *Phalanger vestitus*	Stein's Cuscus	x	x	x	1400-2400	1200-2200	arboreal		nocturnal
18 : *Phalanger orientalis*	Northern Common Cuscus	x	x	x	1600-3500	sea-level - 1500	arboreal	mainly lowland forest; disturbance tolerant	nocturnal
19 : *Phalanger gymnotis*	Ground Cuscus	x	x	x	1500-4850	sea-level - 2700	arb/terr, sleeps in burrow or in cave	500 - 1500 m	nocturnal
20 : *Phalanger* sp. cf. *P. intercastellanus*	Southern Common Cuscus		x	-	1400-2200	sea-level - 1250	arboreal	mainly lowland forest	nocturnal
21 : *Phalanger* sp. indet.	cuscuses	x	x		--		arboreal		nocturnal
Pseudocheiridae									
22 : *Pseudochirulus schlegelii*	Arfak Ringtail	x	x	x	244-305	750 - 1900	arboreal	montane forests only; disturbance intolerant	nocturnal
23 : *Pseudochirulus* sp. cf. *P. canescens*	Lowland Ringtail	x		x	235-380	sea-level - 1300	arboreal	mainly lowland forests	nocturnal
24 : *Petauroides ayamaruensis*	Ayamaru fossil ringtail	x	x	-	no info				

Table 7.3 (continued). List of vertebrates recorded from the combined Kria and Toé faunal assemblages, after Pasveer & Aplin (1998).

scientific name	common name	present in Kria Cave	present in Toé Cave	modern Bird's Head fauna	body weight in g (males and females)	elevational range in metres	habitus	habitat and comments	activity period
25 : *Pseudochirops albertisii*	D'Albertis's Ringtail		x	x	640-875	1000 - 1900	arboreal	montane forests only	nocturnal
26 : *Pseudochirops coronatus*	Reclusive Ringtail		x	x	±1500	1000 - 2250	arboreal	montane, mainly > 1800m; disturbance intolerant	nocturnal
27 : *Pseudochirops* sp. indet.	Ringtail Possums		x		--		arboreal		nocturnal
Petauridae									
28 : *Dactylopsila trivirgata*	Striped Possum	x	x	x	280-470	sea-level - 2300	arboreal	mainly < 1100 m; disturbance tolerant	nocturnal
29 : *Dactylopsila kambuayai*	Ayamaru fossil Striped Possum	x		-	no info		arboreal		
30 : *Dactylopsila palpator*	Long-fingered Triok		x	x	260-550	1200 - 2950	arboreal	mainly > 1500 m	nocturnal
31 : *Petaurus breviceps*	Sugar Glider	x	x	x	69-114	sea-level - 3000	arboreal	disturbance tolerant	nocturnal
Muridae									
32 : *Mallomys* sp. cf. *M. aroaensis*	De Vis's Woolly-Rat		x	x	1410-2000	1100 - 2700	arb/terr, sleeps in burrow	montane forests only	nocturnal
33 : *Xenuromys barbatus*	Rock-dwelling Giant Rat		x	-	900-1100	75 - 1200	terrestrial	always rare?	nocturnal
34 : *Uromys caudimaculatus*	Mottled-tailed Giant Rat	x	x	-	400-695	sea-level - 1925	arboreal	disturbance tolerant	nocturnal
35 : *Uromys*/*Xenuromys*	Giant Rats		x						nocturnal
36 : *Parahydromys asper*	Waterside Rat	x		x	490-590	700 - 2200	terrestrial		???
37 : *Hydromys chrysogaster*	Common Water-rat	x		-	210-500	sea-level - 1900	terrestrial	mainly lowland forests	

Table 7.3 (continued). List of vertebrates recorded from the combined Kria and Toé faunal assemblages, after Pasveer & Aplin (1998).

scientific name	common name	present in Kria Cave	present in Toé Cave	modern Bird's Head fauna	body weight in g (males and females)	elevational range in metres	habitus	habitat and comments	activity period
38 : *Pogonomelomys bruijnii*	Large Pogonomelomys	x	x	x	no info	60	arboreal	lowland forests only	
39 : *Pogonomelomys* sp. A	Ayamaru fossil *Pogonomelomys*	x	x	-	no info				
40 : *Pogonomys* sp. cf. *P. loriae*	Large Tree-mouse	x		x	53-128	240 - 3000	arboreal, sleeps in burrow	disturbance tolerant; mainly < 1000 m	nocturnal
41 : *Paramelomys* sp.	small Paramelomys	x		?	no info	900 - 3000			
42 : *Paramelomys platyops*	Lowland Paramelomys	x		x	65-102	sea-level - 900	terrestrial	lowland forests only; disturbance tolerant	nocturnal
43 : *Rattus praetor*	Large Spiny Rat	x	x	x	164-228	sea-level - 1900	terrestrial	mainly lowlands; very disturbance tolerant	nocturnal
44 : *Rattus* sp. cf. *R. verecundus*	Slender Rat	x		x	88-133	150 - 2750	terrestrial	mainly montane forests	nocturnal
45 : Muridae gen. & sp. indet.	small rats or mice	x	x		--				nocturnal
Pteropodidae									
46 : *Dobsonia* sp. cf. *D. magna*	Great Bare-backed Fruit-bat	x	x	-	360-440	sea-level - 2700	roosts in caves or trees	mainly higher elevations; cave roosters; colonial	nocturnal
Hipposideridae									
47 : *Hipposideros diadema*	Diadem Horseshoe-bat	x		-	33.5-43	50 - 1210	roosts in caves	cave roosters; colonial	nocturnal
Suidae									
48 : *Sus scrofa**	pig	x	x	x	60,000*		terrestrial		diurnal

Table 7.3 (continued). List of vertebrates recorded from the combined Kria and Toé faunal assemblages, after Pasveer & Aplin (1998).

scientific name	common name	present in Kria Cave	present in Toé Cave	modern Bird's Head fauna	body weight in g (males and females)	elevational range in metres	habitus	habitat and comments	activity period
Birds									
49 : *Casuarius bennettii*	Dwarf Cassowary	x	x	?	no info	400 - 2000?	terrestrial	mainly montane forests	diurnal
50 : *Casuarius* sp. indet.	cassowary	x	x		± 36,600		terrestrial		diurnal
51 : *Aepypodius arfakianus*	Wattled Brush Turkey	x		x	± 1500	750 - 2700	terrestrial	mainly montane forests; disturbance intolerant	diurnal
52 : *Tyto* sp. cf. *T. tenebricosa*	Sooty Owl		x	?	± 500	sea-level - 3500?		mainly montane forests	nocturnal
53 : Aves gen. & sp. indet.	small birds	x	x						diurnal
Reptiles and frogs									
54 : Serpentes, incl. Boidae gen. & sp. indet.	snakes (incl. pythons)	x	x		± 5000	sea-level - 2000	arb/terr.	mainly lowland forests	diurnal
55 : *Varanus* sp. indet.	monitors	x	x		± 3000	sea-level - 2000	arb/terr.	mainly lowland forests	diurnal
56 : Agamidae gen. & sp. indet.	Dragon lizards	x	x		± 200	sea-level - 1500	terrestrial		diurnal
57 : Anura gen. & sp. indet.	frogs	x	x		± 30		arb/terr.		nocturnal
Fishes									
58: Pisces gen. & sp. indet.	small fishes	x	x		± 15		aquatic	lakes, rivers and streams	

Note: Common names, weight and elevational ranges are taken from Flannery (1995a) for mammals, and Coates (1985) and Prange et al. (1979) for birds; the details on mammals are modified in some cases to accommodate additional observations by Aplin. The modern status of each species on the Bird's Head is also indicated. x = present; - = absent; x? = possibly present; ? = unknown status. *Pogonomelomys* sp. A is an unknown and possibly new species which has not yet been formally described. *New Guinean pigs are here regarded as belonging to the species *Sus scrofa*; Groves (1981) has suggested that New Guinean pigs are of hybrid origin but this requires genetic confirmation. The estimated weight of *Sus scrofa* is taken for a juvenile animal, as was found in the top levels of Kria.

8. Prehistoric Exploitation of the Brown Dorcopsis

8.1. INTRODUCTION

In the last chapter it was shown that the basic pattern of exploitation of animal resources in the Ayamaru area was remarkably stable throughout the greater part of the Holocene, at least insofar as this can be judged archaeologically. The central activity through this period was the hunting and consumption of the Brown Dorcopsis, *Dorcopsis muelleri* (locally known as *Djief*), a medium-sized mammal but one of the largest terrestrial prey items available to the prehistoric occupants of the caves.

The absence of any obvious change in animal exploitation patterns from the early Holocene to within the last few thousand years might not be regarded as a surprise given the relative climatic stability of the lowland New Guinean environment over this period (see Chapter 9). However, such a view ignores the rapidly growing ethnographic and archaeological evidence that even human groups organised into small-scale societies are capable of bringing significant changes to their local environment, including the depletion of key prey resources and eventual species loss (Smith & Wishnie 2000; Grayson 2001). Indeed, as Grayson (2001: 16) remarks 'virtually wherever archaeologists have sought such effects, they have found them'. In this context, the apparent stability of animal exploitation through the Holocene in the Ayamaru region is an important finding, and one that warrants closer examination.

8.2. OPTIMAL FORAGING THEORY AND ITS IMPLICATIONS

Over the last three decades ethnographers and archaeologists have sought to understand patterns of use of plant and animal resources by hunter-gatherer societies in terms of 'optimal foraging theory', as borrowed from general ecology (Pyke et al. 1977; Krebs 1978; Martin 1983). The optimal foraging theory is founded on the assumption that animals, including people, make choices about which resources to pursue on the basis that the choice will maximise individual or group fitness and minimise costs. In the case of human activities, this can be thought of in terms of costs (e.g. effort expended in travel, pursuit, extraction; use of raw materials) and benefits (kilojoules or some other dietary value; other materials obtained; social benefit from distribution of catch).

Among ethnographically documented hunting societies, the general rule appears to be that the highest return prey items tend to be among the larger animals within a local vertebrate community (Alvard 1995; Fa et al. 1995; FitzGibbon et al. 1995; Bodmer et al. 1997; Peres 2000). High return prey items are those that return greater amounts of kilojoules per unit of energy expended in their pursuit, capture and processing. If encounter rates with the preferred targets drop because they have become locally scarce, the hunter is faced with a number of choices. Hunting effort may shift, either temporarily or permanently, to lower return resources. Alternatively, the hunter may choose to travel further in search of the preferred prey item. A third choice is to move the base camp to a new area where the balance of prey items is consistent with the original hunting preferences. The latter is of course a societal rather than an individual decision, and options for both travel and relocation may be limited by population pressure in surrounding areas.

Grayson (2001) recently reviewed ethnographic and archaeological examples of resource depression – the situation in which overhunting leads to reduced capture rates. He concluded that resource depression is an almost inevitable outcome of exploitation of a preferred resource. In fact, he postulates only two circumstances under which the outcome might be different: where human population densities are not high enough to have any impact; and where available technology is not effective against a particular prey item.

As was shown in Chapter 7, remains of the Brown Dorcopsis (henceforth called *Dorcopsis muelleri*, or simply *Dorcopsis*) are abundant throughout the Kria Cave deposit at least until the end of Unit II, c. 4000 years ago, when the occupation appears to have changed quite abruptly in character. In Toé Cave, *Dorcopsis* becomes more abundant in the assemblage at the end of the Pleistocene. It reaches dominance by the early Holocene and remains abundant from then until the last few thousand years. In Chapter 7 it was argued that the abundance of *Dorcopsis* remains in both sites through the Holocene most likely reflects three factors, namely 1) its relatively large body size, making it an attractive prey item; 2) its likely natural abundance as a relatively mobile terrestrial herbivore; and 3) its relative ease of capture as reported by contemporary Ayamaru inhabitants.

Sustained predation evokes the risk of overhunting and pushing a species to become rare if not extinct in a local area. So how did the Ayamaru people manage to hunt *Dorcopsis* for some 8000 years or more without depleting the local populations? More specifically, given that *Dorcopsis* is today quite rare in the vicinity of the Ayamaru Lakes (see Appendix 3), does the archaeological record show any indication of overexploitation of this animal? And if so, can we identify the nature of the human response to any decline in the *Dorcopsis* population?

In the remainder of this chapter various aspects of the *Dorcopsis* remains are examined. First, the intensity of *Dorcopsis* exploitation through time is investigated, based on estimates of numbers of animals at the site (or absolute culling rates). Second, *Dorcopsis* population structure through time is examined, based on evidence from dental eruption and wear. This may provide clues as to the particular hunting strategies employed and indicate any change in population structure through time, a potentially good indication of overexploitation. And third, the mode of exploitation of the *Dorcopsis* carcasses through time is examined. In general, if food is abundant, bones may be left largely intact, and the consumers may even be selective in which parts of the carcass they preferred to eat. Conversely, if food is scarce, people may be more careful to extract as much value from the bones and carcass as possible; the leftover skeletal remains will be highly fragmented (Binford 1981; Gould 1996). Long-distance transport of animal carcasses might also be detected from analysis of body part representation, with selective transport of only the most meat-rich body parts of large animals being worth the effort to carry (O'Connell et al. 1990; Bartram 1993).

8.3. ECOLOGY OF *DORCOPSIS MUELLERI*

Dorcopsis muelleri is recorded from widely scattered localities in the lowlands of western New Guinea and on some of the larger surrounding islands [Misool, Salawati, Japen (Groves & Flannery 1989)]. It is a medium-sized terrestrial mammal, with a maximum body length of some 60-70 cm and a tail length of c. 50 cm; males grow larger than females, with maximum body weights of c. 10-12 kg[1] and 5-6 kg respectively (Flannery 1995a: 148, 1995b: 80). Both sexes are brown to reddish brown in colour, with a white to creamy underside. When resting and moving, the tail is typically held in an arc, only the tip touching the ground; this may be an adaptation to prevent the tail becoming infested with leeches (Archer & Flannery 1985: 39).

Very little is known about the ecology and behaviour of *Dorcopsis muelleri*, but some information is available for other members of the genus. All species of *Dorcopsis* inhabit deep forest habitat (Tate 1948: 319; Van Deusen 1972: 714) and are thus difficult to observe in the wild. They are shy, secretive animals, and are probably most active at dawn and dusk. Their masticatory anatomy suggests a nonabrasive, low-fibre diet (Sanson 1980) and in captivity they are known to prefer soft vegetable foods including leaves, flowers and fruits (Bourke 1989). They have a slow, distinctive gait involving extensive use of the relatively large forearms, but can move briskly when startled. Although nothing is known about their social behaviour in the wild, in captivity, the Grey Dorcopsis (*Dorcopsis luctuosa*), a species of similar size found in similar habitats in the eastern half of New Guinea, tends to aggregate in small groups (Bourke 1989; Menzies 1991).

The most detailed breeding information pertains to *Dorcopsis luctuosa* (Menzies 1989). In this species females become reproductive at 15 months, males a little later. Breeding appears to be aseasonal and, like most kangaroos, a single young is produced. Pouch life of the young is c. 6 months. *Dorcopsis muelleri* is also reported to produce a single young (Collins 1973), although information gathered by Mr Elimas Kambuaya from inhabitants of the Ayamaru region suggests the occurrence of twins (see Appendix 3). Flannery (1995a: 148) reports that a captive-born young of *Dorcopsis muelleri* first ventured out of the pouch after 24 weeks, had vacated the pouch permanently after 28 weeks, but continued suckling at least through into its ninth month. Flannery (1995b: 80)

[1] As far as I know there are no records of the body weight of a male *Dorcopsis muelleri*; however, body weights of other species of *Dorcopsis* suggest that males weigh generally about twice as much as females (Flannery 1995a,b).

also notes that the species is 'common in riverine habitats in northern Salawati, with seven individuals collected in a single afternoon by a group of hunters with dogs'. Females captured on the island of Salawati in October 1992 carried pouch young weighing between 300-1900 g, suggestive of aseasonal breeding (Flannery 1995b: 80). A captive *Dorcopsis muelleri* survived for almost four years (Collins 1973). However, captive *Dorcopsis luctuosa* is found to live for at least 8 years (Menzies 1991), and other similar-sized wallaby species can survive up to 10-12 years in captivity (e.g. *Macropus eugenii* 11-14 years; Smith & Hinds 1995). Few wild animals would be likely to survive as long due to the influence of diseases and general competition for resources with members of the same and other species.

Dorcopsis muelleri today is hunted mostly with dogs, although they can also be captured by hand (Flannery 1995b: 80). Mr Kambuaya (see Appendix 3) reported that nowadays they are most easily caught by torchlight; this is shone into their eyes, which appears to mesmerise them, allowing them to be shot or approached and clubbed. Elmberg (1955: 61) does not specifically mention the capture of wallabies. He does report that the Meybrat people set traps or simply catch prey with their hands. Traps were usually set for smaller animals such as rats. He further noted that the Meybrat displayed no great skill in the use of bow and arrow, which seemed to have been a recent introduction. The blowpipe was sometimes used to catch lizards or bats. On occasion spears were used in hunting larger animals (he specifically mentions pigs), in which process dogs were of great assistance.

Prior to the arrival of humans and dogs, *Dorcopsis muelleri* did not have any large terrestrial predators. The New Guinean Harpy Eagle (*Harpyopsis novaeguineae*), and sometimes pythons or quolls, are the only species that have ever posed a threat to the young (Menzies 1991). A Harpy Eagle is capable of carrying off piglets, young dogs and young wallabies, and is said to have carried off small children (Majnep & Bulmer 1977: 139). A serious threat to *Dorcopsis* is the destruction of its habitat by humans and pigs.

8.4. INTENSITY OF EXPLOITATION OF *DORCOPSIS MUELLERI*

Although *Dorcopsis muelleri* was brought in more often than any other prey item at both Kria and Toé Caves during the Holocene period, this fact alone does not justify the conclusion that the species was subject to intensive exploitation that would potentially lead to extinction. To support any such claim, it first needs to be investigated whether the archaeological record can reveal anything about actual consumption rates of *Dorcopsis* over time.

In this section, two methods are used to investigate the intensity of exploitation of *Dorcopsis*. The first is the familiar method of estimating the Minimum Numbers of Individuals (MNI) represented in the faunal assemblages based on counts of bones and teeth. The second is a less frequently used but nonetheless interesting method based on the absolute weight of *Dorcopsis* bone in the deposits.

8.4.1. *Estimating the Minimum Number of Individuals based on body parts*

The usual procedure for deriving an MNI value is to count various identified body parts for a particular species within an assemblage and then calculate the smallest number of individuals of that species that would be needed to account for the observed remains (e.g. Klein & Cruz-Uribe 1984; Reitz & Wing 1999). However, the actual process can be made more or less elaborate by taking (or not taking) into account the age and/or sex of the remains, and by attempting to match left and right elements from individuals (Chaplin 1971: 69-75). The method is often applied as a means of standardising the estimates of relative abundance of taxa within and between assemblages, but it is in particular situations, such as burials, also used as a means of estimating the actual number of individuals contained within an assemblage (Grayson 1984).

The MNI method is heavily criticised in the literature (see for the most important criticisms Chapter 7). Where the aim is to estimate actual numbers of individuals, the most critical issue is the extent to which an assemblage is to be subdivided or aggregated (Grayson 1984). For example, in the present context, the MNI for *Dorcopsis* in the Kria Cave assemblage will vary enormously depending on whether each level is treated as an independent sample or whether the levels are grouped into units or even pooled for the entire site. In addition, there are issues related to the horizontal sampling of the site. If the excavated area represents 20% of the area of the site, what proportion of the total number of *Dorcopsis* in the site have been sampled by the excavation? We cannot assume that all individuals have been detected, nor can we justify a simple fivefold multiplication of the MNI values.

As will be demonstrated here, the calculation of MNI values does have some interesting applications. For the present analysis, MNI values were calculated for Kria *Dorcopsis* on the basis of all specifically identified body parts (i.e. dental, cranial and all limb elements). Individual age was also taken into account. For post-cranial bones this was simply scored as adult or sub-adult, based on epiphyseal fusion. For dental remains, three age classes based on tooth eruption stages were defined as follows:

- juvenile: deciduous teeth retained;
- subadult: deciduous teeth lost but last molar not fully erupted;
- adult: all teeth fully erupted.

Less complete dental specimens often lack information on eruption stage; these were allocated to either juvenile/subadult or subadult/adult categories based on tooth wear (see below for further details). Symmetry was always taken into consideration for dental remains. However, because this was not recorded for other body parts, the count per bone for each age class was divided by 2 and rounded up to the nearest number.

MNI estimates were made for the entire Kria Cave assemblage and for each of the five stratigraphic units defined in Chapter 3. The results for Kria are shown in Table 8.1. MNI values were not calculated for Toé Cave because the complex and compacted stratigraphy made it more difficult to aggregate samples in a meaningful way.

The *Dorcopsis* MNI estimates based on body parts (MNI_{body}) and teeth (MNI_{teeth}) are remarkably close for all units ($R^2 = 0.9968$) and there is no consistent pattern of deviation of the estimates (i.e. neither body parts nor teeth give consistently higher

Table 8.1. Kria Cave, counts of fragments of identified body parts for *Dorcopsis muelleri* and estimates of Minimum Number of Individuals (MNI).

body part	Unit I	Unit II	Unit III	Unit IV	Unit V	all units
cranium	4	283	69	59	459	874
humerus		43	20	16	99	178
ulna		27	12	9	67	115
radius		10	2	5	17	34
scapula		12	6	4	23	45
pelvis		11	11	8	54	84
femur		51	49	12	111	223
tibia		53	25	23	100	201
fibula		56	35	24	112	227
4th metatarsal		34	20	8	38	100
5th metatarsal	1	59	23	10	77	170
astragalus	1	15	1	1	8	26
calcaneum		29	10	10	20	69
cuboid		10	1		16	27
claws		36	16	16	20	88
phalanges		19	6	5	8	38
MNI based on all body parts	1 sa 1 ad	13 sa 18 ad	6 sa 6 ad	4 sa 4 ad	22 sa 20 ad	43 sa 41 ad
total	2	31	12	8	42	84
MNI based on teeth	1 juv	4 juv 2 juv/sa 2 sa 5 sa/ad 15 ad 6 ?	2 juv 1 sa 1 sa/ad 5 ad 2 ?	1 juv 1 sa 1 sa/ad 6 ad 2 ?	10 juv 2 juv/sa 2 sa 5 sa/ad 19 ad 3 ?	15 juv 2 juv/sa 2 sa 9 sa/ad 39 ad 5 ?
total	1	34	11	11	41	72
MNI based on body parts and teeth	1 ad 1 juv	18 ad 13 juv/sa 3 ?	6 ad 6 juv/sa	6 ad 4 juv/sa 1 ?	20 ad 22 juv/sa	41 ad 43 juv/sa
total	2	34	12	11	42	84

Note: Codes for age classes: juv = juvenile; sa = subadult; ad = adult; ? = age not determined but specimens represent additional individuals. The counts for some body parts is more than twice the calculated MNI because the table does not present information on bone portion (e.g. distal vs. proximal femur) that was taken into account when calculating MNIs. 'Cranium' includes all fragments, including isolated teeth.

estimates). This result gives some faith that the methods are producing consistent and repeatable MNI estimates for these samples. For the five units combined, the total MNIs for body parts is 95, and for teeth is 98; the total MNI for body parts and teeth combined is 101 individuals.

The MNI estimates based on the entire assemblage are 10-20% lower than the combined value for each of the individual units (Table 8.1). These values also show a larger but unexplained discrepancy between the MNI calculated from all body parts (84 individuals) and that based on teeth (72 individuals).

The bones of individual *Dorcopsis* brought into the cave were most likely spread widely across its floor area. Many individuals might thus be represented in both excavated squares, and indeed in many of the unexcavated squares as well. However, given that only 12% of the floor area in Kria Cave was excavated, it is probably fair to assume that the MNI would increase with the excavation of additional squares. But how great would this increased MNI be? A maximum value can be obtained simply by scaling up the current MNI to the larger floor area. This would give a maximum MNI for the site of 842 individual *Dorcopsis* if calculated per unit, and 700 if calculated for the entire deposit. However, these values clearly represent extreme maximum estimates.

8.4.2. *Estimating the Minimum Number of Indivduals based on bone weight*

The minimum number of individuals of a particular species can also be estimated from the absolute weight of its remains within an assemblage. Given the approximate skeletal weight of one individual of the taxon, the method is simply to ask how many individuals would be needed to account for the recovered quantity of bone. Anderson (1983: 42-43) applied this method to the heavy fragmented moa bone material at the Hawksburn site in New Zealand and showed that a traditional MNI based on body part counts would have grossly underestimated the actual number of moas processed at this site. Correction might also be needed to account for differences in bone density between modern and fossil bones. This method is applied here to derive an independent estimate of the numbers of *Dorcopsis* represented in each of Kria and Toé Caves.

A total of 3090 g of *Dorcopsis* bone was identified in the Kria Cave assemblage (Table 8.2). A further 674 g was identified as 'macropodid' because it could not be ruled out that it might belong to a species of tree-kangaroo (*Dendrolagus*). However, *Dendrolagus* appears to be rare in the deposit and the bulk of the 'macropodid' material is almost certainly referable to *Dorcopsis*. The large quantity of 'unidentified mammal' bone (which in Table 8.2 includes the category 'vertebrae and ribs') also consists

Table 8.2. Kria Cave, distribution of *Dorcopsis* and possible *Dorcopsis* bone in total weight (g), summarised by unit and for the entire assemblage.

taxon/category	Unit I	Unit II	Unit III	Unit IV	Unit V	all units
1. *Dorcopsis muelleri*	3.3	808.7	304.7	251.0	1722.5	3090.2
2. unid. macropodids	2.9	223.3	66.9	62.0	319.2	674.3
3. min. *Dorcopsis* (1 + 2)	6.2	1032.0	371.6	313.0	2041.7	3764.5
4. unidentified mammal	41.0	2372.0	711.5	575.2	4039.2	7738.9
5. max. possible *Dorcopsis* (1 + 2 + 4)	47.2	3404.0	1083.1	888.2	6080.9	11,503.4

predominantly of fragments of larger bones, the majority of which presumably belong to *Dorcopsis*. The quantity of *Dorcopsis* bone in the assemblage thus lies somewhere between 3090 g and 11,503 g, with the true value probably being closer to the latter figure than the former.

No complete reference specimens of *Dorcopsis muelleri* were available from which to determine actual skeletal weight, nor was the archaeological material sufficiently intact to estimate the weight of one complete skeleton from the excavated sample itself. However there are several options for estimating this value, based either on general regression equations for mammals as a whole or for other members of the family Macropodidae. As noted earlier, adult male and female *Dorcopsis* weigh around 10 kg and 5 kg respectively (Flannery 1995a, b). The Kria Cave assemblage appears to contain roughly equal numbers of male and female animals (see section 8.5.3 below) and a mixture of juveniles, subadults and adults, hence it seems reasonable to infer an average body weight for the prehistoric *Dorcopsis* of around 6 kg.

Based on the general allometric relationship between body weight and skeletal weight in mammals [Skeletal Mass = $0.061(\text{Body Mass})^{1.09}$; Prange et al. 1979], the skeleton weight of a 6 kg mammal is estimated at 381g. Although this relationship is non-linear for mammals as a whole (Prange et al. 1979), it appears to be essentially linear for macropodids (see Fig. 8.1), as expressed by the regression Skeletal Mass = 0.0628 (Body Mass) – 3.35 (R^2= 0.9884). For the smaller species, such as the Brush-tailed Bettong (*Bettongia penicillata*) the skeleton makes up c. 7% of the body weight. For larger wallabies and kangaroos, such as an Agile Wallaby (*Macropus agilis*) this proportion is reduced to around 4%. For a 6 kg macropodid a skeletal mass of 374 g is predicted, remarkably close to the estimate based on the wider mammalian relationship.

As noted earlier, it may be necessary to make an adjustment to account for any difference in bone density between the prehistoric and modern bones. Prehistoric bone might be less dense if it has undergone any loss of organic or mineral content. It might be more dense if it has been mineralised or severely burnt. Variation in bone density through each of Kria and Toé Caves were described in Chapters 3 and 4. Although some fluctuations are present, the values are remarkably consistent at around 2 g/cc^3, suggesting little if any change in bone density through diagenetic change in either site.

Binford & Bertram (1977) have shown that measuring bone density by submerging it into a measuring jug filled with water can lead to considerable error, on top of the variation between species, age, body part and size of the animal. The values that these authors give for modern bones of deer and sheep of varying ages range from 0.58 to 1.76 (Binford & Bertram 1977: Table 3.9). These figures are well below the average value of 2 g/cc^3 as established for the archaeological bone samples from Kria and Toé Caves. While this might indicate an increase in density, perhaps as a result of fossilisation, the values of Binford & Bertram are also clearly affected by the inclusion of air pockets within the bones, which in their analysis were undamaged and/or sealed off by paraffin to exclude the possibility that spongy bone would absorb more water than dense cortical bone (Binford & Bertram 1977: 107-108). This would have increased the volume relative to the weight of the bone, which in turn would decrease the density values.

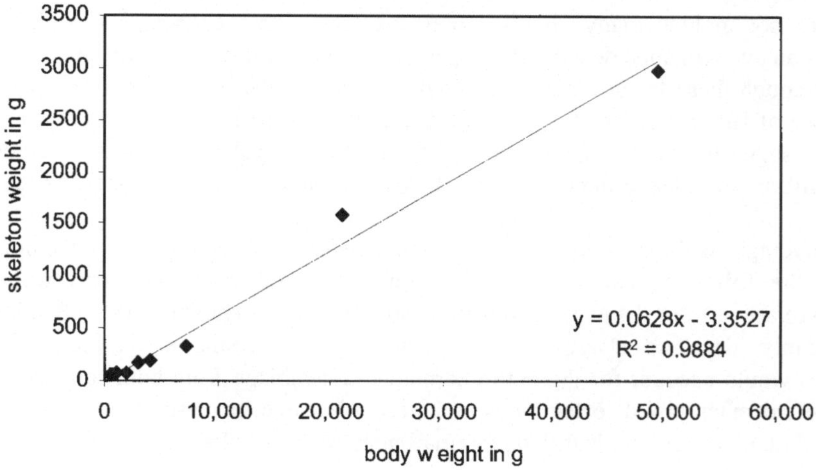

Figure 8.1. Correlation between body weight and skeletal weight for a number of Australian macropodids. Specimens originate from the Australian National Wildlife Collection (CSIRO); codes starting with CM and M are the specimens' reference numbers. Included are *Bettongia penicillata*: CM11452: body weight (bw) 500 g, skeletal weight (sw) 53 g; M18016: bw 780 g, sw 59 g; CM11456: bw 1000 g, sw 71 g; CM11455: bw 1080 g, sw 78 g; CM11453: bw 1972 g, sw 78 g; *Macropus eugenii*: CM18020: bw 3000 g, sw 168 g; CM18013: bw 4000 g, sw 199 g; CM18017: bw 7250 g, sw 322 g; *Macropus rufogriseus*: M18011: bw 21,000 g, sw 1576 g; *Macropus robustus*: M11042: bw 49,000 g, sw 2985 g.

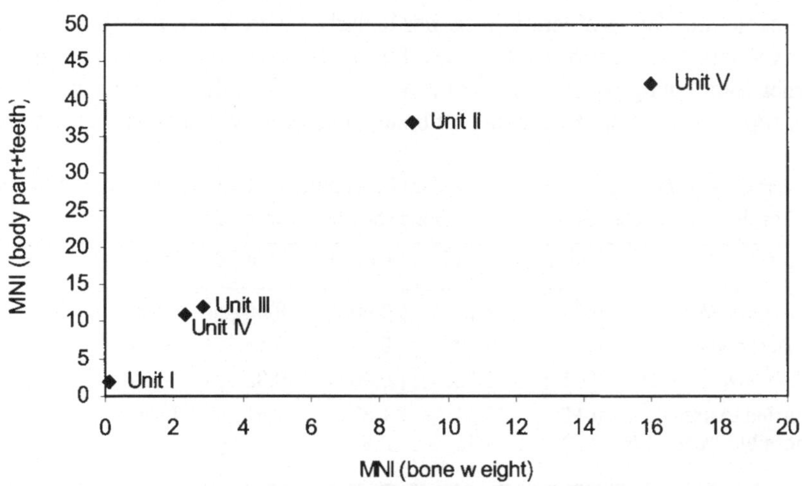

Figure 8.2. Kria Cave, *Dorcopsis muelleri*. Correlation between values for MNI based on body parts and MNI based on bone weight.

The bone material from Kria and Toé was similarly submerged in water (without a paraffin coating), but most of the total bone assemblage is in a fragmentary state and thus does not include many air-filled bone cavities. The archaeological values also represent an average bone density of all species, body parts and age groups present at the sites. Although these factors make it difficult to compare the results from Kria and Toé with those of Binford & Bertram, the stability of the densities between and through the two sites suggests that any adjustment of the archaeological bone densities would be fairly uniform and thus unlikely to affect the overall pattern of skeletal weights through time.

An average skeletal weight for *Dorcopsis muelleri* of 380 g will be employed through the following calculations. The minimum number of *Dorcopsis* wallabies required to account for the excavated bone from Kria Cave is shown in Table 8.3. If only the certainly identified *Dorcopsis* bone is taken into account, a minimum of nine wallabies would account for the entire combined assemblage from both squares. If the 'unidentified macropodid' category is included, this value increases to 10 individuals. By including all of the 'unidentified mammal' bone as potential *Dorcopsis*, the total rises to 31.

These numbers are very much lower than the 84 to 101 individual *Dorcopsis* that are required to account for all of the identified skeletal parts and teeth. Two factors probably contribute to this difference. The first is that the MNI based on identified skeletal parts and teeth is calculated in such a way as to give the highest possible 'minimum' number, taking account of age differences between individuals. To illustrate this point, consider the situation where the skeletons of five *Dorcopsis* wallabies of various ages are scattered randomly across a surface of 10 m^2. All of the five individuals are likely to be represented in each of the squares by some skeletal elements, and the MNI estimate for any one of the squares will be at least one and might well be three or more. However, the average quantity of bone in each square will be only 190 g, half the weight of one *Dorcopsis* skeleton. This problem will be particularly severe when the total quantity of bone is very small, and proportionally less so for increasingly larger samples of bone.

The other contributing factor would be the loss of bone from the site through removal or post-depositional destruction. Destruction of bone might be expected to have a particalary

Table 8.3. Kria Cave, *Dorcopsis muelleri*. Minimum Number of Individuals required to account for the various classes of identified bone shown in Table 8.2.

taxon/category	Unit I	Unit II	Unit III	Unit IV	Unit V	all units
1. *Dorcopsis muelleri*	0.01	2.13	0.80	0.66	4.53	8.13
2. unid. macropodids	0.01	0.59	0.18	0.16	0.84	1.77
3. min. *Dorcopsis* (1 + 2)	0.02	2.72	0.98	0.82	5.37	9.91
4. unidentified mammal	0.11	6.24	1.87	1.51	10.63	20.37
5. max. possible *Dorcopsis* (1 + 2 + 4)	0.12	8.96	2.85	2.34	16.00	30.27

Note: The MNI calculations are based on an average skeletal mass of 380 g. Values are given as fractions of individuals to illustrate the small quantities of *Dorcopsis* bone present in Unit I.

strong impact owing to the fact that the more durable parts of a skeleton are generally the same pieces that are used for identifications – teeth, foot bones etc. An assemblage that has lost a substantial proportion of the original bone through weathering or post-depositional destruction might still yield an MNI estimate based on body parts and teeth close to its original value, despite the fact that the total bone weight is greatly reduced.

Although the two estimates are very different for each unit, the paired values nonetheless show a strong positive correlation and a suggestion of a curvilinear relationship (see Fig. 8.2). The apparent curvilinearity is consistent with the previous suggestions regarding the relationship between total sample size and MNI estimates based on body parts and teeth, but it must be noted that it is based on relatively few data points (N=5).

Unlike the MNI based on body parts or teeth, the MNI based on bone weight can be scaled up to give realistic estimates of how many individual *Dorcopsis* wallabies might be represented in the unexcavated portion of Kria Cave (88% of total). This is because the bone excavated from each square provides a discrete MNI estimate derived from the simple mass of that sample. This is not true for MNIs based on skeletal or dental counts, where one individual might be counted repeatedly from different squares if they are each treated as independent samples. The estimated MNIs for the entire site based on bone weights are shown in Table 8.4.

The various MNI estimates can now be compared and contrasted (Table 8.5). The MNI estimate based on body parts and teeth for the whole site ranges from an absolute minimum of 84 (assuming that every individual was represented in the two excavated squares; all units pooled) to 842 (assuming separate units and that each square provides and independent sample). Clearly, neither of these assumptions is realistic, hence the true MNI almost certainly lies somewhere between these values.

The MNI estimates based on bone weight are not affected by assumptions about the vertical or spatial structure of the assemblage because they simply reflect the mass of *Dorcopsis* bone contained within the deposit. Nevertheless, the MNI estimates based on bone weight for the entire site vary from 68 to 253 individuals, the lower value based on the bone identified specifically as *Dorcopsis*, and the larger value based on all bone that could potentially be derived from *Dorcopsis*, including the large 'unidentified mammal'

Table 8.4. Kria Cave, *Dorcopsis muelleri*. Estimated Minimum Number of Individuals for the entire deposit, based on the weight (g) of the various classes of identified bone.

taxon/category	Unit I	Unit II	Unit III	Unit IV	Unit V	all units
1. *Dorcopsis muelleri*	0.07	17.73	6.68	5.50	37.77	67.77
2. unid. macropodids	0.06	4.90	1.47	1.36	7.00	14.79
3. min. *Dorcopsis* (1 + 2)	0.14	22.63	8.15	6.86	44.77	82.55
4. unidentified mammal	0.90	52.02	15.60	12.61	88.58	169.71
5. max. possible *Dorcopsis* (1 + 2 + 4)	1.04	74.65	23.75	19.48	133.35	252.27

Note: Values are given as fractions of individuals to illustrate the small quantities of bone found in Unit I. The excavated area represents 12% of the total site area.

Table 8.5. Kria Cave, *Dorcopsis muelleri*. Summary of Minimum Number of Individuals estimates for the excavated squares and extrapolated for the whole site, based on the two methods using body parts and teeth, and bone weight.

	excavated squares only		whole site	
MNI method	min.	max.	min.	max.
body parts and teeth	84 (pooled units)	101 (separate units)	84	700 (pooled units) 842 (separate units)
bone weight	9	31	68	253

Note: Separate MNI estimates based on body parts and teeth are shown for the whole assemblage with units pooled and for the assemblage divided into its five stratigraphic units.

category. As argued above, the true value in this case almost certainly lies closer to the upper estimate than to the lower one, for the simple reason that both of the 'unidentified macropodid' and 'unidentified mammal' categories are almost certainly made up predominantly of *Dorcopsis* remains.

If all of the *Dorcopsis* bone that was brought into the site were actually preserved in the deposit, then the estimate of 253 individuals based on bone weight would not be a minimum number of individuals, but rather a maximum number. However, there are strong grounds to suspect that a significant proportion of the original bone assemblage was variously destroyed in hearths, discarded outside of the site area, scavenged, weathered away whilst exposed on the surface, or broken down completely after burial within the deposit.

How much of the assemblage might have been lost through these combined processes is difficult to say. If we take the highest of the various estimates as a starting point, 842 individual *Dorcopsis* would produce 319.96 kg of bone. The maximum possible quantity of *Dorcopsis* bone in the site is only 95.86 kg, even including the less certainly identified categories, and this would imply the loss of at least 70% of the original bone. This figure seems excessively high for Kria given the generally good condition of the bone in the site. A more reasonable figure for bone loss might be around 30-50%, which would imply an original bone weight of 137-192 kg and corresponding MNI for *Dorcopsis* of 361-506.

From the above it should be clear that the number of individual *Dorcopsis* wallabies in the Kria deposit is by no means huge. This point is even more striking when the MNI values are converted into animals per year. The lower four units in Kria Cave, containing the vast bulk of the *Dorcopsis* bone, accumulated over a total period of just over 3000 years (see Chapter 3). Even with the highest MNI estimate of 842 individual *Dorcopsis*, this amounts to one wallaby every 3-4 years. Given that *Dorcopsis* appears to have been a major source of animal protein in peoples' dietary intake over this period, the conclusion is unavoidable that the site was visited only very rarely and probably by very small groups of people.

The MNI of *Dorcopsis muelleri* was not calculated for Toé Cave on the basis of body part or tooth counts. However, a rough estimate of the MNI can be produced from bone weights, as demonstrated above. In the case of Toé Cave this figure is more likely to be an overestimate because a larger proportion of the 'unidentified macropodid' category in this site probably belongs to tree-kangaroos (*Dendrolagus* spp.) and the smaller forest wallaby *Dorcopsulus* sp., and because the 'unidentified mammal' category probably comprises a more even mix of species than in Kria Cave. The total quantity of bone material that could potentially come from *Dorcopsis* is only 3506 g. Approximately 40% of the site was excavated, bringing the total weight of potential *Dorcopsis* bone in the whole site up to around 8765 g. This can be accounted for by as few as 24 individual wallabies. Even if this number is restricted to the Holocene period, the implications for *Dorcopsis* consumption rates in Toé Cave are remarkable – perhaps one wallaby every 200-400 years. Although all of these figures might be out by quite large factors, there would seem little doubt that Toé Cave was visited much less frequently than Kria, or at least not in a way that produced remains of the visitors' meals.

8.5. POPULATION STRUCTURE OF *DORCOPSIS MUELLERI*

The population structure of any species can be described in terms of its age and sex composition at a particular place and time (Krebs 1978). To understand why the population has this particular structure, it is necessary to know various basic facts about the animals' life-history, namely its birth rate (fecundity), death rate (mortality) and dispersal rate (emigration and immigration). It is also necessary to understand how each of these parameters varies in relation to the abundance or population density of the species and its main competitors and predators (Caughley 1977).

Considerable effort has been put into understanding the impact of different hunting or harvesting regimes on the population structure and abundance of different kinds of animals (e.g. Caughley 1977; Getz & Haight 1989; Milner-Gulland & Akçakaya 2001). These studies have produced a number of principles and general models which have been usefully applied to the analysis of subsistence hunting activities by modern hunter-gatherer groups (Alvard et al. 1997; Robinson & Bodmer 1999). However, there have been relatively few attempts to interpret patterns of prehistoric exploitation and/or their impact on prehistoric animal populations from archaeological remains. The recent analysis of moa extinctions in New Zealand stands out as one major contribution in this area (Holdaway & Jacomb 2000), and there have been other significant analyses of various African (Binford 1984; Klein & Cruz-Uribe 1984) and European assemblages (Chase 1986; Stiner 1990a, b, 1991).

The paucity of archaeological analyses of this kind is due in part to the fact that archaeological sites with abundant, well-preserved faunal remains suitable for this kind of analysis are the exception rather than the rule; but also to the still very incomplete state of ecological knowledge of many prey species. The large collection of *Dorcopsis* remains from Kria Cave provides an opportunity to explore the impact of long-term prehistoric hunting of this species on its population structure and perhaps by inference, its local abundance. This study follows several earlier attempts to estimate and interpret the age structure of archaeological wallaby samples (e.g. Ross 1977), although not previously of a member of the genus *Dorcopsis*.

Methods to determine individual age and sex are not available for *Dorcopsis*, but are available for various other species of kangaroos and wallabies (Kirkpatrick 1964, 1965, 1970; Sharman et al. 1964; Kirkpatrick & Johnson 1969; Newsome et al. 1977; McArthur & Sanson 1988; Jarman 1989). However, these were generally developed for use with live animals or for complete museum specimens, and use features such as body weight or measurements of body parts (most often length of the hind foot) that are not available for archaeological material. The criteria employed in this study were therefore developed specifically for this purpose, using reference skulls and mandibles of various species from the modern mammal collection of the Western Australian Museum. A methodology was developed that can be universally applied to macropodids. The full details of this study will be published separately; here are presented only the data and results that are necessary to demonstrate the validity of the methods that are used to quantify patterns of dental eruption and wear in the archaeological *Dorcopsis* sample. First, however, the question of size variation within *Dorcopsis* through time will be briefly addressed.

8.5.1. *Size variation in* Dorcopsis muelleri *through time*

Sustained predation can lead to changes in body size (Reitz & Wing 1999: 315; Stiner et al. 1999; Stiner 2001). This is especially true where the predator actively selects individuals of a species on the basis of body size. Considering the strong indications for sustained hunting of *Dorcopsis muelleri* on the Ayamaru Plateau, the possibility of size change in this species was investigated. This was based on measurements of teeth, which are the most complete remains, are the least affected by weathering, and are not overly influenced by age (see below). It is acknowledged that changes in tooth size may lag behind or be uncoupled from changes in body size; however, few bones were sufficiently complete to provide useful measurements and many bones were affected by weathering. Moreover, all bone measurements are strongly dependent on individual age.

The macropodid dentition consists, on each side, of three permanent upper incisors and one permanent lower incisor (the deciduous precursors of these teeth are resorbed before eruption), two upper and two lower deciduous cheekteeth [one a sectorial premolar (dP3), the other a molariform premolar (dP4), both of which are replaced by the much larger, single permanent premolar (P4)], and four upper and lower molars (M1-4; by definition, lacking deciduous precursors). The present study focuses on the deciduous and permanent cheekteeth (premolars and molars).

With few exceptions, it was possible to allocate even isolated archaeological *Dorcopsis* teeth to a position within the mandible or maxilla. Two measurements were taken on premolars – crown length and maximum posterior width; and two on molars – crown length and width of the posterior half or 'loph' (see Fig. 8.3). Posterior loph width was measured with the arms of the calliper held vertically with respect to the base of the crown. Molar crown length was measured 'from cingulum to cingulum' in a horizontal plane along the long axis of the tooth, on the buccal side for upper molars and the lingual side for lower molars. Tooth length measurements are subject to progressive interdental

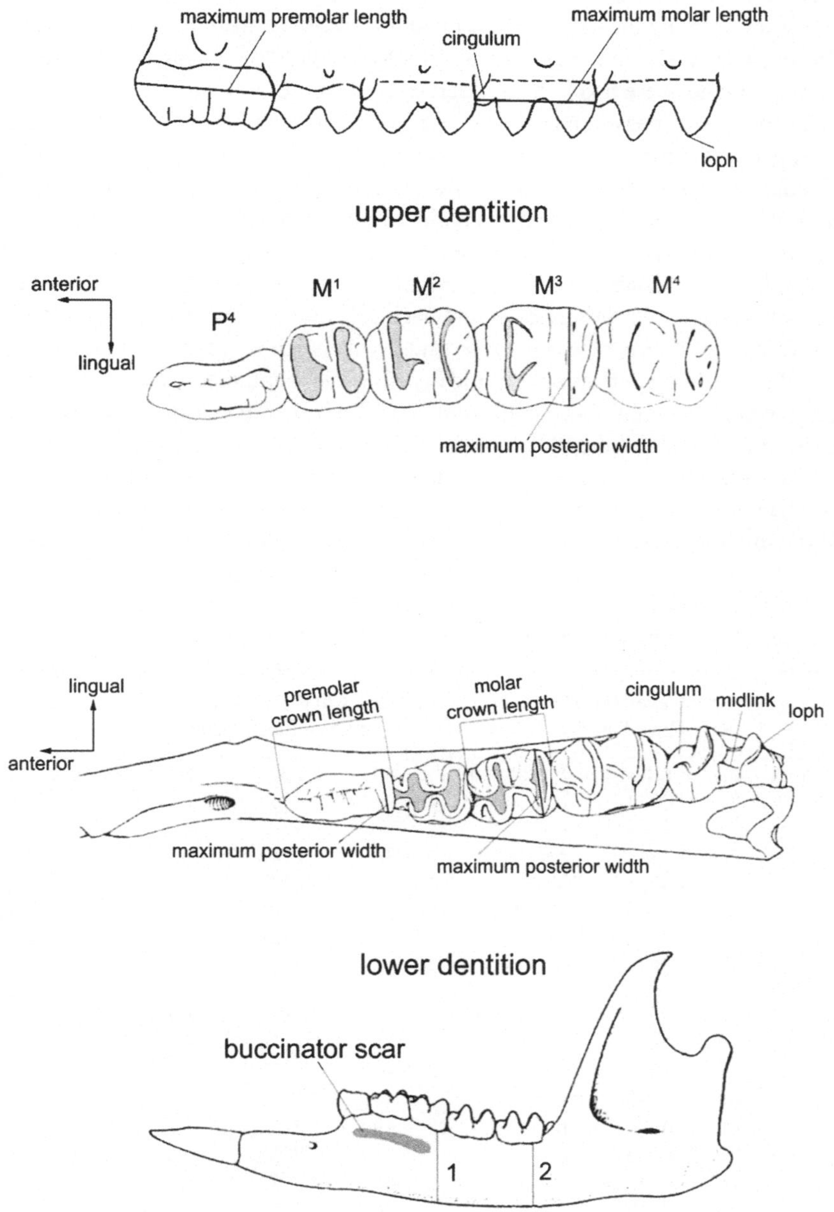

Figure 8.3. *Dorcopsis muelleri* upper (top) and lower (bottom) cheekteeth with tooth measurements. Mandibular depth measurements: 1: buccally, between the lower border of the horizontal ramus and the upper margin of the alveolar bone at the first root situated behind the posterior end of the muscle scar; 2: buccally, between the lower border of the horizontal ramus of the mandible and the upper margin of the alveolar bone at the posterior root of the first fully-erupted molar anterior to the ascending ramus.

wear. No attempts were made to correct for this bias, however, this fact was taken into account when interpreting the results. Tooth width measurements in macropodids are not affected by occlusal wear until very advanced stages where the enamel crown itself is lost. All measurements were taken to an accuracy of 0.1 mm.

Table 8.7 shows the measurements of lower *Dorcopsis* teeth (providing larger sample sizes than upper teeth) from the two large samples from Unit II and V in the Kria Cave deposit. Statistical comparisons of these two samples are summarised in Table 8.6. Although some variables show significant differences between the two samples, the majority show no difference. The measurements that show significant differences are mostly crown lengths, which are generally slightly longer in Unit II. As will be shown below from studies of molar progression and occlusal wear, this is most likely related to differences in age structure between these units, with slightly younger individuals on average in Unit II. Younger animals typically show less interdental wear, hence the slightly greater molar lengths. The small sample of *Dorcopsis* teeth (maximum N = 4 for any measurement) from the Late Pleistocene levels of Toé Cave is consistent in size with the Kria Cave samples (Table 8.7). The dental measurements thus demonstrate no significant change in tooth size within the Ayamaru *Dorcopsis muelleri* population over the period in which this species was hunted from Toé and Kria Caves.

Table 8.6. Statistical comparison of lower *Dorcopsis* tooth measurements from Kria Cave and Toé Cave. One Way Analysis of Variance: n.s. = not significant (p > .05).

	Kria Cave Unit II vs. Kria Cave Unit V			Kria Cave Unit V vs. Toé Cave Pleistocene		
crown length	d.f.	F ratio	F probability	d.f.	F ratio	F probability
dP_3	8	.0534	n.s.	-	-	-
dP_4	9	.5758	n.s.	-	-	-
P_4	54	.2943	n.s.	38	.0146	n.s.
M_1	84	8.3221	.0050	50	.0145	n.s.
M_2	86	6.5998	.0119	55	.0176	n.s.
M_3	75	5.7923	.0186	44	.3146	n.s.
M_4	44	.1560	n.s.	26	.3264	n.s.
posterior width	d.f.	F ratio	F probability	d.f.	F ratio	F probability
dP_3	10	1.6863	n.s.	-	-	-
dP_4	14	.1613	n.s.	-	-	-
P_4	69	.0996	n.s.	48	.0223	n.s.
M_1	85	.0995	n.s.	49	.2424	n.s.
M_2	91	2.0066	n.s.	58	.2055	n.s.
M_3	78	5.2715	.0244	45	.0972	n.s.
M_4	46	.0318	n.s.	26	5.1776	.0317

Table 8.7. Summary of lower tooth measurements of *Dorcopsis* from three stratigraphic time periods.

	Kria Cave, Unit II				Kria Cave, Unit V				Toé Cave, Late Pleistocene levels			
crown length	N	range in mm	mean	σ	N	range in mm	mean	σ	N	range in mm	mean	σ
dP_3	2	7.7-7.8	7.75	.0707	7	7.3-8.6	7.83	.4572	--			
dP_4	2	5.7-5.8	5.75	.0707	8	5.5-6.8	5.96	.3777	--			
P_4	18	10.6-15.0	12.90	1.0273	37	11.1-14.5	12.77	.7509	2	11.8-13.6	12.70	1.2728
M_1	36	4.9-6.2	5.56	.2832	49	4.8-6.1	5.38	.2926	2	5.1-5.6	5.35	.3536
M_2	35	5.6-6.5	5.98	.2315	52	5.3-6.5	5.83	.2689	4	5.5-6.1	5.85	.3000
M_3	34	5.9-7.1	6.57	.2962	42	5.8-7.1	6.41	.2591	3	6.4-6.7	6.50	.1732
M_4	20	5.9-7.1	6.62	.3468	25	6.0-7.3	6.66	.3453	2	6.6-7.0	6.80	.2828
posterior width	N	range in mm	mean	σ	N	range in mm	mean	σ	N	range in mm	mean	σ
dP_3	2	3.4-3.4	3.40	.0000	9	3.0-3.5	3.26	.1509	--			
dP_4	4	3.2-3.5	3.38	.1258	11	3.1-3.6	3.34	.1748	--			
P_4	24	2.1-4.1	3.53	.4208	46	3.0-4.2	3.56	.3015	3	3.1-4.0	3.53	.4509
M_1	37	3.4-4.2	3.81	.2134	49	3.3-4.1	3.79	.1847	1	--	3.70	--
M_2	37	3.8-4.7	4.22	.2006	55	3.7-4.7	4.17	.1816	4	4.0-4.3	4.13	.1500
M_3	35	4.5-5.5	4.81	.2011	44	4.3-5.2	4.70	.2241	2	4.7-4.8	4.75	.0707
M_4	21	3.9-4.8	4.40	.2598	26	4.0-4.8	4.39	.2189	1	--	4.90	--

8.5.2. *Methods of sex determination*

The sex ratio of an archaeological sample is of potential interest for several reasons. Firstly, evidence of selective hunting of males or females of a prey species may have different consequences for the hunted population. And secondly, in the event that rates of tooth eruption or wear differ between the sexes, a bias in sex ratio might affect the analysis of the age structure of the population.

Many species of animals are sexually dimorphic, either in gross size or in combinations of size and shape. In some species (e.g. humans) the sex of a skeleton can be determined from the morphology of particular bones, such as the pelvis and the skull. More frequently, a series of measurements must be taken to compare an unknown individual with samples of known sex. Although this can be done on any sexually dimorphic bone in the skeleton, many bones are very prone to fragmentation and decay in archaeological contexts. Teeth are much less prone to breakdown and have the additional advantage that the tooth crown generally does not grow once it has formed, although it will be affected by wear.

Sexual dimorphism in kangaroos is mainly a matter of size, with males growing larger than females in almost all species (Jarman 1989). As mentioned earlier, males of *Dorcopsis* can grow to approximately twice the body weight of females. Sharman et al. (1964) found that differences in growth rate in Red Kangaroos (*Macropus rufus*) were not evident until after one year of age. Sexual dimorphism is thus unlikely to be prominent in the skeleton of very young animals, although it may be present in the dentition of even the youngest age groups.

Sexual dimorphism in dental measurements has been demonstrated for various species of macropodids including the Grey Kangaroo, *Macropus giganteus* and the Agile Wallaby, *M. agilis* (Bartholomai 1971), and the Red Kangaroo, *M. rufus* (Sharman et al. 1964). Although a detailed comparative study has not been attempted, my own observations of a variety of species suggest that sexual dimorphism in the macropodid dentition may take two different forms. The first is where all of the teeth are larger in males than in females, but with no differences in the relative size of anterior vs. posterior molars within the dentitions (e.g. in *Macropus giganteus*). The second is where the anterior molars are not appreciably different in size between the sexes, but the posterior molars are larger in males than females, thus producing a difference in relative posterior molar size between the sexes. The latter pattern is evident in the data provided by Bartholomai (1971) for the Agile Wallaby, *Macropus agilis*. The one adult specimen of each sex of *Dorcopsis muelleri* that was available for study from the Australian Museum show the latter kind of sexual dimorphism, with enlarged third and fourth molars in the male specimen.

The reference specimens of known sex of *Dorcopsis muelleri* also appeared to differ in the depth of the mandibular ramus, and similar contrasts are apparent in other macropodids. Mandibular depth is strongly related to age, hence the relation of age and sex will need to be assessed for any mandibular measurements. Mandibular depth was measured at two locations on the archaeological specimens: Location 1: buccally, between the lower border of the ramus and the upper margin of the alveolar bone at the first root situated behind the posterior end of the muscle scar (see Fig. 8.3); and Location 2: buccally, between the lower border of the ramus of the mandible and the upper margin of the alveolar bone at the posterior root of the first fully-erupted molar anterior to the ascending ramus (Fig. 8.3). Like teeth, the mandible (or at least the ramus) has a high chance of survival in archaeological deposits.

8.5.3. *Sex ratio of the prehistoric* Dorcopsis *samples*

Tooth and jaw measurements are displayed as simple histograms and examined for evidence of bimodality (Fig. 8.4). Particularly in the case of the premolar and molar widths, measurements that are not affected by interdental wear or by growth, any significant bimodality can be assumed to be a reflection of sexual dimorphism (Reitz & Wing 1999: 189).

To minimise the effect of age, the histograms for the measurements of mandibular depth include only those specimens with a fully erupted dentition. The histogram for mandibular depth 1 contains a suggestion of unequal bimodality, with the dominant peak being at the small, 'female' end of the range. However, the second mandibular measurement does not show an equivalent pattern, but rather shows a normal distribution.

Histograms were examined for all individual tooth measurements. Plots for the anterior molars consistently show no indication of bimodality (graphs not provided), in keeping with observations of modern macropodids. Among the more posterior teeth, the histograms show variable patterns, from distinctly bimodal distributions, e.g. in the posterior width of the M_4 (see Fig. 8.4), to others that appear to be normally distributed (e.g. crown length of M_3, see Fig. 8.4). The lack of clear bimodality is consistent with dental measurements presented by Groves & Flannery (1989) who state that *Dorcopsis muelleri* is only weakly dimorphic in its teeth. Where a bimodal distribution is present in the archaeological sample (e.g. in M_4 width), the 'male' and 'female' peaks are of equal height, suggesting an approximate even number of males and females.

Bivariate plots of crown length against posterior width were also prepared for the two posterior lower molars (Fig. 8.5). The plot for M_4 approximates the 'dumbbell'-shaped scatter of the kind expected in a sexually dimorphic species (Reitz & Wing 1999: 190); in contrast, the M_3 plot shows no such pattern.

Overall, the prehistoric *Dorcopsis* sample shows only weak evidence for sexual dimorphism. However, the vague indications of sexual dimorphism do point to a fairly even sex ratio within the sample. For the following analysis of age structure, it will therefore be assumed that the sex ratio is even; and further, that any sexual dimorphism in the population will not impact on the estimates of relative age based on molar eruption, progression and wear.

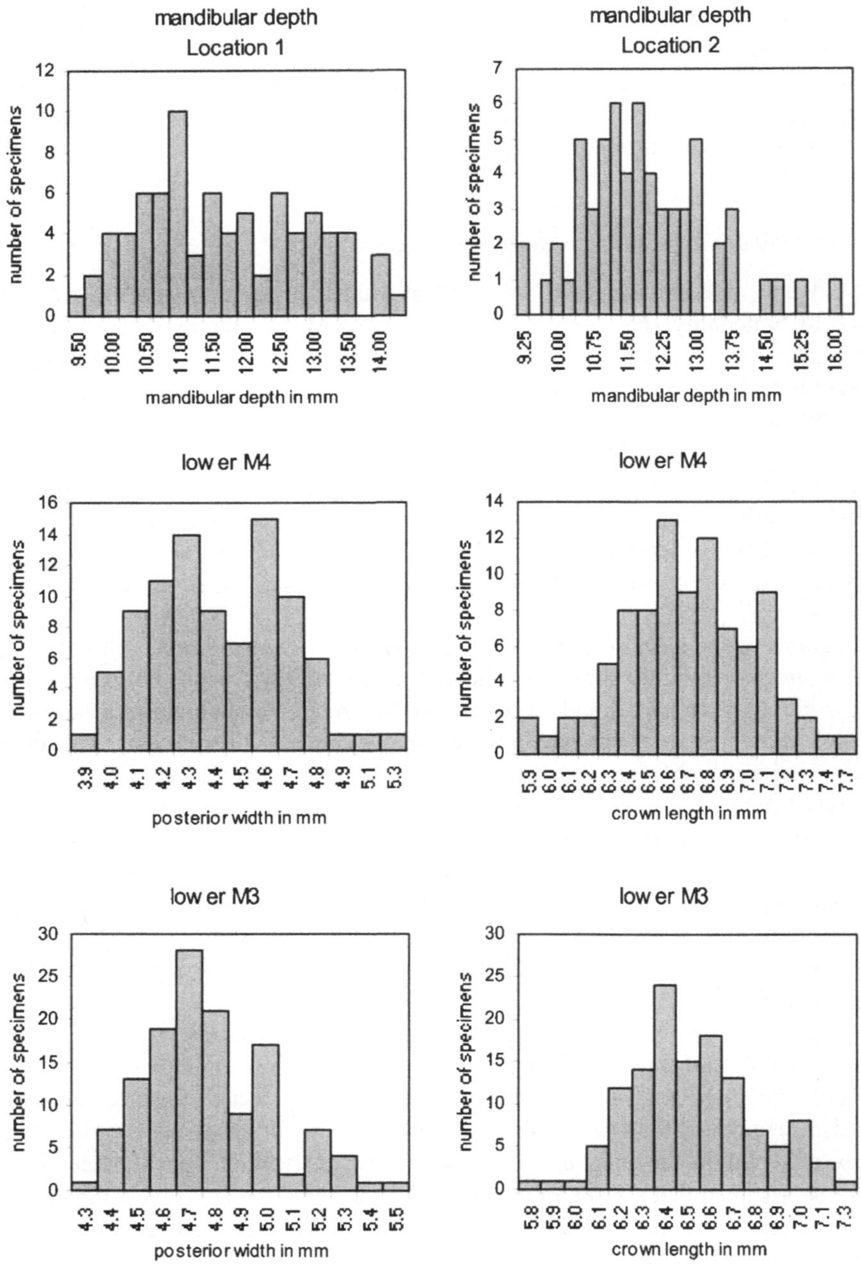

Figure 8.4. Kria Cave, *Dorcopsis muelleri*. Top left and right: histograms of mandibular depths measured at two different locations on the mandible (for definitions see Figure 8.3). X-axis values are midpoints of intervals with standard range of 0.25. Middle and bottom left and right: histograms of posterior width and crown length of the two most sexually dimorphic molars.

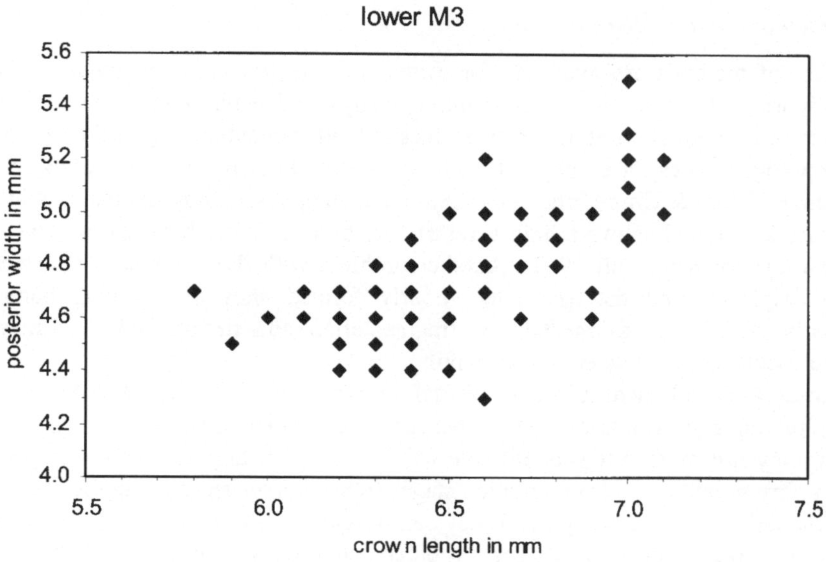

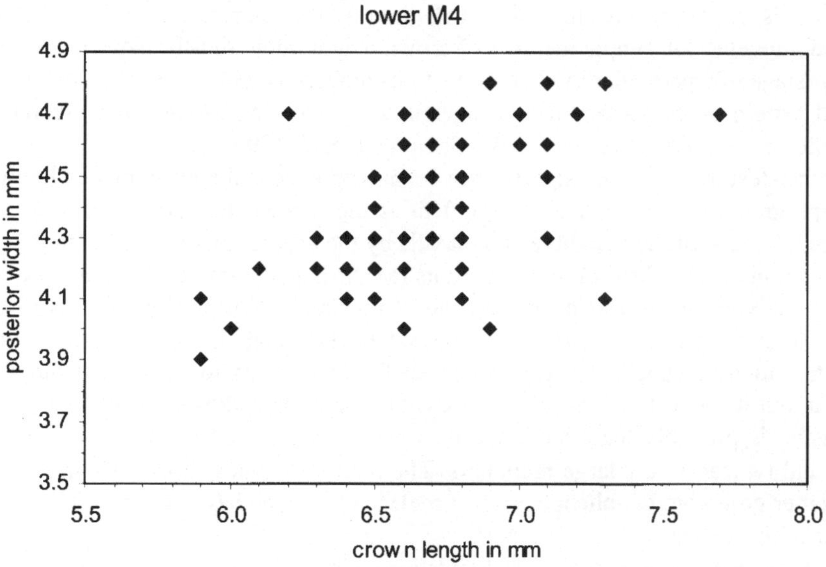

Figure 8.5. Kria Cave, *Dorcopsis muelleri*. Bivariate plot of posterior width and crown length in each of the most sexually dimorphic lower molars.

8.5.4. *Methods of age determination*

A number of methods are available for estimating age from skeletal remains. These include scoring of epiphyseal fusion patterns; analysis of dental development, eruption and wear; and examination of incremental lines in tooth cementum (e.g. Aitken 1975).

Chronological ages for epiphyseal fusion are poorly known for all but a few species of mammals (Klein & Cruz-Uribe 1984). Among marsupials, epiphyseal fusion typically occurs late in life and shows a high level of variation in timing between skeletal parts (Washburn 1946; Kingsmill 1962). Another problem with this method is that weakly ossified epiphyses and the associated, poorly formed ends of the long bone are particularly susceptible to mechanical fragmentation and decay, and thus may be underrepresented in an archaeological sample.

In most archaeological contexts, dental remains are likely to provide the best information on age structure. This is because teeth often survive better than other remains, they are more often identifiable to species level, they can generally provide precise information on age from eruption stages (which, however, only applies to a short part of the animal's life history), and they can provide a less precise or at least relative measure for age from wear stages in older individuals. Members of the family Macropodidae are particularly amenable to this kind of analysis for several reasons. Their teeth are relatively large and robust, and preserve well in archaeological sites. The molars are relatively high-crowned but do not show continuous growth and eruption of the crown, such as occurs in cheekteeth of a horse or some rodents. As wear progresses, the crown is gradually lowered until such time as the enamel crown and dentine is removed altogether. A simple measure of crown height might thus be expected to reflect the wear stage of a particular tooth, relative to its unworn height. During this process, the occlusal pattern of the molars also changes in a regular way, which can be scored as a series of discrete stages (Lentle et al. 1998; Hartzell et al. 1999).

Another feature of special significance in macropodids is the phenomenon known as 'molar progression' (Kirkpatrick 1964). This is the gradual forward movement of the cheekteeth in each of the mandible and maxilla by the absorption of bone in front of, and the deposition of bone behind the tooth roots (McArthur & Sanson 1988), an adaptation to keep relatively unworn teeth in the zone of maximal occlusal effectiveness (Sanson 1980). Molar progression is most pronounced in the larger living kangaroos, among which the anterior cheekteeth are moved so far forward as to be pushed out at the diastema, but it seems to occur to some extent in all macropodids (Sanson 1980). Molar progression is probably least pronounced in browsing macropodids with low-crowned molars and typically very large premolars. The forest wallabies, *Dorcopsis* spp., belong to this latter group but the phenomenon of molar progression has not been studied in any member of this genus.

Although molar progression is an adaptation to improve the occlusal effectiveness of the cheekteeth, the rate of progression does not appear to be linked to rates of tooth wear (McArthur & Sanson 1988). Rather, molar progression appears to be determined mainly by age (Kirkpatrick & Johnson 1969; Newsome et al. 1977), with a slowing of the process through individual life. Newsome et al. (1977) demonstrated very slight sexual dimorphism in the rate of molar progression (and eruption) in the Agile Wallaby, with males showing an earlier eruption and thus slightly more advanced stage of molar progression than females.

For the present study, patterns of tooth eruption, molar progression and tooth wear are investigated for the prehistoric *Dorcopsis* sample. The combination of these three aspects will provide a complementary picture of the age structure of this population. Because no information is available on the timing of dental eruption or molar progression for any species of *Dorcopsis*, the analysis lacks the advantage of a precise chronological framework. Another potential limitation is that molar progression and wear have not previously been studied in detail for any member of the 'browser' grade of macropodids to which *Dorcopsis* belongs (Sanson 1980). Unfortunately, reference material of all *Dorcopsis* species is rather scarce in museum collections (Groves & Flannery 1989), hence it was necessary to look for a suitable analogue. The Quokka (*Setonix brachyurus*) was chosen, despite its significantly smaller body size, because it has a similar dental morphology to *Dorcopsis* and shows a similarly limited degree of molar progression (Sanson 1980). Unfortunately, only limited information is available on dental eruption times in the Quokka. For this reason, two other species were investigated, namely the Agile Wallaby (*Macropus agilis*) and the Red Kangaroo (*Macropus rufus*); both of these taxa have been the subject of detailed studies on the timing of dental eruption and molar progression (*M. agilis*: Kirkpatrick & Johnson 1969; Newsome et al. 1977; *M. rufus*: Sharman et al. 1964; Kirkpatrick 1965, 1970) and they represent different stages along the continuum from low-crowned (brachyodont) to high-crowned (hypsodont) cheektooth morphology.

For each of the reference species samples, the relationship between tooth eruption, molar progression and tooth wear (the latter as reflected by crown height and morphological wear stages) was investigated. The first objective was to determine whether molar progression was present in a browsing macropodid (the Quokka). The second was to determine whether either measure of tooth wear (crown height or occlusal wear stage) could be used as an indicator of individual age among macropodids. In total 19 male, 24 female, and four specimens of unknown sex were studied of the Red Kangaroo; 28 male and 23 female specimens of the Agile Wallaby; and 15 male, 21 female, and two specimens of unknown sex of the Quokka.

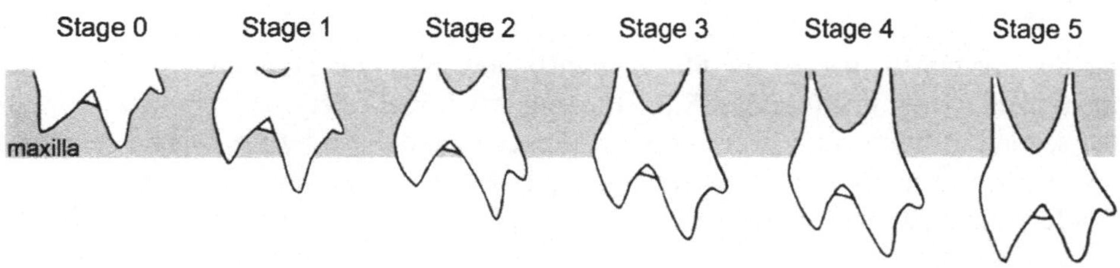

Figure 8.6. Eruption stages, applicable to both upper and lower molars. Modified after Sharman et al. (1964).

Dental eruption

The stage of dental eruption for each molar was scored according to the system of Sharman et al. (1964; see Fig. 8.6):

- Stage 0: no part of the crown is erupted above the jawbone;
- Stage 1: only the anterior loph of the molar has emerged partly above the jaw bone;
- Stage 2: both the anterior and posterior loph of the molar have emerged partly above the jaw bone, but not the body of the tooth;
- Stage 3: both anterior and posterior lophs have emerged fully above the jaw bone, yet not the body of the tooth;
- Stage 4: virtually all of the crown has emerged, although the posterior end may still be just below the jawbone;
- Stage 5: fully erupted: the crown and some of the roots have emerged above the jawbone.

The total developmental stage of the upper or lower dentition was scored as the last fully erupted molar plus the eruption stage of the molar posterior to this tooth (for example, stage 3.2 would mean that the third molar was fully erupted while the fourth was at eruption stage 2; a fully developed dentition (all four molars in place) is simply scored as 4.0). As will be shown below, Stage 3 was never recorded, presumably because its definition is too close to that of Stage 2; these stages could be united to simplify future analyses.

For archaeological specimens, eruption stage could be scored according to the same system for the more complete maxillae and mandibles. For isolated, unworn teeth without roots, or without fully developed roots, the eruption stage could not be scored unless some of the jawbone was still present so that its level above the bone could be established. Likewise, the eruption stage could not be established for specimens that lacked the distal parts of the tooth-bearing section of the jaw, while any present anterior teeth were fully erupted.

Molar progression

For each of the samples of the macropodid reference species, upper molar progression was estimated using Kirkpatrick's (1964) method. This requires that a reference line is drawn across the anterior margins of the orbits, with the skull held in palatal view (see Fig. 8.7). The Upper Molar Progression Index or UMI is defined as the number of whole molars positioned anterior to this reference line, plus the proportion of the molar intersected by the line (see Fig. 8.7). The proportion was calculated from the length of the section of the tooth in front of the reference line, divided by the average length of that particular tooth within the entire sample. For example, in a specimen with two whole molars forward of the reference line, plus 3 mm of the third molar (with an average length of 7 mm), the UMI is 2 + 3/7, or 2.43. This precise method may not always be applicable to archaeological specimens because of the fragmentary state of the skull; however, provided the base of the zygomatic arch is preserved on the buccal side of the maxilla, Kirkpatrick's line can be drawn fairly accurately from this point through the upper molars, remaining parallel to the median valley between the lophs.

The relationship between molar eruption and upper molar progression is shown for each of the three modern reference species in Figure 8.8. As expected, the relationship is linear and particularly strong in both *Macropus* species. More surprisingly, there is also a strong linear relationship in *Setonix brachyurus*, a browsing species that, according to Sanson's (1980) general observations on macropodid masticatory adaptations, might be expected to show little or no molar progression. However, the maximum observed values for molar progression in *Setonix* (< 2.5) are much lower than those observed in *Macropus* spp. (> 4), suggesting an earlier or more dramatic slowing of molar progression in the browsing species.

Equivalent methods have not been developed to score molar progression in the lower dentition. This is clearly desirable for archaeological assemblages where mandibles are more often preserved than maxillae. For this study, two landmarks were identified on the mandible that could potentially yield a Lower Molar Progression Index (LMI), comparable to the UMI. The first landmark is a vertical line drawn at the posterior end of the prominent buccinator muscle scar on the buccal side of the mandible (LMI-1 in Fig. 8.9) and a similar procedure was followed for the calculation of LMI as for the UMI. The second landmark is the anterior rim of the ascending ramus (LMI-2 in Fig. 8.9); the horizontal distance between the ascending ramus and the top of the posterior loph of the last fully erupted molar was measured. This value therefore measures the proportion of the incompletely erupted molar that lies forward of the ascending ramus (see Fig. 8.9). The value was converted into a fraction of the average length of the particular molar in question, as for the UMI. In a specimen where the fourth molar is fully erupted and has progressed some distance anterior to the ascending ramus, then this value would be divided by the average fourth molar length to provide a progression index. For example: if the distance between the posterior loph of the fourth molar and the ascending ramus is 3 mm, then the LMI-2 is 4 + 3/7 mm (7 mm being the average length of M_4), or 4.43.

For the museum specimens, molar progression was scored for both the upper and lower dentition. This allowed the relationship between Upper and Lower Molar Progression Indices to be assessed across a number of species. The results of these comparisons (Table 8.8) show that for all three species lower molar progression is best measured at the ascending ramus (LMI-2). The consistently lower correlations obtained with the buccinator muscle landmark may reflect a greater degree of variability in the development of this muscle or a higher error associated with the delimitation of its posterior margin which was usually sharply defined but sometimes diffuse. The lower molar progression as measured at the ascending ramus will therefore be used in subsequent analyses. Lower molar eruption and progression, not surprisingly, are highly correlated in all species.

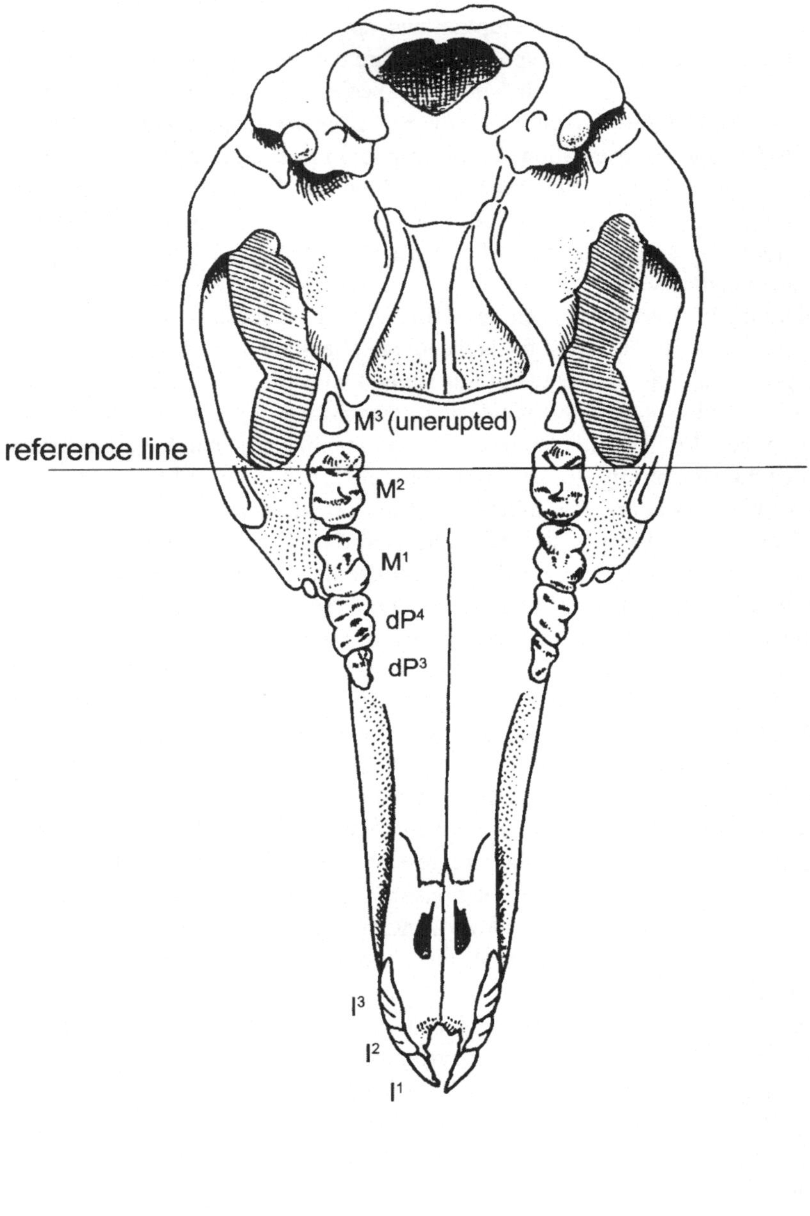

Figure 8.7. Macropodid skull, palatal view. The molar progression reference line to determine the Upper Molar Progression Index (UMI) runs along the anterior rims of the orbits. Modified after Kirkpatrick (1964).

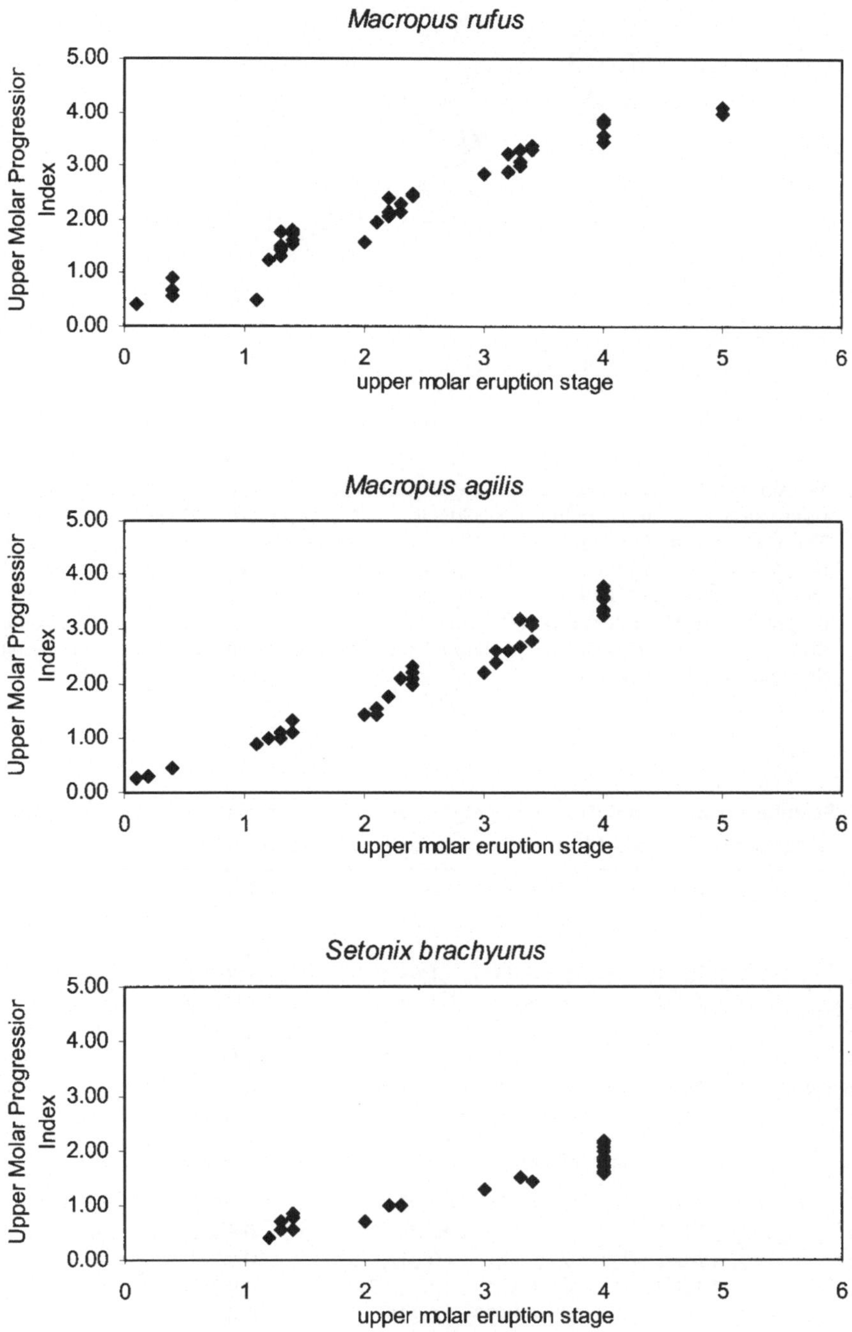

Figure 8.8. Relationship between upper molar eruption and upper molar progression stages in each of the three modern reference series. Eruption stage 5 in *M. rufus* is a code for the eruption of the full dentition (stage 4) plus the loss of the M^1 as a result of molar progression.

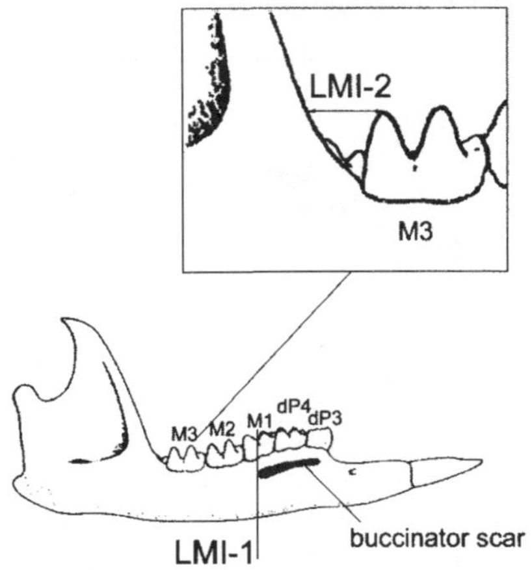

Figure 8.9. Measurement of the Lower Molar Progression Index (LMI). The jaw illustrated in this figure shows the two deciduous cheekteeth, the M_1, M_2 and the M_3 just fully erupted. The reference line at landmark LMI-1 (the posterior margin of the buccinator scar) runs through the M_1; the lower molar progression is therefore 0 + the proportion of M_1 anterior of this line. The lower molar progression at landmark LMI-2 is 3 (=M_3) + the proportion of M_4 anterior of the ascending ramus (as measured by the horizontal distance between the top of the posterior loph of M_3 and the anterior rim of the ramus). See text for full definitions.

Table 8.8. Relationship between upper molar progression and lower molar progression (measured at location LMI-1 and LMI-2) for each of the three reference species.

species	location	regression equation	R^2	N
		males		
M. rufus	LMI-1	y = 0.7927x - 1.0658	0.8670	18
M. rufus	LMI-2	y = 1.06x + 0.4249	0.9764	18
M. agilis	LMI-1	y = 0.8768x - 0.1913	0.9299	28
M. agilis	LMI-2	y = 1.0342x + 0.7958	0.9901	28
S. brachyurus	LMI-1	y = 0.9174x - 0.3339	0.8676	15
S. brachyurus	LMI-2	y = 1.1811x + 1.5263	0.9433	15
		females		
M. rufus	LMI-1	y = 0.7061x - 0.739	0.8352	28
M. rufus	LMI-2	y = 0.9656x + 0.6372	0.9661	28
M. agilis	LMI-1	y = 0.8227x - 0.0113	0.9498	23
M. agilis	LMI-2	y = 0.9852x + 0.8817	0.9793	23
S. brachyurus	LMI-1	y = 0.8723x - 0.2268	0.8416	21
S. brachyurus	LMI-2	y = 1.139x + 1.6268	0.9728	21

Note: LMI-1: lower molar progression measured at the posterior end of the buccinator muscle scar; LMI-2: lower molar progression measured at the anterior rim of the ascending ramus.

Tooth wear

Macropodid molars wear in a very regular and characteristic fashion that allows individual teeth to be readily assigned to discrete 'wear stages'. Tedford (1966) was the first to propose such a scoring system for fossil kangaroos, and his system has been modified and used in various ways by McArthur & Sanson (1988), Lentle et al. (1998) and Hartzell et al. (1999). McArthur & Sanson (1988) investigated the relationship between molar wear and molar progression in the Grey Kangaroo and found significant differences in the maximum amount of wear attained by molar teeth both within and between populations, presumably reflecting differences in diet, but no differences in rate of molar progression. They did not discuss the potential limitations of molar wear as a method of age determination. Lentle et al. (1998) showed that molar wear stage is correlated with body size (= age) within a single population of the Tammar Wallaby (*Macropus eugenii*). Hartzell et al. (1999) scored tooth wear along with molar eruption and progression in their study of modern Bennett's Wallaby (*Macropus rufogriseus*). Although they conclude that 'the use of dental attrition indices alone does not appear to be a reliable means for estimating age of macropods' (Hartzell et al. 1999: 59), no evidence was presented to support this view. Moreover, it should be noted that molar wear has been used with some success as an age indicator in several species of possums (Thomson & Owen 1964; Winter 1980), hence it is worth further exploring the utility of this method in macropodids.

No previous study has explored the utility of crown height measurements for quantifying molar wear in macropodids, despite the well-established use of such measurements in other groups of mammals with hypsodont dentitions (e.g. many ungulates; Klein & Cruz-Uribe 1984). For this study, molar wear was quantified in two ways: according to wear stage; and by direct measurement of crown height. The wear stages are defined in Figure 8.10. Both sets of data were initially recorded for the reference series, in order to test the utility for age determination of tooth wear measures against the more established criteria of dental eruption and molar progression.

Although the three reference species differ markedly in terms of crown height and the extent of molar progression (*Setonix brachyurus* being low-crowned with limited progression and *Macropus rufus* being high-crowned with extreme molar progression), the sequence of progressive wear stages differs only slightly among them. The wear stages (see Fig. 8.10) are believed to be applicable to all macropodid species. The system for the lower jaw is very close to that devised by McArthur & Sanson (1988: Fig. 1). Molar crown height was measured vertically from the base of the enamel to the flat occlusal surface of the loph. It was measured on the buccal side of the anterior loph of the lower molars; and on the lingual side of the anterior loph on the upper molars.

Because molar eruption, molar progression and molar wear are all scored as interval or ordinal values, patterns of correlation between these variables cannot be validly analysed with parametric correlation statistics. Accordingly, Kendall's rank correlation coefficient (*tau*-b) is used here to assess the probability that molar wear and crown height are non-randomly associated, both with each other and with age (as indicated by upper and lower molar progression from the UMI and LMI-2). Table 8.9 gives a summary of the Kendall *tau* values for each of the three reference species, analysed

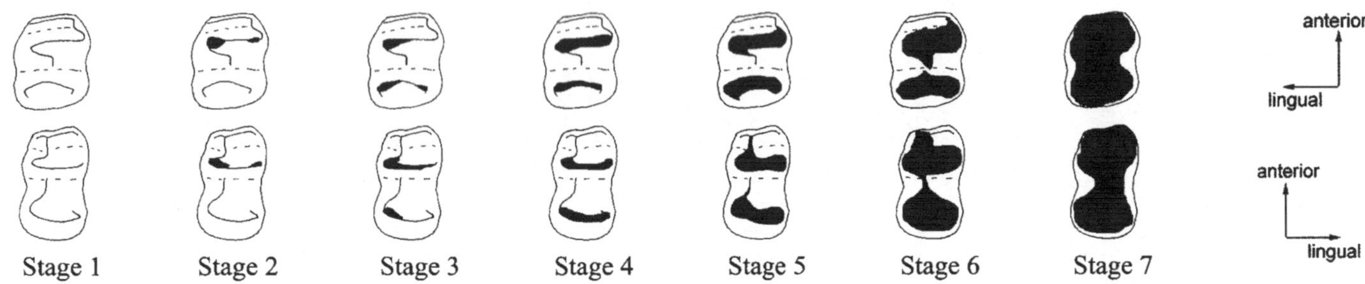

Figure 8.10. Left upper (top row) and left lower (bottom row) molar wear stages in *Macropus rufus*, *Macropus agilis* and *Setonix brachyurus*.

Upper molars

- Stage 1: no dentine is exposed;
- Stage 2: dentine is only exposed on the anterior loph; in *S. brachyurus* potentially also on the posterior loph;
- Stage 3: dentine is exposed at the lingual end of both lophs, occasionally with small areas of dentine on the buccal cusps (or a narrow transverse line of exposure);
- Stage 4: dentine is broadly exposed across at least one of the lophs, with small areas or lines of dentine exposure on the other loph;
- Stage 5: dentine is exposed across both lophs, and both show small dentine spurs extending anteriorly or posteriorly;
- Stage 6: both lophs support broad dentine pools, and the anterior/posterior spurs almost connect the two pools;
- Stage 7: the greater part of the tooth area is covered with a single dentine pool, generally with an enamel rim around the margins. Two separate dentine pools are occasionally retained, but in such cases the enamel rim is removed entirely from part of the tooth margin.

Lower molars

- Stage 1: no dentine is exposed;
- Stage 2: one or two small areas of dentine are exposed on the anterior loph, occasionally (especially in *S. brachyurus*) also on the posterior loph;
- Stage 3: the anterior loph shows a continuous band of exposed dentine; discrete areas of dentine are often exposed on the posterior loph. In *S. brachyurus* both the anterior and posterior lophs may show small discrete areas of dentine rather than a continuous band;
- Stage 4: both lophs show continuous but narrow bands of exposed dentine;
- Stage 5: large but separate pools of dentine are exposed on the anterior and posterior lophs, often with dentine spurs extending onto the fore- and mid-links;
- Stage 6: the two dentine pools are narrowly connected via the midlink, with the anterior pool broadly extended onto the anterior cingulum;
- Stage 7: the greater part of the tooth area is covered with a single dentine pool, generally with an enamel rim around the margins.

Table 8.9. Comparison between tooth wear (wear stage and crown height) and age (Lower Molar Progression Index LMI-2) of the three reference species.

Macropus rufus			M_1	M_2	M_3	M_4
males	crown height vs. wear stage	*tau*-b	-.5331	-.8261	-.8819	-.3162
		p (N)	.003 (17)	.000 (17)	.002 (8)	n.s. (5)
females	crown height vs. wear stage	*tau*-b	-.8240	-.7293	-.8439	-.8578
		p (N)	.000 (15)	.000 (18)	.000 (14)	.001 (10)
males	LMI-2 vs. wear stage	*tau*-b	.7936	.8838	.8584	.8367
		p (N)	.000 (18)	.000 (18)	.000 (13)	.008 (7)
females	LMI-2 vs. wear stage	*tau*-b	.6774	.7973	.7892	.8118
		p (N)	.000 (16)	.000 (20)	.000 (16)	.001 (10)
males	LMI-2 vs. crown height	*tau*-b	-.5509	-.7970	-.5926	-.1054
		p (N)	.001 (17)	.000 (17)	.022 (8)	n.s. (5)
females	LMI-2 vs. crown height	*tau*-b	-.4533	-.6733	-.8269	-.7361
		p (N)	.011 (15)	.000 (18)	.000 (14)	.002 (10)

Macropus agilis			M_1	M_2	M_3	M_4
males	crown height vs. wear stage	*tau*-b	-.8348	-.9082	-.8895	-.8154
		p (N)	.000 (27)	.000 (22)	.000 (13)	.001 (10)
females	crown height vs. wear stage	*tau*-b	-.7821	-.8302	-.9032	-.5004
		p (N)	.000 (21)	.000 (18)	.000 (13)	.049 (8)
males	LMI-2 vs. wear stage	*tau*-b	.7345	.8628	.8563	.7939
		p (N)	.000 (28)	.000 (27)	.000 (19)	.001 (12)
females	LMI-2 vs. wear stage	*tau*-b	.7228	.8951	.8821	.6318
		p (N)	.000 (22)	.000 (20)	.000 (17)	.010 (10)
males	LMI-2 vs. crown height	*tau*-b	-.7208	-.7539	-.7020	-.5350
		p (N)	.000 (27)	.000 (22)	.001 (13)	.018 (10)
females	LMI-2 vs. crown height	*tau*-b	-.6917	-.7518	-.8246	-.3282
		p (N)	.000 (21)	.000 (18)	.000 (13)	n.s. (8)

Setonix brachyurus			M_1	M_2	M_3	M_4
males	crown height vs. wear stage	*tau*-b	-.7946	-.6819	-.5782	-.4021
		p (N)	.000 (15)	.000 (15)	.007 (13)	.047 (13)
females	crown height vs. wear stage	*tau*-b	-.6780	-.5978	-.5113	-.1844
		p (N)	.000 (21)	.000 (20)	.007 (16)	n.s (14)
males	LMI-2 vs. wear stage	*tau*-b	.7306	.7811	.7196	.5590
		p (N)	.000 (15)	.000 (15)	.001 (14)	.009 (13)
females	LMI-2 vs. wear stage	*tau*-b	.7686	.7611	.6505	.5021
		p (N)	.000 (21)	.000 (21)	.001 (17)	.014 (15)
males	LMI-2 vs. crown height	*tau*-b	-.5971	-.5529	-.4106	-.3514
		p (N)	.001 (15)	.003 (15)	.028 (13)	n.s. (13)
females	LMI-2 vs. crown height	*tau*-b	-.6718	-.6097	-.4774	-.5006
		p (N)	.000 (21)	.000 (20)	.008 (16)	.010 (14)

Note: Deciduous teeth and permanent premolars are excluded because of small samples and/or too few wear stages. Only lower jaw test results are presented; upper teeth show very similar results. Kendall rank correlation coefficient (*tau*-b), one-tailed test: n.s. = not significant (p >.05). LMI-2 = Lower Molar Progression Index at landmark 2.

separately for each sex. In both sexes of all three reference species, molar wear, crown height and LMI-2 values are all highly associated (at $p<.05$) for all but some comparisons for the M_4. This result is somewhat surprising in the case of *Setonix brachyurus* considering its relatively small size (hence its proportionally greater measurement error) and the lesser degree of molar progression in this browsing macropodid. In all three species the Kendall *tau* values are consistently higher between the LMI-2 values and molar wear categories than between LMI-2 and crown height. Contrary to suggestions by Hartzell et al. (1999), it is clear that all three measures of relative age hold potential for analysis of archaeological assemblages. This analysis will focus on molar wear because of its slightly better statistical performance against LMI-2.

The bivariate relationship between molar wear stage and molar progression is shown in Fig. 8.11a to 8.11c (at the end of this chapter) for the upper and lower molars each of the three reference species. The pattern of molar wear is very similar in all three species. For each tooth, there is a period following eruption in which wear appears to progress at a fairly constant rate with respect to molar progression. However, once a tooth enters wear stage 6, the rate of wear evidently slows and there is a somewhat erratic progression to wear stage 7. Molars at wear stage 6 or above thus indicate the minimum age for the individual, while specimens with a wear stage below 6 can provide a relatively coarse but nonetheless useful indication of individual age. In *Setonix brachyurus* the rate of molar wear is evidently slower relative to molar progression than in the two *Macropus* species. For *Setonix* this probably means that molar wear stage is a less effective predictor of age than in the species with higher-crowned molars. Better results might be expected for *Dorcopsis muelleri*, in which the molars are higher crowned than in *Setonix brachyurus* (though less so than in *Macropus agilis*).

8.5.5. *Relationship of dental age to important life history parameters*

Some basic life history data on each of the reference species and *Dorcopsis muelleri* are given in Table 8.10. Lee & Ward (1989) suggest that much of the variation in life history traits among macropodoids can be accounted for by their scaling with body size. This relationship is particularly strong for age at weaning (Lee & Ward 1989: Fig. 6g), while the period of pouch life and the age at sexual maturity are less clearly related to body size. One factor that is not highlighted by Lee & Ward (1989) is that forest-dwelling macropodoids in general seem to show slower maturation than equivalent-sized open country species. This would make sense in terms of the relative importance of avoidance of predators, and suggests that *Dorcopsis* species may show more prolonged pouch life, delayed weaning and slower sexual maturation than other medium-sized macropodids. However, this interesting observation is based on a fairly limited sample of forest-dwelling species and needs further investigation.

Important events in the life history of kangaroos and wallabies are birth, the end of pouch life, the time of weaning, and the onset of sexual maturity. As outlined in section 8.3, the timing of these events is either reported for *Dorcopsis muelleri* or else it can be reasonably inferred from data on closely related species. In addition, comparative studies

Table 8.10. Life history parameters of the four macropodid species studied, with the equivalent molar eruption and progression (LMI-2) stages. Sexes are pooled.

	Macropus rufus	*Macropus agilis*	*Dorcopsis muelleri*	*Setonix brachyurus*
body weight (range in kg)	37-93	13-30	5-10	2.7-3.3
pouch life (days)	235	219	183	190
pouch life (molar eruption)	0.1	± 0.1	0.1-0.3	1.0
pouch life (LMI-2)	1.18	1.13	?	<2.0
age at weaning (days)	360	328	275	240
age at weaning (molar eruption)	0.4	1.0	1.2	2.0
age at weaning (LMI-2)	1.48	1.50	<2.0	2.55
age at first reproduction (months)	17-30	12-14	15-17	9-13
age at first reproduction (molar eruption)	1.3-2.2	1.1-1.2	3.0-4.0	3.4-4.0
age at first reproduction (LMI-2)	1.80-2.38	1.75-1.81	<3.7	3.38-3.53
life span (years)	16-18	10-12	c. 10	?

Note: The details of life stages for *Dorcopsis muelleri* were taken from Flannery (1995a) and Menzies (1989; 1991); see also footnote 1 in this chapter for comments on body weight. The value for age at first reproduction is taken from the closely related species *D. luctuosa*. For the other three species details were obtained from Jarman (1989) for body weight; Shield (1958) and Lee & Ward (1989) for pouch life, age at weaning, and age at first reproduction; Kirkpatrick & Johnson (1969) and Archer & Flannery (1985) for life span. Some variation exists in the values provided by different authors; in most cases this probably reflects differences between wild and captivity study animals, or values for pooled sexes. Ages of the museum reference specimens were calculated with the equations (based on Upper Molar Progression Index) given by Kirkpatrick (1965) for *M. rufus*, and Newsome et al. (1977) for *M. agilis*. Such equations are not available for *S. brachyurus*, but Shield (1958: 58) gives eruption stages for various life history events. Eruption stages of the three reference species were recorded for all museum specimens involved. Lower Molar Progression Index (LMI-2) in the table is associated with the calculated age in days and/or molar eruption as recorded for the museum specimens. Molar eruption and associated LMI-2 of *Dorcopsis muelleri* is estimated from the patterns found in the reference species.

of life history parameters within the Macropodoidea indicate a high degree of consistency in the timing of these events (Lee & Ward 1989).

For a total of ten kangaroo species, for which their approximate life span was indicated in the literature, the pouch life, the age at weaning and the age at sexual maturity could be established, albeit with varying levels of certainty related to the differences in timing between males and females (not consistently presented in the literature), and between wild and captive animals. The age at which the offspring leaves the pouch falls for most of these macropodoids between c. 4-6% of their life span (see Fig. 8.12).

As indicated by Lee & Ward (1989), the proportion of life spent in the pouch is thus fairly constant within this group and does not vary in relation to the size of the species. The age at which the young is weaned is also relatively constant, falling for most

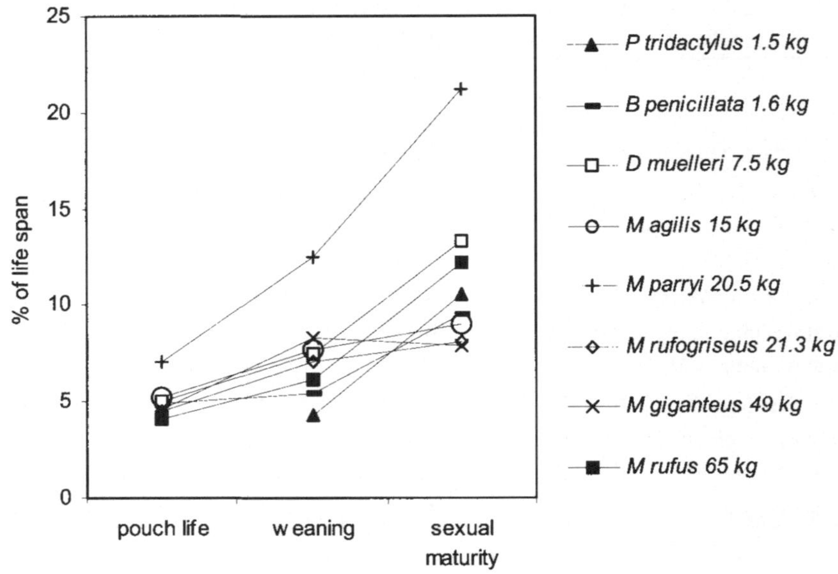

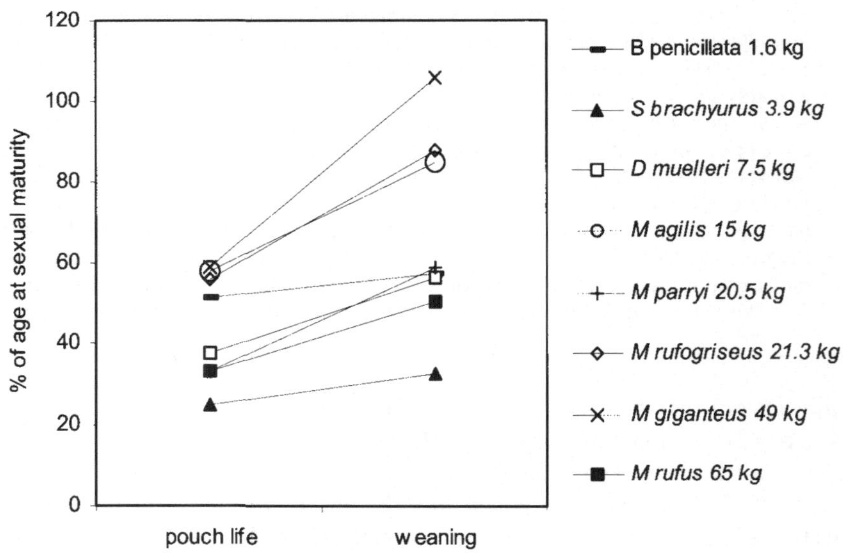

Figure 8.12. Life stages of various kangaroos as proportions of their life span (top) and as proportions of their age at sexual maturity (bottom). Body weights and life stages are averages for males and females. For *M. giganteus* only an age at sexual maturity for females is available, hence this figure is slightly underestimated; its weaning age in the bottom figure is therefore overestimated. Of *P. tridactylus* no data on pouch life are available. Of *S. brachyurus* no data on life span are available.

macropodoids between ages at c. 5-8 % of their life span. The time to sexual maturity seems to be less constant but usually falls between c. 8-13% of their life span. While these patterns are probably reasonably accurate, it has to be noted that life span values may be subject to considerable error since this is difficult to estimate for wild animals, and captive individuals may live much longer than wild ones. This may be the reason, for example, why *Macropus parryi* (the Whiptail Wallaby) appears to be exceptionally slow to reach each of the critical life stages (see Fig. 8.12 top). This species conforms more closely with other macropodoids if the age of pouch exit and weaning are taken proportional to age at sexual maturity rather than to life span (Fig. 8.12 bottom).

Although the overall life history pattern is not clearly linked to variation in body weight, there does seem to be a tendency for larger species (e.g. *M. giganteus*, *M. rufogriseus*) to have a fairly late weaning, but to attain sexual maturity quickly, immediately after (or even before) weaning has been completed. Small species such as the Brush-tailed Bettong (*Bettongia penicillata*) and the Long-nosed Potoroo (*Potorous tridactylus*) tend to have a relatively early weaning but a fairly late sexual maturation after weaning. This tendency is, however, not entirely consistent as one of the largest kangaroos in Australia (*M. rufus*) matches the pattern of the latter group. *Dorcopsis muelleri* fits well within the general life history pattern of the smaller macropodoids but may be slower than average in attaining sexual maturity.

Comparing the molar eruption, progression and wear stages associated with the various life stages of the Red Kangaroo, the Agile Wallaby and the Quokka, it appears that the Quokka differs from the larger kangaroos predominantly in its much faster molar eruption. It is likely that this is related to the smaller teeth (relative to skull size) in this species allowing eruption to accelerate. Interesting is the relatively early stage at which the larger kangaroos are weaned and become reproductively active. In the Red Kangaroo and the Agile Wallaby weaning occurs even before the first permanent molar has fully erupted. In the Agile Wallaby, the time when the animal is first reproductively active is when the second permanent molar is only halfway its eruption. The same applies to the Red Kangaroo, although the timing of this event varies more in this species (within and between males and females) so that in some animals the second molar may have started to erupt. In the Quokka the young leaves the pouch when the first permanent molar is fully erupted, is weaned when the second is fully erupted, and does not become reproductively active until all four permanent molars are in place.

The Lower Molar Progression Index provides a basis for comparison of the timing of the important life history events in *Dorcopsis muelleri* with those in the three reference species. The LMI-2 was used rather than the UMI since the archaeological sample of lower jaws is considerably larger than that of the maxillae. Although *Dorcopsis* molars are larger than *S. brachyurus* (Quokka), they are more similar in relative size and morphology to this species than to either *M. rufus* or *M. agilis*. *Dorcopsis* pouch life is also similar in duration to *S. brachyurus* (see Table 8.10). It is therefore expected that *Dorcopsis* molar eruption will be relatively fast compared to the two *Macropus* species. A *Dorcopsis* young at the time that it first leaves the pouch of its mother is likely to have the first permanent molar in the process of erupting, perhaps between stage 0.1 and 0.3. No specimen in the archaeological sample has such a low eruption value.

The age at weaning for *Dorcopsis* falls between the values given for *M. agilis* and *S. brachyurus*. Considering that relative tooth size and the extent of molar progression in *Dorcopsis* are both more similar to *S. brachyurus* than *M. agilis*, for this life event it seems reasonable to estimate a molar eruption of 1.2-1.3. This is associated with a maximum LMI-2 of 2.0.

As indicated above, *Dorcopsis muelleri* is probably slow to reach sexual maturity, with a relatively long gap (compared to other Macropodoidae) from when it is fully weaned and independent of its mother until the animal becomes reproductively active. While the two *Macropus* species in Table 8.10 are far from having even their third molar fully erupted, *Setonix brachyurus* has all teeth (including still its deciduous teeth) in place when it is sexually mature. With its late maturity and its eruption pattern more or less comparable to the Quokka, *Dorcopsis* may be assumed to have at least an eruption stage of 3.0, if not already a fully developed set of molars. This corresponds to an LMI-2 around 3.7 (the lowest value for eruption stage 4). This pattern of dental development for *Dorcopsis* may be somewhat speculative, but it provides an important basis for the interpretation of the observed distributions of molar eruption and molar progression values for the hunted population (see below).

8.5.6. *Age structure of the prehistoric* Dorcopsis *sample from Kria Cave*

The archaeological *Dorcopsis* sample from Kria Cave includes a total of 340 jaws and isolated teeth. This includes 82 lower jaw fragments and 26 relatively intact maxillae (both allowing calculation of the relevant molar progression index). The bulk of the sample comes from Unit II (118 specimens) and Unit V (172 specimens), with Units I, III and IV contributing one, 26 and 23 specimens respectively. To a large extent the following analysis will focus on the two larger samples.

As shown above for each of the three reference kangaroo species, the most accurate indicators of age are the indices of molar eruption and progressions. The age profile of the prehistoric *Dorcopsis* samples based on the more complete jaws and maxillae will be examined first, and then the more abundant but less precise data obtained from wear stages and molar crown height.

Molar eruption scores for the upper and lower jaws (see Fig. 8.13) show an overwhelming predominance of mature individuals with all four molars fully erupted, and an even spread of younger animals, down to a minimum eruption stage of 1.2 (i.e. first molar fully erupted and both lophs of second molar above the jaw line, see Fig. 8.6). In both the upper and lower jaw samples the dominance of old animals is slightly stronger in Unit V than in Unit II, although the difference is not statistically significant.

Molar progression scores are available for a larger sample of individuals than eruption scores. As in the reference kangaroo species, lower molar progression in Kria *Dorcopsis* is highly correlated with molar eruption stage (see Fig. 8.14). An essentially linear relationship is maintained through until the maximum eruption stage of 4, at which

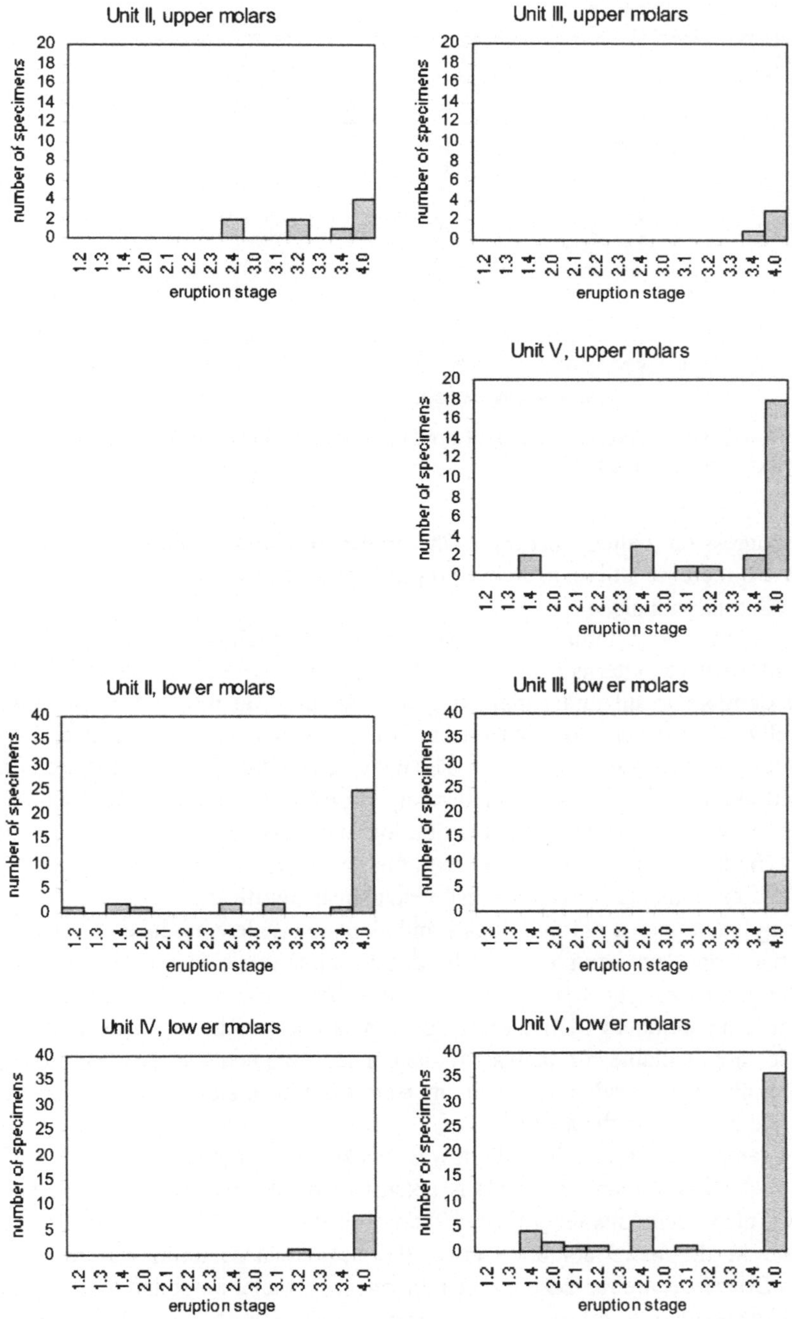

Figure 8.13. Kria Cave, *Dorcopsis muelleri*. Upper and lower molar eruption. Unit IV yielded no specimens of which the upper molar eruption could be determined.

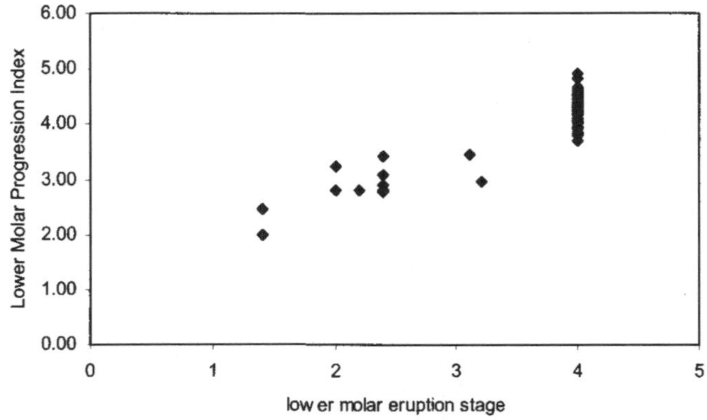

Figure 8.14. Kria Cave, *Dorcopsis muelleri*. Correlation between lower molar eruption and lower molar progression (LMI-2).

point the progression values disperse. The molar progression data might thus be expected to be more sensitive than the eruption data to differences in age structure among the older individuals.

Histograms of lower molar progression do in fact show more complex and potentially informative patterns (Fig. 8.15a). LMI-2 scores are strongly skewed towards values of 4 or more in all units, indicating that not only do most of the individuals possess a fully erupted dentition, but one in which all of the molars have undergone some post-eruptional forward movement. The patterns in Unit V and Unit II are similar but subtly different. Unit V shows a moderately strong secondary peak at around 2.75, while Unit II shows a somewhat flatter distribution, with less clear emphasis on the older individuals. The latter result is consistent with the lower peak at eruption stage 4 in Unit II, but the difference also fails to pass a test of statistical significance.

The strong bias towards old individuals and almost complete absence of very young animals in the Kria Cave sample might be due to taphonomic factors that favour the survival of larger, more robust jaws. Proof that this factor is at work in the site is found in the disparity in numbers of mandibles over maxillae within the collection. Much larger samples are available for analysis if smaller jaw fragments and isolated teeth are included. This also has the advantage that any teeth from the disintegrated jaws of young animals will be included in the analysis.

The total sample size for each of the lower cheekteeth is shown in Table 8.11. If the abundance of the first molar (N = 123) is taken to indicate the overall sample size of original jaws, then deciduous teeth (N = 19) were present in 15.4% of cases. Based on the condition in the more complete jaws, the deciduous sectorial and molariform premolars in *Dorcopsis* are replaced by the permanent premolar at around molar eruption stage 2.4 (corresponding to a Lower Molar Progression Index of 2.77-3.43). Among the 102 relatively intact jaws, a total of 19 (or 18.6%) retain the deciduous teeth. The close agreement of these values shows that the relatively low numbers of young animals is not an artefact of differential destruction, but is a genuine feature of the prehistoric sample.

Table 8.11. Kria Cave, *Dorcopsis muelleri*. Number of lower tooth elements in the sample.

tooth	left	right	total
dP_3	7	8	15
dP_4	10	9	19
P_4	55	42	97
M_1	63	60	123
M_2	64	55	119
M_3	61	52	113
M_4	39	40	79

A very small proportion of isolated deciduous teeth were lost presumably due to a combination of taphonomic factors and excavation methods.

The analysis of modern species demonstrated that tooth wear stages and molar crown height are both valid indicators of individual age. The *Dorcopsis* lower molars are more numerous than the upper molars, therefore the analysis will concentrate on the lower molars.

The relationship between Lower Molar Progression Index (LMI-2) and molar wear stage is shown for each tooth in Figure 8.16. For M_1 and M_2 the molar wear stages are obviously correlated with LMI-2 for as long as the latter value displays any variation. However, by the time these teeth reach wear stage 4, molar progression has effectively ceased, thereby removing any further association between these values. However, it may be reasonably assumed that continued molar wear is at least broadly correlated with advancing age, perhaps up until wear stage 5 or 6 if the reference kangaroo species are taken as valid comparisons.

Histograms of lower molar wear stages are shown for each tooth in Fig. 8.17 (at the end of this chapter), with the sample divided between the four stratigraphic units (Unit I contains only one specimen). The two largest samples from Units II and V show very similar distributions of wear stages for each individual tooth. This impression is confirmed by statistical testing for differences among *k* independent samples, using Kruskal-Wallis one-way analysis of variance by ranks (Table 8.12). In both units the M_1 shows a relatively flat distribution of wear stages with approximately equal numbers in each of wear stages 1 to 5 but a slight peak at wear stage 2. The M_1 of *Dorcopsis* reaches wear stage 2 and 3 at LMI-2 scores of 2.67 and 3.80, respectively, and the fact that 84% of M_1s are at wear stage 2 or higher is consistent with the observation that 93% of LMI-2 values fall above 3.25 (Fig. 8.15a and 8.16). The molar wear plots for M_2, M_3 and M_4 are each progressively more skewed towards the lower wear stages, but with appreciable numbers of very worn teeth, even for the more posterior molars. The overall impression that the *Dorcopsis* assemblage consists predominantly of mature to aged individuals is strongly reinforced by the molar wear data.

Interestingly, the smaller samples of *Dorcopsis* teeth from Units III and IV display a slight shift towards higher wear stage values for all teeth (Fig. 8.17). This difference with Unit II and V is not statistically significant if the data from all four units are compared, but attains significance at the .05 probability level against Unit V under a

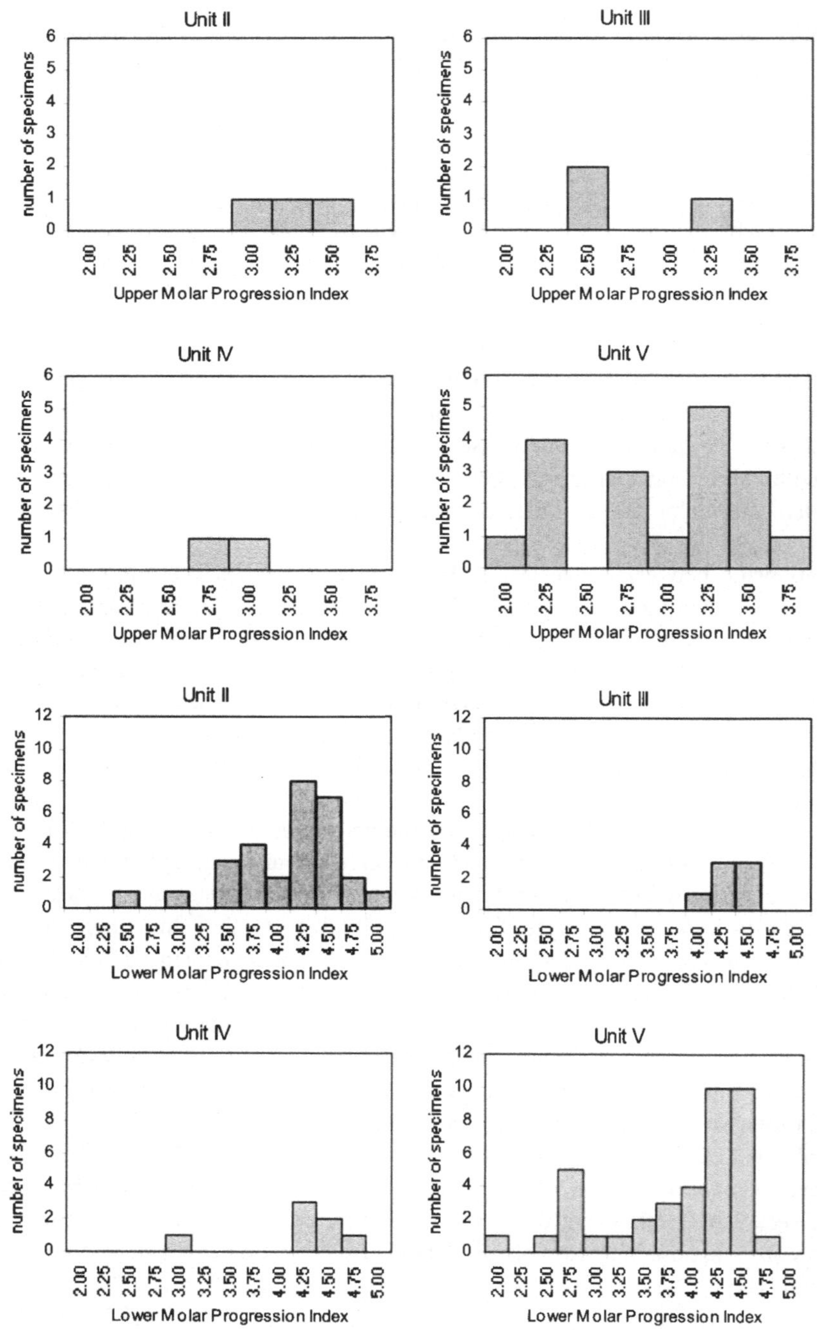

Figure 8.15a. Kria Cave, *Dorcopsis muelleri.* Upper molar progression (top) and lower molar progression LMI-2 (bottom). X-axis values are midpoints of intervals with standard range of 0.25.

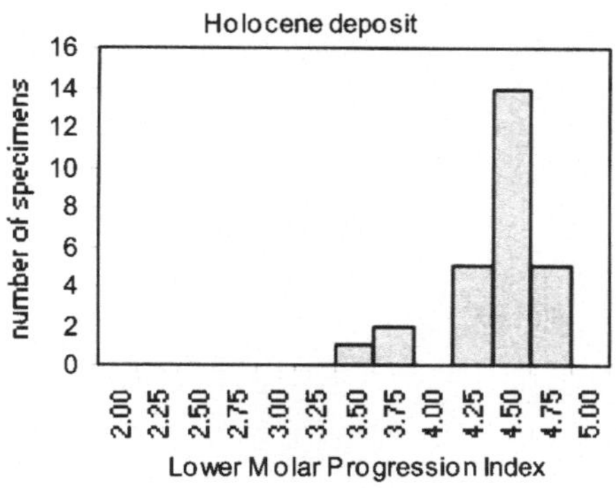

Figure 8.15b. Toé Cave, *Dorcopsis muelleri*. Lower molar progression (LMI-2).

Kruskal-Wallis test for the M_2 if the samples from Units III and IV are combined (Table 8.12). Comparison of combined Units III and IV with Unit II is almost significant (.0536) for the same tooth.

The relationship between Lower Molar Progression Index (LMI-2) and molar crown height is shown for each tooth in Figure 8.18. For each of M_1, M_2 and M_3 the crown height shows a good correlation with LMI-2. However, as with the molar wear stages, the more extreme lowering of crown height evidently takes place after the teeth have stopped any forward movement. In the case of the M_4, all of the tooth wear takes place after this event.

Histograms of molar crown heights (Fig. 8.19, at the end of this chapter) are also remarkably similar for Units II and V; this impression is supported by a lack of statistical discrimination among the means (Table 8.13). Of special interest is the broad spread of crown height values in both units for M_4, once again pointing to the presence of considerable numbers of very mature to aged individuals. Crown height data for Units III and IV again show a slight but consistent shift towards lower values (i.e. more worn teeth); the differences are not statistically significant (Table 8.13).

8.5.7. *Age structure of the prehistoric* Dorcopsis *sample from Toé Cave*

The much smaller sample of *Dorcopsis* remains from Toé Cave includes 77 dental specimens from the Holocene levels and 12 from Late Pleistocene contexts. Unfortunately, the Holocene sample cannot be subdivided without reducing the sample size below the level of meaningful comparisons.

A total of 41 specimens from the Holocene levels of Toé are sufficiently complete to provide an eruption stage. Of these, 39 are at the maximum stage 4. Stage 4 eruption is also observed in six of the seven specimens with eruption stage data from the Late Pleistocene levels. The *Dorcopsis* samples from both time periods in Toé are thus dominated by mature to aged animals, consistent with the situation in the Kria Cave assemblage.

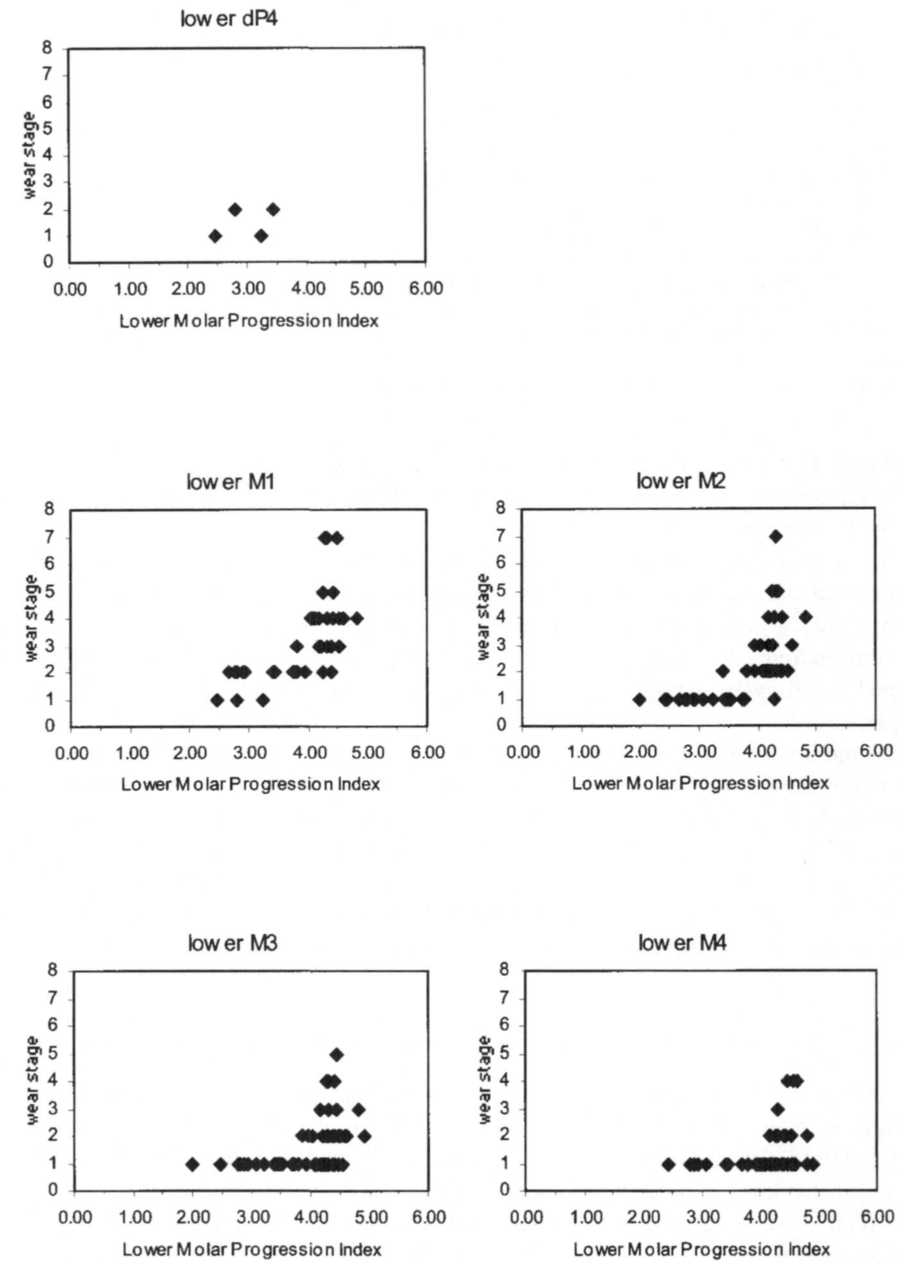

Figure 8.16. Kria Cave, *Dorcopsis muelleri*. Correlation between lower molar progression and lower molar wear.

Table 8.12. Kria Cave, *Dorcopsis muelleri*. Statistical comparison of lower molar wear between various time periods.

lower molar wear		dP_3	dP_4	P_4	M_1	M_2	M_3	M_4
Kria Cave Unit II vs.	X^2	1.5000	5.4000	.5636	.6885	.2579	.4450	.0997
Kria Cave Unit V	p	n.s.	.0201	n.s.	n.s.	n.s.	n.s.	n.s.
Kria Cave Unit II vs.	X^2	1.0000	1.0000	.9457	.9718	3.7252	.7966	.1633
Kria Cave Units III+IV	p	n.s.	n.s.	n.s.	n.s.	n.s.*	n.s.	n.s.
Kria Cave Unit V vs.	X^2	.0000	.0164	3.0527	2.4842	6.3215	2.5639	.4662
Kria Cave Units III+IV	p	n.s.	n.s.	n.s.	n.s.	.0119	n.s.	n.s.
Kria Cave Unit II vs.	X^2	.0000	.0000	1.8252	2.5073	5.0024	3.0869	.9673
Toé Cave Holocene deposit	p	n.s.	n.s.	n.s.	n.s.	.0253	n.s.	n.s.
Kria Cave Unit V vs.	X^2	.8000	3.3000	.8982	5.8125	10.4915	7.0528	1.6990
Toé Cave Holocene deposit	p	n.s.	n.s.	n.s.	.0159	.0012	.0079	n.s.
Kria Cave Units III+IV vs.	X^2	.5000	.5000	4.5714	.0491	.0105	.3941	.2422
Toé Cave Holocene deposit	p	n.s.	n.s.	.0325	n.s.	n.s.	n.s.	n.s.

Note: A comparison with the Pleistocene sample from Toé is excluded because there are too few specimens. Kruskal-Wallis One-Way Anova: n.s. = not significant ($p > .05$); d.f. 1. * value just above .05.

Table 8.13. Kria Cave, *Dorcopsis muelleri*. Statistical comparison of lower molar crown height between various time periods.

lower molar crown height		dP_3	dP_4	P_4	M_1	M_2	M_3	M_4
Kria Cave Unit II vs.	X^2	4.6047	4.1832	.6746	.5065	.0702	.8788	.4040
Unit V	p	.0319	.0408	n.s.	n.s.	n.s.	n.s.	n.s.
Kria Cave Unit II vs.	X^2	1.5000	1.4035	.1930	1.3544	1.0810	2.0082	.5838
Units III+IV	p	n.s.	n.s.	n.s.	n.s.	n.s.	n.s.	n.s.
Kria Cave Unit V vs.	X^2	.5189	.0000	.1972	.4146	.4110	.5345	.2426
Units III+IV	p	n.s.	n.s.	n.s.	n.s.	n.s.	n.s.	n.s.
Kria Cave Unit II vs.	X^2	1.5000	2.0000	.0054	6.2251	7.2693	3.7141	2.2505
Toé Cave Holocene deposit	p	n.s.	n.s.	n.s.	.0126	.0070	n.s.*	n.s.
Kria Cave Unit V vs.	X^2	.2866	.4093	2.5583	4.4150	7.0709	1.6143	.6796
Toé Cave Holocene deposit	p	n.s.	n.s.	n.s.	.0356	.0078	n.s.	n.s.
Kria Cave Units III+IV vs.	X^2	.5000	.0000	1.0876	1.2177	2.0231	.0683	.3029
Toé Cave Holocene deposit	p	n.s.	n.s.	n.s.	n.s.	n.s.	n.s.	n.s.

Note: A comparison with the Pleistocene sample from Toé is excluded because there are too few specimens. Kruskal-Wallis One-Way Anova: n.s. = not significant ($p > .05$); d.f. 1. * value just above .05.

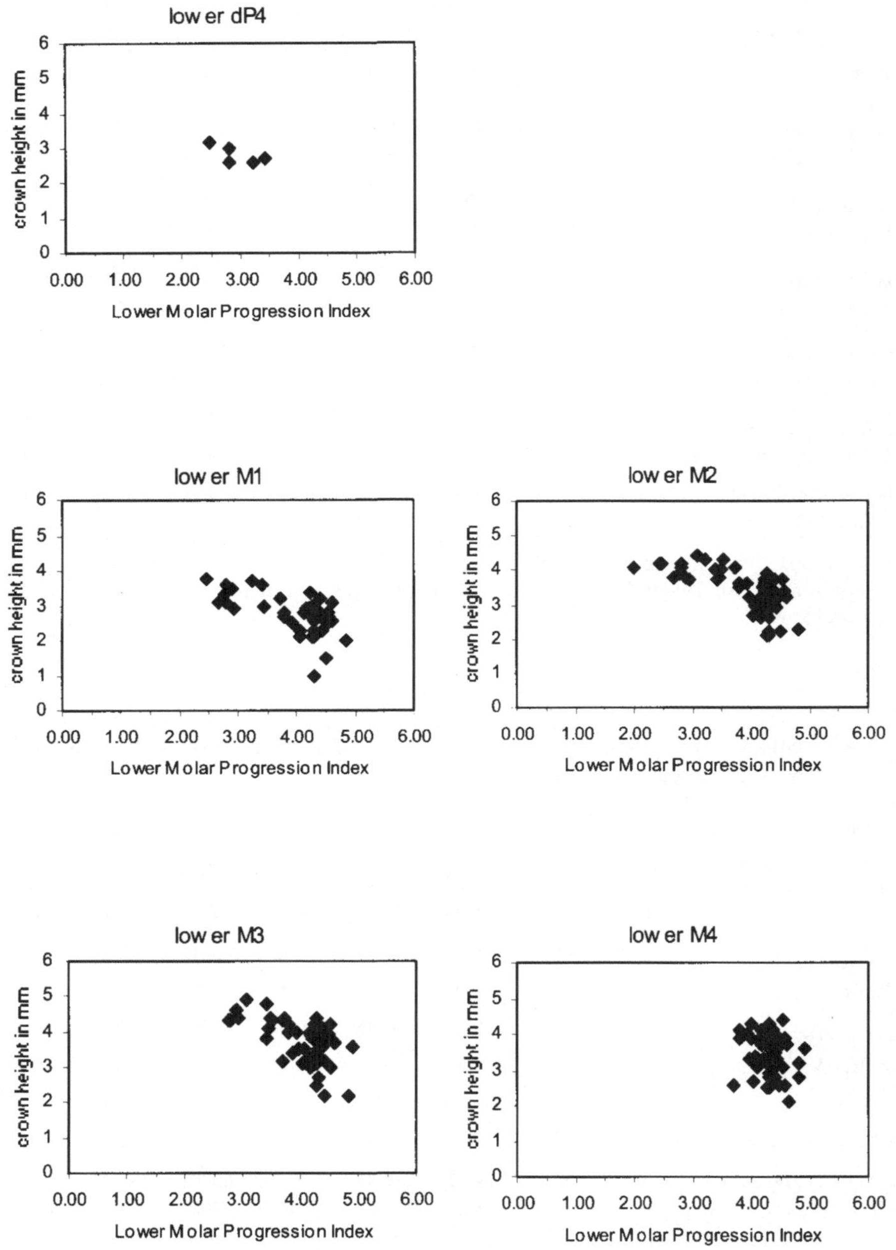

Figure 8.18. Kria Cave, *Dorcopsis muelleri*. Correlation between lower molar progression and lower molar crown height.

Table 8.14. Statistical comparison of *Dorcopsis* lower molar progression (LMI-2) between various stratigraphical time periods in Kria and Toé Caves.

lower molar progression	X^2	p
Kria Cave Unit II vs. Kria Cave Unit V	1.3595	n.s.
Kria Cave Unit II vs. Kria Cave Unit III+IV	1.8889	n.s.
Kria Cave Unit V vs. Kria Cave Unit III+IV	5.6716	.0172
Kria Cave Unit II vs. Toé Cave Holocene	8.4752	.0036
Kria Cave Unit V vs. Toé Cave Holocene	16.8313	.0000
Kria Cave Unit III+IV vs. Toé Cave Holocene	1.8538	n.s.

Note: Kruskal-Wallis One-Way Anova: n.s. = not significant ($p > .05$); df = 1.

More detailed comparisons using data from molar progression give interesting results. These data show that the Holocene *Dorcopsis* sample from Toé Cave consists of significantly older individuals than the samples from Units II and V from Kria Cave (Table 8.14 and Fig. 8.15b). The individuals in the combined Units III and IV are also significantly older than those in Unit V. The contrast between the Holocene Toé Cave sample and that from combined Units III and IV of Kria is less marked.

8.5.8. *Interpretation of the age structures of the prehistoric* Dorcopsis *samples*

The various indicators of age structure in the prehistoric *Dorcopsis* populations from Kria and Toé Caves give a remarkably consistent picture. The great majority of *Dorcopsis* wallabies consumed at both sites were mature to aged individuals and there is little evidence for the consumption of young animals. In Kria Cave, Units II and V, representing periods when most of the bone was deposited at the site, the *Dorcopsis* samples consist of slightly younger (albeit mature) individuals than those consumed during Unit III-IV times. The Holocene sample from Toé Cave displays an older age profile than the samples from Kria Cave Units II and V, and slightly older even than the sample from Units III-IV.

With a minimum eruption stage of 1.2 found in one lower jaw, and an absence of unworn deciduous molars, it is clear (Table 8.10) that remains of pouch young are not present in Kria Cave (nor in Toé). The youngest animal was at an age at which it was just about to become independent of its mother. The time of weaning is potentially the most important in the context of *Dorcopsis* hunting because it is up until this period that the young will be 'at foot' – it has left the pouch but is still suckling – which means that it will generally be captured along with its mother. As a very easy to catch and fairly sizeable prey item, young at this age will be especially vulnerable to predators, including hunters. Disregarding the large number of specimens for which no eruption stage could be established (incomplete jaws and fragmentary teeth), a minimum of nineteen specimens show an eruption stage between 1.4 and 2.4. These are animals which were probably independent of their mother, almost fully grown, but not yet sexually mature. A minimum of eighty-two specimens come from animals that were fully grown and old enough to reproduce. Thus, the *Dorcopsis* population shows an age profile in which all but one individual were independent of their mothers, many were close to fully grown, and the great majority were mature to aged individuals.

To understand what these age profiles might mean in terms of the potential impact of prehistoric hunting on *Dorcopsis* populations it is necessary to look at some theoretical concepts derived from the fields of population ecology (Krebs 1978) and wildlife management (Caughley 1977; Getz & Haight 1989). A useful framework is provided by the general relationship that appears to exist between sustainable yield and population density relative to the carrying capacity of a given environment (Getz & Haight 1989). This relationship is based on population models (reviewed by Milner-Gulland & Akçakaya 2001) and is supported by numerous examples from fisheries and wildlife management (reviewed by Getz & Haight 1989; also Fowler 1981a, b; Roseberry & Woolf 1991).

Three hypothetical situations will be postulated for the prehistoric exploitation of *Dorcopsis* in the Ayamaru area; 1) people are hunting a *Dorcopsis* population which exists at densities very much below the carrying capacity of its environment; 2) people are hunting a *Dorcopsis* population which exists at moderately high densities, but still significantly below the carrying capacity of its environment; and 3) people are hunting a *Dorcopsis* population which exists at densities approaching the carrying capacity of its environment. For each scenario it is possible to make predictions regarding both the age structure of the *Dorcopsis* population and the potential impact of hunting activity. For the sake of simplicity it will be assumed that the prehistoric catch was not strongly biased towards either sex; this assumption is supported by the analysis of tooth and jaw measurements.

Animal populations at densities very much below the carrying capacity of their environment are generally under intense pressure from one or more of the following factors: interspecific competition, predation, and disease or parasitism (Krebs 1978). Any of these factors can lead to increased mortality and/or reduced birth rates, and the factors often work synergistically (i.e. intense predation can cause a decline in fecundity). Populations at densities very much below carrying capacity generally include few breeding adults and a very high proportion of juveniles (see Fig. 8.20), few of which will reach breeding status. The maximum sustainable yield under such circumstances is typically very low, although higher yields can be sustained if young animals are specifically targeted. Importantly, in a population that is already under intense pressure, overexploitation of breeding-age animals can rapidly bring about local extinction. This is especially so where the species in question has a low 'intrinsic rate of increase', itself a function of age at first breeding, litter size, interval between litters and longevity (Caughley 1977: 111-112; Alvard et al. 1997; Bodmer et al. 1997). *Dorcopsis muelleri*, with a relatively slow rate of maturation, a litter size of one, an interval between litters of at least 6 months, and a likely maximum life span of around 10 years, qualifies as a relatively vulnerable species in these terms.

At higher population densities, but still well below carrying capacity, the expected population structure of *Dorcopsis* should approach a typical 'pyramid' form, with

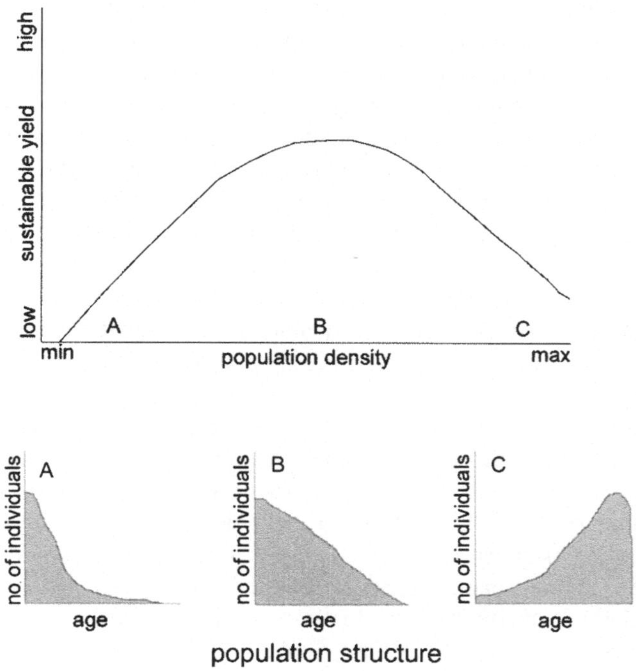

Figure 8.20. Relationship between sustainable yield and population density relative to the carrying capacity of a given environment. The population structure at three different levels of population density are illustrated in A, B and C. Population structure A, with a large proportion of very young animals, is typical for low density populations; random culling has to remain low to save the population from extinction as few animals have reached breeding stage. Structure B allows for higher levels of random culling with yields consisting of primarily young to mature individuals. Hunting will remove a large proportion of the breeding females but the high numbers of subadult individuals allow for a rapid recovery. Structure C is characteristic for high population densities; the already saturated environment leaves little room for recruitment. Even low levels of culling, consisting of predominantly mature to old animals, will result in 'thinning' of the population, increased recruitment, and a shift towards the structure illustrated in B.

relatively large numbers of young animals, fewer adults and even fewer aged individuals (Odum 1971). Population densities in this range may still be influenced by competition, predation, etc., however, others factors such as behavioural or social interactions (e.g. territory defence) may also be important. The maximum sustainable yield value will attain its own peak values within this zone (see Fig. 8.20) where population densities are well below carrying capacity. Some optimisation of the maximum sustainable yield value is possible within this zone by varying the age structure of the catch. However, excessive hunting of either adults or juveniles can be expected to lead to a decrease in population density and hence in the maximum sustainable yield itself.

Population densities approach carrying capacity only where population growth is limited solely by the productivity of the natural environment. For *Dorcopsis*, as one of the few relatively large mammals in the Ayamaru region and with no evidence of large predators prior to the arrival of humans, attainment of carrying capacity is theoretically possible, although the impact of disease and parasites cannot be assessed. Populations at or near the carrying capacity are usually dominated by mature individuals (see Fig. 8.20), with limited opportunities for recruitment of young animals and correspondingly high natural mortality rates among young individuals. Somewhat surprisingly, the maximum sustainable yields decline markedly as population density approaches carrying capacity, at least under a random culling regime. This is because random culling will remove proportionally large numbers of breeding females thereby having an immediate impact on the reproductive potential of the population as a whole.

The age structure of the prehistoric *Dorcopsis* samples can now be considered in the light of these hypothetical harvesting regimes. As noted above, the samples from both Kria and Toé Caves consist predominantly of mature to aged individuals, with relatively few immature animals. If this heavy bias towards older animals is purely a reflection of selectivity in hunting or in transport of the largest carcasses to the caves, then little can be said regarding hunting or its impacts. However, such strong, sustained selectivity seems very unlikely in an environment that can be characterised as 'poor' in terms of soil quality and primary productivity (see Elmberg 1955: 8; Miedema 1986: 1; Polak 1998; and Chapter 2). For the present, it thus seems more reasonable to assume that the age structure of the prehistoric *Dorcopsis* is a reasonably close reflection of the age structure of the natural population at the time.

A hunting catch consisting largely of mature to aged wallabies is difficult to explain in either of the first two scenarios of harvesting from population densities below carrying capacity. If *Dorcopsis* population densities were already depressed well below carrying capacity through predation or other factors, random hunting would be expected to produce a strong bias towards young animals. Furthermore, sustained selective hunting of mature to aged individuals in this circumstance would be likely to result in the local extinction of the species. Non-selective hunting of a population closer to, but still well below carrying capacity is also expected to produce a catch with relatively large numbers of young animals and with relatively few aged individuals. Sustained selective hunting of mature to aged animals in a population of this kind would also be likely to result in population decline, although this may be compensated for by increased survival of young animals provided that hunting pressure was not too extreme.

The prehistoric age profile is clearly more compatible with the third scenario where hunters exploited a *Dorcopsis* population at or near the carrying capacity of its environment. Under this scenario, the catch is expected to consist predominantly of mature to aged individuals, hence there is no need to postulate any selective hunting to account for the prehistoric age profiles. However, this scenario does have other implications. As indicated above, populations at or near carrying capacity typically have low sustainable harvest levels since females have a low reproductive potential, and are sensitive to overexploitation, with changes in age structure being one likely consequence of excessive harvesting. The fact that the prehistoric age structure shows only minor variations through time thus implies that the harvesting rate remained very low throughout the prolonged period of occupation of both sites. This conclusion is consistent with the estimates of very low consumption rates of *Dorcopsis* at both sites.

Two final observations should be made. First, it is worth noting that the slight variations in age structure between the prehistoric *Dorcopsis* samples appear to correlate well with changes in the overall quantities of faunal remains. This pattern is present in the Kria Cave sequence, where Units III and IV, with the lowest quantities of faunal remains per unit time, show a slightly 'older' age profile than either of Units II or V. Similarly, the Holocene unit in Toé Cave, with a lower quantity of faunal remains than any unit in Kria, shows an even more 'aged' *Dorcopsis* population. This pattern of correlated changes, however weak, is consistent with the notion that more or less intensive exploitation of the local *Dorcopsis* population did in fact cause slight changes in age structure and by inference, in its local abundance.

Another point that needs to be made is that while the Ayamaru *Dorcopsis* population may have been at or close to the carrying capacity of its environment, this does not necessarily imply that it was common, only that it was possibly as common as the area would allow. Relatively low prey densities may have always been a feature of the karst environment around the Ayamaru Lakes.

8.6. BODY PART REPRESENTATION AND FRAGMENTATION

Several studies have examined the processing and transport of prey by modern hunters (Binford 1978; Bunn et al. 1988; Bartram 1993; Lupo 1995). These, in turn, have brought about a rapidly growing theoretical literature on the impact of this behaviour on archaeological assemblages (e.g. Binford 1981; O'Connell et al. 1988; Lyman 1992). In general, the likelihood of transport of an entire carcass depends on prey body size, distance of the capture point from a base camp, and the size of a hunting party (Binford 1978; O'Connell et al. 1990; Bartram 1993). Small prey items are generally transported whole, but are also more likely to be consumed during travel than larger items. Medium-sized animals may be carried whole or less useful or less edible parts (e.g. gut contents) may be removed and the carcass brought back to a campsite. Very large prey items may require dismemberment to be transported at all; in such cases selective transport of body parts is more likely to occur, with selection perhaps based on several factors including the nutritional value of the part and the further usefulness of its components (e.g. skin, bone, sinew).

The extent to which prey items are utilised for their potential food value can also vary. If protein is in abundant supply, bones may be left largely intact, and people may even be selective about which parts of the carcass they prefer to eat. Conversely, if protein is scarce, people may go to great lengths to extract every scrap of value from the carcass; the leftover skeletal remains will typically be highly fragmented (Bulmer 1976;

Binford 1981; Gould 1996). Bulmer (1976: 180) describes this situation for the Kalam people of the Papua New Guinea Highlands: 'The extent to which bone is eaten reflects the amount of meat available at the time, and the hunger of those present. While most Kalam are, for example, quite capable of chewing up the wing- and leg-bones of a domestic fowl, they do not bother to do so if food is plentiful and they are replete'.

Ethnographic observations on the transport and processing of game by New Guinean peoples are relatively few in number. Studies among the Gadio Enga (Dornstreich 1973), Rofiafo (Dwyer 1974), Kalam (Bulmer 1976), and Etolo (Dwyer 1980, 1983) peoples of Papua New Guinea suggest a 'standard' treatment of game that involves transport of the entire carcass (or as live animals, as was observed during the fieldwork in the Ayamaru area; see also Brongersma 1958: 11), the removal of the gut (which is either discarded or roasted and consumed after removal of its contents), followed by cooking of the carcass in an earth oven. Smaller prey items are sometimes roasted inside plugged sections of bamboo (Elmberg 1955: 52; Pospisil 1963; Bulmer 1976: 177; Sillitoe 1988: 195). For small animals, Dwyer (1980: 110) also stated that 'all the bones, and the skull but not the mandibles, are usually eaten'. Bulmer (1976: 180) also found that the Kalam do not eat 'jaws of small mammals, reptiles and frogs, and beaks and feet of small birds'. All other bones, of animals up to the size of a pigeon, including smaller and softer bones of larger animals, are consumed. Brongersma wrote in 1958, 'In the Papuan village of Mefkadjem (near Ajamaroe in the Vogelkop Peninsula) I had the experience of eating a small piece of cuscus meat, and it proved quite palatable. The animal had been stewed in its skin to prevent the loss of any of the juice.' (1958: 11). The same treatment is given to 'animals smaller than pig and cassowary' by the Kalam (Bulmer 1976: 180). Dismemberment prior to cooking is more common for carcasses of these larger animals (Bulmer 1976: 180) that also tend to be distributed among households within a community (Rappaport 1968; Dwyer & Minnegal 1991).

Dorcopsis muelleri, with an individual body weight up to 10 kg, does not represent a significant load for a hunter to carry, unless of course multiple individuals were obtained on a single hunting trip. In this circumstance, some differential transport of body parts might be expected, especially if carcasses are being transported over any significant distance prior to consumption. More importantly, these practices might be expected to change in the event that *Dorcopsis* became locally less abundant, forcing hunters to travel increasing distances in search of prey. Longer distance travel might also increase the likelihood that hunters would be returning with more than just a single *Dorcopsis*, unless of course the hunt was unsuccessful.

To investigate these possibilities, body part representation of *Dorcopsis* was analysed for the various stratigraphic subsamples within the Kria Cave assemblage. The Toé Cave *Dorcopsis* remains were not subjected to a comparable analysis on account of the overall smaller sample size and the lesser degree of stratigraphic control. Various indices of fragmentation of the Kria Cave *Dorcopsis* remains were also calculated to determine whether there was any change through time in the treatment of the carcasses.

8.6.1. *Body part representation*

Body part information was recorded in considerable detail for the Kria Cave assemblage (see Chapter 7 for details of methods). This allows fairly precise estimates to be made of the relative contribution of different body parts to the total assemblage. For each stratigraphic unit, total counts (left + right) were extracted from the data for each of the following categories:

- the most abundant individual tooth (e.g. lower first molar);
- lower jaw fragments;
- periotic bones (containing the inner ear), representing the cranium;
- proximal and distal fragments for each major limb bone;
- scapular fragments (glenoid fossa);
- pelvic girdle fragments (ilial, acetabular, pubic and ischial);
- calcaneum, astragalus and cuboid, representing the proximal foot;
- metatarsals IV and V (proximal and distal fragments separately), representing the toes.

The totals include specimens specifically identified as *Dorcopsis muelleri* and those identified as 'macropodid', the vast majority of which undoubtedly also belong to this taxon. Only very small numbers of individuals of the comparable-sized tree-kangaroos, *Dendrolagus* spp., were found in Kria Cave.

Table 8.15 shows the distribution of the *Dorcopsis* and macropodid body parts for each of the five stratigraphic units in Kria Cave. Several of the selected body parts are poorly represented in all units, most notably the distal ulna, distal radius and acetabulum. The low numbers of acetabular fragments is due in part to frequent breakage of the pelvis across this structure, leaving portions attached to the ilium or ischium. The small quantities of *Dorcopsis*/macropodid remains from Unit I are listed in Table 8.15 but will not be mentioned further.

To examine the proportional representation of the different body parts, first a mean body part count for each unit was calculated (see Table 8.15). Each count was then recalculated as a ratio relative to the mean value. A score of 1 indicates that a body part is of 'average' abundance. A value below 1 indicates that it is relatively under-represented. A value above 1 indicates that it is overrepresented. Higher or lower values indicate the strength of under- or overrepresentation. These 'standardised' values are plotted in Fig. 8.21. A similar method was introduced by Binford (1978, also 1984: 56-58) and adapted subsequently by Boyle (1990) and others for studies of Upper Palaeolithic faunas from Europe.

The pattern of body part representation is very similar in all units, but with some subtle differences. Certain elements are poorly represented in all units, most notably the periotic, scapula, proximal humerus, pubis, ischium, proximal femur, distal tibia and cuboid, In each case, the underrepresentation of these elements contrasts with an overrepresentation of other elements that occur in the same general region of the body. For example, the lower jaw and teeth are both well represented whereas the periotic, also derived from the head, is not. Similarly, the calcaneum and metatarsals are well represented, while the cuboid, which is interposed between these elements in the foot,

Table 8.15. Kria Cave, counts of various skeletal fragments for the combined categories *Dorcopsis* and 'macropodid', by major stratigraphic unit.

	Unit I	Unit II	Unit III	Unit IV	Unit V
periotic	-	15	3	1	19
lower jaw	-	29	7	7	39
most common tooth	1	39	8	9	53
scapula	-	15	7	4	27
proximal humerus	-	3	1	3	13
distal humerus	-	14	7	10	44
proximal ulna	-	20	8	8	53
distal ulna	-	-	-	-	-
proximal radius	-	16	3	9	13
distal radius	-	1	-	-	1
ilium	-	15	8	6	41
acetabulum	-	1	2	-	11
pubis	-	3	1	1	12
ischium	1	5	1	4	23
proximal femur	-	18	6	3	47
distal femur	-	42	14	7	70
proximal tibia	-	43	12	10	41
distal tibia	-	14	3	3	20
proximal fibula	-	12	7	9	35
distal fibula	-	34	15	9	58
calcaneum	-	31	10	10	27
astragalus	-	31	6	4	20
cuboid	-	12	1	1	21
proximal metatarsal IV	-	49	17	10	43
distal metatarsal IV	-	30	14	8	38
proximal metatarsal V	-	40	13	8	39
distal metatarsal V	-	26	10	6	34
mean count	-	23.2	7.6	6.3	34.6

Note: The counts include both left and right elements. The average count (bottom row) excludes the values for distal ulna, distal radius and acetabulum, which are underrepresented in all units.

is scarce. With the exception of the periotic, all of the poorly-represented elements consist of relatively soft cancellous bone. These differences in abundance are thus most likely due to the destruction of the softer bones, either during consumption or during and after burial in the deposit. The poor representation of the periotic cannot be explained this way, as it is one of the densest bones in the entire skeleton. As it is also quite easily identified, the reason for its scarcity is obscure.

If the obvious poorly-preserved elements are excluded from consideration, the Unit V sample shows a remarkably even representation of major body parts. Elements from the head (teeth), forelimb (distal humerus, proximal ulna), upper hindlimb (proximal and distal femur, proximal tibia) and lower hindlimb (distal fibula, metatarsals), and axial body (ilium) are all equally abundant. One element only stands out as being particularly overrepresented, the distal femur; this may be because this element is quite large and remains identifiable even if quite highly fragmented. This example illustrates the potential influence of fragmentation on the analysis of body part representation and highlights the care that must be taken in interpreting data of this kind.

The sample from Unit IV shows an almost identical pattern to Unit V, differing only in the higher score for the proximal radius and the lower scores for the proximal and distal femur. These differences are probably not of any great significance in terms of representation of carcass elements; for example, the lower value for the distal femur is contradicted by the higher value for the anatomically adjacent proximal tibia.

The samples from Unit III and Unit II appear to show a reduction in the forelimb elements and a corresponding and consistent increase in all hindlimb values. In the Unit III sample the head elements also appear to be underrepresented, but this is not the case for Unit II where the value for teeth is very close to those for elements from the hindlimb and the foot. A Chi-Squared test based on a subset of the elements chosen to represent these major body parts (Table 8.16) indicates that this difference between the units is statistically significant.

Although these data point to some subtle changes in body part representation through time in the Kria Cave assemblage, it is not immediately clear what this might mean in terms of human behaviour. Unit III appears to have fewer elements from the head and forelimb, and this might point to the selective transport of the wallaby hindquarters during this period, or perhaps the selective removal of the forelimb bones and cranium for use as tools or as trophies. The former option might be the least likely since prey animals are often transported live to prevent early decay of meat. Unit III and IV also showed indications of less intensive hunting in the area, with a *Dorcopsis* population structure that seems to suggest that prey was not depleted. Transport of carcasses over longer distances during this period was probably not necessary.

Table 8.16. Kria Cave, *Dorcopsis muelleri.* Counts of a representative subset of skeletal elements.

skeletal element	Unit II	Unit III	Unit IV	Unit V	totals
tooth	39	8	9	53	109
distal humerus	14	7	10	44	75
proximal ulna	20	8	8	53	89
distal femur	42	14	7	70	133
proximal tibia	43	12	10	41	106
distal fibula	34	15	9	58	116
proximal metatarsal IV	49	17	10	43	119
total	241	81	63	362	747

Note: Result of Chi-Squared analysis: $X^2 = 30.54$; df = 21; p = 0.033.

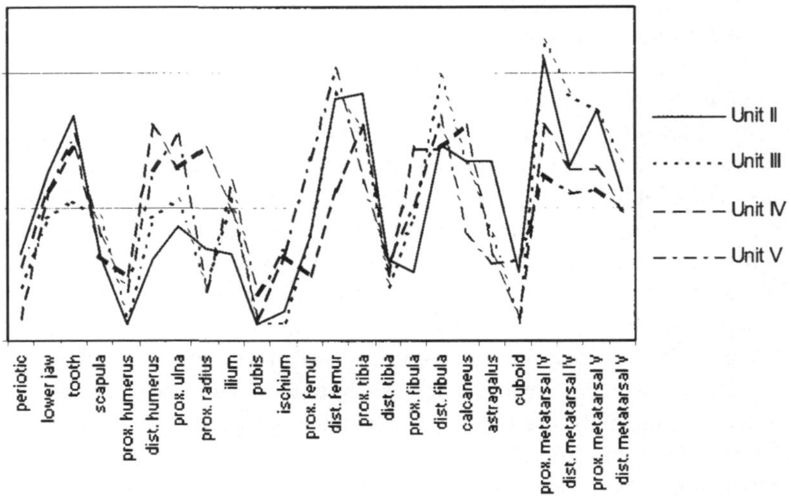

	Unit II	Unit III	Unit IV	Unit V
periotic	0.65	0.40	0.16	0.55
lower jaw	1.25	0.92	1.12	1.13
tooth	1.68	1.06	1.44	1.53
scapula	0.65	0.92	0.64	0.78
proximal humerus	0.13	0.13	0.48	0.38
distal humerus	0.60	0.92	1.60	1.27
proximal ulna	0.86	1.06	1.28	1.53
proximal radius	0.69	0.40	1.44	0.38
ilium	0.65	1.06	0.96	1.19
pubis	0.13	0.13	0.16	0.35
ischium	0.22	0.13	0.64	0.67
proximal femur	0.78	0.79	0.48	1.36
distal femur	1.81	1.85	1.12	2.02
proximal tibia	1.86	1.58	1.60	1.19
distal tibia	0.60	0.40	0.48	0.58
proximal fibula	0.52	0.92	1.44	1.01
distal fibula	1.47	1.98	1.44	1.68
calcaneus	1.34	1.32	1.60	0.78
astragalus	1.34	0.79	0.64	0.58
cuboid	0.52	0.13	0.16	0.61
proximal metatarsal IV	2.11	2.24	1.60	1.24
distal metatarsal IV	1.29	1.85	1.28	1.10
proximal metatarsal V	1.73	1.72	1.28	1.13
distal metatarsal V	1.12	1.32	0.96	0.98

Figure 8.21. Kria Cave, *Dorcopsis muelleri*. Proportional representation of body parts in each unit. A score of 1 represents the 'average' abundance of a body part; any value above and below this score represents an over- and underrepresentation, respectively.

In contrast, the Unit II assemblage does not show any deficit of head elements but only of the forelimb, which seems too small to be worth removing prior to transport, but might have been removed from the site for some special use. No evidence was found of forelimb bones being used for bone tool manufacture at either site (Chapter 6), but this cannot be ruled out as a possibility. Perhaps all that can be said is that some changes in the treatment of the carcasses or their products seem to have occurred through the period of occupation of Kria Cave, but the meaning of these changes is obscure.

8.6.2. *Burning patterns of the* Dorcopsis *remains from Kria Cave*

Analysis of burning patterns among discarded animal remains can reveal useful information on both cooking methods and the subsequent mode of discard of waste material. However, as emphasised in earlier chapters, it is essential that the impact of preservational factors on an assemblage is first taken into account. In Chapters 3 and 4 it was argued that any such impact is relatively minor and subtle. In this situation, we can thus validly examine the evidence of burning frequencies through the sequence.

Table 8.17 gives a very general overview of the burning condition of the *Dorcopsis* remains (as represented by the combined *Dorcopsis* and macropodid categories). In all units the great bulk of the remains are unburnt (or lightly cooked at most) and there is only a very small proportion of calcined bone. The proportion of burnt bone is somewhat higher in Units III-V than in Unit II; this may reflect a slightly better preservational state of bone in this unit. However, the differences are minor and are unlikely to alter any subsequent interpretations.

The distribution of burning by skeletal element and bone portion (e.g. distal, proximal) is shown in Table 8.18 for each of Units II and V, being the two largest samples. The values have been transformed to show the percentage difference from the overall burning frequency for all bones, as given in Table 8.17. A positive value thus means that a particular bone portion is more often burnt than would be expected from the average; a negative value means that it is less often burnt than the average.

The burning data for both units show some interesting patterns that are broadly consistent with the notion that the *Dorcopsis* were cooked as complete carcasses as occurs commonly throughout New Guinea today for small to medium-sized game, and was also experienced by Brongersma (1958: 11) when he visited the Ayamaru area in the 1950s. Cooking in this fashion is likely to result in charring of lightly-fleshed bones near

Table 8.17. Kria Cave, proportion of *Dorcopsis* + macropodid remains in each of three burning classes, based on number of specimens.

	unburnt %	burnt %	calcined %	total number
Unit II	84.1	15.6	0.3	782
Unit III	80.0	19.8	0.2	420
Unit IV	77.5	21.8	0.7	298
Unit V	80.3	19.0	0.7	1164

Table 8.18. Kria Cave, distribution of burning by skeletal element and bone portion for *Dorcopsis* + macropodid remains.

		Unit II				Unit V			
skeletal part	bone part	unburnt %	burnt %	calcined %	total number	unburnt %	burnt %	calcined %	total number
skull	jaws/teeth	2.8	-2.5	-0.3	259	0.8	-0.6	-0.2	412
skull	periotic	-4.1	4.4	-0.3	15	-11.9	12.6	-0.7	19
ulna	proximal	-9.1	9.4	-0.3	20	-8.6	9.3	-0.7	53
radius	proximal	3.4	-3.1	-0.3	16	-18.8	19.5	-0.7	13
humerus	proximal	15.9	-15.6	-0.3	3	12.0	-11.3	-0.7	13
humerus	distal	-5.5	5.8	-0.3	14	1.5	-0.8	-0.7	44
scapula	glenoid fossa	2.6	-2.3	-0.3	15	-2.5	-4.2	6.7	27
pelvis	all	-17.4	17.7	-0.3	21	0.9	-0.2	-0.7	69
femur	proximal	10.6	-10.3	-0.3	19	4.8	-4.1	-0.7	47
femur	distal	6.4	-6.1	-0.3	42	14.0	-13.3	-0.7	70
tibia	proximal	8.9	-8.6	-0.3	43	0.2	0.5	-0.7	41
tibia	distal	15.9	-15.6	-0.3	14	4.7	-9.0	4.3	20
fibula	proximal	-9.1	9.4	-0.3	12	-23.2	21.0	2.2	35
fibula	distal	-16.5	16.8	-0.3	34	-4.4	5.1	-0.7	58
astragalus	all	9.4	-9.1	-0.3	31	9.7	-14.0	4.3	20
calcaneum	all	-2.5	2.8	-0.3	38	6.8	-6.1	-0.7	31
cuboid	all	15.9	-15.6	-0.3	12	14.9	-14.2	-0.7	21
metatarsal IV	proximal	-6.3	6.6	-0.3	45	-14.4	12.7	1.7	41
metatarsal IV	distal	-7.2	3.6	3.5	26	-5.3	6.0	-0.7	36
metatarsal V	proximal	-7.6	7.9	-0.3	34	0.5	0.2	-0.7	26
metatarsal V	distal	-9.1	4.4	4.7	20	-4.1	4.8	-0.7	21

Note: The raw percentage values for each bone or bone portion have been transformed by subtracting the overall percentage for each burning class as shown in Table 8.17. Very small samples (such as the distal radius and ulna) have been omitted.

the ends of the limbs, such as the bones of the feet and forelimbs. These elements (e.g. the metatarsals, ulna, radius, distal humerus) are indeed more commonly burnt in both units than many of the more protected elements or parts of bones (e.g. the proximal humerus, femur, bones of the ankle joint). Interestingly, in Unit II only two of the elements listed attain a calcined state; these are the distal ends of metatarsals IV and V, bones that often exposed during the roasting process.

However, several values are clearly at odds with any simple interpretation of the burning pattern. For example, the pelvic fragments in Unit II show the highest percentage burning of all elements, despite the highly protected environment for this bone if an animal is roasted whole. In Unit V a similar anomaly is observed in the high burning rate of the periotic bone, surely one of the most protected bones within the entire skeleton. Other anomalies emerge if we compare the burning rates for anatomically adjacent bones; for example, the high burning rate of the proximal ulna and radius in Unit V but not of the distal humerus with which they articulate. These examples amply demonstrate that the pattern of burning of the remains cannot be explained solely in terms of cooking practices, but must also include a strong signal derived from the subsequent disarticulation and disposal of the carcasses. As noted in earlier contexts, some burning of bone also undoubtedly occurs after discard, as hearths are constructed in the cave floor. However, burning through this process cannot explain the clear patterning that does exist within the Kria assemblage.

8.6.3. *Fragmentation of the Kria Cave* Dorcopsis *remains*

The intentional pounding up of bones in times of stress to extract all available protein and fat resources has been recorded ethnographically (Gould 1996), and it is reasonable to expect that it may have occurred also in the past. Gould (1996) and others (e.g. Davis & Fisher 1990: 269; O'Connor et al. 1998) have argued that various indices of fragmentation of archaeological bone can serve as indicators of nutritional stress of prehistoric populations. However, others have argued that evidence of such intentional fragmentation would be most likely obscured in archaeological contexts by the numerous post-depositional processes that affect bone prior to and subsequent to burial (e.g. Walshe 2000). Scavenger activity and post-discard burning are two important factors that can cause fragmentation after humans have broken the bones to extract the marrow.

Gould (1980) describes Western Desert Australian Aborigines pounding up small animal carcasses to break the bones prior to it being consumed whole. Dwyer (1980: 110) states for Etolo people of the Southern Highlands of Papua New Guinea that small animals are generally consumed whole, except perhaps for the mandible bones. Ken Aplin (pers. comm.) has also observed people in various parts of New Guinea consume entire animals up to the size of small possums, leaving little or no archaeological trace. This practice may be especially common in areas where animal protein (or fat) is scarce, but there is no hard evidence to support this conjecture. As also argued in Chapter 7, ingested bone fragments of small animals will probably not be recovered on site, and, if so, they may be beyond identification (see also Walshe 2000: 79).

Here the analysis will focus on the extent of fragmentation of the *Dorcopsis* remains in Kria, and in particular on the question of whether the remains show any evidence of dietary stress over the c. 3000 years that *Dorcopsis* was consumed at the site.

Table 8.19 compares the fragmentation of the combined macropodid remains with that of other identified mammal and snake remains. The macropodid category shows a consistent decrease in the average fragment size throughout the deposit, with somewhat less reduction in Unit II. This trend is not seen in the other faunal categories, which appear to fluctuate at random. This fluctuation can be partly explained by the very small samples in Units I, III and IV. If we compare only Units II and V, then it becomes clear that the overall fragmentation remains stable through time for the mammals and snakes, but the size of the macropodid bone fragments reduces significantly. This result points to a specific change over time in the treatment of the macropodid remains, rather than the general operation of a taphonomic factor such as increased trampling or burning.

An independent test of the fragmentation of the macropodid remains can be made by calculating the number of fragments of each major skeletal element, standardised in each unit against the calculated MNI. This test can be performed only because it has already been demonstrated that the major body components (e.g. forelimb, hindlimb) were brought to the site in approximately equal numbers, presumably through the usual transport of whole carcasses. The results of the test are shown in Table 8.20.

On initial examination the results appear somewhat inconsistent and surprising. For the various major limb bones (humerus, pelvis, tibia and fibula), there are consistently higher numbers of fragments in Units III to V than in Unit II. The same contrast is also observed for the scapula and the ulna. Among the foot bones, different elements variously show virtually no differences between units (e.g. cuboid) or a higher number of fragments in Units II-IV than in Unit V (calcaneum, phalanges and claws).

However, these seemingly contradictory indications can be understood if the impact of fragmentation on identifiability is taken into account. In the case of the more complex foot bones, fragmentation does not generally reduce the identifiability of the fragments, unless of course this is extreme. Overall, the degree of fragmentation of these bones is low compared with the major limb bones, as indicated by their lower values for NISP/MNI. In contrast, there is considerable fragmentation in all units of the major limb bones and the pelvis and cranium. With further fragmentation, especially of the shaft sections of these bones, the ease of identification declines rapidly, resulting in a proportional fall rather than rise in NISPs. Hence, Unit II with its more fragmented macropodid remains, has more identifiable foot bone fragments per individual but fewer identifiable fragments of the major limb bones. The result is thus entirely consistent with the trend observed in the general fragmentation index and provides independent confirmation that the *Dorcopsis* remains from the upper units of the Kria Cave deposit are more heavily fragmented than those from Unit V.

Table 8.19. Kria Cave, fragmentation of the combined macropodid remains compared with other identified mammals and snakes.

	all macropodids	other mammals	snake
Unit I	0.78	0.50	0.73
Unit II	0.96	0.47	0.47
Unit III	0.89	0.32	0.57
Unit IV	1.05	0.30	0.55
Unit V	1.22	0.50	0.49

Note: The values are the combined weight divided by the NISP per category.

Table 8.20. Kria Cave, fragmentation (NISP/MNI) of *Dorcopsis* bone by body part.

skeletal part	Unit II	Unit III	Unit IV	Unit V
cranium	8.3	5.8	5.4	10.9
humerus	1.3	1.7	1.5	2.4
ulna	0.8	1.0	0.8	1.6
radius	0.3	0.2	0.5	0.4
scapula	0.4	0.5	0.4	0.5
pelvis	0.3	0.9	0.7	1.3
femur	1.5	4.1	1.1	2.6
tibia	1.6	2.1	2.1	2.4
fibula	1.6	2.9	2.2	2.7
metatarsal IV	1.0	1.7	0.7	0.9
metatarsal V	1.7	1.9	0.9	1.8
astragalus	0.4	0.1	0.1	0.2
calcaneum	0.9	0.8	0.9	0.5
cuboid	0.3	0.1	0.0	0.4
claws	1.1	1.3	1.5	0.5
phalanges	0.6	0.5	0.5	0.2
MNI (body parts + teeth)	34	12	11	42

8.7. SUMMARY AND CONCLUSION

The Brown Dorcopsis (*Dorcopsis muelleri*) was clearly the preferred prey animal through time in the Ayamaru area. Following the principles of optimal foraging theory, it would seem likely that the species was both relatively abundant and easy to catch and process. Based on experience elsewhere in tropical forests, such sustained hunting behaviour might be expected to lead to overpredation and to one of three alternative scenarios: 1. a shift in hunting focus to different resources; 2. hunters travelling further to find their preferred prey; or 3. greater mobility with people moving more frequently in search of game. Each of these outcomes might be visible archaeologically; for scenario 1 as increased bone fragmentation, followed by a shift to lower return faunal groups [increase in diet breadth (Stephens & Krebs 1986; Nagaoka 2001)]; for scenario 2 from evidence of selective transport of body parts; and for scenario 3 from episodic abandonment of the site.

Contrary to these predictions, there is no obvious evidence that *Dorcopsis* suffered any decline in the Ayamaru area ever since its rise to abundance in the early Holocene. What, then, were the factors that protected this species against local extinction or decline? One possible clue is that both Kria and Toé Caves were visited infrequently, and probably by fairly small groups of people. For Kria, it was calculated that, on average, no more than a few forest wallabies were consumed at the site per decade. Although this is almost certainly a considerable underestimate, it is clear that Kria was only occupied on an occasional and probably short term basis. Toé Cave, on the southern margin of the present lake, produced even fewer remains, attributable to one or a few wallabies per two to four centuries. This site was presumably visited even less frequently than Kria Cave, or perhaps, if the quantity of stone-working debitage is any indication, for somewhat different reasons.

Some studies have shown that hunting or harvesting pressure may lead to a reduction in size of the species in question (e.g. Stiner 2001). The teeth of *Dorcopsis* show no sign of a reduction in size through time. While this does not completely rule out the possibility of a body size reduction, it is at least consistent with the notion that *Dorcopsis* was not hunted to dangerous levels.

Much stronger evidence comes from the analysis of the population profile of *Dorcopsis* through time. All age indicators (molar eruption, molar progression, tooth wear and crown height) point unambiguously to an archaeological sample of predominantly mature to aged individuals in both sites. This age structure suggests a culling of mainly older wallabies and a continuation of this pattern over long periods of time. Minor fluctuations in Kria Cave towards still more aged animals are found during the periods of less intense occupation (Unit III and IV), and a slight shift is seen towards somewhat younger (but nonetheless mature) individuals during the formation of Unit II. Such a population structure is typical for an animal population that thrives at or near the carrying capacity of its environment. This does, however, not necessarily mean that the animals are present in large numbers. The limestone Ayamaru area today is fairly poor in natural resources for human populations and its intrinsic productivity would also have limited populations of animals in the area.

Typically, populations at or near their environment's carrying capacity with such a mature age profile, have little 'space' for recruitment of younger animals. Populations in this condition are actually rather prone to overexploitation, and will undergo a rapid decline in population density (and a change in age structure) upon culling of even a relatively small proportion of the population. The lack of change in age profile of the *Dorcopsis* population through time suggests that culling rates were always relatively low and that human hunting had little impact on the *Dorcopsis* population density.

The lack of hunting pressure on the *Dorcopsis* population is further underlined by the stability through the Kria Cave deposit in body part representation of the *Dorcopsis* remains, and the lack of change in diet breadth (see Chapter 7). Evidently, hunters did not need to travel any greater distances to catch their preferred prey (and hence more efficiently butcher the carcass), and they did not need to change their diet to achieve adequate animal protein levels.

The degree of fragmentation of *Dorcopsis* bone increases slightly through Units V-II. This could be interpreted as an indication of dietary stress, especially because this pattern is not seen in the remains of other mammals or snakes. However, the increased fragmentation in Units IV and III is difficult to reconcile with the slight increase in the average age of *Dorcopsis* in these units. Overall, Units IV and III appear to be periods when Kria Cave was less intensively occupied than before or after. The increased maturity of the *Dorcopsis* population over this period suggests that the entire area was less populated with humans, such that the wallaby population could grow even closer to the carrying capacity of its environment. The same argument can be made for rarely-visited Toé Cave, where the *Dorcopsis* sample shows an even older age profile than seen at any point in time in Kria Cave.

Unit II shows a slightly higher proportion of younger animals than any other unit in Kria Cave. Hunting pressure during this period may have caused the population density to fall below carrying capacity, leading to an increased proportion of younger animals in the population. The higher fragmentation of *Dorcopsis* bones in Unit II may thus reflect a somewhat less abundant *Dorcopsis* population, although not to the point where the occupants of Kria had to turn to a different array of animal species for their food. Interestingly, this pattern is not observed at any point in time in Toé Cave.

Toé Cave also fails to register the dramatic change in faunal abundance and composition that characterises Unit I in Kria. In this most recent period in Kria, *Dorcopsis* remains make up only a minor proportion of the total sample. While this seems to correspond to the scarcity of *Dorcopsis* in the area today, there is no equivalent change in the upper levels of Toé Cave. As suggested in Chapter 7, this may be related to different patterns of use of the natural environment on either side of the lake, perhaps with a more 'settled' occupation on the northern side and less frequently visited areas to the south, with consequently much less pressure on *Dorcopsis*.

Taking all the evidence into consideration, it can be concluded that *Dorcopsis* was hunted preferentially, but most likely infrequently, with the result that the species was able to maintain its population at or close to the carrying capacity of the local environment. This was most certainly the case around Toé Cave, but the evidence from Kria suggests a slight decline in the *Dorcopsis* population density, possibly associated with increased levels of dietary stress between c. 5000 and 4000 BP (Unit II). This shift in human behaviour may mark the onset of a process that eventually led to the situation documented ethnographically and reported by local inhabitants (see Appendix 3), characterised by an apparent scarcity of *Dorcopsis* in the area today. Local decline of this key species over the last few millennia might have occurred through a combination of factors, including increased human population densities and hunting pressure, increased forest disturbance and clearance, the introduction of the pig and its establishment as a feral animal (competition for forest floor resources), and the use of the dog in hunting.

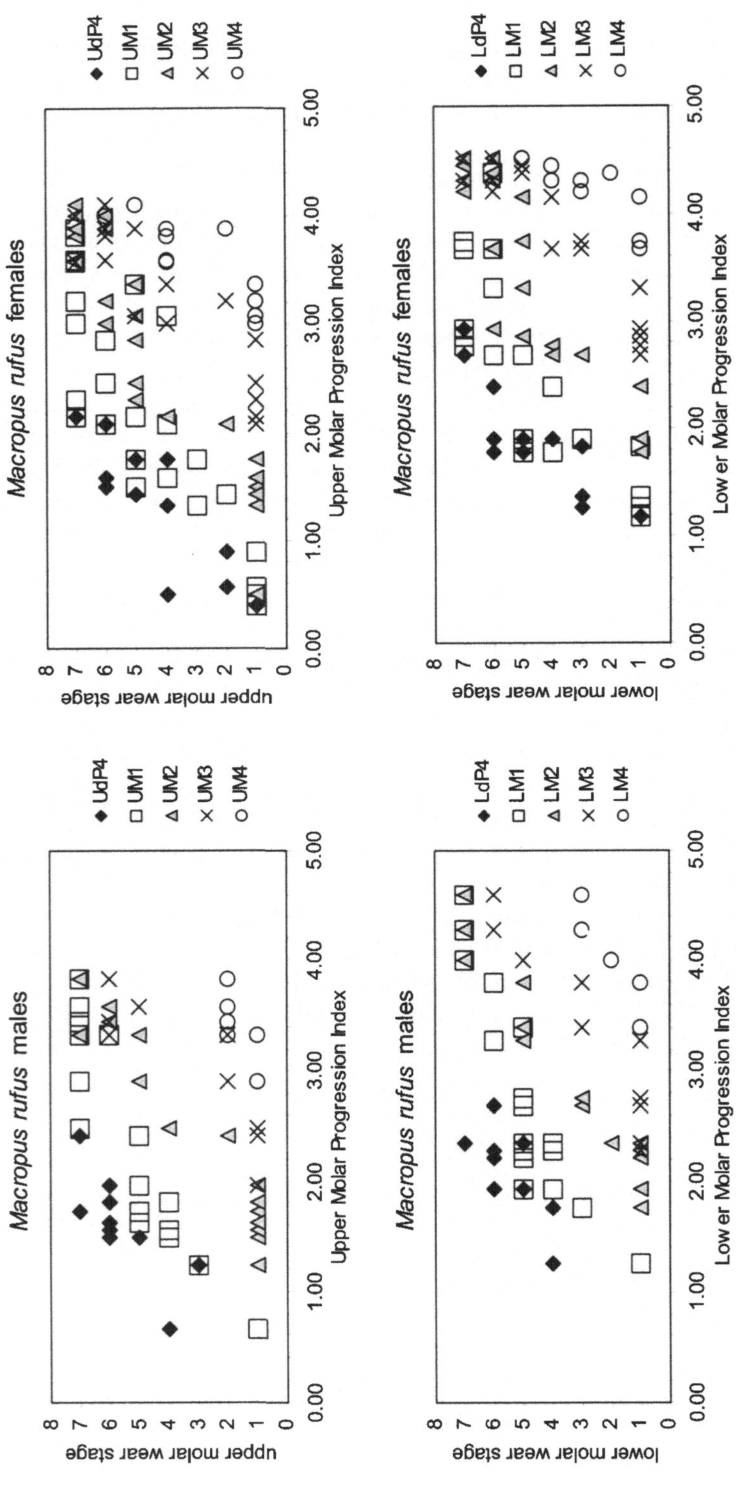

Figure 8.11a. Correlation between molar wear and molar progression in *Macropus rufus*.

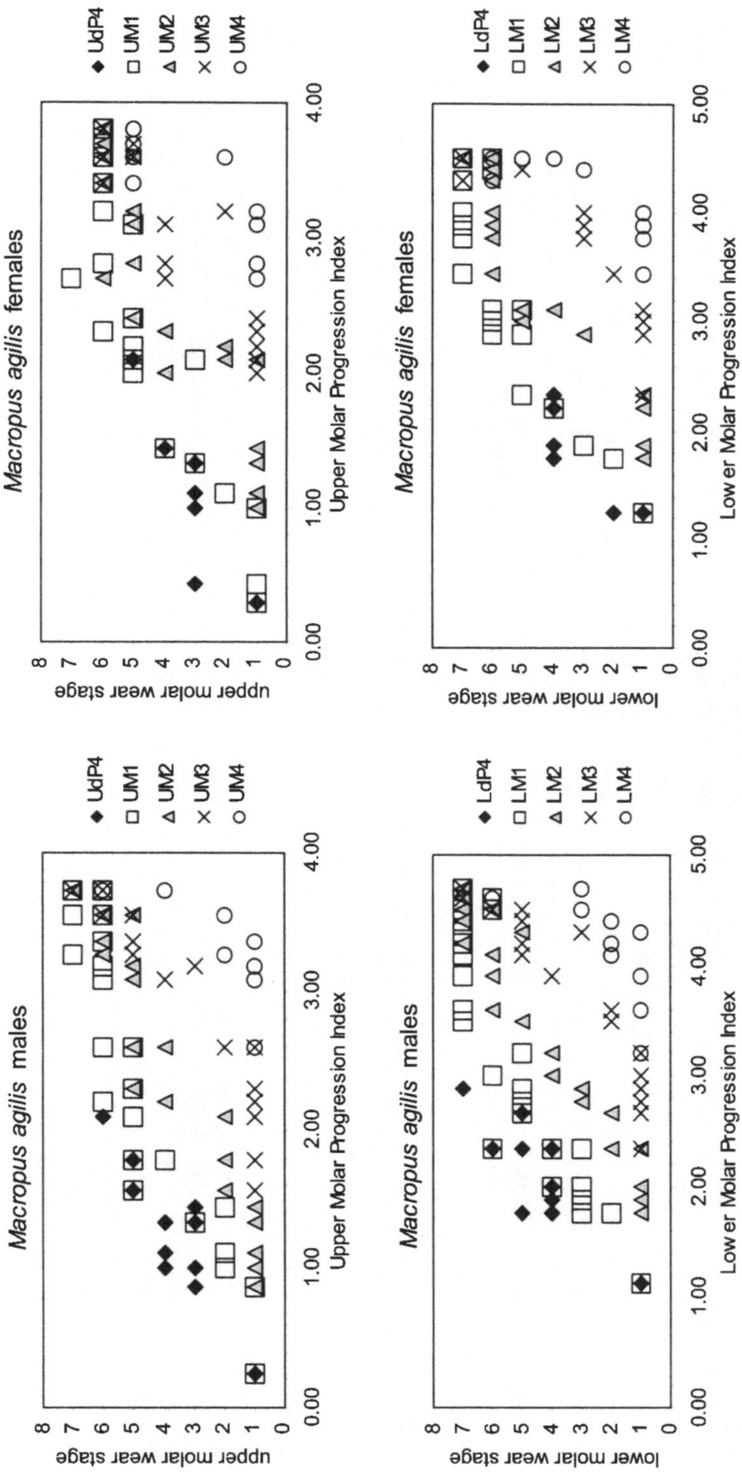

Figure 8.11b. Correlation between molar wear and molar progression in *Macropus agilis*.

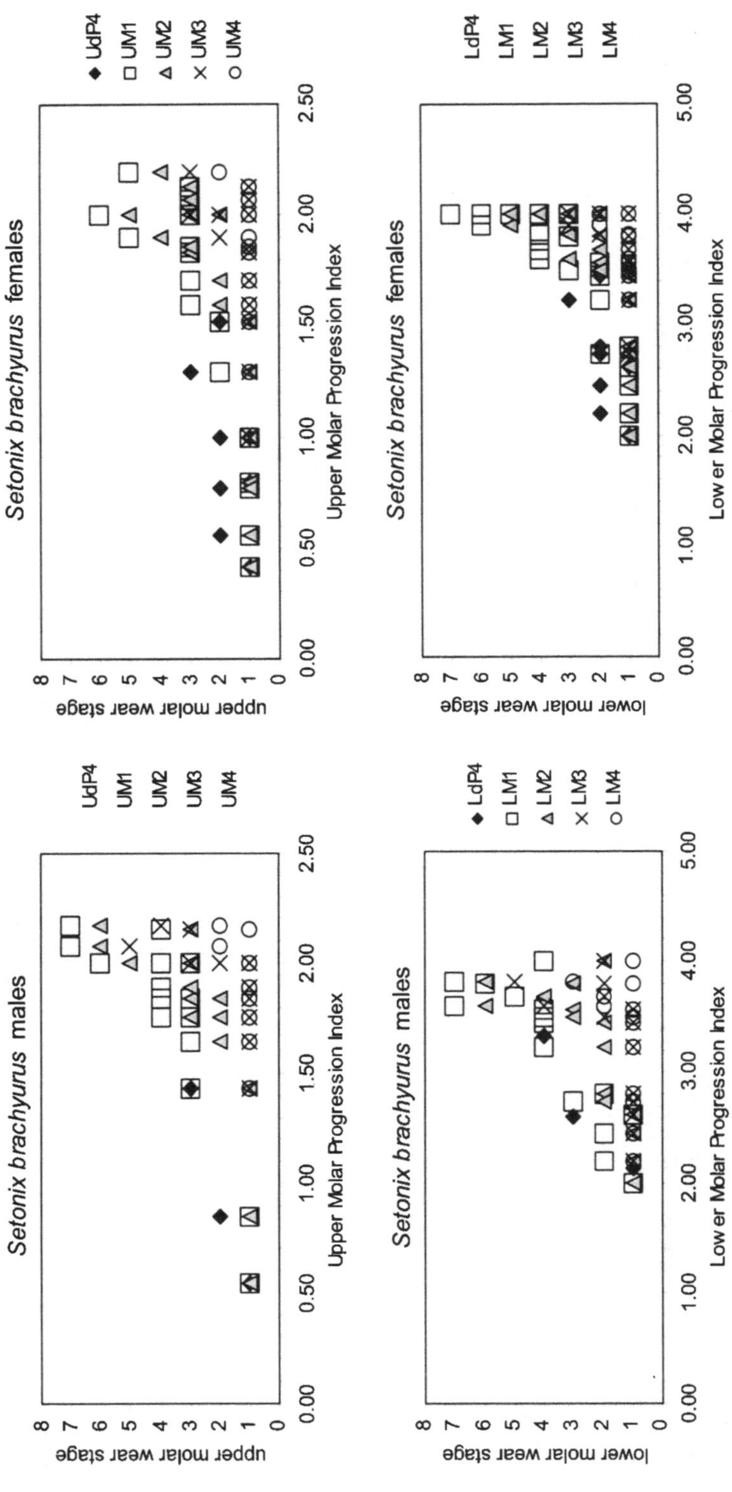

Figure 8.11c. Correlation between molar wear and molar progression in *Setonix brachyurus*.

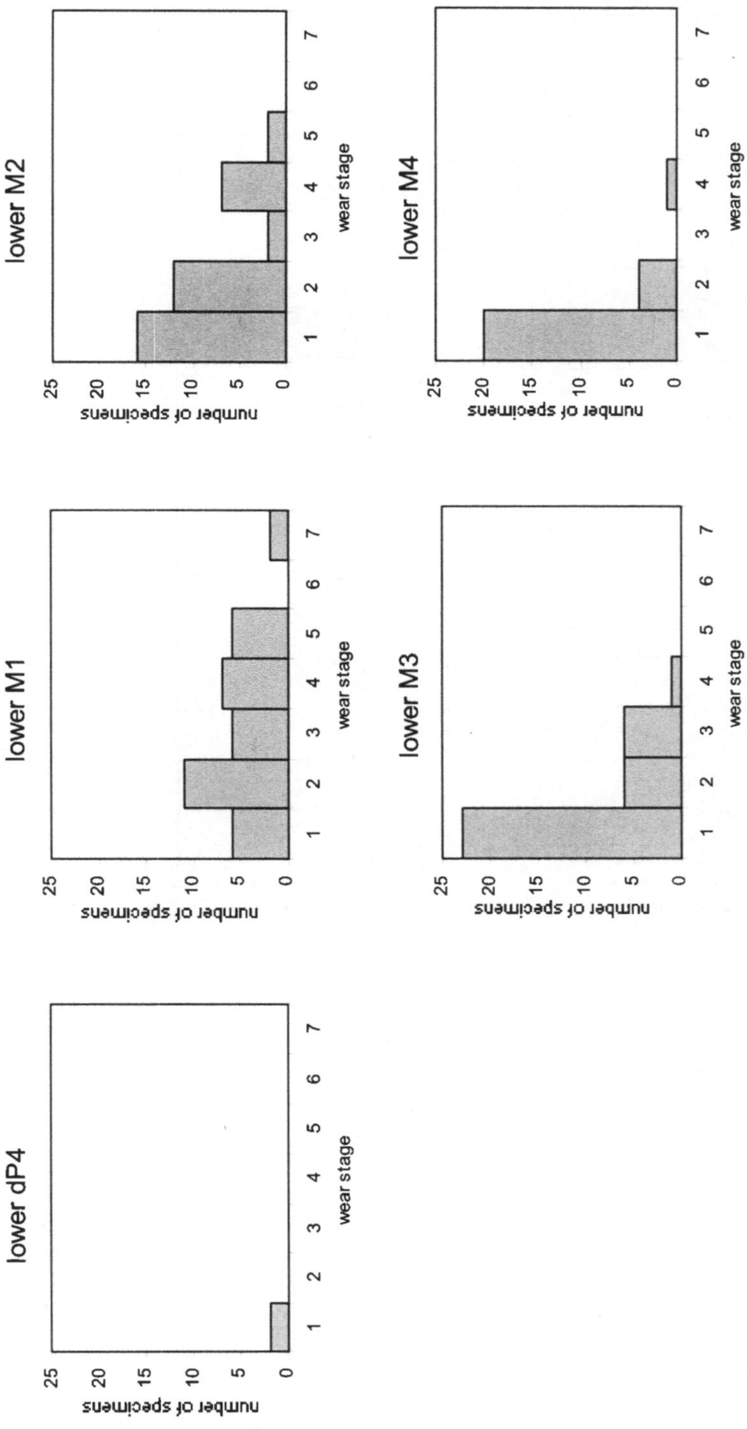

Figure 8.17. Kria Cave, *Dorcopsis muelleri*. Lower molar wear in Unit II.

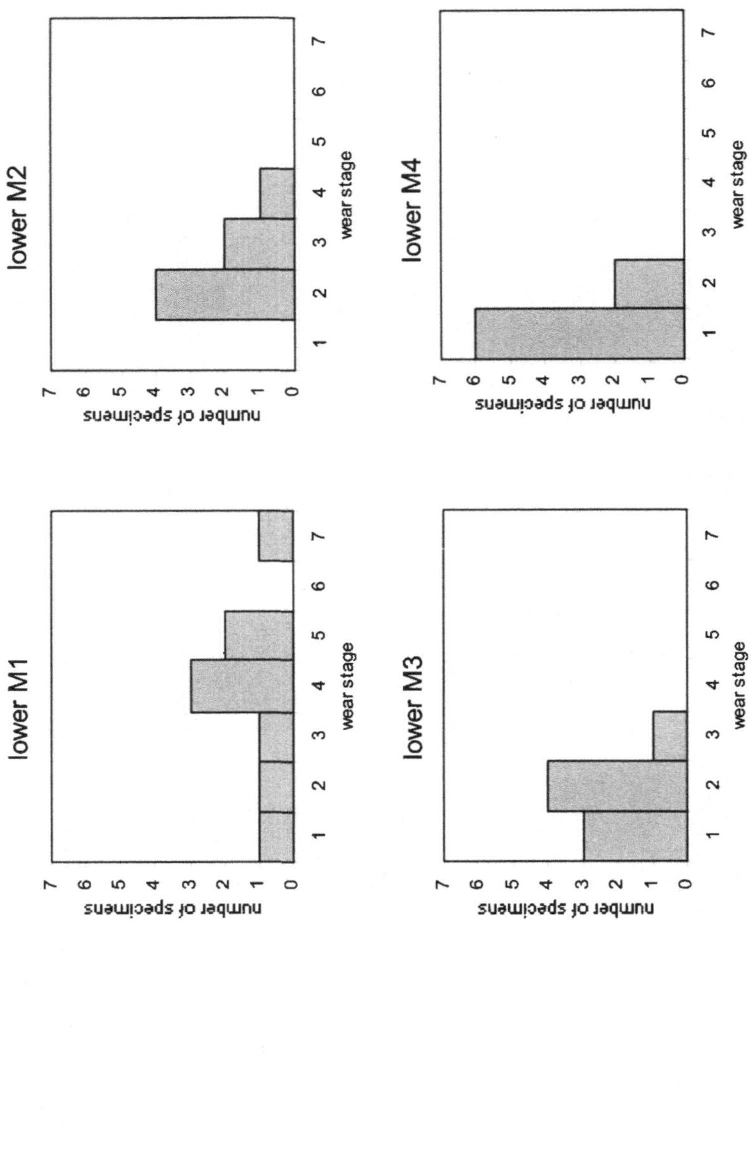

Figure 8.17 (continued). Kria Cave, *Dorcopsis muelleri*. Lower molar wear in Unit III. No deciduous molars were found in this unit.

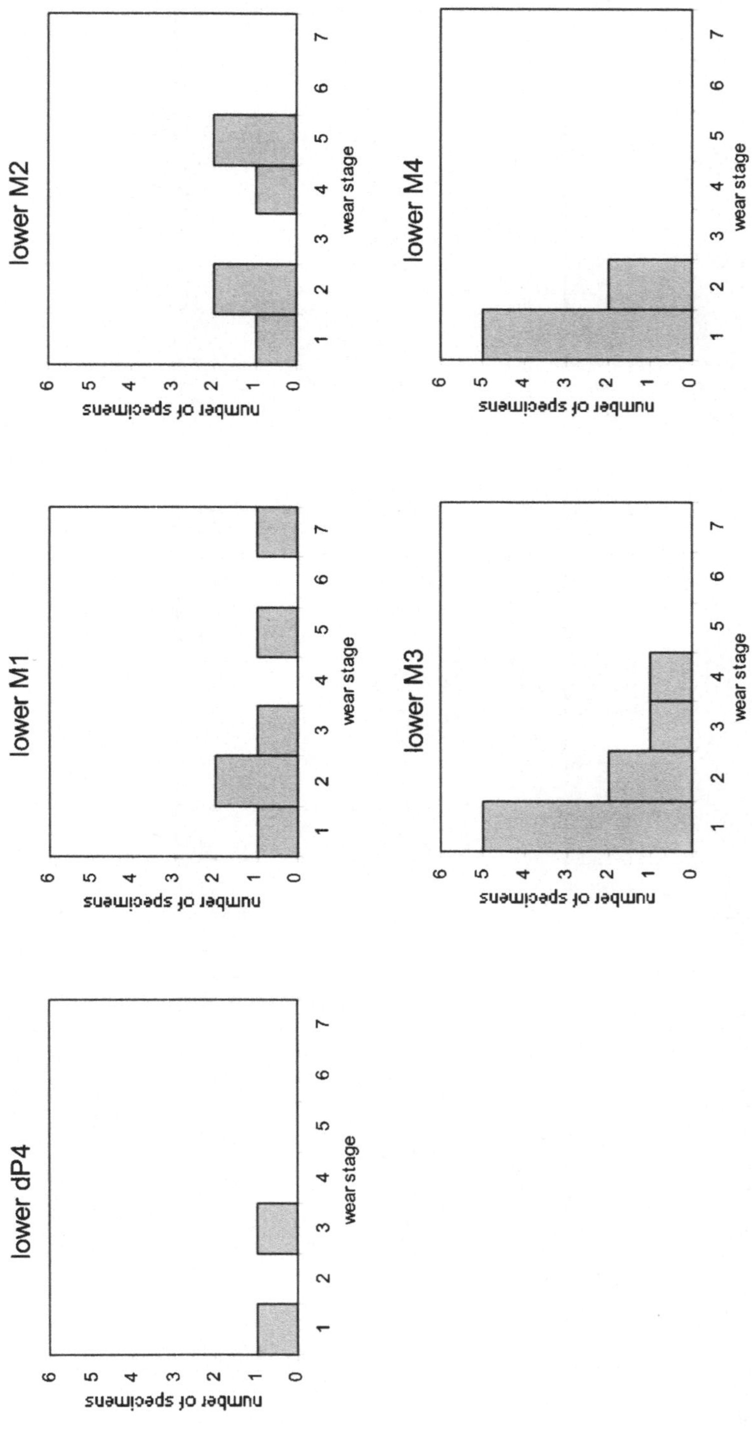

Figure 8.17 (continued). Kria Cave, *Dorcopsis muelleri*. Lower molar wear in Unit IV.

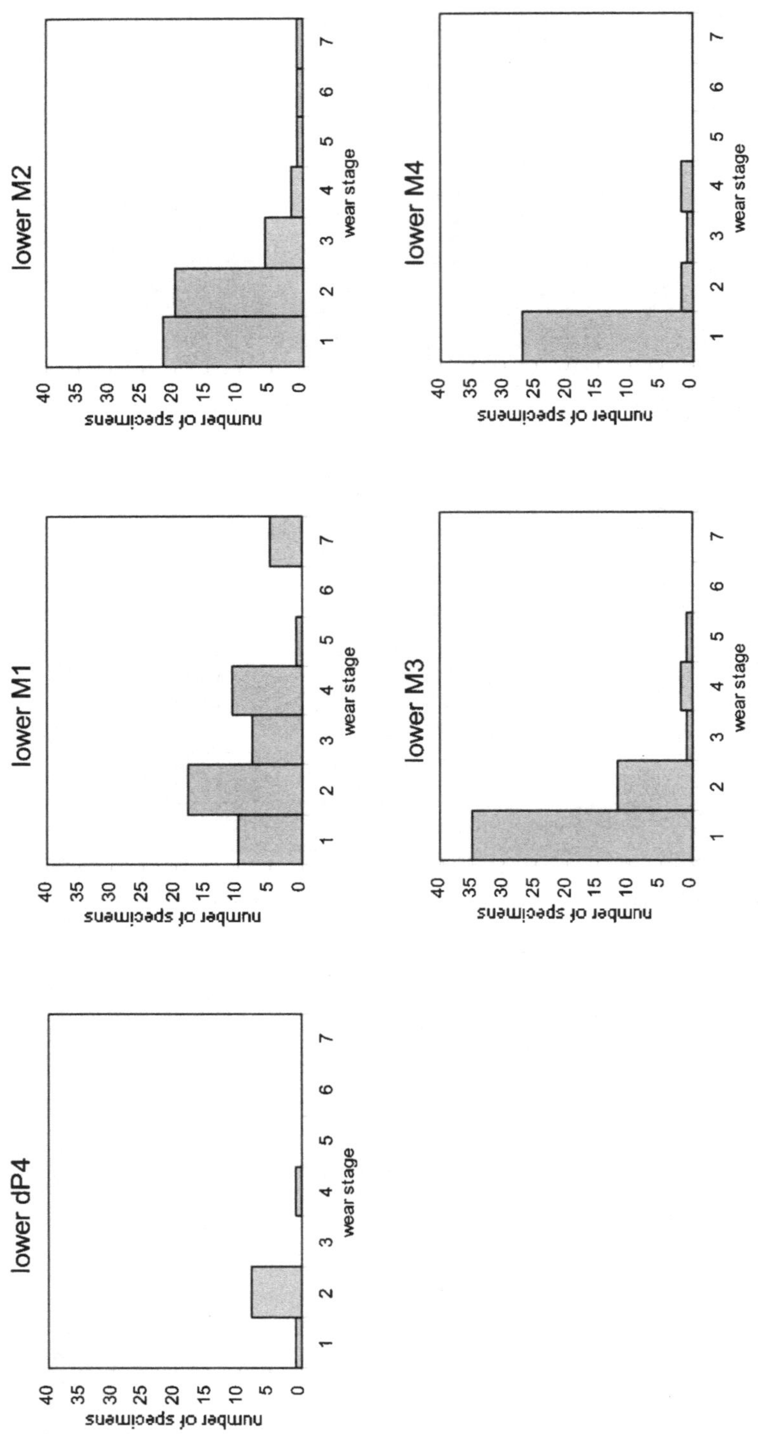

Figure 8.17 (continued). Kria Cave, *Dorcopsis muelleri*. Lower molar wear in Unit V.

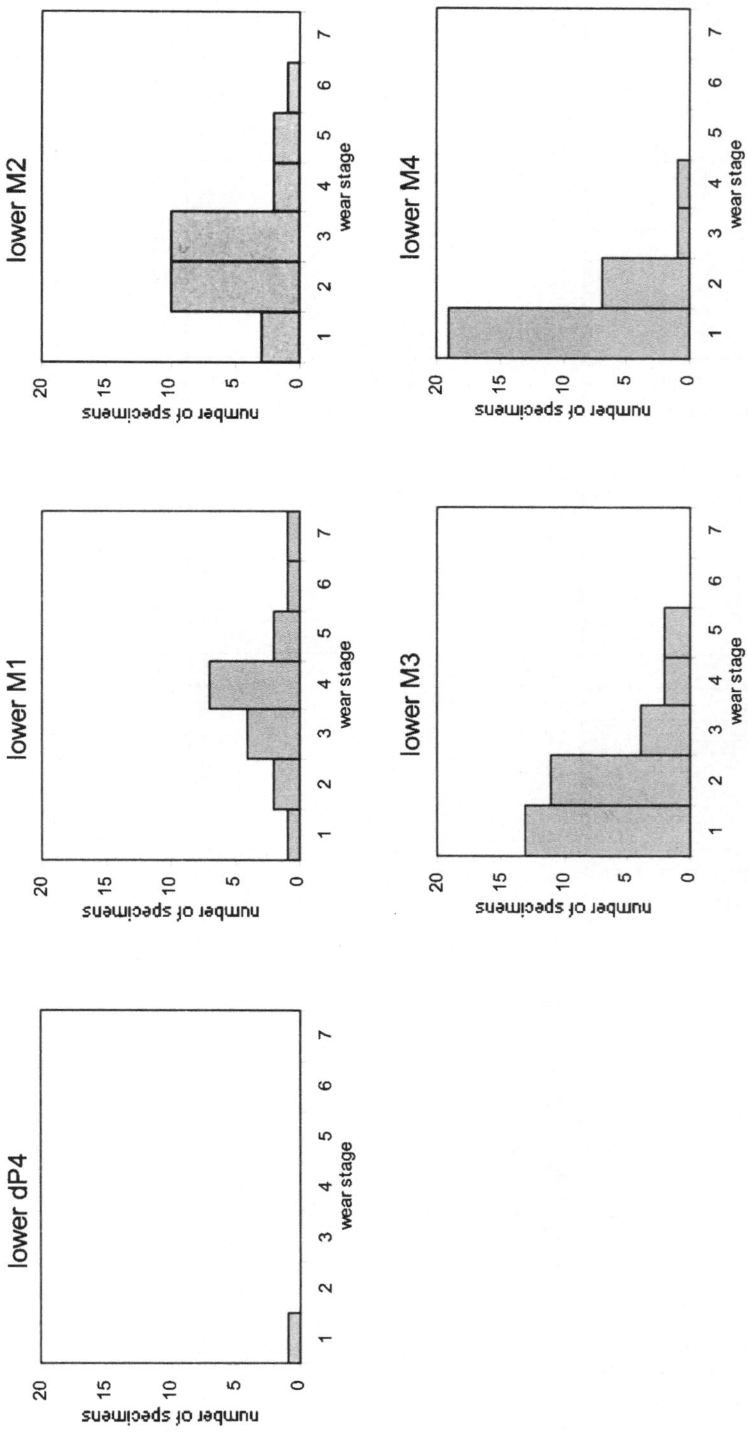

Figure 8.17 (continued). Toé Cave, *Dorcopsis muelleri*. Lower molar wear, Holocene sample.

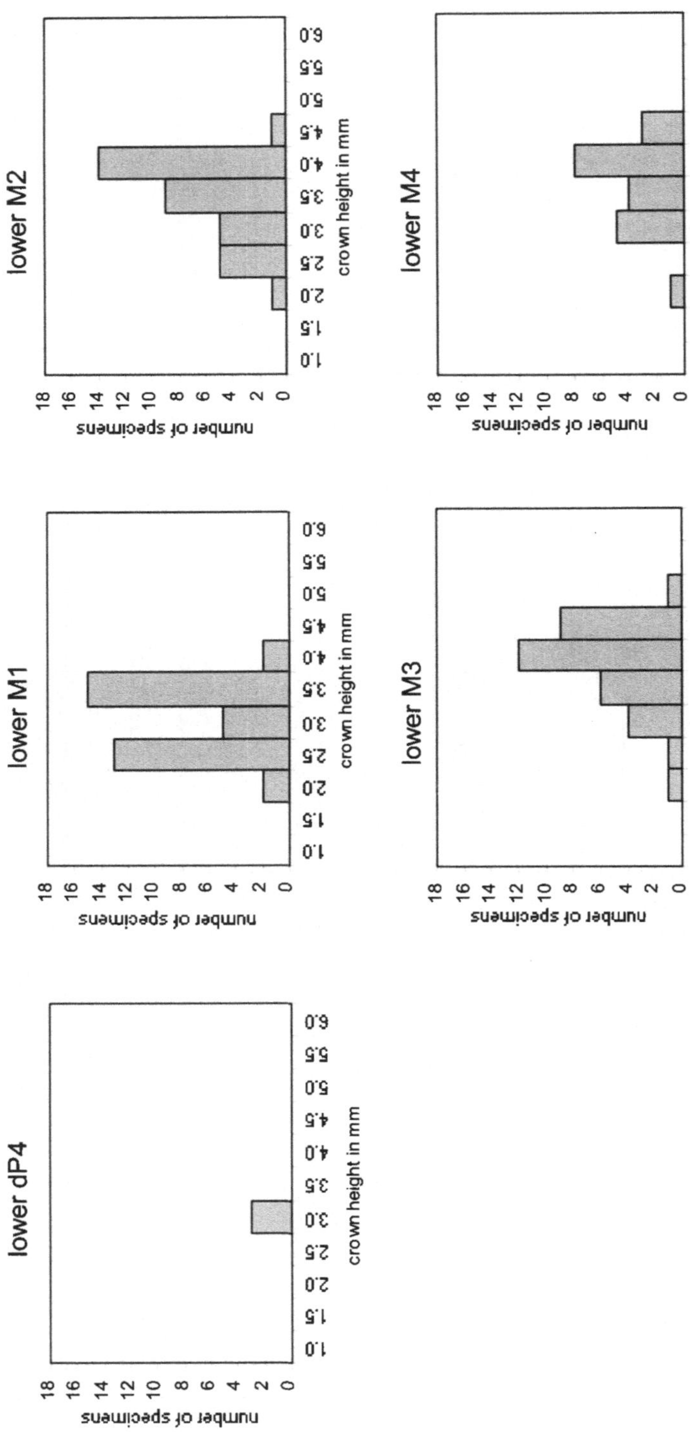

Figure 8.19. Kria Cave, *Dorcopsis muelleri*. Lower molar crown height in Unit II. X-axis values are midpoints of intervals with standard range of 0.5.

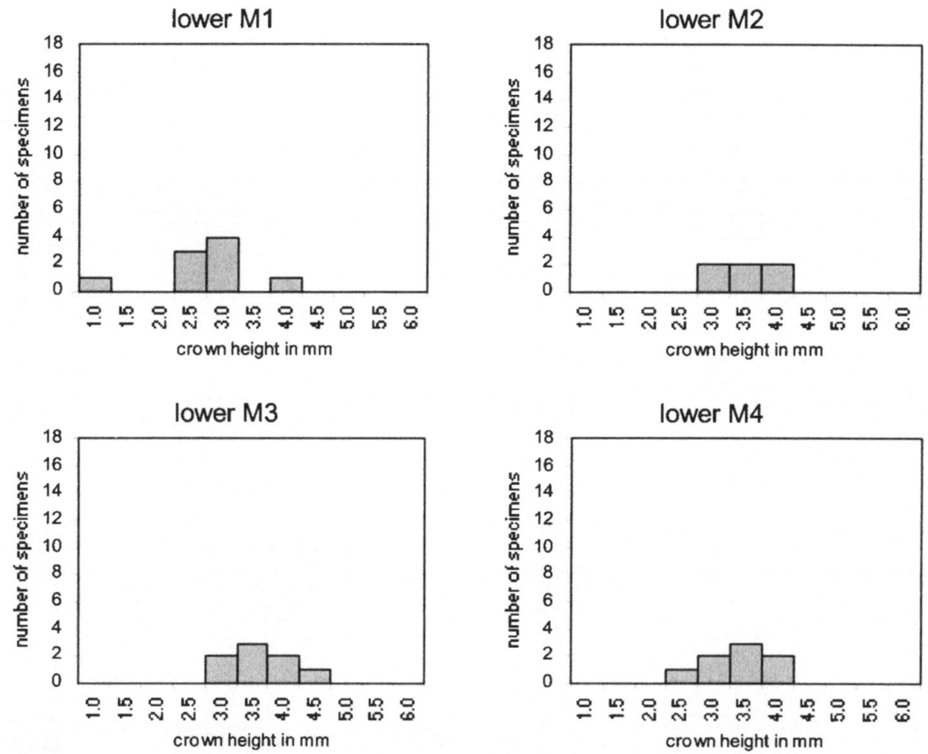

Figure 8.19 (continued). Kria Cave, *Dorcopsis muelleri*. Lower molar crown height in Unit III. No deciduous molars were found in this unit. X-axis values are midpoints of intervals with standard range of 0.5.

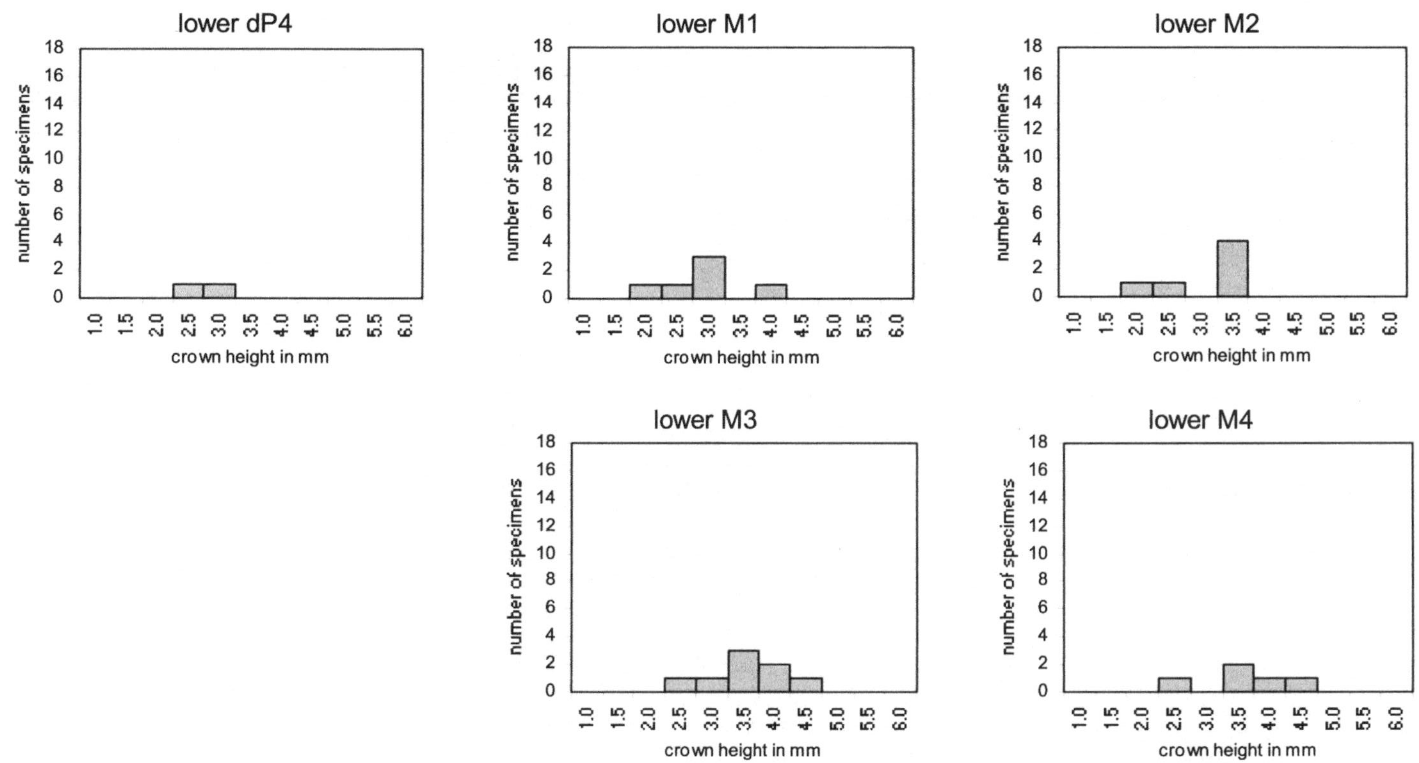

Figure 8.19 (continued). Kria Cave, *Dorcopsis muelleri*. Lower molar crown height in Unit IV. X-axis values are midpoints of intervals with standard range of 0.5.

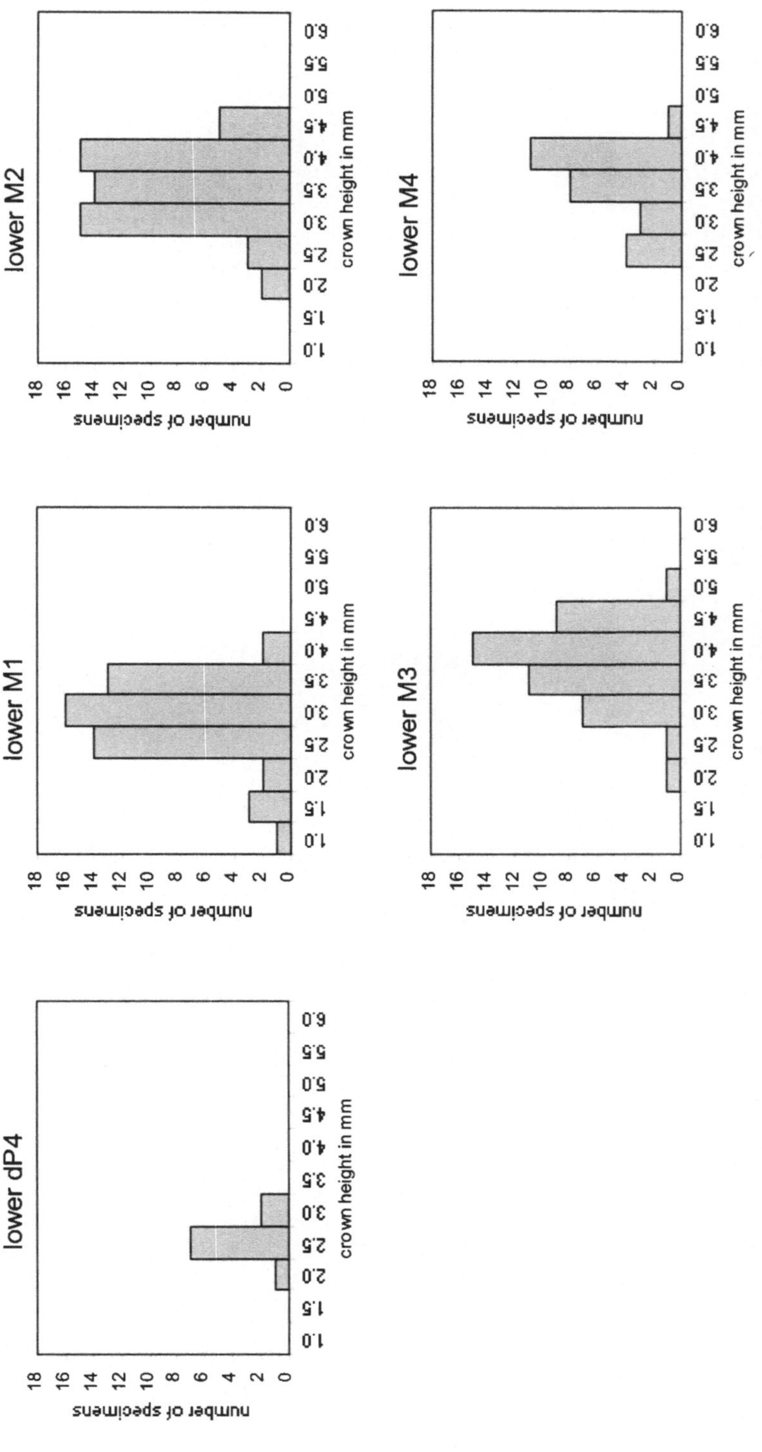

Figure 8.19 (continued). Kria Cave, *Dorcopsis muelleri*. Lower molar crown height in Unit V. X-axis values are midpoints of intervals with standard range of 0.5.

9. Vertebrate Faunal Succession and Environmental Change in Lowland New Guinea

9.1. INTRODUCTION

The impact of Pleistocene glacial episodes was felt in most, if not all, regions of the world. On the equatorial island of New Guinea glacial conditions undoubtedly had their most dramatic effect at higher altitudes where ice sheets expanded to cover 2000 km^2 and treelines fell by almost 2000 m altitude (Hope 1996). At lower elevations, the most obvious effect was the emergence of substantial areas of continental shelf that resulted in the connection of New Guinea with northern Australia and the interruption of oceanic circulation between the Pacific Ocean and the northeast Indian Ocean (Nix & Kalma 1972).

The most detailed and complete record of late Quaternary environmental change in the region comes from long pollen records from localities in Papua New Guinea (Hope & Peterson 1975; Walker & Flenley 1979; Hope 1989; Hope & Golson 1995; see also Haberle 1994). Fewer sites have been examined in Papua, although one pollen record from Lake Hordorli (Hope & Tulip 1994) at 780 m in the Cyclops Mountains and another from a marine core in the East Banda Sea near the Bird's Head Peninsula (Van der Kaars 1991, 1995) are of special significance. A pollen record covering the last 4000 years is available from the western basin of the Ayamaru Lakes; this record will be briefly discussed below, based on information kindly supplied by Dam and Van der Kaars (unpubl. Data).

Profound changes have also taken place over the same period in the vertebrate fauna of New Guinea, including the extinction of a suite of giant marsupials (Flannery 1994). However, long faunal records are scarce and the preliminary analyses of several archaeological faunas from sites in Papua New Guinea have failed to provide clear discrimination between the impact of climatic change as against other factors such as forest disturbance and/or hunting or cultural bias (White 1972; Gillieson & Mountain 1982). Hope et al. (1993) and Flannery (1999) presented more compelling evidence of environmental changes at high altitude in Papua, based on undated faunal material from cave sites in the Baliem Valley.

Note: The data and conclusions presented in this chapter were published in part by Pasveer & Aplin (1998). This version incorporates new information on dating, stratigraphy and palynology.

Both the palynological and the faunal records share a strong bias towards montane localities. In contrast, comparatively little is known regarding the nature and extent of environmental changes at lower elevations. Kria and Toé Caves together provide a long-term record of vertebrate faunal changes in a lowland environment, dating from 26,000 BP until the last few thousand years. Both assemblages are dominated by the remains of terrestrial animals, with far smaller quantities of fish bone, molluscan shell, avian eggshell and crustacean carapace. The present chapter deals only with the terrestrial vertebrate remains.

9.2. FAUNAL CHANGE IN THE KRIA CAVE SEQUENCE

As outlined in Chapter 3, the Kria Cave deposit consists of five 'units', distinguished on both sedimentological and cultural criteria. Unit I, forming the surface layer, contains a concentration of human remains and a small shell midden, but virtually no stone or bone artefacts, and few animal bone fragments. Only in this unit were pottery sherds and pig bone fragments found. A pig bone fragment from this unit was dated to a minimum of 1840 BP.

A possible hiatus in occupation, perhaps starting around 4000 BP, divides this unit from the underlying Units II to V. These lower units are essentially the same in composition but vary in quantity of especially bone remains and stone artefacts. Units II and V contain the highest quantities of bone and stone. However, bone points were abundant in all of the lower units. No discrete features, such as were found in Unit I, were encountered in the lower part of the deposit.

Unit V, as will be discussed below, differs again from the overlying units in composition of its faunal remains. Occupation of the site probably commenced around 8000 BP (Chapter 3). The termination of Unit V is dated to around 6000 BP.

Figure 9.1 shows the vertical distribution of the various identified animal species through the deposit in each square; corresponding changes in quantitative composition are summarised in Figure 9.2. The numerical key to species for Figure 9.1 is given in Table 9.1. The systematics and nomenclature are discussed in full by Aplin et al. (1999).

As noted above, Unit I is distinguished by the presence of small quantities of bone, some of which probably come from a single pig; two other mammals, a species of ringtail possum (*Pseudochirulus* sp. cf. *P. canescens*) and the Large Spiny Rat (*Rattus praetor*), are also restricted to Unit I and the uppermost level of Unit II (Fig. 9.1). The small quantities of bone from these levels most likely explain both the dramatic fluctuations in quantitative composition (Fig. 9.2) and the low diversity of species. A small proportion of the bone in the upper levels of the site is probably the prey remains of a raptor and/or small mammalian carnivore.

The fauna of Units II-IV is remarkably consistent in overall composition. Remains of the forest wallaby, *Dorcopsis muelleri*, make up the bulk of the material in all levels, with a major, secondary contribution from pythons (Boidae; see Chapter 7). Most other groups of animals are represented only sporadically, with the fluctuations probably reflecting small sample sizes rather than real changes in the local environment or in human activities.

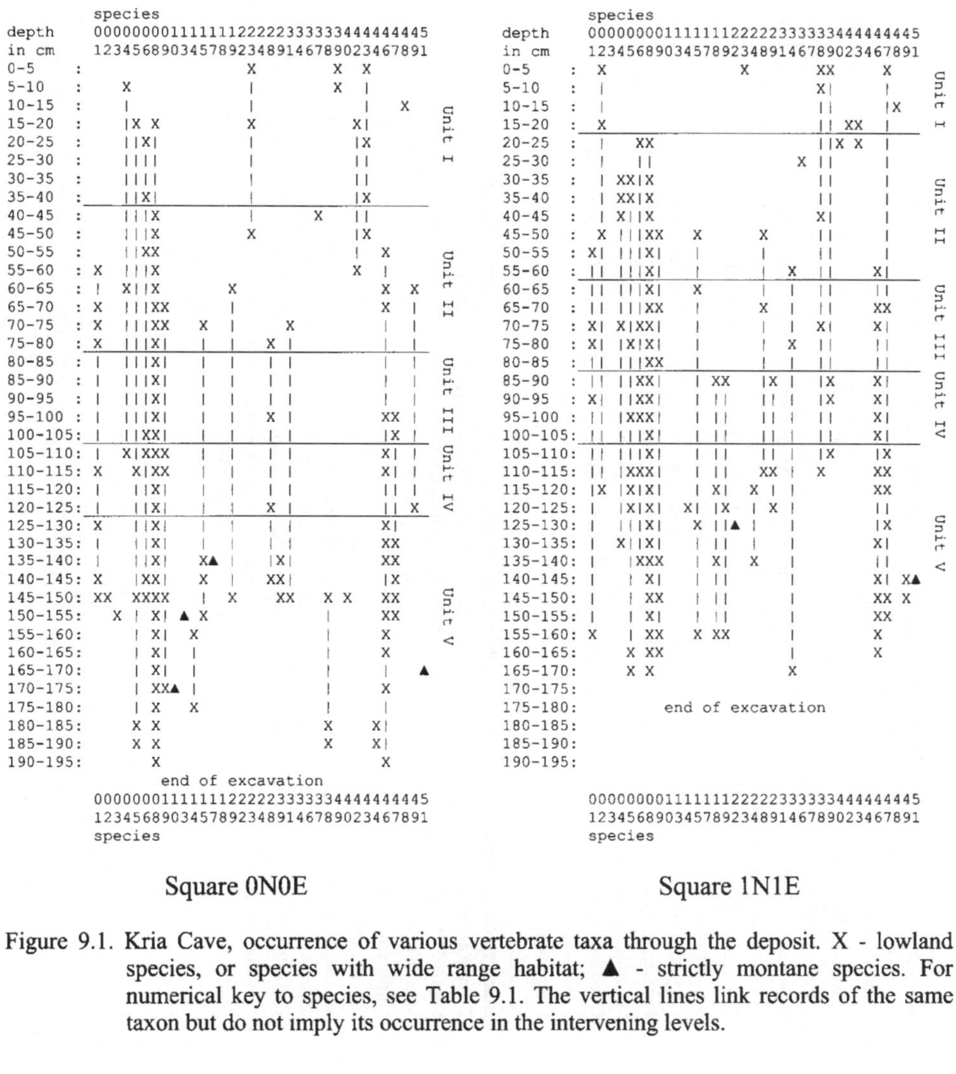

Square 0N0E

Square 1N1E

Figure 9.1. Kria Cave, occurrence of various vertebrate taxa through the deposit. X - lowland species, or species with wide range habitat; ▲ - strictly montane species. For numerical key to species, see Table 9.1. The vertical lines link records of the same taxon but do not imply its occurrence in the intervening levels.

The majority of species present in Units II-IV are recorded from present lowland rainforest habitats either locally on the Ayamaru Plateau or elsewhere on the Bird's Head. Examples include *Dorcopsis muelleri*, *Dendrolagus inustus*, *Spilocuscus maculatus*, *Phalanger gymnotis*, *Dactylopsila trivirgata* and *Pogonomelomys bruijnii*.

Several other taxa found in these levels have not previously been recorded from the Bird's Head but are present in lowland rainforests elsewhere in New Guinea. *Hydromys chrysogaster*, the Common Water Rat, is widely distributed from sea level to 1900 m in New Guinea and on many of the nearby continental shelf and oceanic islands (Flannery 1995a). Although the archaeological specimens constitute the first evidence of this species on the Bird's Head, it probably still occurs around the Ayamaru Lakes. A species of tree mouse (*Pogonomys* sp. cf. *P. loriae*) is also new to the lowland fauna of the

Table 9.1. List of vertebrate species and families from Kria and Toé Caves. Numbers corresponding to those given in Figures 9.1 and 9.3.

Scientific Name	Common Name	Scientific Name	Common Name
1 : *Zaglossus bruijnii*	Long-beaked Echidna	30 : *Dactylopsila palpator*	Long-fingered Triok
2 : *Myoictis melas*	Three-striped Dasyure	31 : *Petaurus breviceps*	Sugar Glider
3 : *Dasyurus albopunctatus*	New Guinea Quoll	32 : *Mallomys* sp. cf. *M. aroaensis*	De Vis's Woolly-Rat
4 : *Echymipera rufescens*	Long-nosed Echymipera	33 : *Xenuromys barbatus*	Rock-dwelling Giant Rat
5 : *Echymipera kalubu*	Common Echymipera	34 : *Uromys caudimaculatus*	Mottled-tailed Giant Rat
6 : *Microperoryctes longicauda*	Striped Bandicoot	35 : *Uromys/Xenuromys*	Giant Rats
7 : *Echymipera* sp. indet.	Echymiperas	36 : *Parahydromys asper*	Waterside Rat
8 : *Dorcopsis muelleri*	Brown Dorcopsis	37 : *Hydromys chrysogaster*	Common Water-rat
9 : *Dendrolagus inustus*	Grizzled Tree-kangaroo	38 : *Pogonomelomys bruijnii*	Large Pogonomelomys
10 : *Dendrolagus* sp. cf. *D. goodfellowi*	Goodfellow's Tree-kangaroo group	39 : *Pogonomelomys* sp. A	Ayamaru fossil Pogonomelomys
11 : *Dendrolagus* sp. indet.	Tree-kangaroos	40 : *Pogonomys* sp. cf. *P. loriae*	Large Tree-mouse
12 : *Dorcopsis/Dendrolagus*	Large wallabies or tree-kangaroos	41 : *Paramelomys* sp.	Small Paramelomys
13 : *Dorcopsulus* spp. (incl. *D. vanheurni*)	Small Dorcopsis	42 : *Paramelomys platyops*	Lowland Paramelomys
14 : *Spilocuscus rufoniger*	Black-spotted Cuscus	43 : *Rattus praetor*	Large Spiny Rat
15 : *Spilocuscus maculatus*	Common Spotted Cuscus	44 : *Rattus* sp. cf. *R. verecundus*	Slender Rat
16 : *Spilocuscus* sp. indet.	Spotted Cuscuses	45 : Muridae gen. & sp. indet.	Small rats or mice
17 : *Phalanger vestitus*	Stein's Cuscus	46 : *Dobsonia* sp. cf. *D. magna*	Great Bare-backed Fruit-bat
18 : *Phalanger orientalis*	Northern Common Cuscus	47 : *Hipposideros diadema*	Diadem Horseshoe-bat
19 : *Phalanger gymnotis*	Ground Cuscus	48 : *Sus scrofa*	Pig
20 : *Phalanger* sp. cf. *P. intercastellanus*	Southern Common Cuscus	49 : *Casuarius bennettii*	Dwarf Cassowary
21 : *Phalanger* sp. indet.	Cuscuses	50 : *Casuarius* sp. indet.	Cassowary
22 : *Pseudochirulus schlegelii*	Arfak Ringtail	51 : *Aepypodius arfakianus*	Wattled Brush Turkey
23 : *Pseudochirulus* sp. cf. *P. canescens*	Lowland Ringtail	52 : *Tyto* sp. cf *T. tenebricosa*	Sooty Owl
24 : *Petauroides ayamaruensis*	Ayamaru fossil ringtail	53 : Aves gen. & sp. indet.	Small birds
25 : *Pseudochirops albertisii*	D'Albertis's Ringtail	54 : Boidae gen. & sp. indet.	Pythons
26 : *Pseudochirops coronatus*	Reclusive Ringtail	55 : *Varanus* sp. indet.	Monitors
27 : *Pseudochirops* sp. indet.	Ringtail Possum	56 : Agamidae gen. & sp. indet.	Dragon lizards
28 : *Dactylopsila trivirgata*	Striped Possum	57 : Anura gen. & sp. indet.	Frogs
29 : *Dactylopsila kambuayai*	Ayamaru fossil Striped Possum		

Note: *Petauroides ayamaruensis* (24) and *Dactylopsila kambuayai* (29) are new species, described in Aplin et al. 1999. *Pogonomelomys* sp. A (39) is an unknown and possibly new species which has not yet been formally described.

Bird's Head; the same or a similar species has been recorded previously from the Arfak Mountains (Dennis & Menzies 1979; see Fig. 9.5). *Pogonomys loriae* is recorded down to 240 m above sea level in northern New Guinea (Flannery 1995a).

Several species are more commonly associated with higher elevation forests but may well occur in lowland rainforests. The Waterside Rat, *Parahydromys asper*, is recorded from numerous localities throughout New Guinea, between 700 and 2200 m altitude. On the Bird's Head Peninsula, the species has been recorded only once – in the Arfak Mountains. As its remains occur at the top of Unit II in Kria Cave, this species is probably still present in the Ayamaru Lakes area; this would represent a slight downward extension of its known altitudinal range.

Microperoryctes longicauda, the Striped Bandicoot, is recorded with confidence only from the lower unit of Toé (see below). However, post-cranial material of a small-bodied peroryctid (bandicoot) is present in various levels of Units V-II in Kria. The Striped Bandicoot is best known from Papua New Guinea where the subspecies *M. l. ornata* is common in high elevation forest and into alpine grasslands (Flannery 1995a). However, the species is recorded down to 1000 m altitude (Flannery 1995a) and may occur even lower in the southern highlands of Papua New Guinea (K. Aplin pers. comm.). The typical race (*M. longicauda longicauda*) has been recorded only once, in the Arfak Mountains of the Bird's Head. It is likely that this species will eventually be found to occur across a relatively wide altitudinal range on the Bird's Head Peninsula.

Two other mammal species found in Units IV-II of Kria Cave have probably never been collected as living animals (Aplin et al. 1999). One of these is a newly named, diminutive striped possum, *Dactylopsila kambuayai*, which may be a lowland relative of the Long-fingered Triok, *Dactylopsila palpator*; the other is an undescribed species of *Pogonomelomys*. The new striped possum occurs to the top of Unit IV but is absent above this point, despite the large quantity of bone from Unit II. The *Pogonomelomys* is confined to the upper levels of the sequence. Little can be said regarding the ecological significance of these previously unknown taxa.

The fauna in Unit V is very similar in overall composition to that from the overlying units. However, it contains a total of ten extra species including at least five whose modern populations occur in montane forest habitats. Of these, two marsupials (*Phalanger vestitus* and *Pseudochirulus schlegelii*), a rodent (*Rattus* sp. cf. *R. verecundus*) and a megapode bird (*Aepypodius arfakianus*) are today found in the Arfak and/or Tamrau Mountains. *Phalanger vestitus* is also found in the central ranges of Papua and Papua New Guinea where it occurs above elevations of about 1000 m above sea level. The Arfak Ringtail Possum, *Pseudochirulus schlegelii*, is restricted to the Arfak Mountains, with records between 750 m and 1900 m above sea level. The *Rattus* species is only tentatively identified as the montane species *R. verecundus* owing to a lack of locally-sourced reference material for direct comparisons (Aplin et al. 1999); an alternative identification could be the Moss Forest Rat, *R. niobe*, which is also recorded exclusively from the Arfak Mountains on the Bird's Head. All small-bodied *Rattus* species in New Guinea are restricted to montane habitats.

Square 0N0E

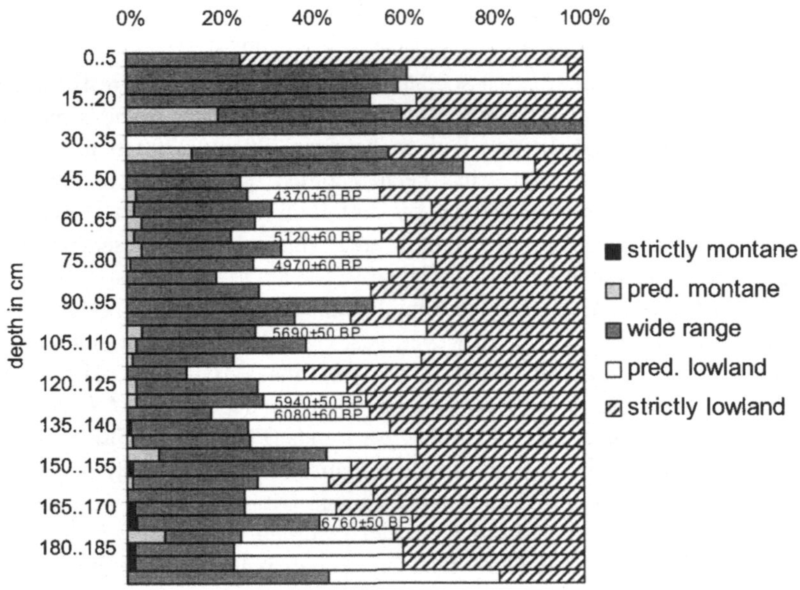

Square 1N1E

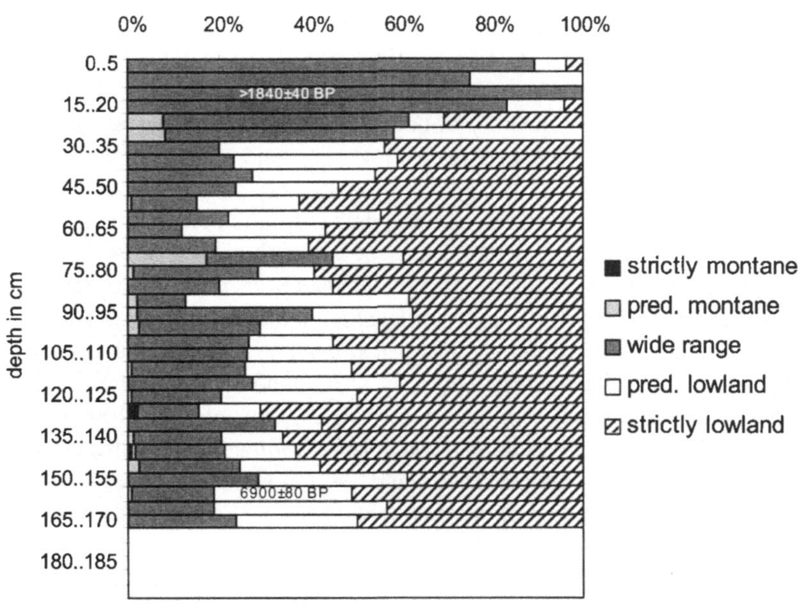

Figure 9.2. Kria Cave. Proportional distribution (by NISP) of vertebrate groups by habitat through the deposit. 'pred.' = predominantly. The radiocarbon dates are positioned according to their depth below the surface; they are not related to any faunal group.

The Wattled Brush Turkey, *Aepypodius arfakianus*, is widespread through montane elevations across New Guinea and generally occurs only above 1000 m above sea level (Rand & Gilliard 1967). An unusually low elevation population of the species is found on Misool Island to the southwest of the Bird's Head (Mees 1965). A related species, *A. bruijnii*, is found on mountainous Waigeo Island off the northwest corner of the Bird's Head (Rand & Gilliard 1967).

Other taxa confined to Unit V in Kria Cave include a previously undescribed, small-bodied ringtail possum (*Petauroides ayamaruensis*), an undetermined species of the exclusively montane wallaby genus *Dorcopsulus*, an as yet undetermined species of tree-kangaroo belonging to the *Dendrolagus goodfellowi*-group, and the Black-spotted Cuscus, *Spilocuscus rufoniger*. The *Dorcopsulus* species is represented in Kria by a single upper incisor; however, more complete material was obtained from the lower levels of Toé Cave (see below). All *Dorcopsulus* species are restricted to montane forests and are relatively intolerant of disturbance. No member of the genus has been previously recorded from the Bird's Head; the closest record of the widespread *D. vanheurni* is from the upper Idenburg River in north-central Papua.

The Black-spotted Cuscus and a tree-kangaroo of the *goodfellowi*-group also represent new records for the Bird's Head. The cuscus is distributed along the north coast from the Huon Peninsula in Papua New Guinea west to the Idenburg River in Papua, and from near sea level to about 1200 m above sea level (Flannery 1995a). Flannery (1995a) has noted the widespread decline and disappearance of this species as a consequence of hunting pressure and habitat disturbance. In contrast, the closely related Spotted Cuscus, *Spilocuscus maculatus*, remains abundant throughout the same region; the latter species is remarkably tolerant of disturbance and is often encountered in secondary forest and even active gardens.

The *goodfellowi*-group of tree-kangaroos is represented elsewhere in New Guinea by three species with differing habitat associations: *Dendrolagus spadix* is restricted to an area of limestone karst in lowland southern Papua New Guinea; *D. matschiei* ranges from 1000 m to the limit of tree growth on the Huon Peninsula; and the more widely distributed *D. goodfellowi* occupies mid-montane forests along the central ranges and in the north coast ranges (Flannery 1993). As described by Aplin et al. (1999), the archaeological specimens are closest in morphology to *D. g. pulcherrimus* from the Torricelli and Foja Mountains of northern New Guinea; the limited information on this race suggests that it may extend to lower elevations than the other subspecies of *D. goodfellowi*, perhaps even below 680 m above sea level. All members of this group of tree-kangaroos are threatened as a consequence of forest clearance and overhunting.

The last of the taxa recorded exclusively in Unit V is the Northern Common Cuscus, *Phalanger orientalis*, a species with a wide altitudinal range, from sea level to 1500 m above sea level. *Phalanger orientalis* is a characteristic element of the modern lowland rainforest fauna of the Bird's Head, and a specimen was obtained locally by hunters from the village of Suwiam/Mapura during the period of fieldwork. This species seems quite tolerant of disturbance and it is commonly taken in old gardens and regrowth forest. Although the absence of this species from the higher units in Kria Cave may be an artefact of sampling, it is also confined to the lower levels of Toé Cave (see below) and thus may have been more abundant locally during earlier times.

Occasional remains of the Long-beaked Echidna, *Zaglossus bruijnii*, were found in Units V, IV and II, with a peak abundance in Unit V. Although *Z. bruijnii* is often thought of as a montane species, it actually occurs across a very broad altitudinal range and is recorded from as low as 450 m above sea level in southern New Guinea (Flannery 1995a). It has been hunted to extinction in many of the more densely populated regions of New Guinea. Local information indicates that the species may still be obtained in the more remote parts of the Ayamaru Plateau (see Appendix 3).

The presence of such a wide range of montane forest species and disturbance-sensitive taxa in Unit V stands in sharp contrast with the much less diverse faunal assemblage from all of the overlying units. However, what is equally noteworthy is the scarcity of these taxa in Unit V, which make up no more than 2.2% (as expressed by NISP values) of the total identified fauna in any level in this unit (Fig. 9.2). As in the overlying Units IV to II, the remains of *Dorcopsis muelleri* are dominant in Unit V, comprising 46% of all identified bone; the category 'macropodid', most of which probably belongs to *Dorcopsis* as well, makes up another 16.6%. Pythons, too, are similarly well represented, with proportional abundance around 22.8%.

Today, suitable habitats for these montane animals occur at a distance of at least 50 km from the Ayamaru Lakes (see Fig. 9.5), well beyond the potential hunting range of the prehistoric people in such rugged karst terrain. It must be inferred, therefore, that these animals lived locally on the Ayamaru Plateau during Kria Unit V times. This conclusion will be further discussed in the next section about the faunal assemblage of Toé Cave, where a montane component is also present.

Thus, while it is clear that a suite of montane and disturbance-sensitive animals must have coexisted with the typical lowland fauna during Unit V times, it is also clear that they were collected only infrequently, presumably reflecting a natural scarcity in the region.

The overwhelming dominance of *Dorcopsis muelleri* through Units V-II suggests that it was locally abundant throughout this time. However, the extreme over-representation of this species in the archaeological assemblage is probably also an indication of an economic preference and selective patterns of food disposal (see Chapter 7 and 8).

The faunal record from Kria Cave documents the presence of more or less continuous lowland rainforest cover within the hunting radius of the site through the period 8000-4000 BP, providing an abundance and diversity of animal resources. However, during the earliest period of occupation, represented by Unit V, it was also possible to procure a number of 'montane' mammal and bird species and a somewhat wider range of disturbance-sensitive taxa in the local area. Considering the differences between Units II-V and the overlying Unit I, it is likely that the cave was abandoned or visited only very infrequently after 4000 BP. Some unknown time later, but before 1840 BP, the site was reoccupied but with a different pattern of site use including the disposal of human remains and accumulation of a small discrete shell midden. The rate of accumulation of animal bones declined markedly over this period.

9.3. FAUNAL CHANGE IN THE TOÉ CAVE SEQUENCE

The occupation sequence in Toé Cave has been outlined in Chapter 4. In summary, the deposit consists mainly of two units: an upper, Holocene unit containing lowland species dominated by *Dorcopsis muelleri*, and a lower, Late Pleistocene unit consisting of a mixed suite of montane and lowland species. The boundary between these units is not clearly defined but follows the slope of the bedrock. The deposit is essentially intact, as is demonstrated by a series of radiocarbon dates, associated A/I ratios from amino acid racemisations, and a good correspondence of events documented in Kria Cave; however, it appears to have been subject to some local truncation, slumping and mixing, especially in the lower levels, and clearly has a complex stratigraphy. The vertical and horizontal distributions of cultural and faunal remains fail to show major disturbances in the deposit.

The results of the study of the faunal material from Toé Cave are presented in Figures 9.3 and 9.4, showing the vertical distribution of species in each square and a summary of the quantitative changes, respectively. For reasons explained in Chapter 7, the quantitative changes are based on NISP values. The information in Figure 9.3 supersedes the preliminary results of a study by Aplin and myself as reported by Jelsma (1998).

The fauna of the lower part of the deposit differs markedly from that of the overlying levels in the presence of a suite of characteristically montane mammal species such as *Mallomys* sp. cf. *M. aroaensis*, *Pseudochirops albertisii, Pseudochirops coronatus*, *Pseudochirulus schlegelii*, *Phalanger vestitus*, *Dactylopsila palpator, Dendrolagus* sp. (*goodfellowi*-group) and *Dorcopsulus* spp. (see Fig. 9.3). Some of these taxa are also present in the lowermost unit of Kria Cave and have been discussed above. Not found in Kria are a giant rat, *Mallomys* sp. cf. *M. aroaensis,* the Long-fingered Triok, *Dactylopsila palpator*, and two ringtail possum species of the genus *Pseudochirops*. All of these taxa are recorded as living animals in the modern forests of the Arfak and/or Tamrau Mountains (Flannery 1995a), and all are characteristic elements of montane forest faunas across their wider distributions in the central and/or north coast ranges of Papua and Papua New Guinea (*Pseudochirops coronatus* is endemic to the Arfak Mountains). As in Kria Cave, the remains of *Spilocuscus rufoniger* and a species of the *Dendrolagus goodfellowi*-group of tree-kangaroos are also restricted in Toé Cave to those levels containing montane fauna.

The upper limit of occurrence of the montane taxa varies across the transect of excavated squares, from c. 85 cm below the surface in Square 2N1E and 75 cm in Square 1N1E, down to 85 cm in Square 0N0E and 95 cm in Square 1S0E. This roughly follows the contour of the bedrock floor, which falls from a depth of 110 cm in Square 2N1E through to a depth of 140 cm in Square 1S0E. Only in Square 1S0E are a few isolated montane taxa represented at significantly higher levels (60-70 cm). However, as will be argued below, this may reflect local disturbance in a small area of the site.

The dramatic change in faunal composition between the lower and upper parts of the deposit is even more clearly demonstrated in Figure 9.4. In all squares, the upper part of the deposit is dominated by the remains of *Dorcopsis muelleri*, a typical inhabitant of lowland rainforest, with a significant secondary contribution from snakes and fluctuating

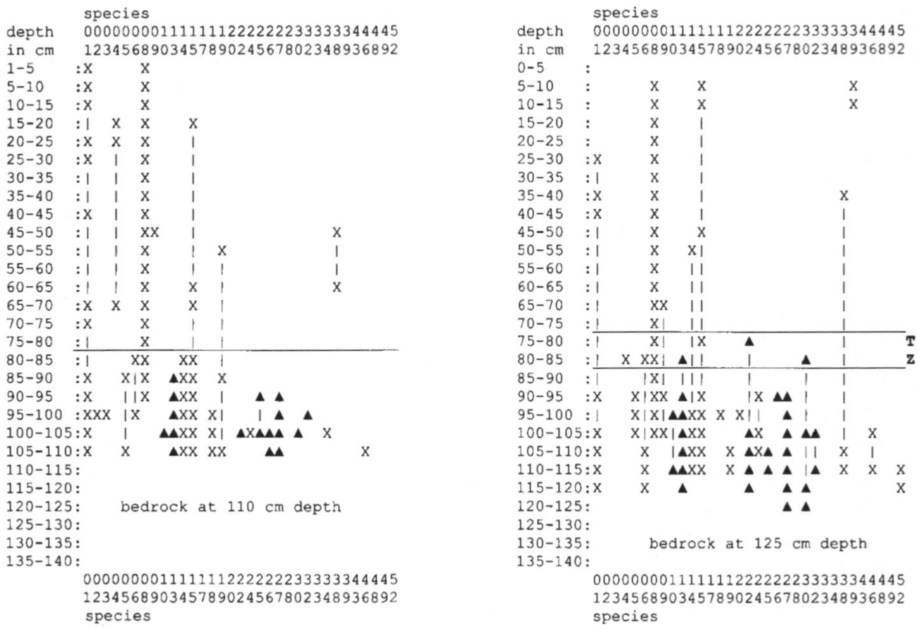

Square 2N1E

Square 1N1E

Figure 9.3. Toé Cave, occurrence of various vertebrate taxa through the deposit. X - lowland species, or species with wide range habitat; ▲ - strictly montane species. For numerical key to species, see Table 9.1. Horizontal lines indicate approximate upper limit of occurrence of montane species and subsequent transition zone (TZ) to lowland species. Square 2N1E was truncated at 80 cm depth. The vertical lines link records of the same taxon but do not imply its occurrence in the intervening levels.

but invariably small quantities of other taxa including bandicoots, cuscuses and rodents (see also Chapter 7). In contrast, the lower levels of the deposit show significant quantities of various strictly or primarily montane taxa including several ringtail possums (*Pseudochirops* spp., *Pseudochirulus* spp.*)*, cuscuses (*Spilocuscus* spp., *Phalanger* spp.), tree-kangaroos (*Dendrolagus* spp.) and Echidna (*Zaglossus bruijnii*), with a much lower proportion of *Dorcopsis muelleri*.

The stratigraphic position and nature of the transition varies across the transect of excavated squares. In Square 2N1E the transition from a montane fauna to one dominated by *Dorcopsis muelleri* is abrupt, with a sharp change between the 75-80 cm and 80-85 cm levels. However, this transition appears to be more gradual in the other three squares with a progressive increase in *Dorcopsis muelleri* and corresponding decline in montane taxa over a 10-15 cm zone. This is best seen in Square 1S0E, which shows the highest excavated bone densities throughout this part of the deposit, perhaps reflecting its 'downslope' location and position against the sloping cave wall, protected from human trampling. The more abrupt transition in Square 2N1E, together with the slightly deeper position of the boundary relative to the adjacent Square 1N1E, suggests

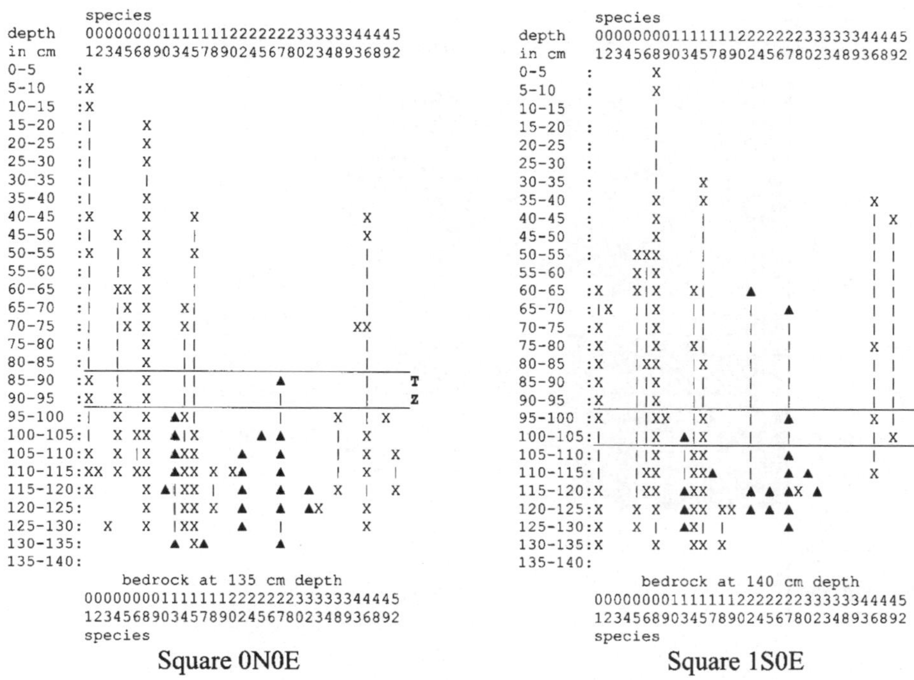

Figure 9.3 (continued). Toé Cave, occurrence of various vertebrate taxa through the deposit. X - lowland species, or species with wide range habitat; ▲ - strictly montane species. For numerical key to species, see Table 9.1. Horizontal lines indicate approximate upper limit of occurrence of montane species and subsequent transition zone (TZ) to lowland species. The vertical lines link records of the same taxon but do not imply its occurrence in the intervening levels.

that the lower unit in 2N1E may be slightly truncated, perhaps through local water erosion or human disturbance along the more vertical cave wall.

As noted above, two of the montane taxa (*Pseudochirulus schlegelii* and a *Pseudochirops* sp.) reappear at 60-70 cm depth in Square 1S0E, well above the general upper limit of their occurrence. In each case, these taxa are represented by a few fragments that make only a very small contribution to the total bone assemblage. The high position of these specimens is perhaps the result of some local disturbance within the deposit; however, they could also reflect the local survival of very small populations of the taxa, similar to the isolated montane elements found in Unit V in Kria Cave.

Jelsma (1998) concluded that the stratigraphy in Toé Cave had suffered significant disturbance during its history. However, the integrity of the faunal sequence and its close relationship to the contours of the original cave floor suggest that the deposit is essentially intact. As noted above, some evidence points to small-scale, local disturbance of the deposit; however, this is to be expected in any habitation site, especially one with limited habitable floor area such as Toé Cave.

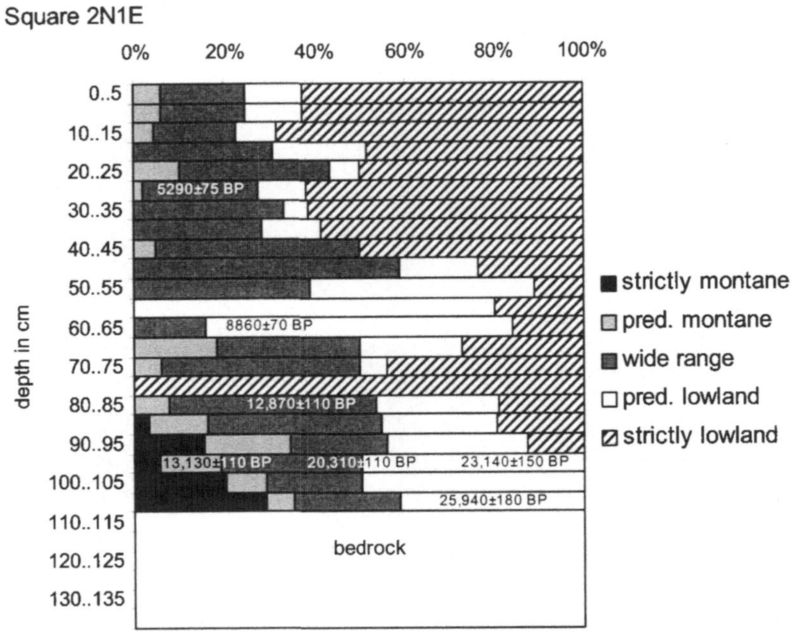

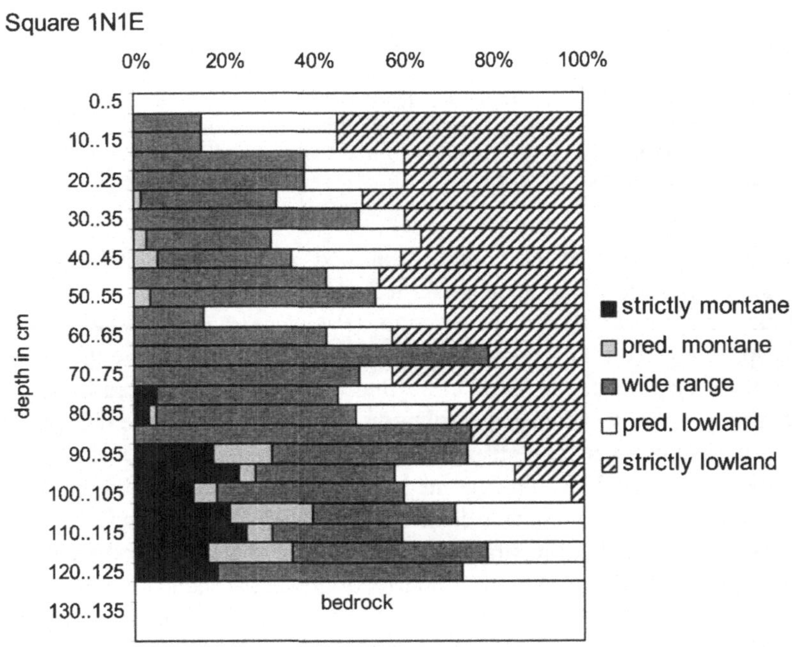

Figure 9.4. Toé Cave, proportional distribution (by NISP) of vertebrate groups by habitat through the deposit. 'pred.' = predominantly. The radiocarbon dates are positioned according to their depth below the surface; they are not related to any faunal group.

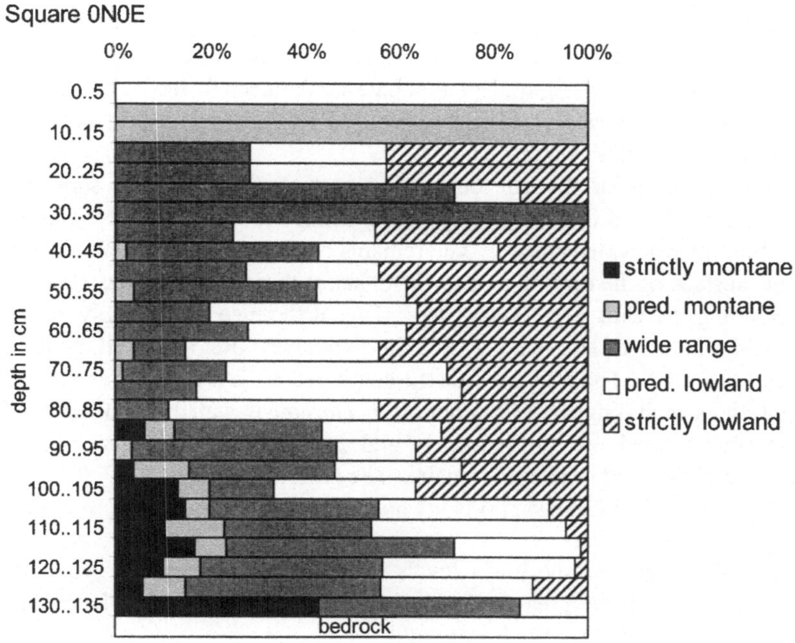

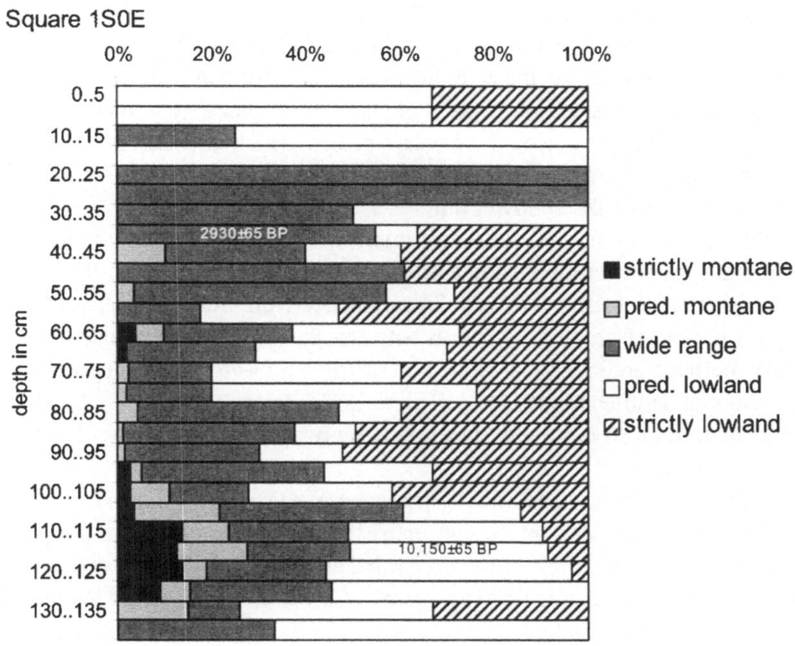

Figure 9.4. (continued). Toé Cave, proportional distribution (by NISP) of vertebrate groups by habitat through the deposit. 'pred.' = predominantly. The radiocarbon dates are positioned according to their depth below the surface; they are not related to any faunal group.

The upper part of the Toé Cave deposit appears to be broadly comparable to Units IV-II of Kria Cave, both showing an overwhelming predominance of *Dorcopsis muelleri* remains and an absence of a significant montane component to the fauna. The AMS radiocarbon date on *Casuarius* eggshell of 5290 ± 75 (25-30 cm depth; see Fig. 9.4) from Toé is consistent with this interpretation.

The uppermost levels of the Toé deposit, as in Kria Cave, contain some pottery sherds. Pig bone was found in a disturbed context on the surface of the deposit, but was not found within the excavated squares. The remains of *Dorcopsis muelleri* are abundant through to the surface of the deposit in Toé Cave, hence there is no obvious stratigraphic unit corresponding to Unit I of Kria Cave. This difference is probably related to differences in the nature of site use between the two caves after 4000 BP, as has been postulated in Chapter 8: Toé Cave evidently continued in use as a focus for general economic activities including the consumption of *Dorcopsis* wallaby, while Kria Cave was by that time visited only infrequently and perhaps used for more specialised functions including disposal of human remains. The radiocarbon date of 2930 ± 65 BP at a depth of 35-40 cm in Toé Cave supports the notion of a more continuous use of the site over the last few thousand years.

Jelsma (1998) argued that the lower levels of the deposit were most likely of Late Pleistocene age, based on the great variety of montane animals in these levels. This is now well supported by a series of radiocarbon dates on cassowary eggshell running back to 26,000 BP (see Fig. 9.4 and Chapter 4).

The present analysis has demonstrated that montane elements together make up a significant proportion of the fauna in the lower part of the deposit. The presence of a full range of post-cranial bones of these montane species further indicates that whole animals, rather than just selected body parts or 'hunting trophies' (e.g. mandibles) were brought to the site. The presence of various species which are not currently found anywhere on the Bird's Head also points to substantial, regional environmental changes since the time of deposition of the lower unit.

As indicated above, the nature of the transition in Toé Cave between the lower unit, with its significant proportion of montane fauna, and the upper unit, with its *Dorcopsis* (lowland) fauna, is not yet fully resolved. The evidence from Squares 1N1E, 0N0E and 1S0E hints at a period of gradual transition between the two units (see also Chapter 7, e.g. Fig. 7.5), perhaps indicating a slow but steady build-up of the deposit without significant breaks in sedimentation. Only Square 2N1E appears to show a break in the stratigraphical pattern at 80 cm depth, but this is more likely to be the result of a local truncation of the deposit.

One final point of interest concerns the presence of *Dorcopsis muelleri* in the lower unit of Toé Cave. Although this implies an unusual pattern of sympatry in the past (see below for further discussion), the record is regarded as genuine and not a consequence of disturbance. One reason for this is that specimens of *Dorcopsis* from these levels display the same degree of mineralisation as those of the various montane species, thereby distinguishing them from the much fresher *Dorcopsis* material from the upper levels.

9.4. FAUNAL SUCCESSION AND ENVIRONMENTAL CHANGE ON THE AYAMARU PLATEAU

As discussed in Chapter 3, the stratigraphy and dating of the Kria Cave sequence is now reasonably well established, with minor uncertainty only over the date of initial occupation of the site (c. 8000 BP by extrapolation) and the nature of the transition from Unit II to Unit I (depositional hiatus vs. continuous but much slower sedimentation rate). For Toé Cave, based on the basic integrity of the faunal sequence, the deposit is regarded as essentially *in situ* and only minimally disturbed. This is confirmed by the radiocarbon dates and the intact sequence of A/I ratios (see Chapter 4), both obtained from cassowary eggshell.

The combined evidence from the two caves suggests that there are three distinct periods in the faunal and wider environmental history on the Ayamaru Plateau. For the sake of simplicity, these will be termed the 'glacial', 'transition' and 'modern' periods.

The 'glacial' period

The 'glacial' period is represented by the Late Pleistocene lower unit of Toé Cave. The fauna from this unit is characterised by the remains of montane, closed forest taxa, along with some additional disturbance-intolerant species (*Spilocuscus rufoniger*, *Dendrolagus inustus*). However, elements of lowland rainforest fauna are also present in these levels, most notably the Brown Dorcopsis (*Dorcopsis muelleri*). Significantly, no grassland or savannah species are present at all, such as seen in Late Pleistocene fauna on the Aru Islands (O'Connor et al. 2002), suggesting that the present closed rainforest habitat has a long, continuous history in this area.

As argued earlier, the abundance and other characteristics of the montane animal remains from these levels indicate that the animals were obtained in the immediate environment of the cave, rather than being carried in from any great distance. Today none of the montane species occurs within 50 km of the site (across rugged karst country; see Fig. 9.5) and several appear to be regionally extinct on the Bird's Head (Aplin 1998).

The former occurrence of montane forest species in the vicinity of the Ayamaru Lakes, located centrally on the Ayamaru Plateau, implies wide-ranging changes in environmental conditions. Today, the montane species are confined to elevations above 800-1000 m above sea level. The Ayamaru Plateau has an average elevation around 350 m above sea level (Visser & Hermes 1962). To bring montane fauna into the vicinity of the lakes would require a broad-scale incursion of montane faunal (and presumably floral) elements across a large area of the southwestern Bird's Head.

The presence of *Dorcopsis muelleri* in the same levels as the montane fauna is noteworthy. Today, there is nowhere in mainland New Guinea where one could observe sympatry between any species of *Dorcopsis* and such mammals as *Dorcopsulus vanheurni*, *Dactylopsila palpator* and *Mallomys aroaensis*. The 'glacial' period fauna from the lower unit of Toé Cave thus represents a good example of what are commonly known as 'disharmonious' or 'non-analogue' associations of animals. Similarly unusual combinations of species from different habitats or climatic zones have been reported in

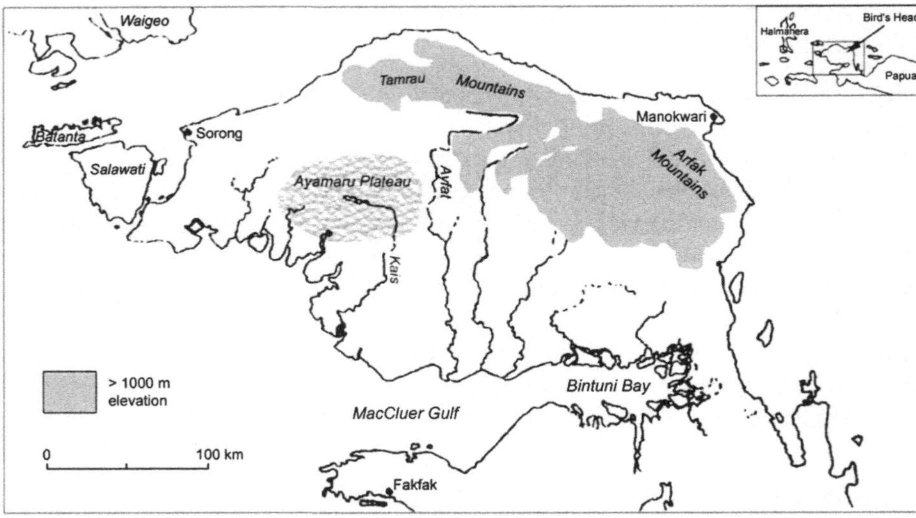

Figure 9.5. Location of the Ayamaru Plateau in relation to areas above 1000 m elevation.

Late Pleistocene deposits in other parts of the world including Australia (Graham 1985; Lundelius 1989). A similar admixture of lowland and montane flora has been noted in glacial age pollen spectra from lowland Amazonia (Colinvaux et al. 1996); in this case the admixture is attributed to a significant cooling which allowed species currently restricted to higher elevations by temperature to coexist with more temperature tolerant taxa (this issue is discussed further below).

An interesting feature of the 'glacial' period fauna is the high abundance of various disturbance-intolerant species (e.g. *Zaglossus bruijnii*, *Dendrolagus* spp., *Dorcopsulus* spp., *Spilocuscus rufoniger* and *Pseudochirops coronatus*). From this it can be inferred that the region was relatively undisturbed through this period, with no forest clearance and only low levels of hunting pressure on the larger mammals.

The 'transition' period

The 'transition' period is best known from Unit V of Kria Cave, dated from around 7000 (or possibly earlier) to c. 6000 BP. It may also be represented in the Toé Cave sequence by the zone especially visible in Squares 1N1E, 0N0E and 1S0E (see Fig. 9.3), as the montane 'outliers' recorded up to some 15 cm above the bulk of montane fauna. In Square 1S0E isolated montane elements occur as high as the 60 to 70 cm levels. As argued above, this may be a product of local disturbance, or, it may in fact represent the upper limit of a wider transition zone in this square. Although the stratigraphy in Square 1S0E is less well controlled, an age of c. 6000 BP for these levels in this square is not impossible (see Fig. 9.4, and Fig. 4.25 in Chapter 4).

As described earlier, the fauna from Kria Unit V is dominated by the remains of lowland mammal species among which *Dorcopsis muelleri* is overwhelmingly predominant, but it also includes a selection of montane forest and disturbance-intolerant species. Together, these make up only a small proportion of the total bone. The 'transition' period is so named because it is presumed to record a phase of replacement of the 'glacial' montane fauna by a modern lowland one over the Ayamaru Plateau. Exactly how long this process would have taken is not known, although the evidence from Kria Cave suggests a process taking millennia rather than centuries. In general, the process was probably one of range fragmentation followed by local extinctions, the latter process possibly accelerated by an expanding human presence in the area.

More direct evidence of human impact is provided by the local extinction of *Spilocuscus rufoniger* and the tree-kangaroo of the *Dendrolagus goodfellowi*-group, as well as the decline in abundance of *Zaglossus bruijnii* after this transition period. These changes were most likely caused by an increase in hunting activities. However, the detailed study of *Dorcopsis* remains from Kria Cave (see Chapter 8) suggests that habitat disturbance associated with small-scale gardening activities did not have an appreciable impact on this species, at least until c. 5000 BP, a thousand years or so after the last montane species was recorded in Kria Cave.

The 'modern' period

The 'modern' period is represented by Units IV-I of Kria Cave and by the upper part of the Toé Cave deposit. These sediments in Kria range in age from c. 6000 BP through to the recent past and contain a vertebrate fauna which is essentially the same as that recorded by contemporary zoological records from lowland habitats on the Bird's Head Peninsula. Typical elements of this widespread lowland fauna are the mammals *Dorcopsis muelleri*, *Spilocuscus maculatus*, *Phalanger gymnotis*, *Dactylopsila trivirgata* and *Rattus praetor*, along with several taxa with more restricted distributions such as *Dendrolagus inustus* (Bird's Head and north coast ranges) and *Pogonomelomys bruijnii bruijnii* (Bird's Head endemic).

Changes in faunal composition through the modern period are relatively minor and unlikely to be of any regional significance. One possible exception is the appearance of the Large Spiny Rat, *Rattus praetor*, in the uppermost unit of Kria Cave. Today, this species is found across a wide altitudinal range and in a wide variety of habitats; however, it is best known as the common garden rat of northern New Guinea and eastward through the Bismarck and Solomon Islands. Its conspicuous abundance in the upper levels of Kria Cave may reflect an increase in gardening activities and other forest disturbance in the immediate vicinity of the site during the most recent phase of deposition. In Toé Cave, one jaw fragment of *Rattus praetor* was collected only from 70-75 cm below the surface in Square 0N0E.

The great abundance of *Dorcopsis muelleri* remains through this period contrasts with the reported scarcity of this species in the area today. According to Elimas Kambuaya (local school teacher in Ayamaru), this species (known locally in Meybrat language as *Djief*) is known to everyone but is locally rare; it is generally obtained through the local markets (presumably caught in more remote areas). Local hunters noted that it is most easily killed at night 'when it cannot see very well' (see Appendix

3). The local scarcity of *Dorcopsis muelleri* populations may be partly a consequence of overhunting, but may also reflect an increase in agricultural activity and associated habitat destruction throughout the region in recent times. All *Dorcopsis* species are shy, secretive animals that are intolerant of excessive human presence.

Other comments from Elimas Kambuaya (Appendix 3) suggest that many other game species have also undergone a significant decline in the Ayamaru region. Specifically, he mentioned the present-day scarcity of *Spilocuscus maculatus* (white form: *Swi,* spotted form: *Tifdjah*) and *Zaglossus bruijnii* (*Gi*) in the area, and the local disappearance of *Dendrolagus inustus* (*Kadjo*; still present in remote country to the northeast and southwest of the Ayamaru Lakes). In contrast, a Striped Possum (*Siwejah*), presumably *Dactylopsila trivirgata* but also possibly including the newly described *Dactylopsila kambuayai,* is stated to be widespread and common in the local area; this accords well with observations of *Dactylopsila trivirgata* as a common inhabitant of both primary forest and regrowth vegetation (Flannery 1995a).

The last c. 4000 years, covering Unit I in Kria Cave, show a change in the use of the cave and in the frequency of occupation. Unfortunately, few faunal remains are available from this period in Kria. Occupation in Toé Cave appears to have continued more or less into (sub-)modern times, however there are no radiocarbon dates available for the uppermost levels of this deposit. While the sample of faunal remains from Kria from the late Holocene is too meagre to document environmental change through the last 4000 years, some information is available from pollen cores taken from the Ayamaru Lakes.

The Ayamaru basin was probably formed as a result of local tectonic movement along a faultline, which generated a small (c. 15 km long and 1-3 km wide), laterally widening depression in the limestone. Harder limestone formations in the southeast of the depression prohibit any deeper erosion by the rivers and rainwater, causing the plain to flood (R. Dam pers. comm.). While the two easternmost lakes are fed by the river Isme, which comes in from the southeast and then curves sharply to the northeast to drain into the two basins of Lake Hain (see Fig. 2.1, p. 14), the larger westernmost Lake Ayamaru appears to be mainly fed by rainwater, which runs down from the surrounding karst hills and disappears locally underground. At the northwest edge of the lake, close to the village Segior, are a series of wells that drain into the lake (Reynders 1961; Miedema 1986). These wells are presumably the source of some local myths, which relate to how the lakes came into existence:

> 'Once upon a time there was a man who had a dog. One day he went out hunting with the dog. The dog smelled a rat and went chasing it. The rat disappeared into a hole between the rocks. [...] The man came and tried to dig out the rat with a sharp stick. [...] At some stage he reached a root. When he pulled the root away water came spurting out of the ground. That was somewhere around Segior. So much water came flowing out of the hole that the man got scared and walked away. Four days later he returned and saw that the water that had flown out of the hole had formed a large lake.' (Translated by the author from Reynders 1961: 19; also described in Elmberg 1968: 256-257 (M6), and Miedema 1986: 3-5).

The westernmost Ayamaru Lake was cored in 1998 by Rien Dam, currently of CICAT, Delft University of Technology (Netherlands), and a preliminary pollen diagram was prepared from one core hole by Sander van der Kaars of Monash University, Melbourne. Information on the core logs, two radiocarbon dates and pollen spectra from one core, were kindly supplied by Dam and Van der Kaars. Assistance with interpretation of the pollen spectra was provided by Geoff Hope of the Dept. of Archaeology and Natural History, the Australian National University. The lake basin contains around 5 m of sediments, overlying a hard limestone unit (probably bedrock) and a thin organic soil. Basal organic muds in two cores returned ^{14}C ages around 4000 BP, indicating that the western lake and associated swampland was formed during the mid-Holocene and has been in existence only for the last c. 4000 years of the period of occupation of Kria and Toé Caves. The other two lakes might be older.

The sediments from a core taken only a few hundred metres from Toé Cave show a basal laminated clay, overlain by humic, carbonaceous muds. Pollen spectra from the laminated clays show low values for grasses (Poaceae) and sedges (Cyperaceae), suggesting a high-water phase with less fringing swampland than is found around the lake today. A sample from above the laminated clays contains abundant Cyperaceae, marking a change to swampier conditions, perhaps associated with a fall in lake levels. Samples from closer to the surface show an initial fall of Cyperaceae, suggesting a return to higher lake levels, followed by a more recent expansion of sedge and grass communities. The lake today has a central zone of permanent standing water, fringed by seasonally inundated sedgeland and grassland.

The terrestrial pollen groups indicate a mixed primary and secondary forest throughout the core with no change in the proportion of disturbance-tolerant taxa. This suggests that there has been no dramatic change in the extent or nature of the forest cover over the last 4000 years or so. The area today supports a mosaic of gardens, low primary rainforest and secondary forest. One curious feature of the pollen record is the consistent presence of *Pandanus* pollen in samples from the laminated clay, but their absence from the overlying sediments. *Pandanus* was also scarce at the time of Elmberg's work in this region the 1950s (1955: 15). In the highlands *Pandanus* also appears to have been much more widespread during the Pleistocene and early Holocene than it is today (Haberle 1994: 188); its nuts may have been a focus for seasonal harvesting in prehistory (White et al. 1970; Groube 1989; Golson 1991a).

9.5. LATE PLEISTOCENE TO HOLOCENE ENVIRONMENTAL CHANGE IN WESTERN NEW GUINEA

Environmental conditions prior to the last period of intense glaciation are not well known from the New Guinea region, although they may be inferred from generalised sea level curves and nearby oceanic records (e.g. Chappell 1994; Linsley 1996). These records indicate that a phase of climatic warming commencing around 60,000 BP ended abruptly around 30,000 BP with a rapid cooling and expansion of glacial ice world-wide (Lambeck & Chappell 2001). In the New Guinea mountains, glaciers expanded to reach their all-time maximum extent by around 18,000 BP (Hope 1989; Hope & Golson 1995).

Global warming evidently commenced around 16,000 BP (Miller et al. 1997), but significant impact in the New Guinea region is not observed in pollen records until 13,500 BP (Hope 1989; Van der Kaars 1995). The retreat of glaciers and upward spread of forest cover in New Guinea continued until about 10,500 BP and even later in some areas. By 9000 BP, present-day treelines had been attained at most localities (Hope 1989).

The nature and extent of environmental change at lower elevations is less well known. Hope (1976) observed minor vegetational changes across the Pleistocene to Holocene transition at 1400 m on Mt Wilhelm in the central ranges of Papua New Guinea and suggested that thermal impact was greatly reduced at mid-montane elevations. Similar results have been obtained elsewhere in wider Malesia, most notably the Sumatran highlands (Flenley 1984; Newsome & Flenley 1988) and montane Java (Stuijts 1993) where sites between 1300 and 1600 m above sea level show very little change in forest structure and only minor changes in forest composition through the Late Pleistocene to Holocene.

More pronounced evidence of climatic change was found by Hope & Tulip (1994) in a long pollen record from Lake Hordorli at 780 m in the Cyclops Range of northeast Papua. Here, higher altitude forest taxa are well represented between 25,000 BP and 10,500 BP, indicating a depression of altitudinal limits by at least 800 m. After 10,500 BP lowland taxa increase briefly but decline again until about 8500 BP. Reinvasion of the site by lower altitude forest taxa was not completed until about 7000 BP.

Van der Kaars (1995) reported a detailed record of glacial to post-glacial events from deep-sea cores taken from the northeast Banda Sea close to Fakfak on the Bomberai Peninsula, south of the Bird's Head. In this record, a significant downward expansion of montane forest communities is indicated prior to 13,500 BP, with some evidence pointing to slightly drier as well as colder conditions. After 13,500 BP, mid- to upper montane forest taxa decline in importance, presumably indicating a general retreat of these taxa to higher elevations. An expansion of mangrove vegetation at 11,750 BP coincides with an increase in the rate of sea level rise (Fairbanks 1989) and also with the establishment of an open connection between the Arafura Sea and the formerly land-locked Lake Carpentaria situated between New Guinea and northeast Australia (Torgersen et al. 1988). On theoretical grounds the presence of such a large expanse of shallow sea in this area could be expected to result in regional warming and humidification (Nix & Kalma 1972). The Banda Sea pollen record indeed records a further decrease in upper montane taxa and a rise in sub- to lower montane elements between 11,750 BP and 10,500 BP, consistent with a phase of slightly warmer and wetter conditions. The latter part of this period overlaps with the so-called 'Younger Dryas' (11,000-10,000 BP), which saw a global return to colder conditions and a significant dry event in the tropical north of Australia (Kershaw 1986; De Deckker et al. 1991; Van der Kaars 1991).

From further afield, Van der Kaars and Dam (1995) recorded a significant downward displacement of montane forest communities during the glacial maximum at 655 m above sea level at Bandung, Java, while Hope (1996) mentions evidence of colder and drier conditions during the period 35,000-10,000 BP and a lowering of montane forests at Soroaka at 440 m elevation in South Sulawesi. Barmawidjaja et al. (1993) inferred a downward shift of montane forests by about 270 m on Halmahera for the period 28,000-14,000 BP, with a rapid retreat to higher elevations after that time. They posited that conditions of glacial aridity were primarily restricted to the eastern Indonesian islands and northern Australia, compared with more stable conditions further to the west, and suggest a weakening of the wet northwesterly winds during glacial times.

By 9500 BP, climatic conditions were similar to those of the present day in all parts of the wider Indo-Pacific region. However, a subsequent rise in the representation of *Nothofagus* and *Phyllocladus* in the Banda Sea core until 7000 BP points to a return to cooler or possibly drier conditions during the early Holocene (Van der Kaars 1995). Significantly, the eastern connection between Lake Carpentaria and the Coral Sea was established at c. 8000 BP (Torgersen et al. 1988), and this is likely to have brought about a lowering of water temperatures across southern New Guinea.

The faunal evidence from the Ayamaru Plateau of the Bird's Head provides the first direct evidence of Late Pleistocene and early Holocene environmental conditions at lowland elevations in western New Guinea. As detailed in the preceding sections, the cave faunas document the former occurrence of numerous montane mammal species in the immediate vicinity of the Ayamaru Lakes at c. 350 m above sea level. Today, these taxa are generally restricted to mountainous regions above 1000 m elevation and, for most species, the nearest extant populations are at least 50 km away across the rugged Ayamaru Plateau to the Arfak and Tamrau Mountains to the east and north. Furthermore, in several cases, the taxa in question are no longer extant anywhere on the Bird's Head and the nearest living populations are probably in the Weyland and Charles Louis Mountains more than 500 km to the east in Papua. The scale of the changes required to bring these montane taxa down onto the central portion of the Ayamaru Plateau convincingly demonstrates that the identified changes are of a regional rather than purely local nature.

Exactly what the expansion of montane fauna on to the Ayamaru Plateau means in terms of palaeoclimatic regime is difficult to say. Hope (1986) has discussed the problems of inferring climatic changes from complex biotic responses in plant communities, even where the distributions and modern associations are well known. For animals, these difficulties are accentuated on account of the scarcity of detailed distributional data and the complete lack of ecological data for any species.

Terborgh (1971) presented a theoretical framework for the interpretation of distribution patterns in a Peruvian montane bird fauna, backed up by detailed mapping of each species' altitudinal range. He suggested that the altitudinal range of each species is determined by three factors, namely 1) physiological fitness along an environmental gradient; 2) competitive interaction with other species; and 3) the location of environmental discontinuities (ecotones). Subsequently, by going to a smaller and lower mountain range with fewer competing species, Terborgh and Weske (1975) were able to demonstrate that the second factor, competitive exclusion, was perhaps the most important determinant of a species' altitudinal range, or at least of its upper altitudinal limit.

Within the New Guinean vertebrate fauna there are many examples of altitudinally replacing species and it is clear that here, too, competition is of great importance (Donnellan & Aplin 1989; Flannery 1995a). However, Ken Aplin (pers. comm.) has observed various montane species being transported alive by hunters to lower elevation villages, and it is clear that individuals of many montane species possess a profound physiological intolerance of high temperatures and will rapidly succumb to heat stress when transported to warmer conditions. Assuming that certain critical physiological parameters have remained the same (e.g. coat thickness), then the successful altitudinal displacement of various species which today occur at altitudes in excess of 1000 m above sea level, down to 400 m above sea level to the vicinity of the Ayamaru Lakes, implies a fall in local ambient temperature of around 4-5°C (assuming a lapse rate of 0.6-0.8°C/100 metres; Flenley 1984). However, this calculation also needs to take account of any difference in humidity levels, for, as a rule, thermal tolerance thresholds are lower under conditions of high humidity and higher under drier conditions. Under less humid conditions, such as have been posited by Van der Kaars (1995) and others for late glacial times, montane animals may thus have been able to descend to lower elevations with a less pronounced fall in temperature.

As a final word on the possible role of competition, it is worth noting that the 'glacial' period fauna contains a wider range of species owing to the 'non-analogue' association between what are today exclusively montane and lowland species. The presence of the montane species at lower elevation is thus not readily explained by the absence of a potential competitor during glacial times (e.g. *Dorcopsulus* spp. excluded by *Dorcopsis muelleri*).

The maximum age of the lower unit of Toé Cave, the source of the 'glacial' period fauna, is now established at around 26,000 radiocarbon years. A minimum age for the unit is indicated by the well-dated sequence from Kria Cave in which the oldest unit, dated to c. 8000-6000 BP, contains an essentially modern lowland fauna, similar to that in the upper part of the deposit in Toé, but with rare, late-surviving elements of the once abundant montane fauna.

By 6000 BP, climatic conditions on the Ayamaru Plateau were comparable with those of the present time. Lowland rainforest communities formed a more or less continuous forest cover throughout the region, and a typical lowland Bird's Head fauna was established. However, compared with the modern environment, the major game animals of the lowland fauna were evidently more abundant. More recently, a probable combination of overhunting and widespread forest disturbance and destruction appears to have reduced the local populations of many species, some even to the point of regional extinction on the Bird's Head. The pollen record from the Ayamaru Lakes suggests that forest disturbance had commenced at or prior to 4000 BP.

9.6. WIDER COMPARISONS AND IMPLICATIONS

The nature of climatic and vegetational change in the tropical lowlands remains a controversial issue. Early theories of environmental stasis in the tropical zone (e.g. Fischer 1960) were bolstered by the publication of the sea-surface temperature reconstruction for the Last Glacial Maximum (CLIMAP 1981), which suggested negligible changes in sea-surface temperature in tropical oceans. This evidence appeared contrary to indications from tropical mountains in areas such as New Guinea of major glacial expansion and tree line depression (e.g. Hope & Peterson 1975), and led to the development of various theoretical models including the notion of steeper lapse rates than found today (Walker & Flenley 1979; Barmawidjaja et al. 1993) or radical changes in patterns of atmospheric circulation (Webster & Streten 1978; Rind & Peteet 1985; Markgraf et al. 1992). Later, more detailed isotopic records for the New Guinea region (Thunell et al. 1994) supported the original predictions but amended the inferred change in sea-surface temperature to less than 2°C. This figure was still far short of the 6-7°C change estimated to be necessary for the major advance of glacial ice over the Last Glacial Maximum.

Another body of literature has grown up around the notion that drier conditions at the Last Glacial Maximum led to the conversion of large tracts of tropical rainforests into grassland and savannah habitats with rainforest taxa surviving in refugia (Haffer 1969; Vanzolini 1973). This view attained great popularity amongst evolutionary biologists as a mechanism for tropical speciation but always remained thin on primary evidential support (Flenley 1993).

Detailed pollen evidence of vegetational changes in lowland environments has recently become available for various parts of the world. By far the most detailed records come from the Americas where Last Glacial Maximum pollen spectra record montane forest elements existing some 800-1000 m below their current lower elevational limits (Bradbury et al. 1981; Liu & Colinvaux 1985; Bush et al. 1992; Colinvaux et al. 1996). These records have been taken to indicate a climatic cooling of 5-6° C, more in line with estimates of temperature changes at high elevations. As noted earlier, the far less complete pollen evidence from lowland sites in southwest China, Java, New Guinea and the north Moluccan islands is consistent with the American evidence. In all of these areas, Last Glacial Maximum pollen spectra show a marked increase in montane forest taxa consistent with the local occurrence of cool, moist forest (Liu Jinling et al. 1986; Barmawidjaja et al. 1993; Van der Kaars & Dam 1995; Hope 1996). Interestingly, recent isotopic work on tropical corals (Guilderson et al. 1994) gave Last Glacial Maximum cooling estimates of c. 5°C, and it has been suggested that the lower estimates based on foraminifera might be a consequence of isotopic fractionation in this latter group of organisms.

Taken in this broader context, the Bird's Head faunal record can be seen as broadly consistent with the emerging picture of vegetational change in tropical lowland habitats around the world. At the Last Glacial Maximum, a significant climatic cooling of the order of 5-6°C allowed the downward migration of numerous species of plants and, not surprisingly, animals that were no longer restricted to upland areas on account of heat intolerance. While temperature changes or new competitors at lower elevations might also have forced some lowland taxa down to even lower elevations, in at least some cases (e.g. *Dorcopsis muelleri*) this appears not to have happened, with the result that new patterns of species association emerged – the so called 'non-analogue' or 'disharmonious' assemblages.

In most parts of the tropics, retreat of montane elements back to their current altitudinal positions generally seems to have occurred rapidly in response to global warming [e.g. in Panama (Bush et al. 1992); Halmahera (Barmawidjaja et al. 1993); Java (Dam 1994)]. As noted earlier, evidence from the Cyclops Range in northeast Papua (Hope & Tulip 1994) and from the northeast Banda Sea (Van der Kaars 1995) off the Bomberai Peninsula (close to Fakfak) suggests a rather later retreat of montane elements, with 'modern' conditions not established until 7000 BP. The Ayamaru faunal evidence also records the late persistence of certain montane faunal elements at low elevation, with some surviving through to about 6000 BP.

Hope & Tulip (1994) and Van der Kaars (1995) both mention the possible role of extended cloud lie during the early Holocene in delaying the regional impact of global warming. There is no question that the impact of dense cloud cover on local climate and vegetation can be dramatic, with potential reduction of irradiance to 10-50% of that under sunny conditions and with other possible impacts on aspects of leaf physiology including photosynthesis and nutrient relations (Bruijnzeel & Veneklaas 1998). In this regard, it is of great interest to repeat an observation made by Rand & Brass (1941: 370) in their report on the 1936-1937 Archbold Expedition regarding the local environment at Black River Camp, at 100-250 m above sea level at the hillward margin of the Fly Plateau in Papua New Guinea:

> 'As with plants, a number of birds which over most of New Guinea occur only in the mountains, as *Clytoceyx rex* and *Diphyllodes magnificus*, come to near sea level on the Upper Fly River. The prevalence of mist and cloud conditions, lowering the light intensity, may have an effect in producing the conditions recalling cloud-shrouded mountain forest.'

A significant increase in cloud lie over southern and western New Guinea might have occurred at around 12,000 BP as a consequence of the initial conversion of the Arafura Shelf into a closed, warm-water embayment. Such conditions may have mimicked the cooler conditions of the Last Glacial Maximum and delayed the retreat of montane elements from the adjacent lowlands. Eventual breach of the Arafura embayment to the east occurred around 8000 BP and the influx of deep basin water presumably led to cooler sea surface temperatures along southern New Guinea. A progressive retreat of montane flora and fauna from the lowland habitats after that time is consistent with the as yet limited palynological and faunal records.

9.7. CONCLUSION

The faunal record from Toé and Kria Caves provides important new evidence on the nature and extent of environmental change in lowland New Guinea. The former occurrence of a montane forest vertebrate fauna at c. 350 m above sea level on the Ayamaru Plateau indicates that glaciation had a significant impact on lowland environments in New Guinea. This evidence runs counter to evidence from isotopic studies that glacial episodes resulted in little or no change in sea surface temperatures in this part of the world, but is consistent with the emerging palynological evidence of significant cooling of lowland rainforest habitats around the world. The resultant 'glacial' period communities of plants and animals included different combinations of species than those observed today. Evidence of a relatively late retreat of montane fauna from the Ayamaru Plateau is also consistent with the limited palynological record from this region.

As suggested by Van der Kaars (1995), initial inundation of the Arafura Sea and the MacCluer Gulf may have led to conditions of increased cloudiness over southern and western New Guinea and a prolongation of cool conditions despite global warming. The final retreat or local extinction of montane fauna on the Ayamaru Plateau is dated to c. 6000 BP and thus appears to postdate the eventual opening of the Arafura Sea to the east. Once more, local factors related to patterns of circulation through Torres Strait may account for the late establishment of fully modern conditions in this area.

Certain aspects of the faunal record are not readily interpreted in terms of climatological factors. Various mammal species may have declined through time as a consequence of increasing human presence in the area; in several cases this has apparently led to local extinction of the species from the wider Bird's Head region. The pressure on larger mammals appears to have become especially pronounced in the recent past and several, formerly common species are now quite rare in the Ayamaru region.

10. Discussion and Conclusion

...until the present day, the structureless society as New Guinea is called, has no past of its own; cannot provide a continuous, comprehensive history because of its large difference in development. Up into the 20th century, life of the Papuans remained static, archaic, so little adapted to advanced development elsewhere, that one could speak of this land as 'the back corner of earth'.[1]

Galis 1953: 16

This thesis presents an analysis of two prehistoric sites excavated in the Bird's Head of Papua and provides the first systematically gathered archaeological information for the entire western half of the island of New Guinea. The results have important bearing on a number of issues, including the early settlement and exploitation of tropical rainforest, independent of agriculture; human impacts on the natural environment and fauna including the prolonged hunting of one major prey animal; and the influence of glacial conditions on tropical lowland environments. Each of these issues will be discussed below in relation to the changing intensity and nature of the occupation of the two sites.

Did humans in the Bird's Head ever live in rainforest without access to cultivated plants or domesticated animals?

In 1989, Bailey et al. put forward the suggestion that people are unable to live in tropical rainforest in any part of the world without access to cultivated foods, and were not able to do so until ways were developed to change the distribution and abundance of edible resources (see also Headland & Bailey 1991). This hypothesis was based on the notion that rainforest habitats are generally too poor in wild food stuffs to sustain viable populations of foragers, with protein resources either located too high above ground, or too unreliable and dispersed. Moreover, even where sufficient protein might be available from animal sources, human occupation might be prohibited by a lack of reliable carbohydrate sources without the use of cultivars (Headland 1987). This hypothesis brought about a lively debate [e.g. *Human Ecology* 19 (2) 1991] and a wide response from a variety of fields (e.g. Hladik et al. 1993; Mercader 2003). Headland & Bailey (1991) suggested that archaeology would play a leading role in testing their hypothesis because it might provide insights into the use of 'true' pre-agricultural tropical rainforest

[1] Original text: '...de verbrokkelde samenleving, welke Nieuw Guinea genoemd wordt, [heeft] tot heden ten dage geen eigen geschiedenis [...]; [kan] geen lopende, allesomvattende historie [...] geven wegens de grote differentiatie in ontwikkeling. Tot in de 20e eeuw toe bleef het leven van de Papoea's statisch, archaeïsch, zo weinig aangepast aan de elders voortgeschreden ontwikkeling, dat men spreken kon van dit land als van "de achterhoek der aarde".'

habitats unaffected by recent or historic forest clearance and plant cultivation, especially where occupation could be dated to before c. 9000 BP.

Bailey et al. (1989) were unable to find convincing archaeological evidence of early rainforest occupation. However, subsequent authors have reported prehistoric sites that appear to contradict their hypothesis. In particular, sites in central Malaysia show that hunters and gatherers occupied this rainforested region from c. 10,000 BP, seemingly without altering their environment to increase the quantities of food or the reliability of resources (Endicott & Bellwood 1991). Headland & Bailey (1991) accepted this as a valid exception and conceded that other examples might be found in this part of the world. Shortly thereafter, Pavlides & Gosden (1994) provided evidence of 35,000-year habitation of an inland, lowland area of West New Britain. Although no direct evidence of past habitat was recovered, rainforest cover through the Late Pleistocene can be reasonably assumed in this case.

The two excavated sites in the Bird's Head provide undeniable evidence for prolonged occupation of lowland rainforest, presumably without access to cultivated foods for much of this time. Kria Cave contains a two-metre thick archaeological deposit that is well stratified and securely dated from sub-recent times back to c. 7000 BP, with an extrapolated date around 8000 BP for initial occupation. The archaeological deposit in Toé Cave, with a maximum depth of 140 cm, is essentially intact but its stratigraphy is complicated due to the sloping bedrock that has caused local slumping and erosion of the sediment. AMS radiocarbon dates were obtained from mammal bone, charcoal and cassowary eggshell, and the latter material also provided associated A/I ratios that increase in proportion to the ^{14}C ages and so confirm the integrity of the ages. The dated cassowary eggshell from Toé Cave provides direct evidence that humans were visiting the Ayamaru area, at least sporadically, since at least 26,000 BP. Today the area is a remote and rugged lowland plateau situated at c. 350 m above sea level. At the time of the Last Glacial Maximum the sites were located some 160 km inland from the southwestern coastline, and c. 90 km from the northwestern coast but with intervening mountain ranges. Human activity in such a remote part of the Bird's Head during Last Glacial Maximum times surely implies a more widespread and significant level of human activity in lowland habitats within New Guinea.

The presence of abundant and well-preserved faunal remains in both sites allowed for a detailed reconstruction of the environmental history of the area, spanning the entire period of human occupation of the sites. Both faunal sequences show that the Ayamaru Plateau supported a continuous rainforest cover throughout the Late Pleistocene and through until the late Holocene. Palynological data for the area confirm the local presence of lowland rainforest for the last c. 4000 years, albeit with some human disturbance.

Faunal remains from the lower levels of the Toé deposit, and to a lesser extent from the basal unit of Kria Cave, indicate that during Late Pleistocene times the Ayamaru Plateau supported a mixed community of lowland and montane mammal species, presumably reflecting lower temperatures than at present and a rainforest vegetation with characteristics of modern lower montane forests. Montane habitats are found today at least 50 km from the sites. Bulmer (1976: 181) recorded that the Kalam people of the

Papua New Guinea Highlands obtained their prey from areas within half a day's walk from a site, and occasionally from distances further than that (especially when the carcass is smoke-dried), but in the rugged karst environment of the Ayamaru Plateau it is unlikely that a hunter could reach such remote locations even after a few days walk.

When temperatures rose out of the Last Glacial Maximum, the faunal assemblage gradually changed, with a progressive loss of diversity, especially among the medium-sized mammals, and a corresponding rise in importance of the forest wallaby *Dorcopsis muelleri*. This species is a shy, secretive and disturbance-intolerant animal. It evidently remained common south of Lake Ayamaru, around the area of Toé Cave, until occupation in Toé ceased some time within the last few thousand years or so, but seems to have become slightly less abundant northeast of the lakes from c. 5000 BP, perhaps as a result of hunting pressure and habitat disturbance (see below). *Dorcopsis* gradually became scarce in the area northeast of the lakes, and it appears to be nowhere abundant around the Ayamaru Lakes today. Other disturbance-intolerant species, such as the Echidna (*Zaglossus bruijnii*), and the Grizzled Tree-kangaroo (*Dendrolagus inustus*), are present in both assemblages, but show similar trends to *Dorcopsis*, with evidence for persistence through to more recent times south of the lakes. Interestingly enough, *Rattus praetor*, the Large Spiny Rat, a typical species of gardens and general disturbance, is only found in Kria in the top levels of Unit II and throughout Unit I, therefore dating to around 4000 BP and later. One isolated element of this species was found in early Holocene levels in Toé Cave. The short pollen record from the western basin of the Ayamaru Lakes, provided by Dam (CICAT, Delft University of Technology) and Van der Kaars (Monash University, Melbourne), also suggests that forest clearance had probably commenced regionally by 4000 BP or earlier.

The absence of any typical tree-felling implements such as ground axes (Bulmer 1964) and waisted blades (Groube 1989) in the two sites is noteworthy in this regard, as it suggests that forest clearing to promote the growth of root crops may not have played a significant role in the lives of the Ayamaru people through the greater part of the occupation period. However, burning of forest may have provided an alternative method for opening up the forest canopy (Groube 1989). Evidence for burning of montane and subalpine vegetation is reported at various highland locations in New Guinea (Hope & Hope 1976; Haberle et al. 1991; Haberle 1994), dating back to a maximum of c. 30,000 BP (Hope & Golson 1995). Information about the role of burning as a method of forest modification in lowland areas including the Ayamaru Plateau is currently unavailable.

In summary, the archaeological record from the Ayamaru Plateau shows that humans occupied rainforest habitat on a more or less continuous basis for at least 26,000 years with true lowland rainforest present through the entire Holocene period. There is little, if any, evidence for forest disturbance until mid-Holocene times, hence plant cultivation is unlikely to have played a major role in their economy through at least 21,000 years of regional occupation. The only domesticated animal found in either of the assemblages is the pig, with the earliest example dated between 1840 and c. 4000 BP.

How did humans manage to live in this rainforest habitat?

The Pleistocene-Holocene transition on the Ayamaru Plateau was marked by significant changes in faunal composition, with the decline of many species including *Spilocuscus rufoniger*, *Dendrolagus* sp. cf. *D. goodfellowi*, *Dorcopsulus* spp., the Giant Rat *Mallomys* sp. cf. *M. aroaensis* and various species of *Pseudochirops*, and a rise to dominance of the Brown Dorcopsis, *Dorcopsis muelleri*. These changes probably reflect the general climatic amelioration coming out of the last glacial period with the retreat of many montane faunal elements to higher elevations. *Dorcopsis* subsequently dominated the mammal assemblage of the area for many thousands of years, evidently remaining abundant until 4000 BP in the northeastern part of the area (as represented in Kria Cave) and until more recent times (post 3000 BP) southwest of the Ayamaru Lakes around Toé Cave. Although it seems that *Dorcopsis* was steadily hunted around Kria Cave over a 3000-year period (between c. 7000 to at least 4000 BP), and for at least twice that long around Toé Cave, there is no sign from the wallaby age structure or the pattern of exploitation of the carcasses, that overhunting or other human activities caused any significant depletion of the species during these periods.

Grayson (2001) argued that the prolonged human exploitation of one major prey species inevitably leads to its decline and a shift to alternative resources. He presents numerous examples where prehistoric exploitation of animals resulted in local depletion, caused in most cases by a combination of overpredation, habitat modification and the introduction of non-native animals and plants. Typically, where detailed faunal analyses are performed, there is evidence for the loss of key prey species over time and an increase in diet breadth by the consumer population. In Grayson's (2001: 8) view, only a very low human population density or inadequate hunting technology might result in absence of any such impacts.

The phenomenon of overhunting depends on the maturation and reproductive rates of the particular species, its longevity, the frequency of kills, and the size of the consumer population (Bodmer et al. 1997). *Dorcopsis muelleri* has, relative to other species in the region, a fairly long lifespan (around 10 years in captivity), a slow maturation (c. 15 months), and a low number of offspring per year (1-2). Based on these characteristics, *Dorcopsis* would certainly be susceptible to overpredation. In Chapter 8, suggestions have been made as to why *Dorcopsis* did not decline on the Ayamaru Plateau despite the fact that its remains dominated the fauna accumulations in the caves over many thousands of years. One of the reasons may be related to the low human population densities in the area. The calculations of minimum numbers of consumed animals clearly showed that in spite of its relative abundance in the assemblages, the number of wallabies actually represented in the sites is far fewer than could sustain even a single human individual, even if this person would only visit the cave seasonally. For Kria Cave, the minimum number of killed wallabies was calculated to be one in every 3-4 years. Toé Cave, in which *Dorcopsis* dominates the assemblage ever since the end of the Pleistocene, produced only enough wallaby remains to support a visit by one or a few individuals every 2-4 centuries. Of course, these estimates are based on the *minimum* number of wallabies. However, even if the true numbers were 50 times higher, they would still have been barely enough to sustain even a single consumer on any regular basis.

Given these estimates, it is difficult to avoid the conclusion that the caves were visited infrequently and that hunting in the vicinity of the sites was sporadic and of low intensity. This notion of infrequent visits to the caves further implies that the people living on the Ayamaru Plateau were probably quite mobile and made use of numerous other rockshelters and perhaps also many open campsites. A high level of human mobility would help maintain the *Dorcopsis* population, giving it time to recover from any potentially excessive hunting in certain areas, although the population structure of the consumed animals, dominated by mostly older adult animals, suggests that excessive hunting was unlikely at any time. Some authors have suggested that an age structure of predominantly old animals found at archaeological sites is indicative of scavenging or selectivity in hunting (e.g. Stiner 1990a: 334-336, 1990b). These explanations are not plausible in the present case. Scavenging of carcasses is unlikely in a tropical environment because of the early onset of decay. Selective hunting of older individuals is also unlikely because *Dorcopsis* is a solitary animal. As described by many authors [e.g. Bodmer (1995) for people from Amazonia, and Bulmer (1976), Dwyer (1982) and Sillitoe (2002) for people from the Highlands of Papua New Guinea], hunters may have a preference for certain animals, but they will take anything they encounter on their way. Adult animals must therefore have been abundant in the *Dorcopsis* population, which in turn indicates that the species existed at, or close to, their demographic maximum (see Chapter 8).

The suggestion that *Dorcopsis* occupied the region at its maximum potential abundance does not imply that the prehistoric forests on the Ayamaru Plateau were 'overpopulated' with wallabies. The carrying capacity of the forest itself may have been very low and unable to sustain large numbers of medium-sized ground-dwelling animals. As a rule, most of the available nutrient in rainforests is located in the canopy (Bailey et al. 1989; Stahl 1995: 167) and the few terrestrial species that occupy such forests are typically either quite small and sedentary or quite large and highly mobile, and therefore widely dispersed. Animal population densities in rainforests can vary considerably. For example, Glanz (1990: 296-297) found that the number of sloths in Central and South American rainforests varied from 633 to less than 1 per km^2. Many factors obviously underpin such variation, including fluctuations (seasonal or annual) in climate, abundance of predators, and availability of food resources. Some natural fluctuation in *Dorcopsis* abundance may have occurred in the Ayamaru region through the Holocene, but whether such factors might account for the observed changes through time in the intensity of site usage is unknown.

To conclude, the low MNIs of *Dorcopsis* in both assemblages suggest that both caves were visited infrequently. This is taken as evidence that human population density in the Ayamaru area probably remained very low throughout the occupation period. But was human population density limited by the availability of *Dorcopsis* and other prey species? Or could there be other reasons?

Why were human population densities low in the Ayamaru region?

The two faunal assemblages show that animal food was sufficiently available in the Ayamaru area to sustain small groups of people over many thousands of years of occupation. There is no sign of overpredation of *Dorcopsis*, nor does the processing of the carcass show any changes through time that might be interpreted as evidence of dietary stress, at least not until the mid-Holocene, and even then only in Kria Cave. Moreover, the diet breadth does not shift to include smaller, low-return species. If these small groups of people were not starving, then why did the human populations stay low, and why did they visit the sites so infrequently? Vickers (1988) described how Amazonian Indians living in rainforest tended to move to different areas on a regular basis. One reason why this was done was local food depletion, but other movement was prompted by outbreaks of disease or infestation with pests, or impoverished raw material resources. Fear of raids from neighbouring populations may also have driven people to move regularly within their range. The substantial number of distinct languages in the Bird's Head area suggests that there may have been strong cultural differences between neighbouring groups, and raids and disputes over resources may have been a regular feature of life. Historically, at least, warfare and raiding was for many New Guineans almost a 'life philosophy' and, in extreme cases, even a major source of entertainment (e.g. Marind-Anim in southern Papua[2]; Van Baal 1966). Warfare may have been one reason why mortality, as well as mobility, was high and why the human population density remained very low.

Disease, too, is likely to have been a major cause of mortality in the past and its potential role in the suppression of human population growth should not be underestimated, especially in lowland tropical environments (Froment 2001). Like Amazonia, lowland New Guinean rainforests and its wildlife harbour numerous diseases that are potentially transferable to humans (e.g. typhus, malaria). Although the likelihood of epidemics is low in small, highly mobile human populations, foraging societies are exposed to rare or new viruses through their close contact with wild animals and their ectoparasites. The hot and humid environment of rainforest is also highly favourable to the transmission of chronic, debilitating diseases such as diarrhoea and intestinal parasites, both potentially serious causes of malnutrition. These factors typically make life expectancy of foraging peoples low and childhood mortality high.

The study of the human remains found in both Kria and Toé Caves, carried out by David Bulbeck (Appendix 1), provides evidence of childhood illness or starvation among the Ayamaru occupants from potentially the early Holocene through to the recent past. Enamel hypoplasia, an interruption of enamel formation during growth, was found on various dental elements of three of the five individuals identified from Toé Cave, as

[2] Interesting in this respect is that De Vries (1998) found influences of languages of the Marind stock in the Inanwatan language in the southern Bird's Head. This suggests that, in the past, the Marind-Anim people, famous for their persistent and wide-ranging head-hunting activities, may have occupied an extended territory across all or part of the south coast of Papua, including the south coast of the Bird's Head.

well as in the dentition of the juvenile individual buried in Unit I of Kria. Moreover, this latter individual bears evidence of a possible violent death. These findings testify to a harsh life on the Plateau, at least during Holocene times, in which illness, hunger and violence must have played a significant role.

Headland (1987; see also Sillitoe 2002: 60-62) suggested that access to reliable sources of carbohydrates is very difficult in tropical rainforests and cited this as the primary reason why people would never have been able to live in rainforest without access to cultivated plant foods (Bailey & Headland 1991: 266). Although this factor was clearly not completely prohibitive in the Ayamaru region, it may well have played a role in limiting population density during the pre-agricultural period. It may also have contributed to keeping the inhabitants highly mobile, at least until such times when gardening activities provided more reliable sources of plant foods.

Besides a few carbonised fruits from Kria, and some *Celtis* and banana seeds from Toé, there is no direct information regarding the contribution of vegetable foodstuff to the diet. However, it seems reasonable to conclude that sago was not exploited as a carbohydrate source to any significant extent (if at all) in the past. The 'sago-choppers' with their typical silica gloss, identified in mid- to late Holocene sites in the eastern half of New Guinea, are absent from the Bird's Head sites. Of course it is possible that other raw materials were used instead for the processing of sago. Ethnographic records of sago-choppers are rare for the Bird's Head region, but Crosby (1973: 173) mentions one such implement made of wood from the eastern margin of the peninsula. Wollaston (1912: 90) also reports the use of hardwood for this purpose in the Mimika River area, southern Papua. However, there was no shortage of chert around the Ayamaru Lakes suitable for manufacture of sago-choppers. Moreover, the number of sago trees in the area was negligible when Elmberg (1955: 57) visited the Ayamaru region, and processed sago was usually obtained by trade with coastal people.

Today, as in Elmberg's time, the staple food in the Ayamaru region is taro. This may have been an important carbohydrate source for Ayamaru people in the past, perhaps using wild taro plants growing close to streams and swamps (Spriggs 2000b: 293). Various authors have suggested that edible plants like tubers would have been present in sufficient densities in rainforests to sustain bands of foragers (e.g. Endicott & Bellwood 1991; Bahuchet et al. 1991). Human activity in rainforest might also promote the spread and growth of edible plants, such as along sunlit trails and around old camps (pers. comm. by Greg Laden, cited in Bailey & Headland 1991: 266). Bailey & Headland (1991: 274) argue that such foods are inefficient and costly to process and may contain relatively few calories. However, archaeological evidence for tuber preparation is available from Kilu Cave on the island of Buka in Melanesia by 28,000 BP (Loy et al. 1992), which further suggests that, by this time, people knew how to process taro to make it edible (Gosden 1995: 811).

The Ayamaru region today is infamous for its thin, poor quality soils (Elmberg 1955, 1968; Miedema 1986) and it is possible that the area has always been deficient in key vegetable nutrients. As Spriggs (2000b: 294) suggested in the context of Vanuatu (southeast of New Guinea) and its lack of pre-agricultural settlement, the area may have been a 'green desert' without the introduction of edible plants and animals.

Bailey et al. (1989) are clearly incorrect in their assertion that tropical rainforest could not be inhabited by people prior to the introduction of domestic plants and animals to the region. However, the Ayamaru evidence could be seen as supporting their less extreme claim that life in such environments was difficult in the absence of abundant and reliable sources of carbohydrates and/or protein (see also Sillitoe 2002). On top of that, warfare and disease may have kept people mobile and population densities low.

Are there indications of plant cultivation in the Ayamaru region?

There are indications in the Ayamaru area for some forest clearance during mid-Holocene times. This could be seen as supporting the notion (e.g. Groube 1989; Bahuchet et al. 1991; Spriggs 2000b: 301) that the growth of tubers might have been actively stimulated by opening up the forest canopy. One source of information comes from subtle changes in the age structure and treatment of *Dorcopsis* carcasses in the Kria Cave assemblage. As explained in Chapter 8, the more an animal population approaches the carrying capacity of its local environment, the 'older', or more senescent, the population becomes. The much smaller quantities of *Dorcopsis* material from Units III and IV, periods that appear to register less intense occupation of the cave, show an even higher proportion of very mature animals than those from Unit II and V, periods of higher intensity site use. This association of intensity and age structure is even stronger in the Holocene sample from Toé Cave, with even older wallabies during a period of overall, very low occupation intensity. The consistent nature of these patterns suggests that Units III and IV at Kria represent a phase of lower human population densities to the north of the lake; and further that levels of human activity around Toé were consistently lower, perhaps until very recent times.

Unit II in Kria Cave, dated from c. 5000 to 4000 BP, shows a slight shift toward somewhat younger – although still mature – individuals. This structure is accompanied by a slightly higher fragmentation of *Dorcopsis* bones, not matched in the remains of any other species in the sample. Although the age structure of *Dorcopsis* still suggests that its numbers in the area were fairly close to carrying capacity, culling must have gradually increased, resulting in a shift in age structure towards a higher proportion of younger animals. Higher fragmentation of the bones also seems to point to a more concerted effort to extract more food value from each capture. Either the consumer population had increased during Unit II times, or the availability of wallabies had declined, or perhaps both in combination. Importantly, *Dorcopsis* was probably never so rare that hunters had to travel large distances and be selective in regard to the body parts of their prey they could return.

One factor that may have brought down the numbers of wallabies (but not necessarily altered its population structure) during this period is a reduction in the carrying capacity of the environment. It is possible that the *Dorcopsis* habitat was being increasingly disturbed, or even gradually destroyed. A combination of factors might be responsible for such a situation, including an increase in human population density in the area; forest clearing to stimulate growth of vegetable items such as tubers; and an increase in predators or competitors for food resources from the forest floor. The latter might have occurred through the introduction of non-native animals such as dogs and pigs. No remains of dogs were found in either cave, and pig remains are also relatively scarce in both assemblages; as noted earlier, the earliest pig remains are directly dated to between 1840 and c. 4000 BP. The time of introduction of both pig and dog to the wider New Guinea region remains poorly dated but is assumed to be mid-Holocene or later.

An increase in human population could explain the insufficient protein supplies and hence the higher fragmentation of *Dorcopsis* bones. By smashing up the larger bones, bone marrow and small bone fragments would become available for consumption. Forest clearing to the north of the lakes would have increased the quantity of plant foods at the expense of sources of meat, although availability of animal protein may not have become a serious problem for the Ayamaru people until recent times. As noted earlier, the pollen record from the western basin of the Ayamaru Lakes adds support for this scenario, with indications of relatively low-level forest disturbance back at least to the time of origin of the western lake at 4000 BP. However, even with this apparent intensification in land use, human population density evidently remained low or at least localised. Large game, including several disturbance-intolerant species, was still being hunted from Toé Cave on the southern side of the lakes until quite recent times.

In conclusion, there are indications of decreasing population densities of *Dorcopsis*, possibly related to forest disturbance, in the period 5000-4000 BP, at least to the northeast of the Ayamaru Lakes. This side of the lakes may have seen a gradual emergence of low intensity agricultural activities over this period. These developments would have lagged some 4000 years behind the earliest evidence of cropping systems in the eastern New Guinean Highlands (Golson 1977a, 1991b; Denham et al. 2003).

How were the caves used through time?

The artefacts found in each of the two caves are quite distinct in some respects but only subtly different in others. One of the most striking differences is the large number of bone artefacts and utilised wallaby teeth in Kria Cave (92 and 10 respectively) as opposed to that of Toé (3 and 1 respectively). The bone artefacts from Kria show a high degree of uniformity, with basically only two classes of bone points found throughout Units V-II: the bipointed artefacts made of splinters of cortical bone; and the unipoints made of small-diameter shaft bones. For the latter group, bat bones were mostly used. The bipointed specimens are likely to have been hafted, while the unipoints may have been mostly hand held.

The pattern of polish and damage on the bipoint tips does not suggest a mode of use involving impact with great force, such as for example as spear or arrow tips. Rather, the bipoints were probably hafted and used as borers or perforators, or perhaps in some wood-carving function. The pattern of damage observed on some wallaby incisors in both caves (the great majority in Kria) may have been formed by drilling activities. The unipoints are more likely to have been hand-held, multipurpose tools.

The high numbers of the bone artefacts found in Kria Cave, including broken specimens and isolated tips, suggest that the occupants of Kria Cave used and probably replaced or repaired these implements (or the composite tools of which they were part) on site. A few potentially 'unfinished' specimens suggest some manufacture on site as well. There are no changes in bone artefact morphology through time. The striking contrast of Kria Cave with Toé with regard to the quantities of bone points suggests that the range of activities carried out at each of the two sites may have differed consistently throughout their occupation periods.

Further information about site use was obtained from the lithic assemblages. Not only are there differences between the sites but also evidence of subtle changes through time. During the Late Pleistocene, Toé Cave appears to have been used for activities that did not require manufacture of stone artefacts, but only maintenance of pre-made tools that were brought into the cave. These tools, typically fairly large but relatively thin retouched flakes, seem to have been selected for transport on the basis of their low weight to cutting edge ratio. The lack of very large flakes and cores, as well as any small debitage, suggests that the cave was primarily used in the Late Pleistocene as a temporary shelter in which limited maintenance of existing toolkits was carried out.

The Holocene lithic assemblage in Toé Cave shows a striking contrast with that from earlier times. Large unfinished cores and flake-cores were left at the site amidst retouched flakes and abundant debitage of all sorts and sizes. This collection is highly indicative of stone-working on site. The retouched flakes are significantly thicker than those from the Pleistocene. It is possible that some of the thinner Holocene specimens were removed for transport for use elsewhere, in the same way that thinner artefacts were selected during the Late Pleistocene to be brought into Toé Cave from elsewhere. The large quantities of unfinished cores and flakes left behind in Toé Cave during the Holocene suggest that abundant sources of raw material were located nearby.

The Holocene stone material from Kria, 12 km to the northeast of Toé, shows an interesting deviation from the pattern seen in the southwest. Here, as in the Pleistocene in Toé, cores are virtually absent (except for one exhausted specimen made on exotic silicified tuff) although some larger retouched flakes feature large flake negatives suggesting that these had functioned as cores as well. The number of large retouched flakes is limited, but their thickness is similar to those of the Toé Holocene assemblage. Overall, however, the debitage is significantly smaller. Stone-working was almost certainly carried out on site at Kria, but the raw material was very much more reduced and recycled. Cores and retouched flakes that were still useful were probably taken away from the site to be used elsewhere, and used-up cores may have been discarded somewhere else in the cave or destroyed in fireplaces, leaving nothing but unrecognisable flaked pieces, many examples of which were found in Kria. The large differences between the two Holocene assemblages prompts the speculation that although chert nodules could be present anywhere within the limestone of the Ayamaru Plateau, suitable raw material was perhaps not so readily available on the northeastern side of the lake. The Kria Cave occupants may have had to obtain their raw material from quite some distance away, and accordingly, were far more selective in what they brought into the site, and what they took away with them. Tools were carefully maintained and chert was used optimally and exhaustively.

The evidence for intensive stone-working on site at Kria is consistent with the high quantities of carefully made bone points at this cave. Taken together, these observations suggest that the site was used on an occasional but extended basis by small, mobile groups involved in a variety of manufacturing and maintenance activities. The almost complete lack of bone artefacts in Holocene Toé Cave is perhaps suggestive of a rarely visited cave, in which casual meals were consumed on hunting or foraging trips, but in which activities involving the manufacture or maintenance of bone points were carried out only rarely.

The large lithic assemblage from the Holocene levels of Toé Cave appears to contradict this picture. Intensive production, use and maintenance of stone implements clearly took place in Toé Cave on various occasions through the Holocene, but without any associated manufacture or repair of bone tools. This may seem puzzling, yet there is an important difference between the bone and stone artefacts from the Ayamaru sites in terms of their energy input. All bone points, without exception, are finely-made artefacts, requiring a fair effort to make or repair them. In contrast, the stone artefacts from both sites are fairly crude implements, and although produced according to a consistent reduction sequence, they could be produced without much effort. Unlike the bone points, the sharpness or shape of the edge was probably more important for a stone implement than its overall shape.

While Toé Cave was only occasionally visited, the main manufacturing activities in the cave probably consisted of the casual reduction of some of the chert material left behind from previous visits. Artefacts were probably used on the spot in the preparation of animal or plant foods or in other activities. Some were evidently resharpened and then left behind or taken away. On occasion, new pieces of chert were selected from a nearby quarry site, following some initial reduction at that site. In Kria, on the other hand, the stone artefacts, albeit crude in form, received more attention than those in Toé. Tool manufacturing and maintenance, both of stone and bone, was perhaps of greater importance during the peoples' stay in Kria than it was in Toé. Kria Cave was visited more frequently than Toé, although perhaps no more than once every 3 or 4 years.

The Holocene patterns of use of Toé Cave, involving very rare visits during which forest wallabies and other game were hunted and some fairly coarse stone-working activities carried out. These visits appear to have continued into the last few millennia. In contrast, Kria Cave appears to have become abandoned around 4000 BP. The reasons why this occurred are obscure. There is no indication that raw materials were becoming exhausted, or that *Dorcopsis* had become scarce to the point that emigration from the area would have been the only solution for the occupants. There is nothing to suggest a change in diet as a consequence of a declining wallaby population. Virtually all bone material, including the otherwise abundant snake remains, gradually drops in quantity in the upper levels of Unit II. In contrast, the remains of rodents increase in these levels, probably on account of visits to the abandoned cave by raptorial birds.

The period of abandonment of Kria is unknown, but by 1840 BP activities had certainly resumed in the cave, albeit at very low intensity. Bone and stone artefacts are virtually absent during this most recent phase of occupation, and food remains are present in small quantities only. *Dorcopsis* remains are surprisingly scarce during this last occupation period and the faunal assemblage consists of proportionally large numbers of bones of smaller mammals (mostly rodents, especially *Rattus praetor*, a typical rat of agricultural landscapes in north and west New Guinea), some pig bones, and a collection of gastropods. It is tempting to interpret this as a typical case of main prey depletion and a change in diet breadth (Grayson 2001; Nagaoka 2001). However, the fact that *Dorcopsis* seems to have remained abundant south of the lakes (at only 12 km distance) tends to challenge this view. Unit I contains a human burial and a few sherds of pottery, suggesting a change in the function of the cave rather than a change in diet. It is interesting to note that the abandonment of Kria coincides with the flooding of the basin of the western Lake Ayamaru. However, the easternmost lakes (closer to Kria) may be older, and the two events may be unrelated. Moreover, the development of Lake Ayamaru did not have any obvious effect on the occupation of Toé Cave, even though it caused this site to become almost completely surrounded by water.

How does the rainforest occupation in the Bird's Head compare to other areas in this region?

In many respects the archaeological record from the Bird's Head sites conforms with the general picture of prehistoric settlement of rainforest habitats in the eastern half of New Guinea and elsewhere in Melanesia. Based on the currently accepted dates in Papua New Guinea and Australia (Groube et al. 1986; Turney et al. 2001), it is likely that the first settlers of Sahul, who may well have landed on the Bird's Head Peninsula (Birdsell 1977), arrived at least 40-45,000 years ago, some 20,000 years earlier than the earliest date now available from the Ayamaru area. The time of occupation of Kria and Toé Caves covers a period in which many significant events have been recorded in this region, including post-glacial environmental changes, the first signs of plant cultivation, and the introduction of domestic animals and pottery.

Environmental changes in New Guinea as a consequence of the last glacial cycle are recorded in pollen cores from elsewhere in Papua (Hope & Tulip 1994), Papua New Guinea (Hope & Peterson 1975; Walker & Flenley 1979; Hope 1989; Hope & Golson 1995) and the Banda Sea (Van der Kaars 1991, 1995). However, with the exception of the data from the Banda Sea, these records relate to environmental changes at higher altitudes, and it has been widely assumed that the impact of the Last Glacial Maximum was much less extreme in lowland areas throughout the tropics (e.g. Groube 1989: 293). The faunal assemblages from the Ayamaru Plateau have shown that during Late Pleistocene times the rainforest was sufficiently different from that in subsequent times to harbour various montane animal species amongst an otherwise diverse lowland fauna. Based on current altitudinal ranges for the various montane species, lowland temperatures may have been 5-6°C lower than they are today (see Chapter 9). When temperatures rose, species that were least tolerant of higher temperatures and that could not adapt to a lowland habitat either left the area as the 'montane' conditions receded, or they became locally extinct. Some species apparently persisted in the area until c. 6000 BP, presumably in remnant pockets of cooler 'montane' vegetation. This date is later than expected from general models of deglaciation and vegetational response, but the late persistence of these montane elements was possibly due to increased cloud cover, caused by the flooding of the Arafura Shelf and creation of shallow seas south and southwest of the Bird's Head (Hope & Tulip 1994; Van der Kaars 1995).

The two sites analysed here are complementary in their time spans, and together provide insights into a number of critical time periods from before the Last Glacial Maximum through to the emergence of contemporary agricultural systems and material culture. Moreover, the two sites, situated only 12 km apart within a relatively uniform landscape, form an interesting contrast in spatial terms with numerous minor to major differences in archaeological content between them. Analysis of faunal remains, providing insight in the spatial organisation of hunting behaviour and other activities, and their effect on the population structure of the main prey item, has not been carried out in such detail before for any archaeological site in New Guinea. The results suggest that human population density in the area was always very low, groups were probably locally mobile, and the caves were used only occasionally. This conclusion is contrary to the initial impression gained from the overwhelming dominance of *Dorcopsis* wallaby remains in both sites, and the large quantities of lithic material. The faunal analysis also provides a rare example (contra Grayson 2001) of long-term, sustainable hunting of this medium-sized mammal, the Brown Dorcopsis, with little evidence of human impact on population densities until the advent of more intensive habitat disturbance. Elsewhere in New Guinea, extinctions of medium to large mammals are attributed to direct human predation and/or habitat modification, but these claims are not backed by any rigorous analysis (e.g. Hope et al. 1993).

The lithic assemblages from the Ayamaru sites reveal a stone-working technology that is broadly consistent with those described from elsewhere in New Guinea. The artefacts are crude and there are no formal 'types', but the reduction sequence followed a clear and distinctive pattern. This technology remained overall stable from Late Pleistocene times, but subtle (as yet unexplained) changes from about 5500 BP, especially in the shape of cortex-bearing flakes from Kria, suggest that the stone-working technology was not as static (or erratic) as it seems. The few recurring forms found in many other New Guinean sites, such as waisted blades, ground adzes and axes, and sago-choppers, are absent from the Ayamaru sites. It is possible that the prehistoric occupants of the caves used other materials to carry out the tasks that were presumably done with these specific tools elsewhere in New Guinea. Or, that such activities, especially sago processing and forest clearing with waisted blades or ground axes, were not carried out at all. Although the occupants of Kria were thrifty in their use of stone, it seems most improbable that all evidence of the use of such implements would have been removed from the site. The abundant bone points from Kria Cave show a similar absence of certain well-known 'types' found in neighbouring areas, most notably bone 'awls' and spatulae. The absence of spatulae is particularly intriguing considering the abundant raw material for these tools (wallaby fibulae) available at each site.

Possible indications of limited habitat disturbance, noted from c. 5000 BP onwards, are observed in Kria Cave only. This may point to some form of plant cultivation, or perhaps to a simple process of opening up the canopy to stimulate growth of edible plants. Areas to the north of the Ayamaru Lakes appear to have been a preferred location for this activity, with the southern areas remaining more densely forested and retaining higher populations of disturbance-intolerant prey species. A 5000 BP date for possible agricultural activity is of course much later than the earliest evidence for agricultural activity at Kuk in the eastern New Guinea Highlands, dated to c. 9000 BP (Golson 1977a, 1991b; Denham et al. 2003). However, this mid-Holocene date corresponds to the timing of land clearing associated with agriculture elsewhere in New Guinea and the wider region (e.g. Flenley 1985: 61; Gillieson et al. 1985; Van der Kaars 1991; Stuijts 1993; Sen 1995; Bellwood 1997: 234, 237; Sémah et al. in press).

No signs of metal were found in either site, although fragments of Dongson drums are reported in the Ayamaru area (Elmberg 1959; Bernet Kempers 1988). They could potentially be over 2000 years old (Bernet Kempers 1988). Bernet Kempers (1988: 244) and Spriggs & Miller (1988) suggest various potential ways in which the kettledrums may have been dispersed into eastern Indonesia, including a distribution through 'refugees' around the middle of the first century AD, when the Chinese conquered northern Vietnam. The well-known Kain Timur cloths of Indonesian origin, used up till 1955 in the Ayamaru area as items of ceremonial exchange (Elmberg 1959), also suggest a long history of extensive and far-reaching trade with regions west and south of the Bird's Head, certainly dating to before the time that European explorers entered this region (Elmberg 1959; Sollewijn Gelpke 1994). Unfortunately, these periods are not represented archaeologically in the two excavated sites.

Kria and Toé Caves produced little information on the date or circumstances of the introduction of domestic animals. Dog and chicken remains are absent from the sites, but pig bones were found on the surface in Toé, and in Unit I in Kria. Bone fragments from the latter were radiocarbon dated to a minimum of 1840 BP, and are, to date, the oldest directly dated pig remains from New Guinea. However, this date is later than the appearance of pig remains elsewhere in Melanesia (Spriggs 1997: 95).

The timing of the appearance of pottery at the Ayamaru sites could not be established but the earliest examples post-date 1840 BP. Toé and Kria Caves both produced low numbers of small pottery sherds, only a few of them decorated (see Figures 3.11 and 4.12). As far as can be determined from such a small sample, they do not seem to show any resemblance to the red-slipped forms from eastern Indonesia, the cord-marked pottery of western Indonesia and Sarawak, or any form of Lapita pottery, found predominantly in Island Melanesia and western Polynesia (Bellwood 1997). Nor does there seem to be a correspondence with the decoration patterns on the pottery illustrated by Solheim (1998) from a site close to Makbon, on the northwest coast of the Bird's Head. However, closer comparison with the Makbon sample would be worthwhile. The Makbon pottery shows similarities to traditions found on the southern Philippines and north-central Indonesia (Solheim 1998: 30).

At the time when Galis wrote the text cited at the beginning of this discussion, little was known about the many remarkable events that have occurred in the wider Melanesian region over the last 40,000 years. The ethnobiologist Ralph Bulmer, writing in the 1950s and 60s, said that 'Papua New Guinea prehistory could be written on the back of a postcard' (Bulmer 1991: 470). Fifty years on, we now have some inkling of the complex history of the region, including the earliest settlement of islands anywhere outside

Southeast Asia (Keegan & Diamond 1987); the earliest large-scale maritime colonisation anywhere in the world (Gorecki 1993: 156); the independent development of one of the world's earliest agricultural drainage systems (Golson 1977a, b; 1991b; Yen 1995; Denham et al. 2003); the earliest evidence of animal translocations (Flannery & White 1991: 108; Diamond 1998); and, contrary to seemingly well-argued expectations (Bailey et al. 1989; Headland & Reid 1989), the human occupation of lowland tropical rainforests long before the development of plant cultivation in the region (Pavlides & Gosden 1994; this thesis). The sites from the Ayamaru Plateau in the centre of the Bird's Head add one more facet to the rapidly emerging history of a region that clearly was anything but a 'back corner of earth'.

Summary

The western half of New Guinea, the Indonesian province of Papua, is a much understudied area in many respects, but in particular in relation to its past. In 1995, as part of the Dutch-Indonesian ISIR project, two cave sites, located on the Bird's Head Peninsula, were archaeologically investigated, providing some of the first insights into the prehistory of this part of New Guinea.

The Ayamaru Plateau, surrounding the three connected Ayamaru Lakes, lies at c. 350 m above sea level in the centre of the Bird's Head. Kria Cave and Toé Cave are located to the northeast and the southwest of the lakes, respectively, and contain overlapping occupation sequences from sub-modern times extending back to c. 26,000 BP, i.e. well before the Last Glacial Maximum. The abundant faunal remains and stone artefacts recovered from both sites provided an opportunity to document long-term environmental and settlement trends, as well as more detailed temporal and spatial patterns in human activity on the Ayamaru Plateau, particularly during the Holocene period.

The study concentrated on three aspects:

1. a detailed analysis of the stratigraphy and chronology of the sites to establish a basis for the interpretation of temporal and spatial patterning in the excavated material (Chapters 3 and 4);
2. a description and analysis of two major artefact classes, namely the stone and bone artefacts (Chapters 5 and 6); and
3. an in-depth investigation of the faunal remains focussing on three issues, namely a) the food resources used by the cave occupants (Chapter 7); b) the effects of their hunting behaviour on populations of their main prey item, *Dorcopsis muelleri*, a wallaby locally known as *Djief* (Chapter 8); and c) the nature of environmental change on the Ayamaru Plateau since the Last Glacial Maximum (Chapter 9).

For both sites the excavated material was subjected to detailed taphonomic analyses to understand the circumstances of preservation, the pattern of distribution of the cultural materials, and the stratigraphy and chronology of the deposit. Kria Cave is stratigraphically intact and shows five distinct units, which for the most part differ in the relative quantities of cultural material. The surficial unit also contains qualitatively different kinds of material, including pottery and the remains of an introduced animal, the pig. Human occupation of this site commenced around 8000 BP (uncalibrated). Kria was probably abandoned around 4000 BP, and then revisited at lower intensity and perhaps for different kinds of activities, including human interment, over the last few millennia.

Toé Cave, with a sloping bedrock floor, has a more complex stratigraphy due to local slumping and erosion of the deposit. However, detailed analysis of bone and stone, along with an intensive dating programme using AMS combined with amino acid racemisation of cassowary eggshell, has demonstrated that the Toé Cave deposit is essentially intact. Two major units are distinguished, a Late Pleistocene unit and a Holocene unit, that differ primarily in faunal composition, and additionally in the preservational state of chert and bone. Toé Cave was first occupied around 26,000 BP (uncalibrated), and was visited on an occasional basis from that time through into recent millennia.

Large quantities of chert artefacts were found in both sites. The stone-working technology, based on a few simple core types, remained relatively stable over the 26,000-year period covered by the sites. However, the intensity of chert reduction differs markedly between sites and between time periods. During the Late Pleistocene, artefacts were mainly brought to Toé Cave ready-made; there is some evidence that they were used and maintained on site, but there is little indication of active stone-working. In contrast, the Holocene period in Toé Cave saw a change to active stone-working on site; the assemblage is comprised of abundant debitage, including large block- and flake-cores, as well as retouched flakes, resharpening flakes and waste material. The function of Toé Cave evidently changed over time, from a 'transit' site where casual tool maintenance occurred during the Late Pleistocene, to a place where significant stone tool production occurred. While chert is potentially available anywhere in the limestone of the Ayamaru Plateau, a source located somewhere close to Toé Cave may have been one reason why it was visited throughout the Holocene. Stone implements were clearly also manufactured during contemporary times at Kria Cave, but the assemblage contains virtually no cores and, with only a few exceptions, the debitage is relatively small. This suggests that the chert reduction process at Kria was more thorough and economic, such that whatever stone material was brought into the cave was thoroughly recycled with larger pieces perhaps taken away for use elsewhere. Good quality chert may have been not so readily available in the vicinity of Kria Cave, leading to high costs of transport and an overall more thrifty approach to this resource.

Bone artefacts are abundant in Kria Cave but extremely rare in Toé. Besides some utilised wallaby incisors, there are large quantities of artefacts made of fragments of bone. The latter fall into two distinct classes: bipointed specimens made of splinters of cortical bone, and unipoints made of fragments of small, slender shaft bones. These categories are also distinct in their pattern of usewear. Analysis of usewear and damage to these artefacts suggests that the bipoints were most likely hafted, and may have been used as borers. The unipoints were probably used for a variety of tasks and in various ways, but were perhaps most often hand-held. The Kria bone points appear to have been manufactured, used and repaired on site. The absence of such artefacts in Toé further points to differences in the pattern of use of the two caves during the Holocene period: Kria seems to have been a focus for more intensive activity involving the manufacture, use and maintenance of bone and stone tools, while Toé appears to have been visited less frequently and perhaps with more specific purpose – to consume the yield of a hunting trip, or to flake some of the locally available chert, perhaps with a view to transporting selected pieces away to more distant parts of the group's territory.

In contrast to what has long been assumed in regard to lowland tropical rainforests, the consequences of the last Ice Age were strongly felt on the Ayamaru Plateau. The mammal and bird communities were significantly altered during Last Glacial Maximum times, with the descent of a suite of montane species down to lowland elevations, creating a for New Guinea undocumented 'mixed' assemblage of montane and lowland species. The vegetation at this time probably consisted of rainforest with floristic and structural similarities to modern lower montane forests. The montane species presumably began to withdraw or decline as temperatures rose under post-glacial climatic conditions. However, some species persisted in the area until 6000 BP, probably surviving in remnant patches of cooler, mossy forest. Flooding of the Arafura Shelf and the shallow bays south and southwest of the Bird's Head may have generated persistent cloud cover over the Ayamaru Plateau that may have led to locally depressed temperatures until mid-Holocene times.

Throughout the Holocene period, the fauna in both caves is dominated by the remains of one species, *Dorcopsis muelleri*, a medium-sized forest wallaby, known locally as *Djief*. Other animals, including echidnas, pythons, cuscuses, cassowaries, megapodes, and a variety of smaller mammals, were caught occasionally as well. Fish, cassowary and megapode eggs, and molluscs further supplemented the diet. Plant remains are scarce in both sites, hence the vegetable component of the diet is virtually unknown. However, carbonised fruits, possibly of the Elaeocarpaceae family, and *Celtis* and banana seeds, were recovered in small quantities. No evidence was found for the prehistoric production of sago, although this cannot be ruled out.

The prolonged hunting of *Dorcopsis* on the Ayamaru Plateau during the Holocene period raises questions regarding the sustainable exploitation of one particular prey animal, especially in a tropical environment. Analysis of the age profile of wallaby remains from Kria and Toé Caves shows that predominantly mature to senescent animals were consumed at the sites. As is evident from ethnographies of hunting-gathering populations in rainforest environments, hunters can rarely afford to be selective in their pursuit of prey; therefore, the *Dorcopsis* age profile is taken as indicative of the demographic character of the *Dorcopsis* population during this period. An age profile dominated by mature to aged individuals is characteristic of a population close to the carrying capacity of its environment. This implies that hunting pressure on *Dorcopsis* was low, a conclusion that is supported by calculations of the minimum number of wallabies (MNI) represented within the faunal assemblages during the period of occupation of the caves. The *Dorcopsis* sample from the Holocene unit of Toé Cave features an even older age profile and still lower MNI. This unit also produced more abundant remains of disturbance-intolerant animals such as echidnas and tree-kangaroos, and it seems likely that the area to the south of the lakes was used less intensively than the area around Kria Cave to the northeast.

The various lines of evidence all point to the fact that human population densities were very low through the area and throughout much of the period of occupation of the caves. Various factors, perhaps in combination, might have been responsible for this, such as the naturally low carrying capacity of the plateau environment, limited carbohydrate resources prior to the advent of extensive agriculture, and a high prevalence of disease and inter-group conflict. Scattered human remains recovered from both sites, dating from the early Holocene to the last few millennia, show signs of childhood illness or starvation, and one a possible violent death. These remains bear testimony to the harshness of life in lowland New Guinea, even during the recent past. Headland's (1987) hypothesis, that a lack of carbohydrates poses a major restriction on human occupation of tropical rainforests, may find some support from the Bird's Head sites. However, contra Headland & Bailey (1991), these environments clearly did support viable human populations, even in remote and rugged areas, long before the emergence or appearance of cultivated plants and domesticated animals.

Pottery and remains of pig, presumably indicative of a new lifestyle based on agriculture, were found only in the uppermost levels of both sites. A pig bone from Kria Cave is radiocarbon dated to a minimum age of 1840 ± 40 BP with an inferred upper range of c. 4000 BP. However, there are some indications for somewhat earlier changes in land use patterns on the Ayamaru Plateau, perhaps dating back to c. 5000 years. In particular, the wallaby remains in Kria Cave from this time on show a slight shift toward a larger proportion of younger, though already mature, animals and a slight increase in bone fragmentation. These changes suggest that the *Dorcopsis* population had declined below its demographic maximum and that wallaby carcasses were being exploited more fully for nutritional content than in earlier times. Why this decline might have occurred cannot be determined with certainty, but the two most likely causes are an increased rate of culling of *Dorcopsis* and the initiation of a more extensive modification of its habitat, perhaps as a consequence of limited gardening activities. Interestingly, those changes in age profile are registered only on the northeastern side of the lakes. A short pollen sequence from the western lake basin, provided by Dam (Delft University of Technology, Netherlands) and Van der Kaars (Monash University, Melbourne), also shows evidence of limited forest disturbance dating back at least to 4000 BP, with no major change in the intensity of disturbance during successive millennia. These indications for small-scale gardening activity are significantly later than the earliest evidence of plant cultivation in New Guinea, but are consistent with the timing of widespread adoption of agriculture across the region.

Abstrak

Bagian paroh barat New Guinea, Propinsi Papua, Indonesia, merupakan daerah yang jarang diteliti dalam berbagai aspek, terutama menyangkut kehidupan masa lampau. Sebagai bagian dari proyek ISIR Indonesia-Belanda, pada tahun 1995 dilakukan penelitian di dua situs dan hingga sekarang telah memberikan beberapa pandangan awal tentang prasejarah daerah ini.

Dataran tinggi Ayamaru, terletak di bagian tengah semenanjung Kepala Burung, memiliki Danau Ayamaru yang terdiri atas tiga bagian yang saling berhubungan, pada ketinggian sekitar 350 m di atas permukaan laut. Gua Kria dan Toé, masing-masing terletak di sebelah timur laut dan barat daya danau, mengandung rangkaian okupasi masa lampau yang bersamaan, paling tidak sejak 26,000 yang lalu, segera sebelum periode glasial maksimum terakhir. Penemuan sisa fauna dan artefak batu yang sangat padat di kedua situs memberikan gambaran tentang lingkungan dan hunian jangka panjang, serta pola ruang dan waktu yang lebih rinci tentang aktivitas manusia pada periode Holosen di dataran tinggi tersebut.

Studi ini difokuskan pada tiga aspek:

1. analisis yang seksama terhadap stratigrafi dan kronologi situs untuk menetapkan dasar yang mantap bagi interpretasi pola ruang dan waktu (bab 3 dan 4);
2. deskripsi dan analisis terhadap kelas-kelas utama tinggalan budaya dari kedua situs, yaitu artefak batu dan tulang (bab 5 dan 6); dan
3. investigasi mendalam terhadap sisa fauna yang difokuskan pada tiga isu, yaitu a) perilaku subsistensi dari penghuni gua (bab 7); b) pengaruh perburuan terhadap populasi binatang buruan utama, *Dorcopsis muelleri*, atau *Djief* (bab 8); dan c) kondisi perubahan lingkungan dataran tinggi Ayamaru sejak periode glasial maksimum terakhir (bab 9).

Analisis taphonomi yang rinci dilakukan terhadap temuan ekskapasi dari kedua situs untuk memahami kondisi preservasi, pola distribusi tinggalan budaya, dan stratigrafi serta kronologi endapan. Gua Kria pada hakekatnya belum teraduk dan memperlihatkan lima unit yang umumnya mengandung tinggalan-tinggalan yang berbeda dalam hal kuantitas. Unit permukaan mengandung sejumlah material yang berbeda, termasuk gerabah dan sisa binatang peliharaan, babi. Okupasi manusia pada gua ini agaknya telah dimulai sejak 8000 BP dan ditinggalkan di sekitar 4000 BP. Gua ini kemudian dikunjungi lagi dengan intensitas yang berkurang dan tampaknya untuk berbagai kegiatan, termasuk penguburan manusia selama beberapa ribu tahun terakhir. Gua Toé memperlihatkan stratigrafi yang lebih kompleks berkaitan dengan batuan dasarnya yang miring, kemungkinan terjadinya longsor lokal, dan erosi sedimen. Namun analisis rinci terhadap tulang dan batu, serta melalui pertanggalan yang intensif dengan menggunakan AMS dan *amino acid racemisation* terhadap kulit telur burung kasuari, menunjukkan bahwa deposit gua masih tergolong utuh, meskipun kemungkinan terdapat gangguan lokal. Dua unit pokok dibedakan dalam gua ini, unit akhir Plestosen dan unit Holosen. Keduanya dibedakan secara jelas berdasarkan keberadaan spesies binatang pegunungan pada himpunan fauna akhir Plestosen dan melalui perbedaan kondisi preservasi batu rijang dan tulang. Awal hunian gua Toé pasti sebelum 26,000 BP. Hunian gua ini tampaknya berlanjut hingga pada millenium resen.

Artefak dari batuan rijang ditemukan padat di kedua situs. Teknologi pengerjaan batu, yang didasarkan pada beberapa tipe batu inti sederhana, menunjukkan relatif stabil selama lebih dari 26,000 tahun. Tetapi intensitas pengerjaan batu rijang berbeda diantara kedua situs dan diantara kedua periode. Pada periode akhir Plestosen artefak yang dibawa ke dalam gua Toé umumnya sudah siap dikerjakan. Di samping adanya beberapa bukti yang menunjukkan penggunaan dan pemeliharaan artefak di dalam situs, terdapat juga sedikit indikasi tentang pengerjaan batu di dalam gua. Hal yang menonjol, bahwa pada periode Holosen, tampak pengerjaan batu yang lebih aktif di Gua Toé, di samping penggunaan dan pemeliharaan alat yang terus berlanjut. Kegiatan ini meninggalkan artefak yang padat, meliputi blok besar dan batu inti, serpih-serpih diretus, serpih-serpih ditajamkan dan material buangan. Fungsi dari Gua Toé terbukti berubah dalam perjalanan waktu dari situs 'transit' dimana pemeliharaan alat-alat berlangsung selama periode akhir Plestosen, menjadi tempat kegiatan untuk memproduksi alat batu. Sumber batuan rijang diperkirakan berada di sekitar Gua Toé dan hal ini menjadi satu alasan mengapa gua ini sering dikunjungi selama periode Holosen. Alat-alat batu juga dikerjakan pada waktu yang bersamaan di gua Kria, tetapi himpunan artefak di gua ini boleh dikatakan tidak mengandung batu inti dan, dalam beberapa perkecualian, *debitage* relatif kecil. Hal ini mengindikasikan bahwa proses pengerjaan batu rijang di gua Kria berlangsung lebih seksama dan ekonomis. Dalam hal ini, bahan batu apapun yang dibawa ke dalam gua sering didaur ulang secara seksama dengan alat-alat yang lebih besar kemungkinan dibawa untuk digunakan dimana saja. Batu rijang yang berkualitas baik agaknya tidak tersedia di sekitar Gua Kria dan hal ini mengakibatkan biaya transport yang tinggi sehingga mendorong perlunya penghematan bahan.

Artefak tulang tergolong padat di gua Kria, tetapi absen di Gua Toé. Pada umumnya terbagi ke dalam dua kelas: alat lancipan ganda dibuat dari tulang kortikal yang dibelah dan lancipan tunggal yang dibuat dari fragmen tulang yang lebih kecil. Kedua jenis alat ini juga memperlihatkan perbedaan dalam pola bekas pakai. Analisis mendalam tentang jejak pakai dan kerusakan menunjukkan bahwa lancipan ganda tampaknya digunakan dengan tangkai. Lancipan tunggal tampaknya dipakai untuk berbagai keperluan dengan berbagai cara, namun yang lebih sering adalah dengan cara dipegang. Alat-alat tulang dari Gua Kria agaknya dibuat, dipakai, dan diperbaiki di dalam situs. Ketiadaan artefak semacam ini di Gua Toé menunjukkan perbedaan-perbedaan dalam pola pemanfaatan kedua gua selama periode Holosen: Gua Kria agaknya lebih difokuskan pada aktivitas yang lebih intensif melibatkan pembuatan, pemakaian, dan pemeliharaan alat-alat batu dan tulang. Gua Toé tampaknya dikunjungi lebih jarang untuk tujuan yang lebih khusus – untuk mengkonsumsi hasil buruan atau untuk mengerjakan beberapa batu rijang yang tersedia. Kemungkinan alat-alat terpilih ditransportasikan ke tempat yang lebih jauh dari wilayah kelompok tersebut.

Berbeda dengan apa yang selama ini diperkirakan dalam kaitannya dengan hutan hujan tropis dataran rendah, konsekuensi jaman es terakhir sangat kuat terasa di dataran tinggi Ayamaru. Komunitas mamalia dan burung berubah secara signifikan selama waktu periode glasial maksimum terakhir dengan turunnya spesies pegunungan ke dataran rendah. Kondisi ini mengakibatkan hal yang sebelumnya tidak terdokumentasikan, yaitu terjadinya 'campuran' antara spesies pegunungan dan dataran rendah. Vegetasi pada waktu itu kemungkinan terdiri dari hutan hujan yang memiliki kesamaan dalam hal bunga dan struktur dengan hutan gunung modern yang lebih rendah. Spesies gunung agaknya mulai mundur atau menurun ketika temperatur naik pada periode post-glasial. Bagaimanapun juga, beberapa spesies masih bertahan di wilayah ini hingga 6000 BP pada tempat yang lebih dingin dan hutan lembab. Tergenangnya selat Arafura dan teluk dangkal di selatan dan barat daya Kepala Burung mungkin telah menimbulkan keberadaan awan yang tetap meliputi dataran tinggi Ayamaru dan yang menyebabkan penekanan temperatur secara lokal hingga pertengahan Holosen.

Selama periode Holosen fauna di kedua gua didominasi oleh sisa dari satu spesies, *Dorcopsis muelleri* sejenis *wallaby* (Walabi) hutan berukuran sedang, dikenal secara lokal sebagai *Djief*. Binatang lain, termasuk *echidnas* (Babi Duri Moncong Panjang), kus-kus, kasuari, *megapode*, dan beberapa mamalia yang lebih kecil, menjadi binatang tangkapan secara okasional. Ikan, kasuari, dan telur *megapode* serta moluska merupakan diet tambahan. Sisa tanaman jarang di kedua situs, agaknya komponen sayur-sayuran sebagai diet tidak dikenal. Bagaimanapun, buah-buahan yang terbakar, agaknya dari famili Elaeocarpaceae serta biji *celtis* dan pisang, ditemukan dalam jumlah sedikit. Sejauh ini tidak ditemukan bukti-bukti produksi sago meskipun kegiatan semacam itu memungkinkan.

Perburuan panjang terhadap *Dorcopsis* di dataran tinggi Ayamaru selama periode Holosen menimbulkan beberapa pertanyaan umum menyangkut eksploitasi berkelanjutan terhadap suatu binatang buruan khusus, terutama dalam lingkungan tropis. Analisis terhadap profil usia *wallaby* dari gua Kria memperlihatkan binatang yang dikonsumsi pada situs ini lebih didominasi binatang dewasa yang lebih tua. Seperti halnya terbukti pada banyak populasi ethnografi berburu dan mengumpulkan makanan di lingkungan hutan hujan, pemburu jarang memilih binatang buruannya, oleh sebab itu profil usia *Dorcopsis* di Gua Kria dapat dipakai sebagai indikasi karakter demografi selama periode tersebut. Suatu profil usia yang didominasi oleh individu dewasa merupakan ciri populasi yang berkaitan erat dengan daya dukung lingkungannya. Hal ini mengimplikasikan bahwa perburuan terhadap *Dorcopsis* tergolong rendah, sebuah kesimpulan yang didukung oleh kalkulasi jumlah minimum dari *wallaby* yang terdapat dalam himpunan fauna selama periode hunian gua. Contoh *Dorcopsis* dari unit Holosen di gua Toé bahkan memperlihatkan profil usia yang lebih tua. Unit ini juga memiliki sisa jenis binatang yang tidak mau diganggu seperti *echidnas* dan kangguru pohon. Agaknya wilayah di selatan gua digunakan kurang intensif dibandingkan dengan wilayah disekitar gua Kria ke arah timur laut.

Bermacam-macam bukti diatas menunjuk pada fakta bahwa populasi manusia sangat rendah diseluruh wilayah ini selama periode hunian gua. Berbagai faktor yang saling mempengaruhi kemungkinan menjadi penyebabnya, seperti rendahnya daya kapasitas lingkungan dataran tinggi, sumber karbohidrat yang terbatas sebelum berkembangnya agrikultur yang ekstensif, dan tingginya penyakit serta konflik antar grup yang merata di daerah tersebut. Kuburan manusia dari gua Kria yang memiliki tanda-tanda penyakit anak atau kelaparan dan kematian yang disebabkan karena kekerasan menyimpan kesaksian bisu tentang kesulitan hidup diberbagai bagian dari dataran rendah New Guinea pada waktu yang baru berlalu. Bagaimanapun juga, hutan hujan dataran rendah telah ada di dataran tinggi Ayamaru selama periode hunian gua. Dengan demikian situs-situs arkeologi telah memberikan bukti-bukti lebih jauh, berlawanan dengan Headland & Bailey (1991), bahwa hutan hujan tropis dataran rendah dapat dihuni dengan baik selama periode yang panjang tanpa mengenal domestikasi tanaman dan binatang.

Gerabah dan babi, kemungkinan indikasi baru dari model hidup agrikultur hanya ditemukan pada lapisan paling atas di kedua situs. Bagaimanapun, ada beberapa indikasi tentang perubahan pada pola pemanfaatan lahan didataran tinggi Ayamaru, kemungkinan di sekitar 5000 tahun lalu. Secara khusus, sisa *wallaby* di gua Kria memperlihatkan sedikit perubahan ke arah proporsi usia muda yang lebih besar, meskipun masih dewasa, dan sedikit pertambahan pada tulang-tulang yang fragmentaris. Perubahan-perubahan ini memperlihatkan bahwa populasi *Dorcopsis* menurun dibawah daya dukung maksimum dan daging *wallaby* telah dieksploitasi lebih intensif untuk kandungan protein dibandingkan pada waktu sebelumnya. Mengapa hal ini terjadi belum diketahui secara pasti, tetapi kemungkinan oleh disebabkan oleh dua faktor, yaitu kenaikan tingkat pemilihan (*culling*) dan permulaan modifikasi habitasi yang lebih intensif, barangkali sebagai konsekwensi dari terbatasnya kegiatan berkebun. Hal yang menarik, perubahan-perubahan tersebut terdokumentasikan hanya pada sisi timur laut danau. Rangkaian pendek polen dari cekungan sebelah barat danau yang diteliti oleh Ryen Dam (Delft University of Technology, Netherlands) dan Sander van der Kaars (Monash University, Melbourne), juga menunjukkan gangguan hutan yang terbatas paling tidak sejak 4000 BP, tanpa perubahan menonjol dalam intensitas gangguan selama ribuan tahun berikutnya. Indikasi-indikasi ini untuk kegiatan agrikultur skala kecil lebih signifikan kemudian dibandingkan dengan bukti-bukti agrikultur yang lebih tua di New Guinea, tetapi konsisten dengan waktu persebaran agrikultur yang luas di seluruh wilayah tersebut.

References

Agogino, G.A. 1979. Book review of V. Watson and J.D. Cole, 1977. Prehistory of the eastern highlands of New Guinea. *Man* n.s. 14: 756.

Aitken, R.J. 1975. Cementum layers and tooth wear criteria for ageing Roe deer (*Capreolus capreolus*). *Journal of Zoology* 175: 15-28.

Allen, G.R. 1996. Freshwater fishes of Irian Jaya. In D.J. Kitchener & A. Suyanto (eds), *Proceedings of the first international conference on eastern Indonesian-Australian vertebrate fauna, Manado, Indonesia, November 22-26, 1994*: 15-21. Perth: Western Australian Museum.

Allen, J. 1994. Radiocarbon determinations, luminescence dating and Australian archaeology. *Antiquity* 68: 339-343.

Allen, J. 1996. The pre-Austronesian settlement of Island Melanesia: implications for Lapita archaeology. In W.H. Goodenough (ed.), *Prehistoric Settlement of the Pacific*. Published as *Transactions of the American Philosophical Society* 86 (5): 11-27.

Allen, J. 2000. From beach to beach: The development of maritime economies in prehistoric Melanesia. In S. O'Connor & P. Veth (eds), *East of Wallace's Line. Studies of past and present maritime cultures of the Indo-Pacific region. Modern Quaternary Research in Southeast Asia* 16: 139-176. Rotterdam: A.A. Balkema.

Allen, J. & C. Gosden (eds) 1991. *Report of the Lapita Homeland Project*. Canberra: Department of Prehistory, Research School of Pacific Studies, The Australian National University.

Allen, J. & S. Holdaway 1995. The contamination of Pleistocene radiocarbon determinations in Australia. *Antiquity* 69: 101-112.

Allen, J., C. Gosden, R. Jones & J.P. White 1988. Pleistocene dates for the human occupation of New Ireland, northern Melanesia. *Nature* 331: 707-709.

Alvard, M.S. 1995. Intraspecific prey choice by Amazonian hunters. *Current Anthropology* 36: 789-818.

Alvard, M.S, J.G. Robinson, K.H. Redford & H. Kaplan 1997. The Sustainability of Subsistence Hunting in the Neotropics. *Conservation Biology* 11 (4): 977-982.

Anderson, A. 1983. The prehistoric hunting of moa (Aves: Dinornithidae) in the high country of southern New Zealand. In C. Grigson & J. Clutton-Brock (eds), *Animals and Archaeology*: 2. *Shell Middens, Fishes and Birds. BAR International Series* 183: 33-52.

Anell, B. 1955. *Contribution to the History of Fishing in the Southern Seas. Studia Ethnographica Upsaliensia* IX. Uppsala: Almqvist & Wiksells Boktryckeri Ab.

Aplin, K. 1980. Faunal identification procedures. In J.W. Rhoads, Through a glass darkly: present and past land-use systems of Papuan sagopalm users. Appendix 13. Unpublished PhD thesis. Canberra: Department of Prehistory, Australian National University.

Aplin, K. 1981. The Kamapuk fauna. A late Holocene vertebrate faunal sequence from the Western Highlands District, Papua New Guinea with implications for palaeoecology and archaeology. Unpublished BA Honours thesis, Canberra: Department of Prehistory, Australian National University.

Aplin, K.P. 1998. Vertebrate zoogeography of the Bird's Head of Irian Jaya, Indonesia. In J. Miedema, C. Odé & R.A.C. Dam (eds), *Perspectives on the Bird's Head of Irian Jaya,*

Indonesia. Proceedings of the Conference Leiden, 13-17 October 1997: 803-890. Amsterdam: Rodopi.

Aplin, K.P., J.M. Pasveer & W.E. Boles 1999. Late Quaternary vertebrates from the Bird's Head Peninsula, Irian Jaya, Indonesia, including descriptions of two previously unknown marsupial species. *Records of the Western Australian Museum*, Supplement No. 57: 351-387.

Archer, M. & T.F. Flannery 1985. *The Kangaroo*. McMahons Point NSW: Kevin Weldon.

Arndt, S. & M. Newcomer 1986. Breakage patterns on prehistoric bone points: an experimental study. In D.A. Roe (ed.), *Studies in the Upper Palaeolithic of Britain and Northwest Europe. BAR International Series* 296: 165-173.

Bahuchet, S., D. McKey & I. de Garine 1991. Wild Yams Revisited: Is Independence from Agriculture Possible for Rain Forest Hunter-Gatherers? *Human Ecology* 19 (2): 213-243

Bailey, R.C. & T.N. Headland 1991. The Tropical Rain Forest: Is It a Productive Environment for Human Foragers? *Human Ecology* 19 (2): 261-285.

Bailey, R.C., G. Head, M. Jenike, B. Owen, R. Rechtman & E. Zechenter 1989. Hunting and gathering in tropical rain forest: Is it possible? *American Anthropologist* 91: 59-82.

Barmawidjaja, B.M., E.J. Rohling, W.A. van der Kaars, C. Vergnaud Grazzini & W.J. Zachariesse 1993. Glacial conditions in the North Molucca Sea region (Indonesia). *Palaeogeography, Palaeoclimatology, Palaeoecology* 101: 147-167.

Bartholomai, A. 1971. Morphology and variation of the cheek teeth in *Macropus giganteus* Shaw and *Macropus agilis* (Gould). *Memoirs of the Queensland Museum* 16 (1): 1-18.

Barton, H. & J.P. White 1993. Use of Stone and Shell Artifacts at Balof 2, New Ireland, Papua New Guinea. *Asian Perspectives* 32 (2): 169-181.

Bartram, L.E., Jr. 1993. Perspectives on skeletal part profiles and utility curves from Eastern Kalahari ethnoarchaeology. In J. Hudson (ed.), *From Bones to Behavior: ethnoarchaeological and experimental contributions to the interpretation of faunal remains. Occasional Papers* 21: 115-137. Carbondale: Center for Archaeological Investigations (Southern Illinois University).

Bellwood, P. 1991. The Austronesian Dispersal and the Origin of Languages. *Scientific American* (July): 70-75.

Bellwood, P. 1997. *Prehistory of the Indo-Malaysian Archipelago*. Revised edition. Honolulu: University of Hawai'i Press.

Bellwood, P., G. Nitihaminoto, G. Irwin, Gunadi, A. Waluyo & D. Tanudirjo 1998. 35,000 years of prehistory in the northern Moluccas. In G.-J. Bartstra (ed.), *Bird's Head Approaches. Irian Jaya Studies - A Programme for Interdisciplinary Research. Modern Quaternary Research in Southeast Asia* 15, 233-275. Rotterdam: A.A. Balkema.

Berlage, H.P.Jr. 1949 Regenval in Indonesie. *Departement van Verkeer, Energie en Mijnwezen, Meteorologische en Geophysische Dienst, Koninklijk Magnetisch en Meteorologisch Observatorium te Batavia, Verhandelingen* 37.

Bernet Kempers, A.J. 1988. *The kettledrums of Southeast Asia. A Bronze Age world and its aftermath. Modern Quaternary Research in Southeast Asia* 10. Rotterdam: A.A. Balkema.

Binford, L.R. 1978. *Nunamiut Ethnoarchaeology*. New York: Academic Press.

Binford, L.R. 1981. *Bones: ancient men and modern myths*. New York: Academic Press.

Binford, L.R. 1984. *Faunal Remains from Klasies River Mouth*. Orlando: Academic Press.

Binford, L.R. & J.B. Bertram 1977. Bone Frequencies – And Attritional Processes. In L.R. Binford (ed.), *For Theory Building in Archaeology. Essays on faunal remains, aquatic resources, spatial analysis, and systemic modeling*: 77-153. New York: Academic Press.

Bird, C. & C. Beeck 1980. Bone points and spatulae: salvage ethnography in southwest Australia. *Archaeology and Physical Anthropology in Oceania* 15: 168-171.

Birdsell, J.B. 1977. The recalibration of a paradigm for the first peopling of Greater Australia. In J. Allen, J. Golson & R. Jones (eds). *Sunda and Sahul: Prehistoric Studies in Southeast Asia, Melanesia and Australia*: 113-167. London: Academic Press.

Blackwood, B. 1950. *The Technology of a Modern Stone Age People in New Guinea. Occasional Papers on Technology* 3. Oxford: University Press.

Bodmer, R.E. 1995. Managing Amazonian wildlife: biological correlates of game choice by detribalized hunters. *Ecological Applications* 5 (4): 872-877.

Bodmer, R.E., J.F. Eisenberg & K.H. Redford 1997. Hunting and the Likelihood of Extinction of Amazonian Mammals. *Conservation Biology* 11: 460-466.

Boeseman, M. 1959. De vissen uit het zoete water van Nederlands-Nieuw Guinea. *Nederlands Nieuw-Guinea* 7: 10-15.

Bourke, D.W. 1989. Observations on the behaviour of the Grey Dorcopsis Wallaby, *Dorcopsis luctuosa* (Marsupialia: Macropodidae), in captivity. In G. Grigg, P. Jarman & I. Hume (eds), *Kangaroos, wallabies and rat-kangaroos*. Volume 2: 633-640. Chipping Norton, NSW: Surrey Beatty & Sons.

Boyle, K.V. 1990. *Upper Palaeolithic Faunas from South-West France. A Zoogeographic Perspective. BAR International Series* 557.

Bradbury, J.P., B. Leyden, M. Salgado-Labouriau, W.M. Lewis Jr., C. Schubert, M.W. Binford, D.G. Frey, D.R. Whitehead & F.H. Weibezahn 1981. Late Quaternary environmental history of Lake Valencia, Venezuela. *Science* 214: 1299-1305.

Brongersma, L.D. 1958. *The animal world of Netherlands New Guinea*. Groningen: J.B. Wolters.

Bruijnzeel, L.A. & E.J. Veneklaas 1998. Climatic conditions and tropical montane forest productivity: the fog has not lifted yet. *Ecology* 79: 3-9.

Bulmer, R. 1976. Selectivity in hunting and in disposal of animal bone by the Kalam of the New Guinea Highlands. In G. de G. Sieveking, I.H. Longworth & K.E. Wilson (eds), *Problems in Economic and Social Archaeology*: 169-186. London: Duckworth.

Bulmer, S. 1964. Prehistoric stone implements from the New Guinea Highlands. *Oceania* 34 (4): 246-268.

Bulmer S. 1966. The prehistory of the Australian New Guinea Highlands. Unpublished M.A. thesis. Auckland: University of Auckland.

Bulmer, S. 1977. Between the Mountain and the Plain: Prehistoric Settlement and Environment in the Kaironk Valley. In J.H. Winslow (ed.), *The Melanesian environment*: 61-73. Canberra: The Australian National University Press.

Bulmer, S. 1991. Variation and change in stone tools in the Highlands of Papua New Guinea: the witness of Wañelek. In A. Pawley (ed.), *Man and a half. Essays in Pacific Anthropology and Ethnobiology in honour of Ralph Bulmer*: 470-478. Auckland: The Polynesian Society.

Bulmer, S. & R. Bulmer 1964. The Prehistory of the Australian New Guinea Highlands. In J.B. Watson (ed.), *New Guinea: the Central Highlands. American Anthropologist* 66 (4), Part 2: 39-76. Special Publication.

Bunn, H.T., L.E. Bartram & E.M. Kroll 1988. Variability in bone assemblage formation from Hadza hunting, scavenging, and carcass processing. *Journal of Anthropological Archaeology* 7: 412-457.

Bush, M.B., D.R. Piperino, P.A. Colinvaux, P.E. De Oliveira, L.A. Krissek, M.C. Miller & W.E. Rowe 1992. A 14,300-yr palaeoecological profile of a lowland tropical lake in Panama. *Ecological Monographs* 62: 251-275.

Campana, D.V. 1989. *Natufian and Protoneolithic Bone Tools. The Manufacture and Use of Bone Implements in the Zagros and the Levant. BAR International Series* 494.

Cann, R.L. 2001. Genetic Clues to Dispersal in Human Populations: Retracing the Past from the Present. *Science* 291: 1742-1748.

Capelli, C., J.F. Wilson, M. Richards, M.P.H. Stumpf, F. Gratrix, S. Oppenheimer, P. Underhill, V.L. Pascali, T.-M. Ko & D.B. Goldstein 2001. A Predominantly Indigenous Paternal Heritage for the Austronesian-Speaking Peoples of Insular Southeast Asia and Oceania. *American Journal of Human Genetics* 68: 432-443.

Casteel, R.W. 1974. A method for estimation of live weight of fish from the size of skeletal elements. *American Antiquity* 39 (1): 94-98.

Caughley, G. 1977. *Analysis of Vertebrate Populations*. London: John Wiley & Sons.

Chaplin, R.E. 1971. *The study of animal bones from archaeological sites*. New York: Seminar Press.

Chappell, J.M.A. 1994. Upper Quaternary sea levels, coral terraces, oxygen isotopes and deep-sea temperatures. *Journal of Geography* 103: 828-840.

Chappell, J., J. Head & J. Magee 1996. Beyond the radiocarbon limit in Australian archaeology and Quaternary research. *Antiquity* 70: 543-552.

Chase, P.G. 1986. *The Hunters of Combe Grenal: Approaches to Middle Paleolithic Subsistence in Europe. BAR International Series* 286.

Church, T. 1994. *Lithic Resource Studies: A Sourcebook for Archaeologists. Lithic Technology*, Special Publication 3.

Clarke, S.J. forthcoming. Investigations of the biogeochemistry of megapode, *Casuarius* and *Aepyornis* eggshell for palaeoenvironmental reconstructions. Unpublished PhD thesis, School of Geosciences, University of Wollongong.

CLIMAP 1981. *Seasonal reconstructions of the Earth's surface at the last glacial maximum. Geological Society of America, Map and Chart Series* MC-36.

Coates, B.J. 1985. *The Birds of Papua New Guinea*. Volume I: *Non-Passerines*. Alderley, Queensland: Dove Publications.

Codrington, R.H. 1891. *The Melanesians. Studies in their anthropology and folk-lore*. Oxford: Clarendon Press. Reprinted in 1969.

Cogger, H.G. 2000. *Reptiles & amphibians of Australia*. Sydney: Reed New Holland.

Colinvaux, P.A., P.E. De Oliveira, J.E. Moreno, M.C. Miller & M.B. Bush 1996. A long pollen record from lowland Amazonia: forest and cooling in glacial times. *Science* 274: 85-88.

Collins, L.R. 1973. *Monotremes and Marsupials. A reference for zoological institutions*. Washington, DC: Smithsonian Institution Press.

Crabtree, D.E. 1972. *An introduction to flintworking. Occasional Papers of the Idaho State University Museum* 28.

Crabtree, D.E. & B.R. Butler 1964. Notes on experiments in flint knapping. I: Heat treatments of silica minerals. *Tebiwa* 7: 1-6.

Cranstone, B.A.L. 1961. *Melanesia, A short Ethnography*. London: British Museum.

Crosby, E. 1973. A comparative study of Melanesian hafted edge-tools and other percussive cutting implements. Unpublished PhD thesis. Australian National University.

Dam, M.A.C. 1994. The late Quaternary evolution of the Bandung basin, West-Java, Indonesia. Unpublished PhD thesis. Amsterdam: Vrije Universiteit.

Dam, R.A.C. 1998. Cenozoic geological development and environmental settings of the Bird's Head of Irian Jaya. In J. Miedema, C. Odé & R.A.C. Dam (eds), *Perspectives on the Bird's Head of Irian Jaya, Indonesia. Proceedings of the Conference. Leiden, 13-17 October 1997*: 757-781. Amsterdam: Rodopi.

Davis, L.B. & J.W. Fisher 1990. A late prehistoric model for communal utilization of pronghorn antelope in the northwestern Plains region, North America. In L.B. Davis & B.O.K. Reeves (eds), *Hunters of the Recent Past*: 241-276. London: Unwin Hyman.

De Bruyn, J.V. 1962. New bronze finds at Kwadeware, Lake Sentani. *Nieuw-Guinea Studiën* 6: 61-62.

De Deckker, P., T. Correge & J. Head 1991. Late Pleistocene record of cyclic eolian activity from tropical Australia suggesting the Younger Dryas is not an unusual climatic event. *Geology* 19: 602-605.

Denham, T.P., S.G. Haberle, C. Lentfer, R. Fullagar, J. Field, M. Therin, N. Porch & B. Winsborough 2003. Origins of Agriculture at Kuk Swamp in the Highlands of New Guinea. *Science* 301: 189-193.

Dennis, E. & J.I. Menzies 1979. A chromosomal and morphometric study of Papuan tree rats *Pogonomys* and *Chiruromys* (Rodentia, Muridae). *Journal of Zoology, London*, 189: 315-332.

De Vries, L. 1998. Some remarks on the linguistic position of the Inanwatan language. In J. Miedema, C. Odé & R.A.C. Dam (eds), *Perspectives on the Bird's Head of Irian Jaya, Indonesia. Proceedings of the Conference. Leiden, 13-17 October 1997*: 643-653. Amsterdam: Rodopi.

Dharma, B. 1988. *Siput dan Kerang Indonesia I (Indonesian Shells)*. Jakarta: Penerbit: PT. Sarana Graha.

Diamond, J.M. 1988. Express train to Polynesia. *Nature* 336: 307-308.

Diamond J.M. 1998. *Guns, Germs and Steel. A Short History of Everybody for the Last 13,000 Years*. London: Vintage.

Diamond, J.M. 2000. Taiwan's gift to the world. *Nature* 403: 709-710.

Donnellan, S.C. & K.P. Aplin 1989. Resolution of cryptic species in the New Guinean lizard, *Sphenomorphus jobiensis* (Scincidae) by electrophoresis. *Copeia* 1989: 81-88.

Dornstreich, M.D. 1973. An ecological study of Gadio Enga (New Guinea) subsistence. Unpublished PhD thesis. Department of Anthropology, Colombia University.

Downie, J.E. & J.P. White 1978. Balof Shelter, New Ireland – report on a small excavation. *Records of The Australian Museum* 31 (19): 762-802.

Dwyer, P.D. 1974. The price of protein: five hundred hours of hunting in the New Guinea Highlands. *Oceania* 44: 278-293.

Dwyer, P.D. 1980. Edible-waste ratios for some New Guinea mammals. *Science in New Guinea* 7 (3): 109-117.

Dwyer, P.D. 1982. Prey Switching: A Case Study from New Guinea. *Journal of Animal Ecology* 51: 529-542.

Dwyer, P.D. 1983. Etolo Hunting Performances and Energetics. *Human Ecology* 11 (2): 145-174.

Dwyer, P.D. 1986. Living with Rainforest: the Human Dimension. In J. Kikkawa & D.J. Anderson (eds), *Community Ecology: Pattern and Process*: 342-367. Melbourne: Blackwell Scientific Publications.

Dwyer, P.D. 1988. Companion foods, schedules and archaeology. In B. Meehan & R. Jones (eds), *Archaeology with ethnography: an Australian perspective*: 148-155. Canberra: Department of Prehistory, Research School of Pacific Studies.

Dwyer, P.D. & M. Minnegal 1991. Hunting in Lowland, Tropical Rain Forest: Towards a Model of Non-Agricultural Subsistence. *Human Ecology* 19 (2): 187-212.

Egloff, B. 1979. *Recent Prehistory in Southeast Papua. Terra Australis* 4. Canberra: Department of Prehistory, Research School of Pacific Studies, The Australian National University.

Elmberg, J.E. 1955. Field Notes on the Mejbrat People in the Ajamaru District of the Bird's Head (Vogelkop), Western New Guinea. *Ethnos* 20: 3-102.

Elmberg, J.E. 1959. Further Notes on the Northern Mejbrats (Vogelkop, Western New Guinea). *Ethnos* 1-2, vol. 24: 70-80.

Elmberg, J.E. 1965. The Popot Feast Cycle. Acculturated exchange among the Mejprat Papuans. *Ethnos* 30, supplement.

Elmberg, J.E. 1968. *Balance and circulation. Aspects of tradition and change among the Mejprat of Irian Barat. Monograph Series* 12. Stockholm: The Ethnographical Museum.

Endicott, K. & P. Bellwood 1991. The Possibility of Independent Foraging in the Rain Forest of Peninsular Malaysia. *Human Ecology* 19 (2): 151-185.

Essai, B. 1961. *Papua and New Guinea. A Contemporary Survey*. Melbourne: Oxford University Press.

Evans, F.G. 1973. *Mechanical properties of bone*. Springfield: Charles C. Thomas.

Fa, J.E., J. Juste, J. Perez del Val & J. Castroviejo 1995. Impact of Market Hunting on Mammal Species in Equatorial Guinea. *Conservation Biology* 9: 1107-1115.

Fairbanks, R.G. 1989. A 17,000-year glacio-eustatic sea-level record: influence of glacial melting rates on the Younger Dryas event and deep-ocean circulation. *Nature* 342: 637-642.

Fischer, A.G. 1960. Latitudinal variation in organic diversity. *Evolution* 14: 64-81.

FitzGibbon, C.D., H. Mogaka & J.H. Fanshawe 1995. Subsistence Hunting in Arabuko-Sokoke Forest, Kenya, and its Effect on Mammal Populations. *Conservation Biology* 9: 1116-1126.

Flannery, T.F. 1993. Taxonomy of *Dendrolagus goodfellowi* (Macropodidae: Marsupialia) with description of a new subspecies. *Records of the Australian Museum* 45: 33-42.

Flannery, T.F. 1994. The fossil land mammal record of New Guinea: a review. *Science in New Guinea* 20: 39-84.

Flannery, T. 1995a. *Mammals of New Guinea*. Revised and updated edition. Chatswood NSW: Reed Books.
Flannery, T. 1995b. *Mammals of the South-West Pacific & Moluccan Islands*. Chatswood NSW: Reed Books.
Flannery, T.F. 1999. The Pleistocene mammal fauna of Kelangurr Cave, central montane Irian Jaya, Indonesia. *Records of the Western Australian Museum*, Supplement No 57: 341-350.
Flannery, T.F., & J.P. White 1991. Animal translocations. The zoogeography of New Ireland mammals. *National Geographic Research and Exploration* 7: 96-113.
Fleming, T.H. 1973. Numbers of Mammal Species in North and Central American Forest Communities. *Ecology* 54: 555-563.
Flenley, J.R. 1984. Late Quaternary changes of vegetation and climate in the Malesia mountains. *Erdwissenschaftliche Forschung* 18: 261-267.
Flenley, J.R. 1985. Quaternary vegetational and climatic history of Island Southeast Asia. In G.-J. Bartstra & W.A. Casparie (eds), *Papers read at Symposium 1, 12th Congress of the Indo-Pacific Prehistory Association, Philippines, 26th January-2nd February 1985. Modern Quaternary Research in Southeast Asia* 9: 55-63. Rotterdam: A.A. Balkema.
Flenley, J.R. 1993. The origin of diversity in Tropical Rain Forest. *Trends in Ecology and Evolution* 8: 119-120.
Flenniken, J.J. & J.P. White 1983. Heat treatment of siliceous rocks and its implications for Australian prehistory. *Australian Aboriginal Studies* 1: 43-48.
Flenniken, J.J. & J.P. White 1985. Australian Flaked Stone Tools: A Technological Perspective. *Records of the Australian Museum* 36: 131-151.
Fowler, C.W. 1981a. Density dependence as related to life history strategy. *Ecology* 62 (3): 602-610.
Fowler, C.W. 1981b. Comparative Population Dynamics in Large Mammals. In C.W. Fowler & T.D. Smith (eds), *Dynamics of Large Mammal Populations*: 437-455. New York: John Wiley & Sons.
Fredericksen, C., M. Spriggs & W. Ambrose 1993. Pamwak Rockshelter: A Pleistocene site on Manus Island, Papua New Guinea. In M.A. Smith, M. Spriggs & B. Fankhauser (eds), *Sahul in Review. Pleistocene Archaeology in Australia, New Guinea and Island Melanesia. Occasional Papers in Prehistory* 24: 144-152. Canberra: Department of Prehistory, Research School of Pacific Studies, The Australian National University.
Frodin, D.G. & J.L. Gressitt 1982. Biological exploration of New Guinea. In J.L. Gressitt (ed.), *Biogeography and ecology of New Guinea*. Part I: 87-130. The Hague: Dr W. Junk Publishers.
Froment, A. 2001. Evolutionary biology and health of hunter-gatherer populations. In C. Panter-Brick, R.H. Layton & P. Rowley-Conwy (eds), *Hunter-gatherers: an interdisciplinary perspective*: 239-266. Cambridge: Cambridge University Press.
Fullagar, R. 1984. Use-wear on quartz. In G.K. Ward (ed.), *Archaeology at ANZAAS Canberra 1984*: 191-197. Canberra: Canberra Archaeological Society, The Australian National University.
Fullagar, R. 1988. Recent developments in Australian use-wear and residue studies. In S. Beyries (ed.), *Industries Lithiques: Traceologie et Technologie. BAR International Series* 411: 133-145.
Fullagar, R. & J. Field 1997. Pleistocene seed-grinding implements from the Australian arid zone. *Antiquity* 71: 300-307.
Fullagar, R., B. Meehan & R. Jones 1992. Residue analysis of ethnographic plant-working and other tools from Northern Australia. *Préhistoire de l'agriculture. Monographie du CRA* 6: 39-53. Paris: Centre National de la Recherche Scientifique.
Galis, K.W. 1953. Geschiedenis. In W.C. Klein (ed.), *Nieuw Guinea. De ontwikkeling op economisch, sociaal en cultureel gebied, in Nederlands en Australisch Nieuw Guinea*. Part I: 1-65. The Hague: Staatsdrukkerij - en uitgeverijbedrijf.
Galis, K.W. 1957a. De grotten van Jaand. *Nieuw-Guinea Studiën* 1: 14-23.
Galis, K.W. 1957b. De Pinfeloe-grot nabij Tainda. *Nieuw-Guinea Studiën* 1: 118-128.

Galis, K.W. 1964. Recent oudheidkundig nieuws uit westelijk Nieuw-Guinea. *Bijdragen tot de Taal-, Land- en Volkenkunde* 120: 245-274.

Getz, W.M. & R.G. Haight 1989. *Population Harvesting. Demographic Models of Fish, Forest, and Animal Resources*. Princeton: Princeton University Press.

Gibbons, A. 2001. The Peopling of the Pacific. *Science* 291: 1735-1737.

Gillieson, D. & M.-J. Mountain 1982. Environmental history of Nombe rock-shelter, Papua New Guinea Highlands. *Archaeology in Oceania* 18: 45-53.

Gillieson, D., P. Gorecki & G. Hope 1985. Prehistoric agricultural systems in a lowland swamp, Papua New Guinea. *Archaeology in Oceania* 20: 37-39.

Glanz, W.E. 1990. Neotropical Mammal Densities: How Unusual Is the Community on Barro Colorado Island, Panama? In A.H. Gentry (ed.), *Four Neotropical Rainforests*: 287-313. New Haven: Yale University Press.

Glover, I.C. 1976. Ulu Leang Cave, Maros: A preliminary sequence of post-Pleistocene cultural development in South Sulawesi. *Archipel* 11: 113-154.

Glover, I.C. 1979. The effects of sink action on archaeological deposits in caves: an Indonesian example. *World Archaeology* 10 (3): 302-317.

Glover, I. 1986. *Archaeology in Eastern Timor, 1966-67. Terra Australis* 11. Canberra: Department of Prehistory, Research School of Pacific Studies, the Australian National University.

Golson, J. 1977a. No room at the top: agricultural intensification in the New Guinea highlands. In J. Allen, J. Golson & R. Jones (eds), *Sunda and Sahul: Prehistoric Studies in Southeast Asia, Melanesia and Australia*: 601-638. London: Academic Press.

Golson, J. 1977b. Simple tools and complex technology. Agriculture and agricultural implements in the New Guinea Highlands. In R.V.S. Wright (ed.). *Stone tools as cultural markers: change, evolution and complexity. Prehistory and Material Culture Series* 12: 154-161. New Jersey: Humanities Press.

Golson, J. 1991a. The New Guinea Highlands on the eve of agriculture. In P. Bellwood (ed.), *Indo-Pacific Prehistory 1990. Proceedings of the 14th Congress of the Indo-Pacific Prehistory Association, Yogjakarta, Indonesia, 26 August to 2 September 1990. Bulletin of the Indo-Pacific Prehistory Association* 11, volume 2: 82-91.

Golson, J. 1991b. Bulmer Phase II: early agriculture in the New Guinea Highlands. In A. Pawley (ed.), *Man and a half. Essays in Pacific Anthropology and Ethnobiology in honour of Ralph Bulmer*: 484-491. Auckland: The Polynesian Society.

Gorecki, P.P. 1979. Disposal of human remains in the New Guinea Highlands. *Archaeology and Physical Anthropology in Oceania* 14 (2): 107-117.

Gorecki, P. 1993. Saltwater people of the southwest Kimberley coast. In G. Burenhult (ed.), *The Illustrated History of Humankind.* Vol. 2. *Hunter-gatherers and Early Farmers*: 154-159. San Francisco: American Museum of National History, Harper.

Gorecki, P., M. Mabin & J. Campbell 1991. Archaeology and geomorphology of the Vanimo Coast, Papua New Guinea: preliminary results. *Archaeology in Oceania* 26: 119-122.

Gosden, C. 1993. Understanding the settlement of Pacific Islands in the Pleistocene. In M.A. Smith, M. Spriggs & B. Fankhauser (eds), *Sahul in Review. Pleistocene Archaeology in Australia, New Guinea and Island Melanesia. Occasional Papers in Prehistory* 24: 131-136. Canberra: Department of Prehistory, Research School of Pacific Studies, The Australian National University.

Gosden, C. 1995. Arboriculture and agriculture in coastal Papua New Guinea. In J. Allen & J.F. O'Connell (eds), *Transitions. Pleistocene to Holocene in Australia & Papua New Guinea. Antiquity* 69, Special number 265: 807-817.

Gosden, C. & N. Robertson 1991. Models for Matenkupkum: interpreting a late Pleistocene site from southern New Ireland, Papua New Guinea. In J. Allen & C. Gosden (eds), *Report of the Lapita Homeland Project. Occasional Papers in Prehistory* 20: 20-45. Canberra: Department of Prehistory, Research School of Pacific Studies, The Australian National University.

Gould, R. 1980. *Living Archaeology*. Cambridge: Cambridge University Press.

Gould, R. 1996. Faunal Reduction at Puntutjarpa Rockshelter, Warburton Ranges, Western Australia. *Archaeology in Oceania* 31: 72-86.

Graham, R.W. 1985. Diversity and community structure of the late Pleistocene mammal fauna of North America. *Acta Zoologica Fennica* 170: 181-192.

Grayson, D.K. 1984. *Quantitative Zooarchaeology. Topics in the Analysis of Archaeological Faunas*. Orlando: Academic Press.

Grayson, D.K. 2001. The Archaeological Record of Human Impacts on Animal Populations. *Journal of World Prehistory* 15 (1): 1-68.

Green, R. 1991. The Lapita Cultural Complex: current evidence and proposed models. In P. Bellwood (ed.), *Indo-Pacific Prehistory 1990. Proceedings of the 14th Congress of the Indo-Pacific Prehistory Association, Yogjakarta, Indonesia, 26 August to 2 September 1990. Bulletin of the Indo-Pacific Prehistory Association* 11, volume 2: 295-305.

Griffiths, D.R., C.A. Bergman, C.J. Clayton, K. Ohnuma, G.V. Robins & N.J. Seeley 1987. Experimental investigation of the heat treatment of flint. In G. de G. Sieveking & M. Newcomer (eds), *The human uses of flint and chert. Proceedings of the 4th International Flint Symposium held at Brighton Polytechnic 10-15 April 1983*: 43-52. Cambridge: Cambridge University Press.

Groube, L. 1984. Waisted axes of Asia, Melanesia and Australia. In G.K. Ward (ed.), *Archaeology at ANZAAS Canberra 1984*: 168-177. Canberra: Canberra Archaeological Society, The Australian National University.

Groube, L. 1989. The taming of the rain forests: A model for Late Pleistocene forest exploitation in New Guinea. In D.R. Harris & G.C. Hillman (eds), *Foraging and Farming. The Evolution of Plant Exploitation*: 292-304. London: Unwin Hyman.

Groube, L., J. Chappell, J. Muke & D.A. Price 1986. 40,000 year old human occupation site at Huon Peninsula, Papua New Guinea. *Nature* 324: 453-455.

Groves, C.P. 1981. Ancestors for the pigs: taxonomy and phylogeny of the genus *Sus*. *Technical Bulletin* 3. Canberra: Department of Prehistory, Research School of Pacific Studies, Australian National University.

Groves, C.P. & T.F. Flannery 1989. Revision of the genus Dorcopsis (Macropodidae: Marsupialia). In G. Grigg, P. Jarman & I Hume (eds), *Kangaroos, wallabies and rat-kangaroos*. Chipping Norton, NSW: Surrey Beatty & Sons.

Guilderson, T.P., R.G. Fairbanks & J.L. Rubenstone 1994. Tropical temperature variations since 20,000 years ago: modulating interhemispheric climate change. *Science* 263: 663-665.

Guthrie, R.D. 1983. Osseous projectile points: biological considerations affecting raw material selection and design among Paleolithic and Paleoindian peoples. In J. Clutton-Brock & C. Grigson (eds), *Animals and Archaeology*. 1. *Hunters and their Prey*. *BAR International Series* 163: 273-294.

Haberle, S.G. 1994. Anthropogenic indicators in pollen diagrams: problems and prospects for late Quaternary palynology in New Guinea. In J.G. Hather (ed.), *Tropical archaeobotany: applications and new developments*: 172-201. London: Routledge.

Haberle, S.G., G.S. Hope & Y. DeFretes 1991. Environmental change in the Baliem Valley, montane Irian Jaya, Republic of Indonesia. *Journal of Biogeography* 18: 25-40.

Haberle, S.G., G.S. Hope & S. van der Kaars 2001. Biomass burning in Indonesia and Papua New Guinea: natural and human induced fire events in the fossil record. *Palaeogeography, Palaeoclimatology, Palaeoecology* 171: 259-268.

Haffer, J. 1969. Speciation in Amazonian forest birds. *Science* 165: 131-137.

Hale, H.M. & N.B. Tindale 1930. Notes on some human remains in the Lower Murray Valley. *Records of the South Australian Museum* 4: 145-218.

Harris, E. & P.J. Hughes 1978. An early agricultural system at Mugumamp ridge, Western Highlands Province, Papua New Guinea. *Mankind* 11: 437-444.

Hartzell, L.L., K. Gale, M. Mebberson & D. Wines 1999. Methodology for determining the age structure of Pleistocene archaeofaunal remains of Bennett's Wallaby (*Macropus rufogriseus*) using tooth eruption, molar progression and attrition in modern populations. In M.-J. Mountain

& D. Bowdery (eds), *Taphonomy: The Analysis of Processes from Phytoliths to Megafauna*: 55-61. Canberra: ANH Publications, Department of Archaeology and Natural History, Research School of Pacific and Asian Studies, The Australian National University.

Hayden, B. 1977. Stone tool functions in the Western Desert. In R.V.S. Wright (ed.), *Stone tools as cultural markers: change, evolution and complexity. Prehistory and Material Culture Series* 12: 178-188. Canberra: Australian Institute of Aboriginal Studies, Humanities Press.

Headland, T.N. 1987. The wild yam question: How well could independent hunter-gatherers live in a tropical rain forest ecosystem? *Human Ecology* 15: 463-491.

Headland, T.N. & L.A. Reid 1989. Hunter-gatherers and their neighbors from prehistory to the present. *Current Anthropology* 30: 43-66.

Headland, T.N. & R.C. Bailey 1991. Introduction: Have Hunter-Gatherers Ever Lived in Tropical Rain Forest Independently of Agriculture? *Human Ecology* 19 (2): 115-122.

Hill, A.V.S. & S.W. Serjeantson (eds) 1989. *The Colonization of the Pacific. A Genetic Trail.* Oxford: Clarendon Press.

Hiscock, P. 1984. A preliminary report on the stone artefacts from Colless Creek Cave, Northwest Queensland. *Queensland Archaeological Research* 1: 120-151.

Hiscock, P. 1985. The need for a taphonomic perspective in stone artefact analysis. *Queensland Archaeological Research* 2: 82-95.

Hiscock, P. 1986. Technological change in the Hunter River Valley and the interpretation of Late Holocene change in Australia. *Archaeology in Oceania* 21: 29-39.

Hiscock, P. 1990. A study in scarlet: taphonomy and inorganic artefacts. In: S. Solomon, I. Davidson & D. Watson (eds), *Problem Solving in Taphonomy. Archaeological and palaeontological studies from Europe, Africa and Oceania. Tempus. Archaeology and Material Culture Studies in Anthropology* 2: 34-49.

Hiscock, P. 1993. Bondaian Technology in the Hunter Valley, New South Wales. *Archaeology in Oceania* 28: 65-76.

Hladik, C.M., A. Hladik, O.F. Linares, H. Pagezy, A. Semple & M. Hadley (eds) 1993. *Tropical Forests, People and Food. Biocultural Interactions and Applications to Development. Man and the Biosphere Series* 13. Paris: UNESCO & The Parthenon Publishing Group.

Holdaway, R.N. & C. Jacomb 2000. Rapid Extinction of the Moas (Aves: Dinornithiformes): Model, Test, and Implications. *Science* 287: 2250-2254.

Hope, G.S. 1976. The vegetation history of Mt Wilhelm, Papua New Guinea. *Journal of Ecology* 64: 627-664.

Hope, G.S. 1986. Development of present day biotic distributions in the New Guinea mountains. In B. Barlow (ed.), *Flora and Fauna of Alpine Australasia*: 129-145. Melbourne: C.S.I.R.O. & Leiden: E.J. Brill.

Hope, G.S. 1989. Climatic implications of timberline changes in Australasia from 30,000 BP to present. In T. Donnelly & R. Wasson (eds), *CLIMANZ* 3: 91-99. Canberra: C.S.I.R.O., Div. Water Resources.

Hope, G.S. 1996. Quaternary change and the historical biogeography of Pacific islands. In A. Keast & S.E. Miller (eds), *The origin and evolution of Pacific Island biotas. New Guinea to Eastern Polynesia: patterns and processes*: 165-190. Amsterdam: SPB Academic Publishing.

Hope, G.S. & J.A. Peterson 1975. Glaciation and Vegetation in the High New Guinea Mountains. *Bulletin of the Royal Society of New Zealand* 13: 155-162.

Hope, G.S. & J.H. Hope 1976. Man on Mt. Jaya. In G.S. Hope, J.A. Peterson, U. Radok & I. Allison (eds), *The equatorial glaciers of New Guinea. Results of the 1971-1973 Australian Universities' Expeditions to Irian Jaya: survey, glaciology, meteorology, biology and palaeoenvironments*: 225-239. Rotterdam: A.A. Balkema.

Hope, G.S. & J. Tulip 1994. A long vegetational history from lowland Irian Jaya, Indonesia. *Palaeogeography, Palaeoclimatology, Palaeoecology* 109: 385-398.

Hope, G. & J. Golson 1995. Late Quaternary change in the mountains of New Guinea. In J. Allen & J.F. O'Connell (eds), *Transitions. Pleistocene to Holocene in Australia & Papua New Guinea. Antiquity* 69, Special number 265: 818-830.

Hope, G.S., J.A. Peterson, U. Radok & I. Allison (eds) 1976. *The equatorial glaciers of New Guinea. Results of the 1971-1973 Australian Universities' Expeditions to Irian Jaya: survey, glaciology, meteorology, biology and palaeoenvironments.* Rotterdam: A.A. Balkema.

Hope, G.S., T.F. Flannery & Boeardi 1993. A Preliminary Report of Changing Quaternary Mammal Faunas in Subalpine New Guinea. *Quaternary Research* 40: 117-126.

Irwin, G. 1985. *The Emergence of Mailu: as a central place in coastal papuan prehistory. Terra Australis* 10. Canberra: Department of Prehistory, Research School of Pacific Studies, The Australian National University.

Jarman, P.J. 1989. Sexual dimorphism in Macropodoidae. In G. Grigg, P. Jarman & I. Hume (eds), *Kangaroos, wallabies and rat-kangaroos*: 433-447. Chipping Norton, NSW: Surrey Beatty & Sons.

Jelsma, J. 1998. Room with a view. An excavation in Toé cave, Ayamaru district, Bird's Head, Irian Jaya. In G.-J. Bartstra (ed.), *Bird's Head Approaches. Irian Jaya Studies - A Programme for Interdisciplinary Research. Modern Quaternary Research in Southeast Asia* 15: 41-65. Rotterdam: A.A. Balkema.

Jennings, J.N. 1985. *Karst Geomorphology.* Oxford: Basil Blackwell Ltd.

Johnson, E. 1985. Current Developments in Bone Technology. In M.B. Schiffer (ed.), *Advances in Archaeological Method and Theory* 8: 157-235.

Jones, R. & I. Johnson 1985. Rockshelter excavations: Nourlangie and Mt Brockman Massifs. In R. Jones (ed.), *Archaeological Research in Kakadu National Park.* Special Publication 13. Canberra: Australian National Parks and Wildlife Service.

Keegan, W.F. & J.M. Diamond 1987. Colonization of Islands by Humans: A Biogeographical Perspective. In M.B. Schiffer (ed.), *Advances in Archaeological Method and Theory* 10: 49-92. San Diego: Academic Press.

Keeley, L.H. 1982. Hafting and retooling: effects on the archaeological record. *American Antiquity* 47: 798-809.

Kershaw, A.P. 1986. Climatic change and Aboriginal burning in north-east Australia during the two last glacial-interglacial cycles. *Nature* 322: 47-49.

Kingsmill, E. 1962. An investigation of criteria for estimating age in the marsupials *Trichosurus vulpecula* Kerr and *Perameles nasuta* Geoffroy. *Australian Journal of Zoology* 10: 597-616.

Kirch, P.V. 1997. *The Lapita Peoples: Ancestors of the Oceanic World.* Oxford: Blackwell Publishers.

Kirkpatrick, T.H. 1964. Molar progression and macropod age. *Queensland Journal of Agricultural Science* 21 (1): 163-165.

Kirkpatrick, T.H. 1965. Studies of Macropodidae in Queensland. 2. Age estimation in the grey kangaroo, the eastern wallaroo and the red-necked wallaby, with notes on dental abnormalities. *Queensland Journal of Agricultural and Animal Sciences* 22: 301-317.

Kirkpatrick, T.H. 1970. Studies of Macropodidae in Queensland. 8. Age estimation in the Red Kangaroo (*Megaleia rufa* (Desmarest)). *Queensland Journal of Agricultural and Animal Sciences* 27: 461-462.

Kirkpatrick, T.H. & P.M. Johnson 1969. Studies of Macropodidae in Queensland. 7. Age estimation and reproduction in the agile wallaby (*Wallabia agilis* (Gould)). *Queensland Journal of Agricultural and Animal Sciences* 26: 691-698.

Klein, R.G. & K. Cruz-Uribe 1984. *The Analysis of Animal Bones from Archaeological Sites.* Chicago: The University of Chicago Press.

Knecht, H. 1997. Projectile Points of Bone, Antler, and Stone. Experimental Explorations of Manufacture and Use. In H. Knecht (ed.), *Projectile Technology*: 191-212. New York: Plenum Press.

Koetje, T.A. 1994. Intrasite spatial structure in the European Upper Paleolithic: Evidence and patterning from the southwest of France. *Journal of Anthropological Archaeology* 13: 161-169.

Krebs, C.J. 1978. *Ecology. The Experimental Analysis of Distribution and Abundance.* 2nd edition. New York: Harper & Row.

Lambeck, K. & J. Chappell 2001. Sea Level Change Through the Last Glacial Cycle. *Science* 292: 679-686.

Lampert, R.J. 1966. An excavation at Durras North, New South Wales. *Archaeology and Physical Anthropology in Oceania* 1: 83-118.

Lampert, R.J. 1971. *Burrill Lake and Currarong. Coastal sites in southern New South Wales. Terra Australis* 1. Canberra: Department of Prehistory, Research School of Pacific Studies, Australian National University.

Leask, M.F. 1943. A kitchen midden in Papua. *Oceania* 13: 235-242.

Leavesley, M. & J. Allen 1998. Dates, disturbance and artefact distributions: another analysis of Buang Merabak, a Pleistocene site on New Ireland, Papua New Guinea. *Archaeology in Oceania* 33 (2): 63-82.

Lee, A.K. & S.J. Ward 1989. Life histories of macropodoid marsupials. In G. Grigg, P. Jarman & I. Hume (eds), *Kangaroos, wallabies and rat-kangaroos*: 105-115. Chipping Norton, NSW: Surrey Beatty & Sons.

LeMoine, G.M. 1997. *Use Wear Analysis on Bone and Antler Tools of the Mackenzie Inuit. BAR International Series* 679.

Lentle, R.G., K.J. Stafford, M.A. Potter, B.P. Springett & S. Haslett 1998. Incisor and molar wear in the tammar wallaby (*Macropus eugenii* Desmarest). *Australian Journal of Zoology* 46: 509-527.

Lilley, I. 1992. Papua New Guinea's human past: the evidence of archaeology. In R. Attenborough & M. Alpers (eds), *Human Biology in Papua New Guinea*: 150-171. Oxford: Oxford University Press.

Lilley, I. 2000. Migration and ethnicity in the evolution of Lapita and post-Lapita maritime societies in nothwest Melanesia. In S. O'Connor & P. Veth (eds), *East of Wallace's Line. Studies of past and present maritime cultures of the Indo-Pacific region. Modern Quaternary Research in. Southeast Asia* 16: 177-195. Rotterdam: A.A. Balkema.

Linsley, B.K. 1996. Oxygen-isotope record of sea level and climatic variations in the Sulu Sea over the past 150,000 years. *Nature* 380: 234-237.

Liu, K.-B. & P.A. Colinvaux 1985. Forest changes in the Amazon Basin during the last glacial maximum. *Nature* 318: 556-557.

Liu Jinling, Tang Lingyu, Qiao Yulou, M.J. Head & D. Walker 1986. Late Quaternary vegetational history at Menghai, Yunnan Province, southwest China. *Journal of Biogeography* 13: 399-418.

Loughnan, F.C. 1969. *Chemical Weathering of the Silicate Minerals*. New York: American Elsevier Publishing Company.

Loy, T.H. 1983. Prehistoric blood residues: Detection on tool surfaces and identification of species of origin. *Science* 220: 1269-1271.

Loy, T.H. 1987. Recent advances in blood residue analysis. In W.R. Ambrose & J.M.J. Mummery (eds), *Archaeometry: Further Australasian Studies*: 57-65. Canberra: Department of Prehistory, The Australian National University.

Loy. T.H. 1990. Prehistoric organic residues: Recent advances in identification, dating and their antiquity. *Archaeometry* 32: 645-656.

Loy, T.H., M. Spriggs & S. Wickler 1992. Direct Evidence for Human Use of Plants 28,000 Years Ago: Starch Residues on Stone Artefacts from the Nortern Solomon Islands. *Antiquity* 66: 898-912.

Lum, J.K. & R.L. Cann 1998. mtDNA and Language Support a Common Origin of Micronesians and Polynesians in Island Southeast Asia. *American Journal of Physical Anthropology* 105: 109-119.

Lundelius, E.L.Jr 1989. The implications of disharmonious assemblages .for Pleistocene extinctions. *Journal of Archaeological Science* 16: 407-417.

Lupo, K.D. 1995. Hadza Bone Assemblages and Hyena Attrition: An Ethnographic Example of the Influence of Cooking and Mode of Discard on the Intensity of Scavenger Ravaging. *Journal of Anthropological Archaeology* 14: 288-314.

Lyman, R.L. 1992. Anatomical considerations of utility curves in zooarchaeology. *Journal of Archaeological Science* 19: 7-22.

Lyman, R.L. 1994. *Vertebrate Taphonomy*. Cambridge: Cambridge University Press.

Majid, Z. 1982. *The West Mouth, Niah, in the Prehistory of Southeast Asia. Sarawak Museum Journal (N.S.)* 31, Special Monograph 3: 1-200.

Majnep, I.S. & R. Bulmer 1977. *Birds of my Kalam Country*. Auckland: Auckland University Press.

Markgraf, V., J.R. Dodson, A.P. Kershaw, M.S. McGlone & N. Nicholls 1992. Evolution of late Pleistocene and Holocene climates in the circum-South Pacific land areas. *Climate Dynamics* 6: 193-211.

Marshall, B. & J. Allen 1991. Excavations in Panakiwuk Cave, New Ireland. In J. Allen & C. Gosden (eds), *Report of the Lapita Homeland Project. Occasional Papers in Prehistory* 20: 59-91. Canberra: Department of Prehistory, Research School of Pacific Studies, The Australian National University.

Martin, J. 1983. Optimal foraging theory: a review of some models and their applications. *American Anthropologist* 85: 612-629.

Massola, A. 1956. Australian fish hooks and their distribution. *National Museum of Victoria, Memoirs. Anthropology Series* No. 1. 22: 1-16.

McArthur, C. & G.D. Sanson 1988. Tooth wear in eastern grey kangaroos (*Macropus giganteus*) and western grey kangaroos (*Macropus fuliginosus*), and its potential influence on diet selection, digestion and population parameters. *Journal of Zoology* 215: 491-504.

McCarthy, F.D. 1940. The bone point, known as *Muduk*, in eastern Australia. *Records of the Australian Museum* 20: 313-319.

McCormac, F.G., A.G. Hogg, T.F.G. Higham, M.G.L. Baillie, J.G. Palmer, L. Xiong, J.R. Pilcher, D. Brown & S.T. Hoper 1998. Variations of radiocarbon in tree rings: Southern Hemisphere offset preliminary results. *Radiocarbon* 40 (3): 1153-1159.

Mees, G.F. 1965. The avifauna of Misool. *Nova Guinea, Zoology* 31: 139-203.

Menzies, J.I. 1989. Observations on a captive forest wallaby (*Dorcopsis luctuosa*) colony. In G. Grigg, P. Jarman & I. Hume (eds), *Kangaroos, wallabies and rat-kangaroos*: 629-631. Chipping Norton, NSW: Surrey Beatty & Sons.

Menzies, J.I. 1991. *Handbook of New Guinea Marsupials & Monotremes*. Madang: Kirsten Press.

Mercader, J. (ed.) 2003. *Under the Canopy: The Archaeology of Tropical Rainforests*. New York: Rutgers University Press.

Meyer, O. 1907. Funde prähistorischer Töpferei und Steinmesser auf Vuatom, Bismarck-Archipel. *Anthropos* 4: 251-252 and 1093-1095.

Miedema, J. 1986. *Pre-capitalism and Cosmology. Description and Analysis of the Meybrat Fishery and Kain Timur-Complex. Verhandelingen van het Koninklijk Instituut voor Taal-, Land- en Volkenkunde* 120. Dordrecht: Foris Publications.

Miller, G.H., J.W. Magee & A.J.T. Jull 1997. Low-latitude glacial cooling in the Southern Hemisphere from amino-acid racemization in emu eggshells. *Nature* 385: 241-244.

Milner-Gulland, E.J. & H.R. Akçakaya 2001. Sustainability indices for exploited populations. *Trends in Ecology & Evolution* 16 (12): 686-692.

Montet-White, A. & J.T. Williams 1994. Spatial organization at a winter campsite of the Last Glacial Maximum: The case of Grubgraben AL1. *Journal of Anthropological Archaeology* 13: 125-138.

Mook, W.G. & H.J. Streurman 1983. Physical and chemical aspects of radiocarbon dating. *Proceedings of the First International Symposium ^{14}C and Archaeology, Groningen 1981, PACT* 8: 31-55.

Mountain, M.-J. 1983. Preliminary report of excavations at Nombe rockshelter, Simbu Province, Papua New Guinea. *Bulletin of the Indo-Pacific Prehistory Association* 4: 84-99.

Mountain, M.-J. 1993. Bones, Hunting and Predation in the Pleistocene of Northern Sahul. In M.A. Smith, M. Spriggs & B. Fankhauser (eds), *Sahul in Review. Pleistocene Archaeology in Australia, New Guinea and Island Melanesia. Occasional Papers in Prehistory* 24: 123-130.

Canberra: Department of Prehistory, Research School of Pacific Studies, The Australian National University.

Mulvaney, D.J. 1960. Archaeological excavations at Fromm's Landing on the Lower Murray Valley, South Australia. *Proceedings of the Royal Society of Victoria* 72: 53-85.

Mulvaney, J. & J. Kamminga 1999. *Prehistory of Australia.* St Leonards: Allen & Unwin.

Nagaoka, L. 2001. Using Diversity Indices to Measure Changes in Prey Choice at the Shag River Mouth Site, Southern New Zealand. *International Journal of Osteoarchaeology* 11: 101-111.

Newsome, J. & J.R. Flenley 1988. Late Quaternary vegetational history of the central highlands of Sumatra. II. Palaeopalynology and vegetational history. *Journal of Biogeography* 15: 555-578.

Newsome, A.E., J.C. Merchant, B.L. Bolton & M.L. Dudzinski 1977. Sexual Dimorphism in Molar Progression and Eruption in the Agile Wallaby. *Australian Wildlife Research* 4: 1-5.

Nicholson, R.A. 1996. Bone Degradation, Burial Medium and Species Representation: Debunking the Myths, an Experiment-based Approach. *Journal of Archaeological Science* 23: 513-533.

Nix, H.A. & J.D. Kalma 1972. Climate as a dominant control in the biogeography of northern Australia and New Guinea. In D. Walker (ed.), *Bridge and Barrier: the natural and cultural history of Torres Strait*: 61-91. Publication B/G3. Canberra: Department of Biogeography and Geomorphology, Research school of Pacific Studies, Australian National University.

O'Connell, J., K. Hawkes & N. Blurton Jones 1988. Hadza hunting, butchering, and bone transport and their archaeological implications. *Journal of Anthropological Research* 44: 112-162.

O'Connell, J.F. & J. Allen 1998. When Did Humans First Arrive in Greater Australia and Why Is It Important to Know? *Evolutionary Anthropology* 6: 132-146.

O'Connell, J.F., K. Hawkes & N. Blurton Jones 1990. Reanalysis of large mammal body part transport among the Hadza. *Journal of Archaeological Science* 17: 301-316.

O'Connor, S., P. Veth & C. Campbell 1998. Serpent's Glen Rockshelter: report of the first Pleistocene aged occupation sequence from the Western Desert. *Australian Archaeology* 46: 12-22.

O'Connor, S., K.P. Aplin, M. Spriggs, P. Veth & L.K. Ayliffe 2002. From Savanna to Rainforest: changing environments and human occupation at Liang Lemdubu, Aru Islands, Maluku (Indonesia). In P. Kershaw, B. David, N. Tapper, D. Penny & J. Brown (eds), *Bridging Wallace's Line: the environmental and cultural history and dynamics of the SE Asian-Australian region. Advances in Geoecology* 34: 279-306. Reiskirchen: Catena Verlag GMBH.

Odell, G.H. 2000. Stone Tool Research at the End of the Millennium: Procurement and Technology. *Journal of Archaeological Research* 8 (4): 269-331.

Odum, E.P. 1971. *Fundamentals of Ecology.* 3rd edition. Philadelphia: W.B.Saunders.

Olsen, S.L. 1984. Analytical Approaches to the Manufacture and Use of Bone Artifacts in Prehistory. Unpublished PhD thesis. London: University of London.

Olsen, S.L. & I.C. Glover in press. The bone industry of Ulu Leang 1 and Leang Burung 1 rockshelters, Sulawesi, Indonesia, in its regional context. In S.G. Keates & J.M. Pasveer (eds), *Quaternary Research in Indonesia. Modern Quaternary Research in Southeast Asia* 18. Leiden: A.A. Balkema Publishers.

Oppenheimer, S.J. & M. Richards 2001. Slow boat to Melanesia. *Nature* 410: 166-167.

Orliac, M. 1975. Empreintes au latex des coupes du gisement Magdalénien de Pincevent; Technique et premier résultats. *Bulletin de la société Préhistorique Française* 72/9: 274-276.

Paoletti, M.G., D.L. Dufour, H.Cerda, F. Torres, L.Pizzoferrato & D. Pimentel 2000. The importance of leaf- and litter-feeding invertebrates as sources of animal protein for the Amazonian Amerindians. *Proceedings of the Royal Society of London* B 267: 2247-2252.

Pasveer, J.M. 1995. Archeologisch onderzoek in Irian Jaya. Verslag van een oriënterende reis. *Paleo-aktueel* 6: 53-55.

Pasveer, J.M. 1998. Kria cave: An 8000-year occupation sequence from the Bird's Head of Irian Jaya. In G.-J. Bartstra (ed.), *Bird's Head Approaches. Irian Jaya Studies - A Programme for Interdisciplinary Research. Modern Quaternary Research in Southeast Asia* 15: 67-89. Rotterdam: A.A. Balkema.

Pasveer, J.M. 2003. The *Djief* hunters. 26,000 years of lowland rainforest exploitation on the Bird's Head of Papua, Indonesia. Unpublished PhD thesis. Groningen: Groningen Institute of Archaeology, University of Groningen.

Pasveer, J.M. & G.-J. Bartstra 1996. Naalden in de Hooiberg. Verslag van een opgraving in Kria, de Vogelkop, Irian Jaya, Indonesië. *Paleo-aktueel* 7: 30-33.

Pasveer, J.M. & K.P. Aplin 1998. Late Pleistocene to modern vertebrate faunal succession and environmental change in lowland New Guinea: evidence from the Bird's Head of Irian Jaya, Indonesia. In J. Miedema, C. Odé & R.A.C. Dam (eds), *Perspectives on the Bird's Head of Irian Jaya, Indonesia. Proceedings of the Conference, Leiden, 13-17 October 1997*: 891-930. Amsterdam: Rodopi.

Pasveer, J.M. & P. Bellwood in press. Prehistoric bone artefacts from the northern Moluccas, Indonesia. In S.G. Keates & J.M. Pasveer (eds), *Quaternary Research in Indonesia. Modern Quaternary Research in Southeast Asia* 18. Leiden: A.A. Balkema Publishers.

Pasveer, J.M., S.J. Clarke & G.H. Miller 2002. Late Pleistocene human occupation of inland rainforest, Bird's Head, Papua. *Archaeology in Oceania* 37: 92-95.

Pavlides, C. & C. Gosden 1994. 35,000-year-old sites in the rainforests of West New Britain, Papua New Guinea. *Antiquity* 68: 604-610.

Pawley, A.K. 1998. The Trans New Guinea Phylum hypothesis: a reassessment. In J. Miedema, C. Odé & R.A.C. Dam (eds), *Perspectives on the Bird's Head of Irian Jaya, Indonesia. Proceedings of the Conference. Leiden, 13-17 October 1997*: 655-690. Amsterdam: Rodopi.

Pearce, J. & R. Luff 1994. The Taphonomy of Cooked Bone. In R. Luff & P. Rowley-Conwy (eds), *Whither Environmental Archaeology*? *Oxbow Monograph* 38: 51-56. Oxford: Oxbow Books.

Peres, C.A. 2000. Effects of Subsistence Hunting on Vertebrate Community Structure in Amazonian Forests. *Conservation Biology* 14: 240-254.

Peterson, N. 1971. Open sites and the ethnographic approach to the archaeology of hunter-gatherers. In D.J. Mulvaney & J. Golson (eds), *Aboriginal Man and Environment in Australia*: 239-248. Canberra: Australian National University Press.

Pickering, M.P. 1979. Aboriginal bone tools from Victoria. Unpublished BA Honours thesis. Melbourne: La Trobe University.

Piepenbrink, H. 1989. Examples of chemical changes during fossilisation. *Applied Geochemistry* 4: 273-280.

Pigram, C.J. & U. Sukanta 1989. *Geologi Lembar Taminabuan, Irian Jaya.* (*Geology of the Taminabuan Sheet area, Irian Jaya.*) Bandung: Geological Research and Development Centre, Indonesia.

Pöch, R. 1907. Ausgrabungen alter Topfscherben in Wanigela (Collingwood-Bay). *Mitteilungen Der Anthropologischen Gesellschaft in Wien* 37: 137-139 and 67-71.

Polak, M. 1998. Initial results of a botanical species richness study in the Ayawasi area, Irian Jaya. In G.-J. Bartstra (ed.), *Bird's Head Approaches. Irian Jaya Studies – A programme for Interdisciplinary Research. Modern Quaternary Research in Southeast Asia* 15, 111-121. Rotterdam: A.A. Balkema.

Pospisil, L. 1963. *The Kapauku Papuans of west New Guinea.* New York: Holt, Rinehart and Winston.

Prange, H.D., J.F. Anderson & H. Rahn 1979. Scaling of skeletal mass to body mass in birds and mammals. *The American Naturalist* 113: 103-122.

Prasetyo, B. 2002. The Bone Industry. In T. Simanjuntak (ed.), *Gunung Sewu in Prehistoric Times*: 181-194. Yogjakarta: Gadjah Mada University Press.

Purdy, B.A. 1974. Investigations concerning the thermal alteration of silica minerals: an archaeological approach. *Tebiwa* 17 (1): 37-66.

Purdy, B.A. 1975. Fractures for the Archaeologist. In E. Swanson (ed.), *Lithic Technology*: 133-141. The Hague: Mouton Press.

Purdy, B.A. & D.E. Clark 1979. Weathering of thermally altered prehistoric stone implements. *Lithic Technology* 8 (2): 20-21.

Pyke, G., H. Pulliam & E. Charnov 1977. Optimal Foraging: A Selective Review of Theory and Tests. *Quarterly Review of Biology* 52: 137-154.

Rand, A.L. & L.J. Brass 1941. Results of the Archbold expeditions no 29. Summary of the 1936-1937 New Guinea expedition. *Bulletin of the American Museum of Natural History* 77, 1940-1941: 341-380.

Rand, A.L. & E.T. Gilliard 1967. *Handbook of New Guinea Birds.* London: Weidenfeld & Nicolson.

Rappaport, R.A. 1968. *Pigs for the Ancestors: Ritual in the Ecology of a New Guinea People.* New Haven: Yale University Press.

Ratman, N. 1998. Geology of the Bird's Head, Irian Jaya, Indonesia. In J. Miedema, C. Odé & R.A.C. Dam (eds), *Perspectives on the Bird's Head of Irian Jaya, Indonesia. Proceedings of the Conference, Leiden, 13-17 October 1997*: 719-755. Amsterdam: Rodopi.

Reesink, G.P. 1996. Morpho-syntactic features of the Bird's Head languages. In G.P. Reesink (ed.), *Studies in Irian Languages*. Part I. *NUSA* 40: 1-19.

Reesink, G.P. 1998. The Bird's Head as Sprachbund. In J. Miedema, C. Odé & R.A.C. Dam (eds), *Perspectives on the Bird's Head of Irian Jaya, Indonesia. Proceedings of the Conference. Leiden, 13-17 October 1997*: 603-642. Amsterdam: Rodopi.

Reeskamp, G.A. 1959. Verslag van het tournee naar Ajamaroe, Negeri Besar en Sorong. *Mededelingen van de Dienst van Gezondheidszorg in Nederlands Nieuw-Guinea* 6-4: 13-19.

Reitz, E.J. & E.S. Wing 1999. *Zooarchaeology*. Cambridge: Cambridge University Press.

Reynders, J.J. 1961.Volksverhalen uit het Ajamaroese. *Nederlands Nieuw Guinea* 9: 18-19.

Rhoads, J.W. 1980. Through a glass darkly: present and past land-use systems of Papuan sagopalm users. Unpublished PhD thesis. Canberra: Department of Prehistory, the Australian National University.

Rind, D. & D. Peteet 1985. Terrestrial conditions at the Last Glacial Maximum and CLIMAP sea-surface temperature estimates: are they consistent? *Quaternary Research* 24: 1-22.

Roberts, R.G., R. Jones & M.A. Smith 1993. Optical dating at Deaf Adder Gorge, Northern Territory, indicates human occupation between 53,000 and 60,000 years ago. *Australian Archaeology* 37: 58-59.

Roberts, R.G., R. Jones, N.A. Spooner, M.J. Head, A.S. Murray & M.A. Smith 1994. The human colonisation of Australia: optical dates of 53,000 and 60,000 years bracket human arrival at Deaf Adder Gorge, Northern Territory. *Quaternary Geochronology* (*Quaternary Science Review*) 13: 575-583.

Robinson, J.G. & R.E. Bodmer 1999. Towards wildlife management in tropical forests. *Journal of Wildlife Management* 63: 1-13.

Röder, J. 1939. Rock-pictures and prehistoric times in Dutch New Guinea. *Man* 172-173: 175-178.

Röder, J. 1940. Ergebnisse einer Probegrabung in der Höhle Dudumunir auf Arguni, MacCluer-Golf (Höllandisch West Neu-Guinea). *Nova Guinea* (n.s.) 4: 1-10.

Roseberry, J.L. & A. Woolf 1991. A comparative evaluation of techniques for analyzing white-tailed deer harvest data. *Wildlife Monographs* 117: 1-59.

Ross, B. 1977. Agile Wallabies in Recent Papuan Prehistory: an Approach to Faunal Analysis. Unpublished BA Honours thesis. Canberra: Department of Prehistory and Anthropology, Australian National University.

Rowney, M. & J.P. White 1997. Detecting Heat Treatment on Silcrete: Experiments with Methods. *Journal of Archaeological Science* 24: 649-657.

Ryan, P. 2000. *Fiji's Natural Heritage*. Auckland: Exisle Publishing.

Sanson, G.D. 1980. The morphology and occlusion of the molariform cheek teeth in some Macropodinae (Marsupialia: Macropodidae). *Australian Journal of Zoology* 28: 341-365.

Scheans, D.J., K.L. Hutterer & R.L Cherry 1970. A newly discovered blade tool industry from the Central Philippines. *Asian Perspectives* 13: 179-181.

Schrire, C. 1982. *The Alligator Rivers. Prehistory and ecology in western Arnhem Land. Terra Australia* 7. Canberra: Department of Prehistory, Research School of Pacific Studies, The Australian National University.

Schumacher, C. 1954. Exploratie. In W.C. Klein (ed.), *Nieuw Guinea. De ontwikkeling op economisch, sociaal en cultureel gebied, in Nederlands en Australisch Nieuw Guinea.* Part III: 1-120. The Hague: Staatsdrukkerij - en uitgeverijbedrijf.

Sémah, A.-M., F. Sémah, R. Moudrikah, F. Fröhlich & T. Djubiantono in press. A late Pleistocene and Holocene sedimentary record in Central Java and its palaeoclimatic significance. In S.G. Keates & J.M. Pasveer (eds), *Quaternary Research in Indonesia. Modern Quaternary Research in Southeast Asia* 18. Leiden: A.A. Balkema Publishers.

Sen, A. 1995. Borneo's history in a handful of husks. *New Scientist* 1972: 10.

Sharman, G.B., H.J. Frith & J.H. Calaby 1964. Growth of the pouch young, tooth eruption and age determination in the red kangaroo, *Megaleia rufa. CSIRO Wildlife Research* 9 (1): 20-49.

Sheppard, P.J. & L.A. Pavlish 1992. Weathering of Archaeological Cherts: A Case Study from the Solomon Islands. *Geoarchaeology* 7 (1): 41-53.

Shield, J.W. 1958. Aspects of field ecology of the Quokka (*Setonix brachyurus* Quoy and Gaimard). Unpublished PhD thesis. Perth: University of Western Australia.

Shipman, P., G. Foster & M. Schoeninger 1984. Burnt Bones and Teeth: an Experimental Study of Color, Morphology, Crystal Structure and Shrinkage. *Journal of Archaeological Science* 11: 307-325.

Sillitoe, P. 1988. *Made in Niugini. Technology in the Highlands of Papua New Guinea.* London: British Museum Publications.

Sillitoe, P. 2002. Always been farmer-foragers? Hunting and gathering in the Papua New Guinea Highlands. *Anthropological Forum* 12 (1): 45-76.

Simanjuntak, T. 1996. Perspektif prasejarah Irian Jaya. *Kebudayaan* 9: 38-49. Jakarta: Departemen Pendidikan dan Kebudayaan.

Simanjuntak, T. 1998. Review of the prehistory of Irian Jaya. In J. Miedema, C. Odé & R.A.C. Dam (eds), *Perspectives on the Bird's Head of Irian Jaya, Indonesia. Proceedings of the Conference, Leiden, 13-17 October 1997*: 941-950. Amsterdam: Rodopi.

Simanjuntak, T. in press. New insight on the prehistoric chronology of Gunung Sewu, Indonesia. In S.G. Keates & J.M. Pasveer (eds), *Quaternary Research in Indonesia. Modern Quaternary Research in Southeast Asia* 18. Leiden: A.A. Balkema Publishers.

Smith, E.A. & M. Wishnie 2000. Conservation and subsistence in small-scale societies. *Annual Review of Anthropology* 29: 493-524.

Smith, M.A. & N.D. Sharp 1993. Pleistocene sites in Australia, New Guinea and Island Melanesia: Geographic and temporal structure of the archaeological record. In M.A. Smith, M. Spriggs & B. Fankhauser (eds), *Sahul in Review. Pleistocene Archaeology in Australia, New Guinea and Island Melanesia. Occasional Papers in Prehistory* 24: 37-59. Canberra: Department of Prehistory, Research School of Pacific Studies, The Australian National University.

Smith, M.J. & L. Hinds 1995. Tammar Wallaby. In R. Strahan (ed.), *The Mammals of Australia*: 329-331. Chatswood: 329-331. The Australian Museum/ Reed Books.

Smyth, W.B. 1878. *Aborigines of Victoria.* Melbourne: Government Printer.

Solheim II, W.G. 1978. Archaeological survey, Irian Jaya. *Pacific Art Newsletter* 6: 1-5.

Solheim II, W.G. 1979. Irian Jaya origins. *Australian Natural History* 19 (10): 324-327.

Solheim II, W.G. 1985. Korwar of the Biak. In J. Feldman (ed.), *The Eloquent Dead*: 147-160. Los Angeles: Museum of Cultural History, University of California.

Solheim, W.G. II 1998. Preliminary report on Makbon archaeology, the Bird's Head, Irian Jaya. In G.-J. Bartstra (ed.), *Bird's Head Approaches. Irian Jaya Studies - A Programme for Interdisciplinary Research. Modern Quaternary Research in Southeast Asia* 15: 29-40. Rotterdam: A.A. Balkema.

Solheim II, W.G. & A. Ap 1977. Pottery manufacture in Abar, Lake Sentani, Irian Jaya. *Irian. Bulletin of Irian Jaya Development* 6 (1): 52-70.

Solheim II, W.G. & J. Mansoben 1977. Pottery manufacture on Mansinam, Manokwari, Irian Jaya. *Irian. Bulletin of Irian Jaya Development* 6 (1): 46-51.

Sollewijn Gelpke, J.H.F. 1994. The Report of Miguel Roxo de Brito of his Voyage in 1581-1582 to the Raja Ampat, the MacCluer Gulf and Seram. *Bijdragen tot de Taal-, Land- en Volkenkunde* 150: 123-145.

Spriggs, M. 1995. The Lapita culture and Austronesian prehistory in Oceania. In P. Bellwood, J.J. Fox & D. Tryon (eds), *The Austronesians: Historical and Comparative Perspectives*: 112-133. Canberra: Department of Anthropology, Research School of Pacific and Asian Studies, The Australian National University.

Spriggs, M. 1996a. What is Southeast Asian about Lapita? In T. Akazawa & E.J.E. Szathmary (eds), *Prehistoric Mongoloid Dispersals*: 324-348. Oxford: Oxford University Press.

Spriggs, M. 1996b. Chronometric hygiene and colonisation in Island Southeast Asia and the Pacific: new data and an evaluation. In J. Davidson, G. Irwin, F. Leach, A. Pawley & D. Brown (eds), *Oceanic Culture History: Essays in Honour of Roger Green*: 33-50. Dunedin: New Zealand Journal of Archaeology, Special Publication.

Spriggs, M. 1997. *The Island Melanesians*. Oxford: Blackwell Publishers.

Spriggs, M. 1998. *The archaeology of the Bird's Head in its Pacific and Southeast Asian context.* In J. Miedema, C. Odé & R.A.C. Dam (eds), *Perspectives on the Bird's Head of Irian Jaya, Indonesia. Proceedings of the Conference, Leiden, 13-17 October 1997*: 931-939. Amsterdam: Rodopi.

Spriggs, M. 2000a. Out of Asia: The spread of Southeast Asian Pleistocene and Neolithic maritime cultures in Island Southeast Asia and the western Pacific. In S. O'Connor & P. Veth (eds), *East of Wallace's Line. Studies of past and present maritime cultures of the Indo-Pacific region. Modern Quaternary Research in. Southeast Asia* 16: 51-75. Rotterdam: A.A. Balkema.

Spriggs, M. 2000b. Can Hunter-Gatherers Live in Tropical Rain Forests? The Pleistocene Island Melanesian Evidence. In P.P. Schweitzer, M. Biesele & R.K. Hitchcock (eds), *Hunters and Gatherers in the Modern World. Conflict, Resistance, and Self-Determination*: 287-304. New York: Berghahn Books.

Spriggs, M. & D. Miller 1988. A previously unreported bronze kettledrum from the Kai Islands, Eastern Indonesia. *Indo-Pacific Prehistory Association Bulletin* 8: 79-89.

Stahl, P.W. 1995. Differential preservation histories affecting the mammalian zooarchaeological record from the forested neotropical lowlands. In P.W. Stahl (ed.), *Archaeology in the lowland American tropics. Current analytical methods and applications*: 154-180. Cambridge: Cambridge University Press.

Stephens, D.W. & J.R. Krebs 1986. *Foraging Theory*. Princeton: Princeton University Press.

Stiner, M.C. 1990a. The use of mortality patterns in archaeological studies of Hominid predatory adaptations. *Journal of Anthropological Archaeology* 9: 305-351.

Stiner, M.C. 1990b. The Ecology of Choice: Procurement and Transport of Animal Resources by Upper Pleistocene Hominids in West-Central Italy. Unpublished PhD thesis. Albuquerque: University of Mexico.

Stiner, M.C. 1991. An Interspecific Perspective on the Emergence of the Modern Human Predatory Niche: Actualistic and archaeological studies of prey mortality. In M.C. Stiner (ed.), *Human Predators and Prey Mortality*: 149-185. Boulder: Westview Press.

Stiner, M.C. 2001. Thirty years on the "Broad Spectrum Revolution" and paleolithic demography. *Proceedings of the National Academy of Sciences* 98 (13): 6993-6996.

Stiner, M.C., N.D. Munro, T.A. Surovell, E. Tchernov & O. Bar-Yosef 1999. Paleolithic Population Growth Pulses Evidenced by Small Animal Exploitation. *Science* 283: 190-194.

Straus, L.G. 1979. Caves: a palaeoanthropological resource. *World Archaeology* 10 (3): 331-339.

Stuijts, I.-L.M. 1993. *Late Pleistocene and Holocene vegetation of West Java, Indonesia. Modern Quaternary Research in Southeast Asia* 12. Rotterdam: A.A. Balkema.

Stuiver, M. & P.J. Reimer 1993. Extended ^{14}C database and revised CALIB radiocarbon calibration program. *Radiocarbon* 35: 215-230.

Stuiver, M., P.J. Reimer, E. Bard, J.W. Beck, G.S. Burr, K.A. Hughen, B. Kromer, F.G. McCormac, J. Van der Plicht & M. Spurk 1998. INTCAL98 Radiocarbon age calibration 24,000-0 cal BP. *Radiocarbon* 40: 1041-1083.

Summerhayes, G.R., J.R. Bird, R. Fullagar, C. Gosden, J. Specht & R. Torrence 1998. Application of PIXE-PIGME to Archaeological Analysis of Changing Patterns of Obsidian Use in West New Britain, Papua New Guinea. In M.S. Shackley (ed.), *Archaeological Obsidian Studies*: 129-158. New York: Plenum Press.

Swadling, P., J. Chappell, G. Francis, N. Araho & B. Ivuyo 1989. A Late Quaternary inland sea and early pottery in Papua New Guinea. *Archaeology in Oceania* 24: 106-109.

Tate, G.H.H. 1948. Studies in the Peramelidae (Marsupialia). *Bulletin of the American Museum of Natural History* 92: 313-346.

Tedford, R.H. 1966. A review of the macropodid genus *Sthenurus*. *University of California Publications in Geological Sciences* 57: 1-72.

Terborgh, J. 1971. Distribution on environmental gradients: theory and a preliminary interpretation of distributional patterns in the avifauna of the Cordillera Vilcabamba, Peru. *Ecology* 52: 23-40.

Terborgh, J. & J.S. Weske 1975. The role of competition in the distribution of Andeaen birds. *Ecology* 56: 562-576.

Terrell, J.E. & R.L. Welsch 1997. Lapita and the temporal geography of prehistory. *Antiquity* 71: 548-572.

Thomson, J.A. & W.H. Owen 1964. A field study of the Australian ringtail possum *Pseudocheirus peregrinus* (Marsupialia: Phalangeridae). *Ecological Monographs* 34: 27-52.

Thorne, A., R. Grün, G. Mortimer, N.A. Spooner, J.J. Simpson, M. McCulloch, L. Taylor & D. Curnoe 1999. Australia's oldest human remains: age of the Lake Mungo 3 skeleton. *Journal of Human Evolution* 36: 591-612.

Thunell, R, D. Anderson, D. Gellar & Q. Maio 1994. Sea surface temperature estimates for the tropical western Pacific during the last glaciation and their implications for the Pacific Warm Pool. *Quaternary Research* 41: 255-264.

Torgersen, T., J. Luly, P. de Deckker, M.R. Jones, D.E. Searle, A.R. Chivas & W.J. Ullman 1988. Late Quaternary environments of the Carpentaria Basin, Australia. *Palaeogeography, Palaeoclimatology, Palaeoecology* 67: 245-261.

Turney, C.S.M., M.I. Bird, L.K. Fifield, R.G. Roberts, M.A. Smith, C.E. Dortch, R. Grün, E. Lawson, L.K. Ayliffe, G.H. Miller, J. Dortch & R.G. Cresswell 2001. Early human occupation at Devil's Lair, southwestern Australia 50,000 years ago. *Quaternary Research* 55: 3-13.

Tyzzer, E.E. 1935-36. The "simple bone point" of the shell-heaps of the northeastern Algonkian area and its probable significance. *American Antiquity* 1: 261-279.

Ubelaker, D.H. 1978. *Human Skeletal Remains*. Chicago: Aldine Publishing Co.

Van Baal, J. 1966. *Dema. Description and analysis of Marind-anim Culture*. The Hague: Nijhoff.

Van der Kaars, W.A. 1991. Palynology of eastern Indonesian marine piston-cores: a Late Quaternary vegetational and climatic record for Australasia. *Palaeogeography, Palaeoclimatology, Palaeoecology* 85: 239-302.

Van der Kaars, S. 1995. Preliminary palynological results on the Pleistocene-Holocene transition, Seram Trench, offshore Irian Jaya, Indonesia (extended abstract). *Geologie en Mijnbouw* 74: 285-286.

Van der Kaars, W.A. & M.A.C. Dam 1995. A 135,000-year record of vegetational and climatic change from the Bandung area, West Java, Indonesia. *Palaeogeography, Palaeoclimatology, Palaeoecology* 117: 55-72.

Vanderwal, R.L. 1978. Exchange in prehistoric coastal Papua. *Mankind* 11: 416-428.

Van Deusen, H.M. 1972. Marsupials. In P. Ryan (ed.), *Encyclopaedia of Papua New Guinea*: 711-714. Melbourne: Melbourne University Press.

Van Huet, S., R. Grün, C.V. Murray-Wallace, N. Redvers-Newton & J.P. White 1998. Age of the Lancefield megafauna: A reappraisal. *Australian Archaeology* 46: 5-11.

Vanzolini, P.E. 1973. Paleoclimates, relief, and species multiplication in equatorial forests. In B.J. Meggers, E.S. Ayensu & W.D. Duckworth (eds), *Tropical forest ecosystems in Africa and Southern America: a comparative review*: 255-258. Washington: Smithsonian Institute Press.

Veth, P., M. Spriggs, A. Jatmiko & S. O'Connor 1998. Bridging Sunda and Sahul: The archaeological significance of the Aru Islands, southern Moluccas. In G.-J. Bartstra (ed.), *Bird's Head Approaches. Irian Jaya Studies – A programme for Interdisciplinary Research. Modern Quaternary Research in Southeast Asia* 15: 157-177. Rotterdam: A.A. Balkema.

Vickers, W.T. 1988. Game Depletion Hypothesis of Amazonian Adaptation: Data from a Native Community. *Science* 239: 1521-1522.

Visser, W.A. & J.J. Hermes 1962. *Geological results of the exploration for oil in Netherlands New Guinea. Verhandelingen van het Koninklijk Nederlands geologisch mijnbouwkundig genootschap, Geologische serie* 20, special issue.

Walker, D. & J.R. Flenley 1979. Late Quaternary vegetational history of the Enga Province of upland Papua New Guinea. *Philosophical Transactions of the Royal Society of London* Series B 286: 265-344.

Walshe, K. 2000. Carnivores, taphonomy and dietary stress at Puntutjarpa, Serpent's Glen and Intitjikula. *Archaeology in Oceania* 35 (2): 74-81.

Washburn, S.L. 1946. The sequence of epiphyseal fusion in the opossum. *Anatomical Records* 95: 353-363.

Watson, V.D. 1979. New Guinea Prehistory: A model of regional comparison. *Archaeology and Physical Anthropology in Oceania* 14 (2): 83-98.

Watson, V.D. & J.D. Cole 1977. *Prehistory of the Eastern Highlands of New Guinea.* Seattle: University of Washington Press.

Webb, C. 1987. Use-wear on bone tools. An experimental program and three case-studies from south-east Australia. Unpublished BA Honours thesis. Melbourne: La Trobe University.

Webb, C. & J. Allen 1990. A functional analysis of Pleistocene bone tools from two sites in Southwest Tasmania. *Archaeology in Oceania* 25: 75-78.

Webster, P.J. & N.A. Streten 1978. Late Quaternary Ice Age climates of tropical Australasia: interpretations and reconstructions. *Quaternary Research* 10: 279-309.

White, J.P. 1969. Typologies for some prehistoric flaked stone artefacts of the Australian New Guinea Highlands. *Archaeology and Physical Anthropology in Oceania* 4: 18-46.

White, J.P. 1972. *Ol Tumbuna. Archaeological excavations in the Eastern Central Highlands, Papua New Guinea. Terra Australis* 2. Canberra: Department of Prehistory, Research School of Pacific Studies, The Australian National University.

White, J.P. 1977. Crude, Colourless and Unenterprising? Prehistorians and their views on the Stone Age of Sunda and Sahul. In J. Allen, J. Golson & R. Jones (eds), *Sunda and Sahul: Prehistoric Studies in Southeast Asia, Melanesia and Australia*: 13-30. London: Academic Press.

White, J.P. & J. Allen 1980. Melanesian prehistory: some recent advances. *Science* 207: 728-734.

White, J.P. & J.F. O'Connell 1982. *A Prehistory of Australia, New Guinea and Sahul.* Sydney: Academic Press.

White, J.P., K.A.W. Crook & B.P. Buxton 1970. Kosipe: a Late Pleistocene site in the Papuan Highlands. *Proceedings of the Prehistoric Society* 36: 152-170.

Whittaker, J.L., N.G. Gash, J.F. Hookey & R.J. Lacey 1975. *Documents and Readings in New Guinea History. Prehistory to 1889.* Hong Kong: The Jacaranda Press.

Wickler, S. & M. Spriggs 1988. Pleistocene human occupation of the Solomon Islands, Melanesia. *Antiquity* 62: 703-706.

Wing, E.S. & A.B. Brown 1979. *Paleonutrition. Method and theory in Prehistoric Foodways.* New York: Academic Press.

Winter, J.W. 1980. Tooth Wear as an Age Index in a Population of the Brush-Tailed Possum, *Trichosurus vulpecula* (Kerr). *Australian Wildlife Research* 7: 359-363.

Wollaston, A.F.R. 1912. *Pygmies & Papuans. The stone age to-day in Dutch New Guinea.* London: Smith, Elder & Co.

Yen, D.E. 1995. The development of Sahul agriculture with Australia as bystander. In J. Allen & J.F. O'Connell (eds), *Transitions. Pleistocene to Holocene in Australia & Papua New Guinea. Antiquity* 69, Special number 265: 831-847.

Appendix 1: Human Remains from Kria Cave and Toé Cave, Papua, Indonesia

DAVID BULBECK
School of Archaeology and Anthropology, The Australian National University

1. INTRODUCTION

The sites of Kria Cave and Toé Cave, excavated by Juliette Pasveer for her PhD research in Archaeology, both yielded a modicum of human material. Pasveer identified this material as human and, after cleaning, weighing and preliminary anatomical identification, passed it onto me for the purpose of comparative study. Though the sample size is small, human remains from an archaeological context are to my knowledge as yet unknown from Papua, and so this contribution at least has the benefit of novelty. My report will proceed through the steps of describing the adult from Kria, the juvenile from Kria, and the minimum of five individuals from Toé.

2. KRIA CAVE

Two individuals can be unambiguously identified among the Kria Cave material owing to the duplication of particular anatomical elements which include both tibiae, one of the humeri, the right scaphoid, and the petrous part of the right temporal bone. The remains of these two individuals evidently overlapped in the sediments, based on the 'find' numbers, as the individual with the bigger elements is represented by find numbers ranging from 641 to 740, while the smaller individual is represented by duplicated elements with numbers between 643 and 1196. More to the point, elements attributable to the larger individual appear more weathered and mineralised than the fresh status of their smaller counterparts; the latter can clearly be ascribed to an adolescent who had died at about twelve years old as represented by jaws with the teeth in the process of eruption and long bones that lack fused epiphyses. After attributing all of the more weathered fragments that could reasonably be adult to the larger individual, we infer representation of that person between numbers 641 and 1194. Number 693 contained fragments from both individuals, and some other numbers probably do likewise.

The essentially coincidental status of the two Kria Cave individuals in terms of their stratigraphic context suggests the following scenario, especially when the freshness of the bone is also considered. An adult, evidently male (see below), was originally buried in the shelter. Some time later, perhaps after any grave marking had disappeared, the juvenile (who is probably also a male) was buried in the same spot. As a result, the skeletal parts of the two burials are intermingled. There is some evidence that the juvenile might have died from a blow to the face, and so this could have been a clandestine burial that disrupted the original grave.

The Kria Cave adult

The cranium is represented by most of the right temporal bone (number 669), numerous calvarial fragments (numbers 671, 675, 681, 688, 693, 706, and 1194) characterised by very thin inner and outer tables compared to diploe thickness, and a right upper canine (number 702). The mastoid process on the temporal bone fragment measures 21 mm in height, 18 mm in length and 13 mm in breadth (cf. Larnach & Macintosh 1966: 43), which would realise a 'small' mastoid process with its module estimated as 49.1. This feminine indicator (cf. Larnach & Freedman 1964) is outweighed by other hints of a male status (see below). The temporal fragment also contained the external auditory meatus which shows no traces of exostoses, the margin of a glenoid fossa whose shape would seem to have been shallow, and thick bone in the squamous part of the temporal. The scattered calvarial fragments vary between 4.5 and 10 mm thick, with the combined thickness of the inner and outer tables varying between 0.5 mm (on the thinner fragments) and 2 mm (on the thicker fragments), and the diploe making up the rest.

The canine, observed on the basis of a cast produced by Wallace Ambrose (Dept. of Archaeology and Natural History, the Australian National Unversity) from the mould made by Pasveer, is well worn [Smith's (1984) grade 4 – see Hillson 1996] with the slope of the wear plane slanting strongly towards the cervico-enamel junction on the lingual face. It is more worn than would be expected if it had belonged to the juvenile, nor does it fit into the socket on the juvenile's extant palate, so it can be assigned to the adult. Though we should note that the extent of occlusal wear makes morphological observations unreliable, neither shovelling nor a distal accessory ridge would seem to have been present. The tooth is single-rooted and quite large. Its crown diameters (Table 1) reflect a larger crown area than the average observed among males in central Australia, Melanesia or Indonesia (cf. Bulbeck 1981: 29).

Table 1. Kria Cave, diameters of the teeth (mm).

specimen	mesio-distal	bucco-lingual	mesiodistal, cemento-enamel junction	buccolingual, cemento-enamel junction
right C, adult	8.1	9.5	6.4	9.05
right P^1, juvenile	7.5	10.3	5.8	9.3
left P^1, juvenile	7.4	10.2	5.2	9.1
right P^2, juvenile	7.5	10.1	5.3	9.2
left P^2, juvenile	7.5	10.5	5.9	9.5
right M^1, juvenile	11.4	12.6	9.4	11.85
left M^1, juvenile	11.9	12.6*	8.7	12.6*
left M^2, juvenile	10.2	14.0	8.5	12.7
right M_1, juvenile	12.4	12.9	10.7	10.4
right M_3, juvenile	12.4	11.4	10.1	10.2

*note identical measurements

Find number 665 contains a limb-bone shaft fragment which, by a process of exclusion, cannot be ascribed to the Kria adolescent, because every anatomical part that could conceivably be related to the fragment is instead identifiable with a better match (see below). Accepting the status of this fragment as belonging to the adult, we may identify it as the disto-medial shaft of the humerus, between the distal portion of the (moderately rugged) deltoid tuberosity as far as the attachment area for the *infraspinatus* muscle. Number 693 contained a proximal fragment of the right scaphoid of adult size. Other sturdy if comminuted limb-bone skerricks that could relate to an adult femur, or the humerus represented by the fragment with number 665, were noted amongst fragments with numbers 643/646, 671, and 675.

The adult's tibiae (numbers 640, 641 and 650) are its best-represented parts, with much of the left tibia reconstructed from the tuberosity downward, and the right tibia represented by its proximo-medial and disto-medial segments of its shaft as well as a tuberosity or malleolus fragment. The right tibia was not measured because the left member provides a fuller and more reliable suite of measurements, taken according to the definitions in Martin & Saller (1957), and compared with Australian Aboriginal data provided by Rao (1966). As the tibia is not quite complete, its maximum length is estimated through Steele's (1970; Steele & McKern 1969) technique which divides the tibia into measurable segments that are then related back to the complete tibia through sets of regression formulae. Segment 2 of the Kria left tibia measures 183 mm, segment 3 measures 108 mm, and segment 4 measures 11 mm, all of which are of characteristically male rather than female size. They result in estimates of the maximum length of 400.2 ± 10.5 mm (Mesoamerican formula in Steele & McKern 1969), 399.1 ± 10.8 mm and 401.4 ± 9.6 mm (Afroamerican and White American formulae, respectively, in Steele 1970). The reliably estimated maximum length of c. 400 mm and the other tibia measurements are compared with Australian Aboriginal male averages in Table 2.

Rao (1966: 55-58) observed that Australian tibiae are long by the standards of other populations but gracile in their epiphysis and shaft dimensions, and that their extent of platycnemia is moderate. Table 2 shows that these same observations would apply to the Kria tibia which, in relation to its length, would appear slightly more gracile than even the average Australian Aboriginal tibia. Further, the shape of the Kria tibiae near the shaft midpoint is Hrdlička Type 2 on both sides, with the lateral surface hollowed out, a variant found with about 13% of Australian males but only some 3% of Australian females (see Rao 1966: 61).

It would be useful to provide an estimate of the stature of the Kria adult but unfortunately the regression formulae in the literature usually relate stature to total tibia length (Martin's no. 1) rather than maximum tibia length. Accepting that the total length of the Kria tibia would be around 391 mm (cf. Rao 1966: 269), we would estimate the person's stature as around 164.5 cm by Australian standards (Rao 1966: iii), 168.8 cm by Javanese standards (cf. Bergman & The 1955, subtracting 2 cm from the result to allow for the difference between corpse length and stature), or 173.0 cm by Sjøvold's (1990) 'race-generalised' formulae. The last estimate would appear too large; instead a stature of around 165 cm, based on the body build of recent populations in the vicinity of the Kria site, should be employed as our estimate of the individual's body size.

Table 2. Kria Cave, left adult tibia measurements (mm) compared with male Australian means.

measurement	Kria Cave	Australians
maximum length (Martin 1a)	~ 400	383.6 ± 19.8
distal epiphyseal breadth (Martin 4)	44.5	43.2
sagittal diameter at distal epiphysis (Martin 7)	34	36.05 ± 2.0
greatest diameter of shaft at mid-shaft (Martin 8)	31	30.8 ± 2.3
greatest diameter of shaft at nutrient foramen (Martin 8a)	28.5	34.6 ± 2.6
transverse diameter of shaft at mid-shaft (Martin 9)	~ 21	21.2 ± 1.7
transverse diameter of shaft at nutrient foramen (Martin 9a)	19	22.6 ± 2.2
circumference at mid-shaft (Martin 10)	~ 80	83.8 ± 5.8
least shaft circumference (Martin 11)	71	72.6 ± 4.9
platycnemic index (Martin 9a x 100/Martin 8a)	66.7	65.4 ± 5.3

Note: Australian data from Rao (1966: 55, 269-270); standard deviation for Martin 4 not supplied.

Finally, the distal shaft of the left tibia shows a fracture line along which is a series of impressions that, upon SEM scanning, proved to resemble pits left by large sharp teeth during a bite delivered with sufficient force to fracture the tibia. The most likely candidate to have left these tooth impressions would be a dog. This apparently occurred after the bone was already dry given the roughly compressed appearance of the sides of the pits, and the way the pits seem to break through other previously established surface markings such as localised fracture lines (created as the fresh bone dried out?) and possible root marks. Hence it would appear that the bite marks were created after the buried bone had been re-exposed on the surface, and indeed the dog's activity may have alerted its owner(s) to the presence of disturbed human material which was then re-interred with the juvenile. This scenario would be consistent with the presence of a wide suite of the adult's remains, from his head to his shins, including a bone as small as the scaphoid. The original burial may have been primary.

The Kria Cave juvenile

The Kria Cave juvenile is represented by fragments from every part of the skeleton and was surely a primary burial. Some tendency is evident for fragments from the same anatomical element to have come from the same or similar find number (Table 3) but with sufficiently wide variation to indicate the bones had lost their integrity by the time of excavation and, for that reason, were recovered as disjunct finds. The left femur is heavily marked with what appear to be scratch and gnaw marks, suggesting post-depositional molestation of the burial by local fauna, and this may have contributed to the disintegration of the anatomical elements.

The unfused status of numerous long-bone epiphyses that would fuse between the approximate ages of 12 and 22, or over shorter periods within this interval, demonstrate the individual's juvenile status (cf. Brothwell 1981), as does the state of the dentition with the P^2 barely erupted, the M^2 in the process of erupting, the M_3 crown completely formed, and enamel polishing on the other extant teeth (cf. Hillson 1996: 131-145).

Table 3. Kria Cave, probable/definite juvenile identifications (continued on p. 384).

anatomical element	portion	find numbers
calvarial	fragments (29, some with unfused sutures)	628, 643, 643/646, 648, 671
occipital	fragment	671
temporal	fragments (4)	643/646
temporal	right petrous part	1063
sphenoid	fragments (2)	643/646
maxilla	semi-complete	656
right mandible	posterior corpus/lower ramus	679/683 join
atlas	anterior arch + tubercle, superior articular facet	643/646, 671
axis	body + dens; transverse process + foramen	643, 643/646
cervical vertebrae	disc (2 fragments)	643/646, 671
cervical vertebrae	superior articular process	643/646
cervical vertebrae	superior articular process/transverse process	648
cervical vertebrae	articular process/superior articular surface	643/646
cervical vertebrae	articular process (5 fragments)	643
cervical vertebrae	lamina	643/646
cervical vertebrae	posterior tubercle	643/646
cervical vertebrae	spinous process (2 fragments)	643/646, 658
thoracic vertebrae	superior articular process	643/646
thoracic vertebrae	superior articular process/lamina (2 frags)	643/646
thoracic vertebrae	superior articular process/ pedicle (2 frags)	643/646
thoracic vertebrae	costal facet/pedicle	643/646
thoracic vertebrae	pedicle	643/646
thoracic vertebrae	spinous process (3 fragments)	643/646
thoracic vertebrae	transverse process (4 fragments)	643/646, 658, 671
thoracic vertebrae	lateral tubercle	671
thoracic vertebrae	inferior process	671
lumbar vertebrae	superior articular process (4 fragments)	643/646
lumbar vertebrae	superior articular process/mamillary process	671
lumbar vertebrae	lamina (2 fragments)	671, 678
lumbar vertebrae	mamillary process	648
lumbar vertebrae	pedicle	643/646
lumbar vertebrae	transverse process (4 fragments)	643/646, 658
lumbar vertebrae	inferior articular process	643/646
vertebrae	miscellaneous disc (9 fragments)	643/646
vertebrae	miscellaneous arch/process (24 fragments)	643/646, 658, 671
ribs	fragments (133)	637, 643, 643/646, 648, 649, 664, 666, 667, 671, 677, 679
right humerus*	virtually complete	653/663 join, 660, 662/664 join, 664
left humerus*	semi-complete	654, 664/668 join
left radius*	medial shaft, distal two-fifths of shaft	643
right radius*	semi-complete	643, 671, 678
radius	shaft, head	643/646, 693
left ulna*	olecranon and cap	643
ulna	shaft	643/646
right scaphoid	complete	693

Table 3 (continued). Kria Cave, probable/definite juvenile identifications.

anatomical element	portion	find numbers
left capitate	fragment	702
carpal	fragment	648
metacarpal*	2nd to 5th proximal shaft	671
manual phalanx*	semi-complete, possibly 3rd medial	693
manual phalanx*	distal shaft, 2nd or 4th medial	693
humerus/femur	head fragment	693
limb shaft	miscellaneous (72 fragments)	637, 643/646, 648, 671, 677, 693, 1196
metacarpal/metatarsal	shaft (3 fragments)	671
pelvic	right acetabulum + ischial tuberosity, ischial ramus, miscellaneous fragment	643/646, 656, 671
sacrum	fragment	648
right femur*	virtually complete	630, 651, 652/660 join, 656
left femur*	proximo-medial shaft, disto-medial shaft + epiphyseal surface	652, 653/655 join, 654, 661, 666
femur/tibia	proximal shaft	652
left patella	semi-complete	657
right tibia*	shaft at nutrient foramen, distal shaft + epiphyseal surface	643, 652, 666
left tibia*	dorso-lateral part of medial malleolus, posterior shaft, distal shaft + epiphyseal surface	652, 659, 1196
right fibula*	distal shaft + epiphyseal surface	643
left fibula*	head, shaft fragments	1196
right calcaneus*	almost complete	643
left talus	semi-complete	679
tarsal	fragment	678
left 2nd metatarsal*	distal shaft + epiphyseal surface	649
right 2nd metatarsal*	semi-complete	693/1193 join
left 3rd metatarsal*	semi-complete	643
right 4th metatarsal	proximal	637
left 5th metatarsal	proximal	643
metatarsal	shaft	693

Note: *epiphysis specifically observed lacking, or present but unfused with diaphysis.

The crown diameters are similar in overall size to those of the male Walpiri of Australia, exceeding the Walpiri mean by up to 1.2 mm or falling below the Walpiri mean by as much as 0.5 mm, and are usually larger than the male averages of other Australasian populations such as the Bougainville Nasioi and Batawi Indonesians (cf. Table 1, and Bulbeck 1981: 29). This comparison suggests the juvenile was a male. Further, the extant right femur suggests a femoral length of c. 400 mm, a value that would be rarely reached among pre-industrial populations until their late teens (cf. Brothwell 1981: Fig. 4.7). Estimated lengths of the right humerus of at least 270 mm and the right radius of over 200 mm also verge on the adult male range of populations in the region such as Australian Aborigines (cf. Van Dongen 1963: Fig. 1; Rao 1966: 265). Extant length of the 3rd (?) medial manual phalanx of 29.5 mm, and estimated length of the right 2nd metatarsal of 65-70 mm, also compare well with Australian male averages (cf. Rao 1966: Tables 1 and 2). Despite the patently gracile relief of the limb bones, and the small size of many post-cranial elements, the individual's status as a boy of around 12 years old seems secure, and even then he would seem to have been on the path towards growing into a tall man before his untimely death.

The maxilla suggests that a blow to the right side of the mouth might be implicated in the lad's demise. While the right P^2 is present as a loose tooth, the socket is about one millimetre too narrow to admit the root, suggesting the tooth had been expelled while the bone was still malleable, and retained in the mouth. The bone along the buccal border of the roots of all the anterior teeth and the right P^1 is lacking, and the right canine socket is deformed. Dehiscence along the entire arcade from the right P^1 to the left $\underline{C}$ would seem unlikely in a 12 year old with no other signs of oral pathology, especially as it would have to have been advanced dehiscence that had led to the shedding of all upper incisors and canines. More likely, the thin bone along the buccal margins of the roots of these teeth had been damaged through physical impact that had at the same time knocked out all the anterior teeth. The right M_1 is split through its upper crown in a manner that would also be consistent with violent impact, as there is no other possible evidence of the remains having been exposed to heat (the other conceivable cause for the fractured status of the first lower molar crown).

The lad's life prior to death was not entirely without incident either. Linear enamel hypoplasia is detectable about halfway up the crown of the M^1 and the right P^1, and the bottom third of the other three premolars, the first upper molars, and the third lower molar. By assuming slightly earlier eruption of the lower molars compared to the upper molars, these observations can be reconciled to suggest two interruptions to enamel formation corresponding to Hillson's (1996: 175) 'C' and 'G' events, at the approximate ages of two and six years old. The other observation to be made on the teeth is the formation of slight calculus (cf. Patterson 1984) on both interproximal faces of the right P^1 and right M_1 near the cemento-enamel junction, and comparable observations on the right P^2 distal and M^1 mesial crowns, left P^2 distal and M^1 mesial crowns, and right M^2 proximo-lingual crown (calculus formation on the right M^1 mesial crown is medium, and the slight calculus on the M^2 occurs in the upper half of the crown.) Mild calculus development of this nature is not unusual even for an adolescent, while its location reflects the lack of any appreciable atrophy of the gums by the time of death.

The fragments afford certain morphological observations, especially on the teeth (facilitated by the lack of occlusal wear). The right maxilla sports a trace of a palatine torus. An inverted gonion, distinct mylohyoid groove, and mylohyoid ridge of medium development are present on the right mandible fragment (cf. Larnach & Macintosh 1966, 1971). The lower narial margins are of non-anthropine form, the anterior nasal spine is absent (cf. Larnach & Macintosh 1966) and nasal breadth would have been c. 26 mm, assuming symmetry. These anatomical features may have changed had the individual survived into adulthood, whereas the dental features (Table 4), recorded with the Arizona State University system (Scott & Turner 1997), relate to the completed development of the teeth. (Root number was not observable on any of the molars.)

Table 4. Kria Cave, dental morphological observations on the adolescent teeth.

trait	right P^1	left P^1	right P^2	left P^2
accessory ridges	absent	absent	mesial	absent
accessory marginal tubercles	distal	absent	distal	mesial + distal
odontomes	absent	absent	absent	absent
disto-sagittal ridge	absent	absent	—	—
root number	?	single	single	single

trait	right M^1	left M^1	left M^2	right M_1	right M_3
hypocone (<u>M</u>)/hypoconulid	ASU 4	ASU 5	ASU 3	ASU 2	ASU 3
metaconule (<u>M</u>)/entoconulid	ASU 4	ASU 2	ASU 1	ASU 1	ASU 3
mesial marginal accessory tubercle (<u>M</u>)/metaconulid	absent	present	absent	ASU 4	ASU 4
Dryopithecus pattern	—	—	—	Y	X
Carabelli's cusp	ASU 4	ASU 4	ASU 7	—	—
parastyle (<u>M</u>)/protostylid	ASU 1	ASU 1	ASU 5	ASU 1	ASU 2
enamel extension	absent	absent	absent	ASU 1 (l)	ASU1 (b)
deflecting wrinkle	—	—	—	slight	absent
anterior fovea	—	—	—	absent	absent
distal trigonid crest	—	—	—	absent	absent

Note: (l) = lingual; (b) = buccal. ASU=Arizona State University system.

Calvarial thickness was observed to vary between 2 and 6.5 mm, although the larger values could correspond to fragments from the adult skull. A few notes on the post-cranial skeletal anatomy were also made. Specific observations on the bones' gracility include the minimal development of the *linea aspera* on the femora, the weakly developed deltoid tuberosity on the humeri, and the modest dimensions of the extant limb-bone epiphyses. Maximum length (Martin 1) and breadth (Martin 2) of the patella are c. 36 and c. 30 mm, respectively (minimum measurements in both cases).

Measurements on the scaphoid (*multangulum minus*, in the system of Martin & Saller 1957: 548-550) include 23 mm (Martin 1), 15 mm (Martin 2), 14 mm (Martin 3), 9.5 mm (Martin 4), 8 mm (Martin 5), 12 mm (Martin 6), 13 mm (Martin 7), 10 mm (Martin 8), 14.5 mm (Martin 10) and 9 mm (Martin 11).

3. TOÉ CAVE

Five individuals comprising two suspected adult males, a suspected adolescent male, a suspected female adult and a totally unsexable adult are tentatively identified at Toé Cave. One of the male adults (Individual 1) is represented predominantly by a semi-complete mandible with an intact right M_1, (found in a disturbed context outside the excavated area); the left P_1 found on the surface of the site, and the left lower C from Square 0N0E (number 143) have roots that slot into the mandibular sockets and so may belong to the same individual. Pelvic remains from Square 1N1E are clearly female and may well be associated with a relatively small left M^3 from a nearby spit in the same square (Individual 5). The two teeth at a depth of 45 to 60 cm assigned to Individual 2

Table 5. Toé Cave, attempted assignment of the *in situ* human remains to individuals ('Ind.')

depth (cm)	Square 2N1E	Square 1N1E	Square 0N0E	Square 1S0E
5-10		3 teeth – Ind. 2 cranial – Ind. 2?* post-cranial – Ind. 2??	tooth – Ind. 4	
10-15	cranial – Ind. 2?			
15-20		cranial – Ind. 2? post-cranial – Ind. 2?? tooth – Ind. 4?		
20-25	mandible/tooth - Ind.2 cranial – Ind. 2?			
25-30	date: 5290 ± 75 BP	cranial – Ind. 2?*		
30-35		cranial – Ind. 2?		tooth – Ind. 3#
35-40				2 teeth – Ind. 3# date: 2930 ± 65 BP
40-45			tooth – Ind. 1?	
45-50	tooth – Ind. 2??			
50-55		tooth – Ind. 4?		
55-60		tooth – Ind. 2?		tooth – Ind. 3
60-65	date: 8860 ± 70 BP	tooth – Ind. 5		
65-70				
70-75		pelvis – Ind. 5		

Note: * Fragments make a reasonable join. # Involves two molars whose interproximal facets nestle together. The mandible and P_1 of Individual 1 were found outside the excavated area.

(Table 5) evidently do not belong to Individual 5 as that assignment would involve duplication of the left M^3 and irregular tooth wear across the arcade, but they may belong to yet another, unnumbered individual(s). Similarly, the identity of Individual 4 based on three teeth spread over two squares and 50 cm of depth in the deposit is far from reliable. It would not seem to me unreasonable that a disturbed or secondary burial centred on Square 1N1E might have led to the associated human remains reaching from c. 5 to 50 cm depth in the deposit, with a small proportion protruding into Square 2N1E, but other interpretations are also feasible. Possibly we should take a conservative stance and conclude that there would be no solid grounds to infer a dating other than the late to mid-Holocene for the Toé Cave human remains, except possibly Individual 5 who could be of early to mid-Holocene antiquity (Table 5).

Individual 1

The mandible found outside the excavation area extends from the third molar socket on the left side to the gonion and lower ramus on the right. An age at death no more than 20 years may be inferred from the slight occlusal wear involving dentine exposure on only the mesiobuccal cusp (Smith's grade 3) on the *in situ* molar. The P_1 and lower canine that slot into the vacant sockets have similarly minimal wear (Smith's grade 2) and, like the first molar, are similar in size to the average dimensions recorded on Australian Walpiri males (Table 6; cf. Bulbeck 1981: 29). A masculine sex assignment to the mandible would be consistent with its bilaterally medium anterior marginal tubercle, and a bigonial breadth between 90 and 95 mm, even though the *sulcus intertoralis* is bilaterally absent, and symphysis height at only 27.5 mm is small (cf. Larnach & Freedman 1971). The specimen is accordingly treated on the balance of probabilities as a small, gracile mandible for a young adult male.

The mandible's alveolar crests are in fine condition, with nil to minimal blunting resorption, apart from slight atrophy at the canine and left P_1 sites. There are no fenestrations or infrabony pockets, and there is only slight inflammation of the interproximal septa, with surficial pitting observable bilaterally between the premolars, the second premolar and first molar, and the first and second molars. Slight calculus formation on the distal surface of the M_1 is the only trace of calculus or caries on the assigned teeth (see Patterson 1984 for pathology definitions). Macroscopic linear enamel hypoplasia (LEH) was not seen on any of the teeth.

In addition to the main anatomical traits (according to Bulbeck 1981) and measurements mentioned in Table 7, and the sexing features noted above, the following traits as defined by Larnach & Macintosh (1971) were observed. The gracile morphology of the lateral corpora includes bilateral absence of the superior lateral torus, absence of the marginal torus on the right side and its trace development on the left, and bilaterally trace development of the posterior marginal tubercle. The bilaterally single

Table 6. Toé Cave, diameters of the teeth (mm).

specimen	mesio-distal	bucco-lingual	mesiodistal, cemento-enamel junction	buccolingual, cemento-enamel junction
left I^1, Individual 4	(9.8)	8.2	8.4	7.6
left <u>C</u>, Individual 2	9.1	9.4	7.3	8.65
right lower C, Ind. 1	7.7	8.2	5.9	7.95
left lower C, Ind. 2	7.9	7.95	5.9	7.3
right lower C, Ind. 3	7.7	8.0	5.5	7.7
left P^1, Individual 3	7.9	10.4	5.1	9.8
left P_1, Individual 1	7.95	8.4	4.9	7.2
right P_1, Individual 2	7.9	9.0	5.6	7.4
right M^2, Individual 2	(10.2)	13.35	9.0	12.9
left M^3, Individual 2	8.0	10.85	6.3	10.5
left M^3, Individual 5	(7.6)	11.9	~6.5	11.5
right M_1, Individual 1	12.2	11.6	9.1	10.4
right M_1, Individual 3	12.6	11.2	10.5	10.2
right M_2, Individual 2	13.4	12.3	10.8	11.5
right M_2, Individual 3	12.3	10.65	9.8	9.5

Note: Mesio-distal measurements in brackets are appreciably affected by interproximal wear.

Table 7. Toé Cave, measurements (in mm) and main anatomical observations on the mandible.

bigonial breadth (g_og_o, M66)*	~92	sulcus extramolaris	absent L, R
gnathion-gonion length (g_ng_o)	76 R	lateral prominence	medium L, R
chin height (h_1, M69)*	26	trigonum mentale	medium
height at mental foramen (M69(1))*	26.5 L, 27.5 R	anterior mandibular incurvature	slight
thickness at mental foramen (M69(3))*	13.3 L, 14 R	planum alveolare decline	slight
bimental breadth (zz)	47	chin projection	neutral
height at second molar (m_2h)	24 L, 25.5 R	fossa precoronoidea	medium R
height between M_1 and M_2	26.8 L, 27.5 R	mylohyoid groove	medium R
thickness at second molar	18.3 L, 17.8 R	lingula	absent R
thickness between M_1 and M_2	17 L, 17.5 R	mylohyoid ridge	large L, R
molar-premolar chord (m_2p_1)	32.5 L, 31.5 R	superior lateral torus	absent L, R
interobliqual breadth	71		

Note: M-tagged measurements from Martin & Saller (1957: 481-482), including the four asterisked measurements used in the Penrose size and shape analysis. Alphanumerically tagged measurements, from the Biometric School, defined by Morant (1923). Measurements between M_1 and M_2 from Olivier et al. (1966), and interobliqual breadth from Jacob (1967). L = left side, R = right side.

mental foramen is located beneath the P_2 socket and is of 2.5 mm diameter. The symphysis region includes a large submental notch, absence of the *fossa mentalis* on the left side but its slight development on the right, absent to trace expression of the *trigonum basale*, a slight *torus transversus superior*, a double genial pit, a bilaterally slight digastric fossa, and bilaterally distinct *sulcus praedigastricus* (dual *sulci* on the right side). On the interior surface of the corpus we observe medium sublingual and submaxillary fossae on the left side but, on the right side, a large sublingual fossa and small submaxillary fossa, without any mandibular torus development. A sufficient portion of the right ramus is preserved to exhibit the large *sulcus colli*, the small triangular torus, and the slightly elevated *fossa masseterica*. According to the traits described by De Villiers (1968), the mental foramen is directed superiorly on the left side and posteriorly/superiorly on the right, the genial spine (a medium-sized mound) conforms to Type D, and a parasagittal groove is present on the right but there are no genial grooves on the left. The jaw is not a rocker jaw (cf. Houghton 1978), and there is no mylohyoid ridge (cf. Sawyer et al. 1990).

Individual 2

The second Toé individual was defined on the presence of a right mandibular corpus fragment with the M_2 present and margins of the sockets of the first and third molars (Square 2N1E). It is a smaller corpus than Individual 1's, with heights of 24.75 mm and 22.5 mm between the M_1 and M_2 and at M_2 respectively, and a thickness of 15 mm at both locations (cf. Table 7). The extant mylohyoid ridge is huge, and the *sulcus extramolaris* shows trace development. Occlusal wear on the M_2 involves early dentine exposure on the buccal cusps (Smith 3). A right M^2, one of the three teeth found near the top of Square 1N1E, occludes comfortably with the M_2 and, despite slightly greater wear (Smith 4), would seem to belong to the same individual. Linear enamel hypoplasia (LEH) is observable on the cervical half of the crown on both teeth. LEH is also present on the cervical third of the crown of the two canines (upper and lower left) excavated with the right M^2, with all four events corresponding to a Hillson (1996: 175) 'F' event that would have represented an interruption to enamel formation at an age of around 3½ years old. The canines, which show Smith wear 3 (lower) and 4 (upper), also occlude neatly together. These related remains constitute the nucleus of Individual 2, a young adult of around 20 years at the time of death, and probably male from the tooth diameters which essentially match the average crown dimensions of Walpiri males (cf. Table 6 and Bulbeck 1981: 29).

Two further teeth may belong to the same person. Square 2N1E produced a lower premolar, probably the right P_1, from a level in between the middle and early Holocene dates in the square (Table 5). The tooth has been sacrificed for DNA analysis and, as my observations come from the cast made by Wallace Ambrose, there may be insufficient resolution to decide whether the lack of observable, macroscopic enamel hypoplasia is a reliable feature. The extent of occlusal wear (Smith 4), crown dimensions which slightly exceed male Walpiri averages, and lack of caries (none were noted on the teeth described above), would all be consistent with attribution to Individual 2. The second tooth involved here is also represented by a cast, a left M^3 from 55-60 cm depth in Square 1N1E. Its anomalous variation of a single root is consistent with the condition of the right M^2 which, having two roots, also displays a reduction in root number; other morphological observations are also consistent with assigning these two teeth to the same person (Table 8). Its slight occlusal wear (Smith's grade 2) and lack of caries are also consistent with attribution to Individual 2. Its crown dimensions are rather small, and best match those of Bougainville Nasioi females or Batawi Indonesian males (cf. Table 6 and Bulbeck 1981: 29), but as is well known, the third molars are the most variable teeth in the human dentition in their size. Macroscopic LEH is evident approximately one-third way up the extant crown and so, if this tooth is correctly attributed, Individual 2 would have experienced a second interruption to enamel formation at an age of around seven years old (cf. Hillson 1996).

The mandible fragment was associated with cranial fragments which may also be reasonable attributed to Individual 2. One is the right inferior parietal in the region of asterion, broken along its largely fused sutures, lacking any trace of an angular torus, of 3.4-6.5 mm thickness, and with normal development of the diploe and plates in relation to each other. The second is the partial right malar including the malar-maxillary suture which follows a strongly everted course. Malar tuberosity development is medium, consistent with Individual 2's assigned male status. Another anatomical trait coined by Larnach & Macintosh (1966) is the rounded status of the orbital border. An interesting feature involves parallel striations across the suture and at the point of maximum malar convexity, presumably the result of blood-vessel impressions.

Find numbers 49, 53, 60/76, 79, and 204 from the upper spits of Squares 2N1E and 1N1E all include cranial material that could be reasonably ascribed to Individual 2. The left superciliary ridge in 60/76 makes a passable if uncertain join with a left zygomatic

Table 8. Toé Cave, dental morphological observations on the teeth (ASU system).

trait	left $\underline{C}$ (Ind. 2)	right lower C (Ind. 1)	left lower C (Ind. 2)	right lower C (Ind. 3)
shovelling (inc. double shovelling)	absent	absent	absent	absent
tuberculum dentale	small	medium	small	small
distal accessory ridge	absent	absent	absent	absent
mesial canine ridge	absent	–	–	–
root number	single	single	single	single

trait	left P^1, Ind. 3	left P_1, Ind. 1	right P_1, Ind. 2
accessory ridges	mesial + distal	absent	distal
accessory marginal tubercles ($\underline{P}$)/ multiple lingual cusps (lower P)	absent	absent (ASU 0)	?
odontomes	absent	absent	absent
disto-sagittal ridge	absent	–	–
root number	double	single	single

trait	right M^2, Ind. 2	left M^3, Ind. 2	left M^3, Ind. 5
hypocone	ASU 2/3	ASU 0	?
metaconulid (cusp 5)	?	ASU 0	?
Carabelli's anomaly	ASU 0	ASU 0	ASU 0
mesial marginal tubercle	?	ASU 0	?
parastyle	ASU 2	ASU 0	?
enamel extension	ASU 0	ASU 0	ASU 0
root number	two	single	single

trait	right M_1, Ind. 1	right M_1, Ind. 3	right M_2, Ind. 2	right M_2, Ind. 3
cusp 5	ASU 4	ASU 3	?	ASU 4
Dryopithecus	Y	Y	+	+
protostylid	ASU 0	ASU 1	ASU 0	ASU 1
enamel extension	ASU 2 lingually, ASU 1 buccally	ASU 2 lingually, ASU 0 buccally	ASU 0	ASU 1 buccally and lingually
root number	?	two	two	two

Note: Lower molars too worn to retain most ASU occlusal crown traits. The mandible sockets of Individual 1 reflect two-rooted lower molars at every site except right M_1 (unobservable).

trigone fragment in number 49. This would produce a supraorbital morphology that combines medium superciliary ridge and medium zygomatic trigone development, both of which are acceptably masculine traits even by male Australian Aboriginal standards (cf. Larnach & Macintosh 1966), as well as a supraorbital notch with a single foramen in it. The small extent of orbital roof present evinces no trace of *cribra orbitalia*. A maxilla fragment with sockets for the posterior teeth protruding into the maxillary sinus was found in number 204. Of the available Toé upper molars, only Individual 2's right M^2 would fit in the socket, confirming the coherence of the association. The morphology of the fragment suggests a tall palate shape. The right glenoid fossa and root of the zygomatic arch labelled 'find 79' could also be attributed to a young adult male as the glenoid fossa is quite large (c. 22.5 mm long by 12 mm wide) albeit shallow (c. 8.5 mm) and with no traces of erosion or hypertrophy of the surface, and the fragment has not yet fused to the tympanic part of the temporal. In addition we have seven cranial fragments (one occipital, one temporal) in find number 49, seven cranial fragments (one posterior temporal) in number 53, and a thin vault fragment (possibly from the frontal) in number 204. Thickness of these fragments varies between 2.5 and 8.1 mm, and the diploe makes up between 50% and 90% of skull thickness.

The following post-cranial fragments may also be attributable to Individual 2. Number 49: partial patella of adult size; part of the disc of a cervical or upper thoracic vertebra; sacrum fragment; humerus head fragment; femur shaft fragment; possible humerus shaft and tibia shaft fragments; an unidentifiable fragment, possibly from a limb bone; and a metatarsal distal shaft fragment with the articular surface for the epiphysis. The epiphysis would appear not to have fused but, as this may occur at any age between 12 and 22 years old (Brothwell 1981), it is not inconsistent with belonging to Individual 2. Number 53: a femur shaft fragment and a left humerus distal shaft fragment, both of apparently adult dimensions; and a possible forearm shaft fragment. As with the calvarial fragments, there is some variation in the condition of the post-cranial material, ranging from fresh-looking to chalky and weathered, so it is conceivable that a second person (Individual 4?) could also be represented.

Individual 3

The four teeth from Square 1S0E are all compatible with having come from the same dental arcade. Two of them, the right lower canine and left P^1, were used for DNA analysis and so the observations provided here are based on Wallace Ambrose's casts. Root formation would not appear quite completed on the left P^1, which accordingly lacks any trace of occlusal wear, and the minimal wear on the other teeth (Smith 3 on the right M_1, and Smith 2 on the canine and right M_2), would also be compatible with an age at death of about 11-12 years old. Crown dimensions, comparable to the averages of Walpiri males or, in the case of the canine, Walpiri females or Bougainville Nasioi males (Table 6; cf. Bulbeck 1981: 29), hint at a male status. Macroscopic LEH was not observed on either crown nor on the casts. Caries is absent though there is slight calculus formation at the central level of the buccal face of the M_1 crown. The lower teeth do not fit into the sockets of the Individual 1 mandible, nor does the P^1 occlude with the left P_1 assigned to that mandible, and so the separate status of this individual, as well as its coherence, seems a reliable inference.

Individual 4

Three well-worn teeth are very tentatively attributed to an older individual. This tentativeness is particularly the case for the two-rooted lower molar from near the top of Square 0N0E. One of the roots is fully broken off, possibly reflecting a combination of atrophy through oral disease (and suggesting the tooth was shed from a living person), and/or post-depositional damage when the tooth lay on the shelter floor. The occlusal surface was well worn, even though the single extant fragment of enamel would be compatible with Smith's stage 5 or 6 of wear. The rationale for potentially relating it to the left upper central incisor from 50-55 cm depth in Square 1N1E is the red colouration that affects both roots, and the similarly advanced occlusal wear (Smith 6 on the incisor). This colouration could, however, simply reflect staining through length of use and, in particular, staining from betel-nut chewing. The incisor retains enough of the labial surface of the crown to display a red staining pattern that is typical of anterior teeth among habitual chewers of the quid. A third tooth tentatively assigned to Individual 3, a (probable) third upper molar from 15-20 cm depth in Square 1N1E, does not show a stained root, but the third molars are located too posteriorly to be affected by betel-nut colouration.

The third molar has only two roots, and these had formed part of the plane of attrition which slants very strongly from the sole remaining trace of enamel in one corner of the tooth (Smith wear 8). The wear plane on the incisor also slants strongly, in this case directed upwards lingually, related to overbite. The buccal crown clearly displays two macroscopic LEH lines at 4.0 and 5.7 mm above the cervico-enamel junction, corresponding to 'B' and 'C' events in Hillson's (1996: 175) scheme, and evidently reflecting interruption to enamel formation at around two to three years old. The nature of occlusal wear makes observations on the crown morphology difficult but shovelling would seem to have been absent.

Individual 5

The lowest of the excavated teeth would appear to be a left M^3 and, as Individual 2 already has that tooth, a different person is in evidence. Wear is quite advanced (Smith's stage 5) and a medium carious lesion (Patterson 1984) has developed at the proximal cervico-enamel junction. While this is the sole case of caries recorded in the Bird's Head collection, this point may reflect the youthful status of most of the buried persons rather than sound oral health. Slight calculus formation is present approximately 2.1 mm from the cervico-enamel junction on the root, indicative of degree '1' blunting resorption (cf. Patterson 1984). A line of macroscopic LEH is detectable about 1.7 mm from the cervico-enamel junction, suggestive of an interruption to enamel formation at around seven years old (cf. Hillson 1996). The mesio-distal diameter is significantly reduced through mesial and occlusal wear, but the bucco-lingual diameter (Table 6) is similar to the Walpiri female mean (cf. Bulbeck 1981: 29). Therefore the tooth is possible female.

A female status would appear certain for the left pelvic fragment – found about 10 cm beneath the tooth – which incorporates the junction of the auricular facet, the arcuate line as far as the start of the iliopubic eminence, and the acetabulum. The greater sciatic notch is markedly broad, measuring around 105°, a typically feminine trait (Brothwell 1981). Impressionistically the size is adult, but with a gracility (e.g. the arcuate line and inferior iliac spine) that would be compatible with a woman's.

4. POPULATION CHARACTERISTICS OF THE HUMAN MATERIAL FROM KRIA AND TOÉ CAVES

The material described here is mainly, if not entirely, of mid- to late Holocene antiquity. Although the single available estimate of stature, around 165 cm for a male, suggests a relatively tall population, muscular markings on the limb bones are always modest, especially (if expectedly) on the adolescent. The population would appear to have been susceptible to metabolic insults resulting in interruptions to enamel formation at any age between two and seven years old, as the majority of individuals show signs of macroscopic linear enamel hypoplasia, and these physical upsets during development may be related to the gracile post-cranial morphology. The teeth would appear to have been quite large, with dimensions akin to those of the Walpiri, one of the smaller-toothed Australian Aboriginal populations (cf. Bulbeck 1981). Nonetheless, the sole semi-complete mandible seems to be smaller than the average male mandible found in surrounding Australasian populations, although groups like the Atayal of Taiwan and the Biakers of Papua tend to have similarly small mandibles (see below). The observed dental morphological features show slightly greater affinities with southwest Pacific than with Southeast Asian populations, whilst basically conforming to the transitional pattern that would be consistent with the location of Papua near Melanesia's border with Southeast Asia (Table 9).

Table 9. Dental morphology assessed in terms of Scott & Turner's (1997) break points.

tooth/trait	Kria expression	Toé expression	Southwest Pacific	Sunda-Pacific
I^1	—	unshovelled?	+ +	+
$\underline{C}$ mesial ridge	—	absent	+ +	+
C root number	—	single	+ +	+
premolars	no odontomes	no odontomes	+ +	+ +
P^1 root number	single	double	+ +	+ +
P Tomes' root	—	absent	+	+
M^1 Carabelli	absent	—	+	+
M^1 cusp 5	present	—	+ —	—
M^2 cusp 5	—	present	+	—
M^2 hypocone	4-cusped	4-cusped	+ +	+
M^2 root number	—	double	—	—
enamel extension	absent	present on 2/6 molars	+ —	+ +
M_1 cusp 5	present	present	+	+ +
M_1 cusp 6	present	—	+ —	+ —
M_1 cusp 7	present	—	—	— —
M_1 def. wrinkle	present	—	—	—
M_1 trigonid	absent	—	+ +	+
M_1 root number	—	double	+ +	+
M_2 pattern	—	+	+	+
M_2 root number	—	double	+ +	+
overall			26 +, 6 —	19 +, 7 —

Note: ++: strongly characteristic of populations in region. +: compatible with region's populations. + —: ambiguous expression. —: unusual for region's populations. — —: very unusual for region's populations. def. wrinkle: deflecting wrinkle; trigonid: distal trigonid crest.

The skerricks of cranial material display anatomical features fully consistent with those of southwest Pacific populations, including rounding of the malar orbital border, medium superciliary ridge and zygomatic trigone, evidence of palatine torus formation, and (based on the Kria individual) non-anthropine lower narial margins and absent to weak anterior nasal spine development (cf. Larnach & Macintosh 1966; Bulbeck 1981).

The mandibular morphology is also typical of southwest Pacific people in features such as the pronounced mylohyoid ridge and distinct mylohyoid groove development (cf. Larnach & Macintosh 1971; Bulbeck 1981). The latter point can be demonstrated formally on the semi-complete mandible from Toé Cave, which preserves the nine features recommended by Bulbeck (1981) to distinguish Melanesian and Southeast Asian mandibles. Its overall score of 23, based on the expressions listed in Table 7, falls near the recorded Melanesian range, and at the upper limits of the Neogene Southeast Asian range recorded in the Philippines, Sulawesi, Melanesia and Java; only the prehistoric eastern Indonesian specimens from Lemdubu and Golo have a more strongly Melanesian morphology (Figure 1, adapted from Bulbeck 2001, presented here with Peter Bellwood's permission).

An attempt was also made to assess the biometric affinities of the Toé mandible with the Penrose (1954) size and shape statistic, interpreting the shape distances through average-linkage clustering and seriation of the resulting dendrogram (see Bulbeck 2000). As Snell (1938) has measured a sample of around 30 Biaker male mandibles from the same part of the world as the Toé site (Papua), I limited the analysis to the four measurements recorded both by Snell and available for Toé (asterisked in Table 7). I designed a procedure to calculate the Penrose distances using an Approach database combined with an Excel worksheet, and used the square roots of the size distances (retaining information on which sample has the larger mandibles) and of the shape distances too. The comparative samples are male Hainan Chinese and Atayal Taiwan Aborigines (Tokitsu 1960), Javanese and Biakers (Snell 1938), Leang Buidane Iron Age, New Guinea and New Britain (Bulbeck 1981), and Australian Aborigines (Keiter 1934). Sample sizes range from as few as four (Leang Buidane) to as many as 164-217 (Hainan), but generally range between 12 and 45. The results are presented in Table 10 and Figure 2.

All samples had larger mandibles on average than Toé, with their greater size varying between 0.195 (Atayal), 0.274 (Biakers), 0.619 (New Guinea), 0.784 (Leang Buidane), 0.866 (Javanese), 0.947 (Australians), 1.372 (Hainan), and 1.451 (New Britain). In terms of shape, the Toé mandible was more or less equidistant from all comparative samples (Table 10, top right half-matrix). Seriation pushed the specimen to the same part of the dendrogram as Leang Buidane and Atayal Aborigines, but little should be made of this, as it is impossible to distinguish any separation between East Asian and southwest Pacific samples based on the results (Figure 2). The poor goodness of fit (only 72.4% for merely four measurements) further indicates a lot of multi-dimensional variation that cannot be well represented in the single dimension allowed by seriation. This equivocal result may reflect the small sample sizes often available, an unfortunate availability of measurements that are poorly tailored to reflect Australasian population variation, stochastic local population variation, or some combination of these factors.

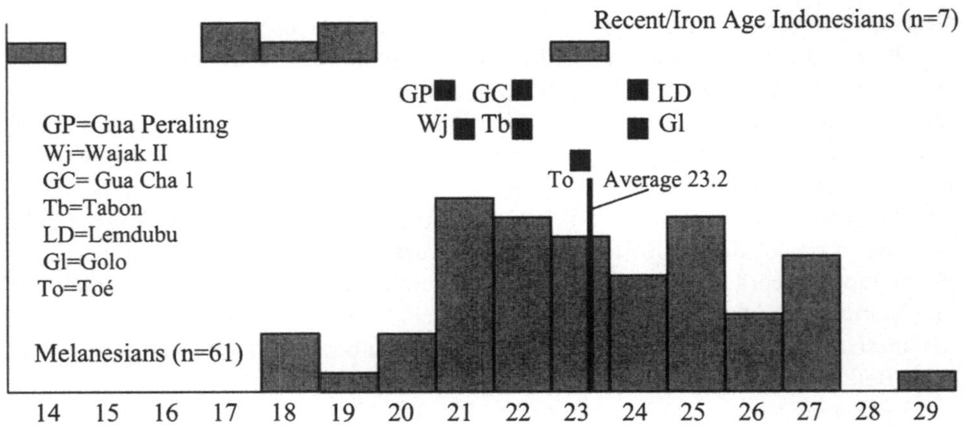

Figure 1. Toé Cave plus Comparative Mandibular Specimens in the Bulbeck 11-trait List.

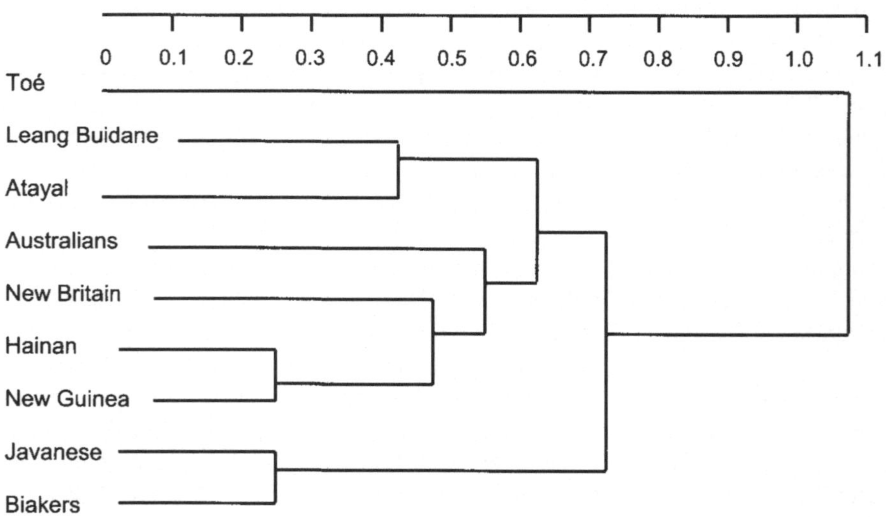

Figure 2. Seriated Average-Linkage Dendrogram of Square Roots of Penrose Shape Distances based on four Mandibular Measurements. Goodness of Fit 72.4%.

Table 10. Seriated square roots of Penrose shape distances based on bigonial width, chin height, and height and thickness at the mental foramen. Goodness of fit 72.4%. See text.

	Toé Cave	L.Buidane	Atayal	Australia	N.Britain	Hainan	N.Guinea	Java	Biaker
Toé Cave	–	1.167	1.172	1.039	0.734	1.079	1.287	1.229	1.053
L.Buidane	1.546	–	0.422	0.758	0.560	1.261	0.514	0.730	0.854
Atayal	1.779	0.788	–	0.467	0.601	0.362	0.529	0.921	0.995
Australia	1.546	1.387	0.979	–	0.522	0.338	0.727	0.877	0.853
N.Britain	1.359	1.276	1.569	1.335	–	0.399	0.555	0.591	0.522
Hainan	1.801	2.589	0.852	0.779	1.145	–	0.266	0.609	0.661
N.Guinea	2.148	1.055	1.244	1.675	1.592	0.688	–	0.477	0.621
Java	1.792	1.295	1.893	1.766	1.481	1.376	1.077	–	0.261
Biaker	1.502	1.499	2.001	1.681	1.280	1.461	1.372	0.504	–

It may be observed that Toé's shape distance from Biakers is rather low given that Biakers also tend to show a divergent shape (Table 10). However, when the shape distances are double standardised, i.e. divided by the product of the average distance for the two specimens/samples in the comparison, Toé and Biakers do not show any observable proximity (Table 10, bottom left half-matrix). The Toé mandible is closest to the New Britain mandibles as is apparent both from the original and the double-standardised shape distances. For instance, when the latter are clustered using average linkage, Toé and New Britain would find each other at a (double-standardised) distance of 1.359, so if these results were seriated, New Britain would be drawn to the side of Toé. The specific similarity between Toé and New Britain would seem to be a relatively broad corpus at the mental foramen (cf. Bulbeck 1981: 385). To draw an affinity between Toé and New Britain on that basis would of course be nonsense.

As the Toé mandible is a single specimen, its small size and rather anomalous shape can be readily attributed to idiosyncratic variation. Rather large tooth size, more of Australian than typically Melanesian dimensions, is the only feature of the material that would seem out of place in a zone near the border between Melanesia and Southeast Asia. A larger sample would be required to confirm the reliability of this observation.

REFERENCES

Bergman, R.A.M. & T.H. The 1955. The length of the body and long bones of the Javanese. *Documenta de Medecina Geographica et Tropica* 7: 197-214.

Brothwell, D.R. 1981. *Digging up Bones*. Third edition. Oxford: British Museum (Natural History)/ Oxford University Press.

Bulbeck, F.D. 1981. Continuities in Southeast Asian Evolution since the late Pleistocene. Unpublished MA thesis. Canberra: The Australian National University.

Bulbeck, D. 2000. Multivariate statistical analysis of the Gua Peraling 4 cranium. In Adi bin Haji Taha, Archaeological Investigations in Ulu Kelantan, Peninsular Malaysia. Appendix 7. Unpublished PhD thesis. Canberra: Australian National University.

Bulbeck, D. 2001. The Golo Burial and its Relationship with other Moluccan Prehistoric Human Remains. Consultancy report prepared for the project 'Archaeological Survey and Excavation in the Halmahera Islands, Moluccas, Indonesia'.

De Villiers, H. 1968. *The Skull of the South African Negro: A Biometrical and Morphological Study*. Johannesburg: Witwatersrand University Press.

Hillson, S. 1966. *Dental Anthropology*. Cambridge: Cambridge University Press.

Houghton, P. 1978. Polynesian mandibles. *Journal of Anatomy* 127: 251-260.

Jacob, T. 1967. *Some Problems Pertaining to the Racial History of the Indonesian Region*. Utrecht: Netherlands Bureau for Technical Assistance.

Keiter, F. 1934. Unterkiefer aus Australien und Neuguinea aus dem Nachlasse Rudolf Pöchs. (Details of journal unfortunately lost.)

Larnach, S.L. & L. Freedman 1964. Sex determination of crania from coastal New South Wales. *Records of the Australian Museum* 26: 295-308.

Larnach, S.L. & N.W.G. Macintosh 1966. *The Craniology of the Aborigines of Coastal New South Wales*. *Oceania Monographs* 13. Sydney: University of Sydney.

Larnach, S.L. & N.W.G. Macintosh 1971. *The Mandible in Eastern Australian Aborigines*. *Oceania Monographs* 17. Sydney: University of Sydney.

Martin, R. & K. Saller 1957. *Lehrbuch der Anthropologie*. Fifth edition. Stuttgart: Gustav Fischer Verlag.

Morant, G.M. 1923. A first study of the Tibetan skull. *Biometrika* 14: 193-260.

Olivier, G., G. Phalippou & H. Tissier 1966. Craniométrie des Indochinois. *Bullétins et Mémoires de la Société d'Anthropologie de Paris* 11: 67-90.

Patterson, D.K. jr. 1984. *A Diachronic Study of Dental Paleopathology and Attritional Status of Prehistoric Ontario Pre-Iriquois and Iriquois Populations*. Ottawa: National Museums of Canada.

Penrose, L.S. 1954. Distance, size and shape. *Annals of Eugenics* 18: 337-343.

Rao, P.D.P. 1966. The Anatomy of the Distal Limb Segments of the Aboriginal Skeleton. Unpublished PhD thesis. Adelaide: Department of Anatomy, University of Adelaide.

Sawyer, D.R., V. Gianfortune, M.L. Kiely & M.J. Allison 1990. Mylohyoid and jugular foramen bridging in pre-Colombian Chileans. *American Journal of Physical Anthropology* 82: 179-181.

Scott, G.R. & C.G. Turner II 1997. *The Anthropology of Modern Human Teeth*. Cambridge: Cambridge University Press.

Sjøvold, T. 1990. Estimation of stature from long bones utilizing the line of organic correlation. *Human Evolution* 3: 431-447.

Smith, B.H. 1984. Patterns of molar wear in hunter-gatherers and agriculturalists. *American Journal of Physical Anthropology* 63: 39-56.

Snell, C.A.R.D. 1938. *Menschelijke Skeletresten uit de Duinformatie van Java's Zuidkust nabij Poeger (z-Banjoewangi)*. Surabaya: G. Kolff.

Steele, D.G. 1970. Estimation of stature from fragments of limb bones. In T.D. Stewart (ed.), *Personal Identification in Mass Disasters*: 85-97. City of Washington: National Museum of Natural History, Smithsonian Institute.

Steele, D.G. & T.W. McKern 1969. A method for assessment of maximum long bone length and living stature from fragmentary long bones. *American Journal of Physical Anthropology* 31: 215-228.

Tokitsu, K. 1960. Anthropological studies on the palatum, the mandible and the teeth of the Atayals (Musha), Taichu Prefecture, Formosa. *Jinruigaku-Kenkyu* 7, supplement: 37-132.

Van Dongen, R. 1963. The shoulder girdle and humerus of the Australian Aborigine. *American Journal of Physical Anthropology* 29: 469-488.

Appendix 2: Archaeobotanical Report on Fruits from Kria Cave, Bird's Head, Papua, Indonesia

KATHLEEN McCONNELL

Formerly: Department of Archaeology and Natural History, Research School of Pacific and Asian Studies, The Australian National University

1. INTRODUCTION

A small collection of archaeobotanical materials were excavated from a rockshelter site named Kria Cave, located on the Bird's Head of Papua, by Juliette Pasveer (then University of Groningen, The Netherlands). Ms Pasveer sent samples to the ANU for possible identification including Kria 991, 1012, 1024, 1025, 1125. Most samples consisted of fragments, but Kria 1125 contained half of one specimen and Kria 1024 contained around a third of one specimen.

2. METHODS

The materials were examined with a low power microscope. Specimen Kria 1125 was coated with gold and examined under the JEOL 6400 Scanning Electron Microscope. Photos by the author.

Plant lists were obtained from other locations in Papua New Guinea, due to the lack of recent, comprehensive plant lists specifically for the Bird's Head area. Sources include an ethnobotanical study of the Jimi Valley (Gorecki 1989) and a review of montane rainforests covering the Fatima Basin and Mt Kerigomna (Grubb & Stevens 1985). Grubb and Stevens' work includes references to, and incorporates the first vegetation review ever done in the Bird's Head by Gibbs (1917) that was otherwise unavailable. An article by Bulmer on foods that could be processed with grinding stones proved useful in providing ethnobotanical suggestions (Bulmer 1964). The flora used was *The Alpine Flora of New Guinea* (Van Royen 1982).

CSIRO Herbarium collections, indicated by prefixes CANB and CBG, were used for comparative purposes. Potential sources were identified from Gorecki's list, using the criteria of small, edible fruits from a highland location. Known edible fruit with multiple carpels, such as Rutaceae and Euphorbiaceae were eliminated. Final sources considered were the *Syzygium* spp., Myrtaceae, known edible Elaeocarpaceae, some Fagaceae and Pandanaceae. Botanical descriptions and terms are taken from Mauseth (1988) and Roth (1977).

3. DESCRIPTION OF FRUITS

The botanical specimens can be described as fruits, 1 - 1.5 cm in diameter and slightly longer, so oval in shape with a slightly pointed apex. The epidermis cannot be distinguished clearly, but the subepidermal cell layers are visible as smaller, thin-walled cells extending around five rows into the fruit (Roth 1977: 374-375; see Fig. 1). The mesocarp comprising the flesh of the fruit consists of tightly packed, larger ovoid cells (Mauseth 1988: 440; see Fig. 2). Kria 1024 shows that a thick network of vascular bundles extends into the mesocarp to the endocarp.

One almost complete endocarp, consisting of one carpel, elliptical in shape and around 2.5 mm in width, was recorded from Kria 1125 (Roth 1977; Mauseth 1988; see Figures 3 and 4). The two margins of the carpel appear not to be joined and, it is this characteristic which suggests the fruit is a pome, although, the search was not limited to pome fruits.

Syzygium sp. fruit were considered but the structure of the fruit differed so markedly that it is clearly not the source. Lynn Craven of the CSIRO Herbarium demonstrated that the fruits of *Syzygium* are dehiscent while the archaeobotanical sample is not (L. Craven pers. comm.). *Syzygium* sp. (species novae) Sankowsky 1592, cultivated on the Atherton Plateau and being studied by Lynn Craven was examined.

Elaeocarpus spp. fruits have been recorded as eaten in the highlands (Bulmer 1964: 148; Paijmans 1976: 110). The Elaeocarpaceae that were examined included *Elaeocarpus bancroftii* F. Mueller & Bailey, CBG 9102499 Sankowsky, and CANB 62014 Carr. The fruits are similar in size, 10-12 mm in diameter, but the internal

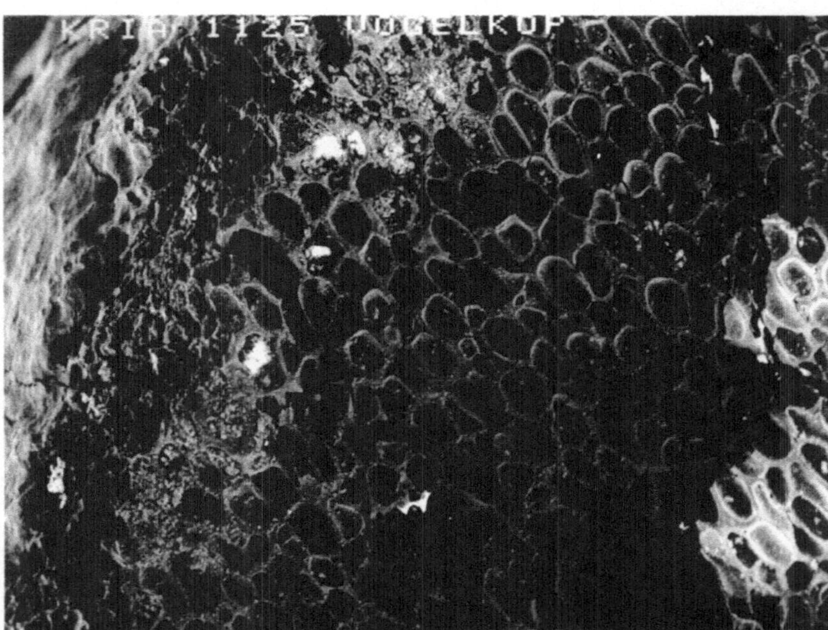

Figure 1. Cross-section of sample Kria 1125 showing subepidermal layers (top left) and mesocarp.

structure differs. In *E. bancroftii* the centre of the fruit is an ovary packed with ovules, the whole surrounded by long fibres or trichomes which surround the endocarp which is enclosed in a fleshy mesocarp (Mauseth 1988: 442). *Sloanea archboldiana* A.C. Smith, CANB 152542 Sayers, suggested by Bulmer's (1964) paper, proved also to be too different from the sample fruits to be considered. The fruits are dehiscent with four carpels and are very (4-5 mm) thick walled. The *Sloanea* spp. seeds are long and narrow in shape.

Castanopsis acuminatissima of the Fagaceae family is another edible fruit (Bulmer 1964: 148; Paijmans 1976: 109). This species is found growing with *Elaeocarpus crenulatus* often in old graveyards, between 5000 and 8000 ft (Powell 1970: 90). It forms almost pure stands on the upper slopes and ridge crests (Paijmans 1976: 88). The fruits contain a large triangular nut, dorsi-ventrally symmetrical, around 7 mm in length, enclosed within a papery exocarp. *Castanopsis acuminatissima* CANB 357287 Kerenga, and CANB 232294 Jackson and Womersley were both examined. The Herbarium specimens were collected from above 7000 ft elevation.

A wide range of Pandanaceae at the CSIRO Herbarium were examined, however, none of the fruits bore the least resemblance to the archaeobotanical samples. A suggestion was made that the unknown fruits could be seeds from the keys of ripe Pandanus, however, experiments burning Pandanus keys and extracting the seeds held within the tips of the compound fruits showed that the archaeobotanical specimens are not *Pandanus*. *Pandanus* sp. CBG 7805188 Verdon, and *P. spiralis* S.T. Blake (collection number) 16433 Blake were the species burnt and cut apart.

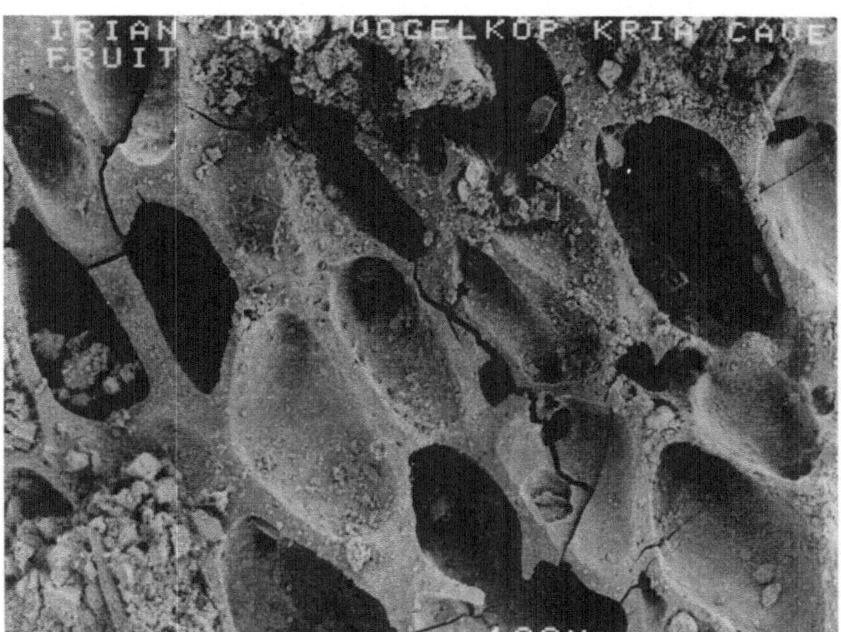

Figure 2. Cross-section of sample Kria 1125 shows the oval cells forming the mesocarp. Some sclerenchymatous development is visible in the thick cell walls.

Figure 3. Endocarp, consisting of a carpel elliptical in shape and around 2.5 mm in width. Sample Kria 1125.

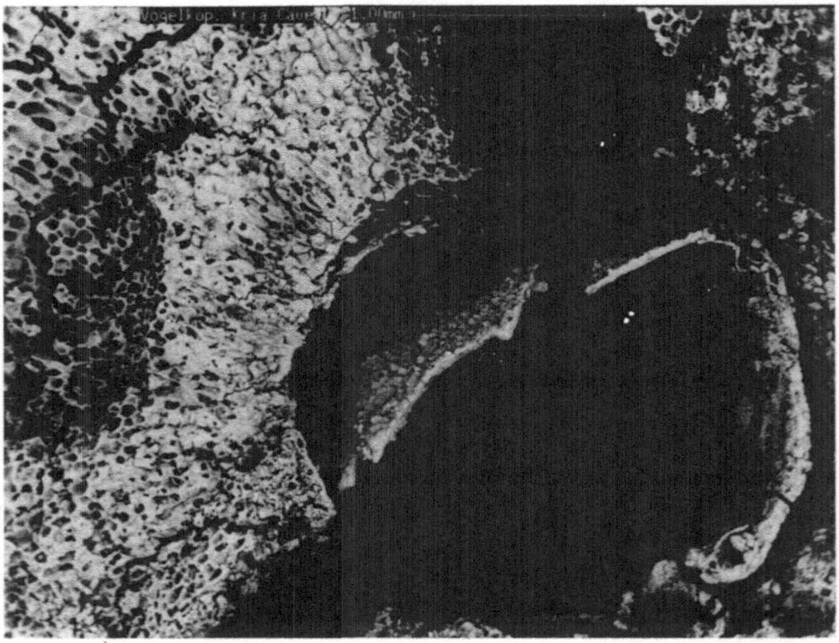

Figure 4. Close-up of sample Kria 1125 showing scale-like cell arrangement on the carpel.

4. CONCLUSIONS

No definite identifications of the fruits can be made at this time. The author's opinion is that the fruits are a member of the Elaeocarpaceae family, but the evidence to date is not conclusive.

REFERENCES

Bulmer, R.N.H. 1964. Edible seeds and prehistoric stone mortars in the highlands of East New Guinea. *Man* 183: 147-150.

Gibbs, L.S. 1917. *A Contribution to the Phytogeography and Flora of the Arfak Mountains etc.* London: Taylor & Francis.

Gorecki, P. 1989. Ethnobotany of the Jimi Valley. In P.P. Gorecki & D.S. Gillieson (eds), *A Crack in the Spine; Prehistory and ecology of the Jimi-Yuat Valley Papua New Guinea.* Townsville: Department of Anthropology and Archaeology, James Cook University.

Grubb, P.J. & P.F. Stevens 1985. *The Forests of the Fatima Basin and Mt. Kerigomna, Papua New Guinea with a review of Montane and Subalpine Rainforests in Papuasia.* Canberra: Research School of Pacific and Asian Studies, The Australian National University.

Mauseth, J.D. 1988. *Plant Anatomy.* New York: Benjamin/Cummings.

Paijmans, K. 1976. *New Guinea Vegetation.* Canberra: The Australian National University.

Powell, J.M. 1970. The impact of man on the vegetation of the Mt. Hagen region, New Guinea. Unpublished PhD thesis. Canberra: Department of Biogeography, The Australian National University.

Roth 1977. Fruits of Angiosperms. In K. Linsbauer (ed.), *Handbuch der Pflanzenanatomie* X, l.

Van Royen, P. 1982. *The Alpine Flora of New Guinea.* Vol. 3, *Taxonomic Part Winteraceae to Polygonaceae.* Hirschberg: Strauss & Cramer.

Appendix 3: Local Knowledge of Various Vertebrate Species in the Ayamaru Region

In 1997, a selection of photographs from Flannery's *Mammals of New Guinea*[1] were sent to Elimas Kambuaya, Junior-Highschool (SMP) teacher in the village of Ayamaru, with the request that he seek information from local people about the abundance and habits of the pictured animals. The following pages of Appendix 3 contain his account, collected on my behalf. His original text in somewhat imperfect Dutch, the language he learned at school during colonial times, is accompanied by my fairly literal English translation. Some remarks related to the translation are put in brackets [..].

The photographs represented some of the larger and more important species in the archaeological deposits of Kria and Toé Caves. The images were selected on the basis of their clarity and overall similarity (but not necessarily exact identity) to the species that occur in the Ayamaru area. For example, although *Dorcopsis hageni* does not occur in the Bird's Head, the pictured animal could easily be mistaken for the locally occurring *Dorcopsis muelleri*, for which a good photograph was not available.

Mr Kambuaya recorded a few interesting and peculiar stories – for example the mysterious 'tears' that are mentioned for *Dendrolagus* (the tree-kangaroo). There is nothing in the known biology of these animals that could possibly explain this phenomenon – assuming that it has actually been observed. Surprising also is the great lengths taken to catch this animal.

Interestingly, the local Ayamaru people make a distinction between all-white individuals of *Spilocuscus maculatus* and spotted ones. They may see the two varieties as different species, but it is unlikely that one or the other animal on the photographs is mistaken for another species, since the accounts of both are very similar.

With regard to the photograph of *Phalanger intercastellanus*, the associated local taxon is probably a composite of two different species, *Phalanger orientalis* and *Phalanger gymnotis*. Both are probably abundant in the area and *Phalanger gymnotis* does live occasionally in caves. *Phalanger intercastellanus* itself is found in the Pleistocene levels of Toé Cave, but does not occur on the Bird's Head today.

---

[1] Flannery, T. 1995. *Mammals of New Guinea*. Revised and updated edition. Chatswood NSW: Reed Books. New Holland Publishers and the photographers of the images have kindly granted permission to reproduce their photographs in this thesis.

The number of offspring claimed for the various species by Mr Kambuaya and his informants deviates somewhat from Flannery's (1995) accounts.

Following Mr Kambuaya's accounts on the various animals, a fable is presented that he provided about *Dorcopsis*, *Dendrolagus*, *Spilocuscus* and *Phalanger* and their horrific encounter with the dog. Again both his original Dutch text and my (fairly literal) English translation are given. Some interesting points are surfacing in this story, again the 'crying' of the tree-kangaroo, and the apparent relation between the cave-dwelling *Phalanger* and the spotted *Spilocuscus maculatus*. The 'hat' that is being referred to in relation to the latter is mysterious. The Ayamaru people are clearly aware of the fact that the forest wallaby and the tree-kangaroo are related species. The dog is recognised as a major cause of disruption of local wildlife.

Notes collected by ELIMAS KAMBUAYA
SMP1 school teacher, Ayamaru

1. *ZAGLOSSUS BRUIJNII* (LONG-BEAKED ECHIDNA)

Naam in Maibrat: *Gi*
Gi leeft in de omgeving van Mapoera/ Soewiam en tussen Mapoera - Soewiam met Kokas (een dorp dichtbij Ajawasi) en in de omgeving van Ajawasi. De andere kant van het meer (Ajamaroe, Kartapoera/Men, Kamboeaja) en de omgeving is zeldzaam, Als de *Gi* de vijand ontmoet, hij steek zijn snavel diep in de grond en hij zet zijn doorn uit op dat de vijand kan niet bij. Ik heb gezien onderweg Sorong – Klamono.
Afstammeling van *Gi* is: 1 jonge *Gi*.

Name in Meybrat: *Gi*
Gi lives in the surroundings of Mapura/Suwiam and between Mapura – Suwiam and Kokas (a village close to Ajawasi) and around Ajawasi. It is rare on the other side of the lake (Ayamaru, Kartapura/Men, Kambuaya). If the *Gi* meets the enemy, it will stick its beak deep into the ground and erects its spines so that the enemy cannot reach it. I have seen it on the way from Sorong to Klamono.
Offspring of *Gi*: one young *Gi*.

Zaglossus bruijnii. From Flannery (1995: 66). Courtesy K. Aplin

2. *DENDROLAGUS INUSTUS* (GRIZZLED TREE-KANGAROO)

Maibrat genoemd: *Kadjo*
Alle mensen in Ajamaroe gebied weten goed dit soort dier. Dit dier heeft een bijzonderheid, dat is dat dit dier kan huilen. Als dit dier met iemand ontmoet, meteen komt de tranen uit en veeg de tranen uit met de hand. Misschien denkt dit dier dat de mensen zullen vangen of dood maken, daarom gaat huilen. Dit dier leeft tussen de omgeving van Mapoera/ Soewiam en Ajawasi (Aifat gebied) en een gedeelte gebied van Aitinjo en Teminaboean. In de omgeving van Ajamaroe, Kartapoera/Men, Kamboeaja en de omgeving van het vliegveld gebied komen ze niet voor.

Als de mensen hebben gezien dat ziet in een boom, ze gaan eerst de omgeving van deze boom schoon maken (haken). Alle bomen gaan ze hakken en de bladeren en de takken van de hakkende bomen maken ze een hek rond die boom waar de *Kadjo* is/zit. En dan gaan ze de boom waar de *Kadjo* zit hakken en als die boom valt vangen ze de *Kadjo* en meteen dood maken en brengen naar huis. Afstammeling van *Kadjo*: 2 jonge *Kadjo*.

Named in Meybrat: *Kadjo*
Everybody in the Ayamaru area knows this type of animal very well. This animal has a peculiarity that it can cry. When this animal meets someone, the tears appear instantly and it wipes them off with its hand. Perhaps the animal thinks that people will catch or kill it and therefore it cries. This animal lives in the area between Mapura/Suwiam and Ajawasi (Aifat area) and in part of the area around Aitinjo and Taminabuan. It does not occur in the area around Ayamaru, Kartapura/Men, Kambuaya or the airstrip.

When it is spotted in a tree, people clear the area around the tree (cut [? or rake?]). They cut all trees and from the cut branches and leaves they make a fence around the tree containing the *Kadjo*. The tree containing the *Kadjo* is cut and when it falls the *Kadjo* is caught and killed immediately and taken home. Offspring of *Kadjo*: two young *Kadjo*.

Dendrolagus inustus. From Flannery (1995: 133). Courtesy J.P. Ferrero/ Auscape International

3. *DORCOPSIS MUELLERI* (BROWN DORCOPSIS)

In Maibrat heet: *Djief*
Djief leeft ook tussen Mapoera/ Soewiam en Ajawasi Aifat, 30 km naar het zuiden van Ajamaroe. Iedereen kent *Djief* en eet *Djief*, hoewel in Ajamaroe is zeldzaam maar de mensen kunnen krijgen in de markt of bij de zwarte markt. In de nacht kan niet goed zien, en gemakelijk dood geslaan in de nacht. Als in de nacht met de zaklantarn in de ogen van de *Djief* doet, kan niet weg lopen maar blijf op de in plaats zitten en kan je gemakelijk schieten of dood maken. Afstammeling van *Djief*: 2 jonge *Djief*.

Called in Meybrat: *Djief*
Djief also lives between Mapura/ Suwiam and Ajawasi (Aifat), 30 km south of Ayamaru. Everyone knows *Djief* and eats *Djief*, although it is rare in Ayamaru but available on the markets or on the black market [private sale?]. It cannot see very well at night, and is easily beaten to death. If at night a torch is shone into the eyes of the *Djief*, it cannot move but stays on its spot and it is easy to shoot it or kill it. Offspring of *Djief*: two young *Djief*.

Dorcopsis hageni. From Flannery (1995: 144). Courtesy I. Craven

4. *SPILOCUSCUS MACULATUS* (COMMON SPOTTED CUSCUS, albino)

In Maibrat heet: *Swi*
Swi komt in de omgeving van Ajamaroe voor, alle mensen van Ajamaroe gebied kennen *Swi*, en hebben zijn vlees gegetan. Nu de *Swi* wordt steeds opzij geduwd door de mensen want de lekkere vlees van *Swi*. De staart van de dodende *Swi* als versiering in de hand van de mensen als armband. *Swi* leeft boven in de bomen. Afstammeling van *Swi*: 2 jonge *Swi*.

In Meybrat called: *Swi*.
Swi is found in the surroundings of Ayamaru, all people from the Ayamaru area know *Swi* and have eaten its meat. Now the *Swi* is being pushed aside [overhunted?] because of the tasty meat of *Swi*. The tail of the dead *Swi* is used as decoration on the hand of the people, as a bracelet. *Swi* lives high up in the trees. Offspring of *Swi*: two young *Swi*.

Spilocuscus maculatus. From Flannery (1995: 181). Courtesy P. German

5. *SPILOCUSCUS MACULATUS* (COMMON SPOTTED CUSCUS)

In Maibrat noemt: *Tifdjah*
Tifdjah woont in de bomen, in de gat van de bomen. *Tifdjah* komt vaak in de omgeving, maar nu wordt geduwd door de mensen want de vlees van *Tifdjah* is lekker. De staart van *Tifdjah* wordt een armband als versiering in de hand van de mensen. Afstammeling van *Tifdjah*: 1 jonge *Tifdjah*.

Named in Meybrat: *Tifdjah*
Tifdjah lives in the trees, in tree hollows. *Tifdjah* is seen often in the area but is now pushed aside [overhunted?] by the people because the meat of *Tifdjah* is tasty. The tail of *Tifdjah* is used by the people as decoration, as a bracelet, around the hand. Offspring of *Tifdjah*: one young *Tifdjah*.

Spilocuscus maculatus. From Flannery (1995: 183). Courtesy G. Steer

6. *PHALANGER* SP.
P. orientalis (Northern Common Cuscus) or *P. gymnotis* (Ground Cuscus)

In Maibrat genoemt: *Ames*
Ames woont in de grot of in de gat van de bomen die niet zo hoog zijn. Als *Ames* in de nacht ontmoet met de vijand, rent direct naar de grot. *Ames* kan niet zo lang klimmen in de bomen net als *Swi* of *Tifdjah*. De staart van *Ames* als armband voor versiering van de mensen. De afstammeling van *Ames*: 2 jonge *Ames*.

Named in Meybrat: *Ames*
Ames lives in a cave or in a tree hollow in a tree that is not very high. If *Ames* meets an enemy at night it runs straight into the cave. *Ames* cannot climb in trees for a long time just as *Swi* or *Tifdjah*. The tail of *Ames* is used as a bracelet, as decoration of the people. Offspring of *Ames*: two young *Ames*.

Phalanger intercastellanus. From Flannery (1995: 169). Courtesy T. Flannery

7. *DACTYLOPSILA TRIVIRGATA* (STRIPED POSSUM)

In Maibrat genoemt: *Siwejah*
Siwejah woont in de gat van de boom. *Siwejah* kan springen van de hoge boom naar de lage boom of van de hoge tak naar de 1age tak. De staart van *Siwejah* vooral de witte gebied, gebruiken de mensen om de trachoom [oogbindvlies-ontsteking] schoon to maken. *Siwejah* komt overal in de omgeving van Ajamaroe en iedereen kent *Siwejah.* De vlees van *Siwejah* smaakt lekker. Afstammeling van *Siwejah*: 1 jonge *Siwejah.*

Named in Meybrat: *Siwejah*
Siwejah lives in a hollow tree. *Siwejah* can jump from a high tree to a low tree, or from a high branch to a low branch. The tail of the *Siwejah*, especially the white area, is used by the people to clean trachoma [conjunctivitis]. *Siwejah* is found everywhere in the area of Ayamaru and everyone knows *Siwejah.* The meat of *Siwejah* is tasty. Offspring of *Siwejah*: one young *Siwejah.*

Dactylopsila trivirgata. From Flannery (1995: 202). Courtesy J.P. Ferrero/ Auscape International

8. *DOBSONIA MAGNA* (GREAT BARE-BACKED FRUITBAT)

In Maibrat genoemt: *Wafoe*
Wafoe is een nacht dier. *Wafoe* komt overal in de omgeving. De vlees van *Wafoe* wordt gegeten door de mensen. De botten van *Wafoe* vroeger gebruikten de mensen als nalt om te naaien. *Wafoe* woont rustig in de grot, onder de klapper blad, onder de hoge Pakis blad of aan de tak van de hoge bomen. De grootste soort van *Wafoe* noemt men: *Wisam*. En ook een paar kleine soorten noemt men: *Tam*. De afstammeling van *Wafoe*: 2 jonge *Wafoe*

Named in Meybrat: *Wafoe*
Wafoe is a nocturnal animal. *Wafoe* is found everywhere in the area. The meat of *Wafoe* is eaten by the people. People used to use the bones of *Wafoe* as a needle to sew. *Wafoe* lives quietly in a cave, underneath a 'klapper' [coconut] leaf, underneat a high 'Pakis' [tree fern?] leaf or hanging on a branch of a tall tree. The largest species of *Wafoe* is called: *Wisam*. There are also a few small species called: *Tam*. Offspring of *Wafoe*: two young *Wafoe*

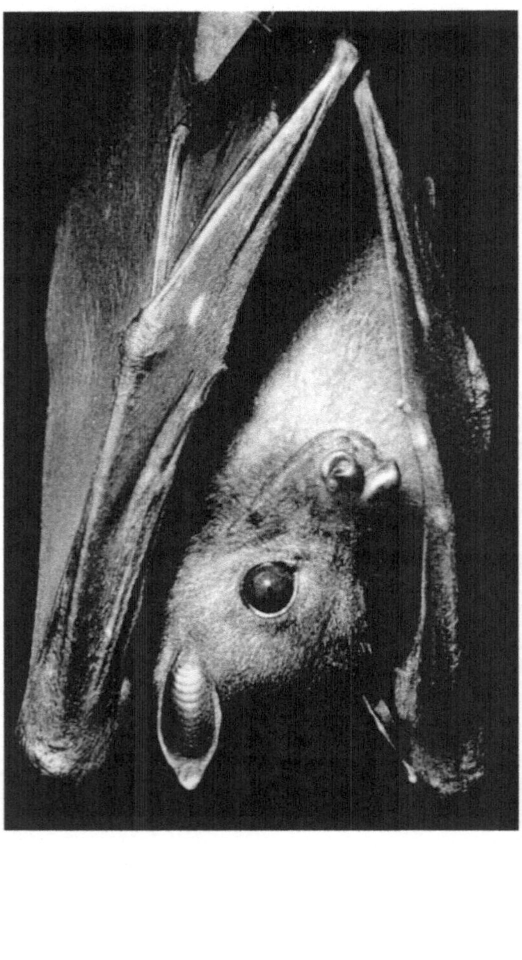

Dobsonia magna. From Flannery (1995: 351). Courtesy S. Donnellan

Een Fabel over de *Djief*, *Swi*, *Ames* en *Kadjo*

A Fable About the *Djief*, *Swi*, *Ames* and *Kadjo*

Elimas Kambuaya
SMP1 school teacher, Ayamaru

Er was een fabel over de dieren. Toen waren de dieren kunnen praten. Op een tijd gingen alle dieren naar een plaats en ze hadden een plan. Ze wouden iets doen. Alle soorten dieren kwamen al, en ze begonnen te praten over zijn plan maar de hond was laat. De hond kwam laat want de hond ging naar een andere plaats om sago te ruilen of barter doen met de andere families.

Alle dieren wachten zo lang en plotseling word de *Djief* boos, en zegt aan de hond. "De hond is onbeleeft! Als de hond at (eet) zijn mond en hoofd altijd bewegen naar links en rechts, omhoog en omlaag en kijkt naar boven en naar beneden."

Daar was ook de muis. De muis luisterde goed wat de *Djief* gezegd heeft. De muis heeft honger. De muis ging weg om de hond afhalen. De muis ontmoette de hond en zegt aan de hond. Ik (de muis) heb honger, mag ik iets eten en ik zal iets aan .jou vertellen, De hond vraagt aan de muis : "Wat is de probleem?" De muis zegt, geef mij sago of mag ik de sago eten en ik zal direct vertellen. De hond draagt de sago op zijn schouder. De hond zegt aan de muis : "Klim maar op en eet de sago aan de ene kant op mijn rug of mijn schouder."

De muis klim op en at de sago rustig aan, en daarna vertelt aan de hond, wat de *Djief* heeft gezegd tegen de hond. De *Djief* zegt : "Als de hond eet, de hond beweegt zijn mond omhoog en omlaag en kijk naar boven of naar beneden."

A fable existed about the animals. It happened at a time when the animals could talk. One day all animals went to a place and they had a plan. They wanted to do something. All kinds of animals had already arrived and started to discuss the plan, but the dog was late. The dog was late because he went somewhere else to barter for sago with other families.

All animals waited very long and suddenly the *Djief* got angry, and said about the dog: "the dog is so rude! When the dog eats his mouth and head always move from left to right and up and down, and he looks up and down."

There was also a mouse. The mouse listened carefully to what the *Djief* had said. The mouse was hungry. The mouse left to go and get the dog. The mouse met the dog and said: I (the mouse) am hungry, if I can eat something I'll tell you something. The dog asked the mouse "what is the problem?" The mouse said: give me some sago, or may I eat some sago and I will tell you immediately. The dog was carrying the sago on his shoulder. The dog said to the mouse: "climb up and have some of the sago on one side of my back or shoulder." The mouse climbs up and started nibbling the sago, and afterwards tells the dog what the *Djief* had said about the dog. The *Djief* had said: "When the dog eats, it moves his mouth up and down and looks up and down."

De hond wordt erg boos, maar zij kwamen verder tot de plaats waar de dieren waren. Alle andere dieren waren in een huis, ze gingen zitten, een paar gingen slapen, de andere met elkaar praten. De *Djief* zat ook, maar de testis / scrutum van de *Djief* hangt beneden de vloer. Dit huis was op pallen gebouwd en de vloer was van de bamboe genaaid, daarom heeft gatjes en daarom hangt de twee testis / scrutum van de *Djief* beneden.

De hond legde de sago neer, de muis springt naar beneden en direct de hond neemt zijn hakmes en snijdt de testis / scrutum van de *Djief* af.

De *Djief* verschrikt heel erg en schreeuwd plotseling met heel harde stem en alle dieren worden verschrikken en heel bang en proberen rennen of springen weg naar de veilige plaats om te verbergen. De *Djief* was erg pijn en sprongen rond in het huis met de achter poten, en de voor poten blijven gewoon vouwen. Het ging zeer verwaard en in rep en roer brengen. De dieren springen tegen elkaar, de andere springen in het vuur waar de keuken plaats, in de as van de keuken, de andere klimen op het boven gebied van het huis, de andere springen naar buiten, de andere blijven stil in de plaats en beven.

Van deze gebeurtenis deze dieren hebben verschillende levensloop vooral over zijn kleur, waar zij gaan leven / slapen, hoe zij lopen, klimmen en wie zijn de vijand en wie de vrienden zijn.

Ik begin èèn voor èèn te vertellen.
1. De *Djief*:
De *Djief* loopt, springt of rent altijd gebruikt zijn twee achter poten. De twee voor poten blijven gewoon vouwen, behalve als gewoon lopen, zitten of rust

The dog got very angry, but they arrived at the place where the animals were. All other animals were inside a house, they sat down, some went to sleep, others started talking. The *Djief* also sat down but its testes / scrotum were hanging below the floor. This house was built on poles and the floor was made of bamboo, sewn together, therefore it had holes and therefore the testes / scrotum were hanging down below the floor. The dog put the sago down, the mouse jumped off and immediately the dog took his machete and cut off the testes / scrotum of the *Djief*.

The *Djief* had a big fright and cried out suddenly with a loud voice and all animals had a terrible fright and were very scared and tried to run or jump to a safe place to hide. The *Djief* was in great pain and jumped around in the house on his hind legs, with his front legs just remaining flexed. It was all very confusing and chaotic. The animals jumped against each other, some jumped in the fire where the kitchen was, in the ash from the kitchen, others climbed up in the upper part of the house, others again jumped outside or stood still, shaking.

Because of this event these animals have varying life histories, especially in regard to their colour, where they live / sleep, how they walk, climb, and who their enemies and who their friends are.

I will tell about them one by one.
1. The *Djief*.
The *Djief* walks, jumps or runs always on both its hind legs. Both front legs remain flexed, except when they walk normally, sit or have a rest.

nemen. De *Djief* kan ook ver springen maar de meeste krachten komen van de achter poten. De vijand van de *Djief* is de hond. De hond is de vijand van alle dieren.

De *Djief* gebruikt overal zijn achter poten als zijn enige kracht en, zijn voor poten alleen voor speciale activiteiten. Dat komt van deze vecht partij waar toen de *Djief* zo erg pijn was en sprong rond om heen met de achter poten en tot nu toe de *Djief* in de meeste activiteiten gebruikt de achter poten.

2. De *Swi* :
Swi was toen ook daar. Toen gebeurde de vecht partij *Swi* wil ook rennen of springen om een veilige plaats te zoeken, want de *Swi* was ook erg bang. De andere dieren springen eerder de as van de keuken in en de as vliegen overal. De meeste as raakt aan *Swi* en *Swi* wordt helemaal wit en nu de *Swi* is wit. want *Swi* wit is *Swi* gaat verbergen hoog in de boom, *Swi* klimt de boom in en gaat rustig zitten in de hoge tak van de boom op dat de hond kan niet zien, en overdag *Swi* kan niet zo maar klimmen heen en weer van de ene boom naar de andere boom of kruipen beneden in het bos want *Swi* is kan gemakelijk gezien door de vijand. De *Swi* maakt zijn eten activiteit in het donker of in de nacht.

3. De *Ames* :
Ames wordt ook bang en springt direct en rent naar buiten van dit huis rent naar het bos. *Ames* wil niet klimmen in de boom anders wordt gezien door de hond. Daarom kruipt de *Ames* het gat in of kruipt in de grot. De familie van *Ames* was *Tifdjah*. *Tifdjah* gebruikt de hoed op zijn kop. Om gemakelijk te kruipen de grot in, de *Tifdjah* vind een moeilijkheid. De *Ames* zegt aan de *Tifdjah*:

The *Djief* can also jump very far but most of its power comes from the hind legs. The enemy of the *Djief* is the dog. The dog is the enemy of all animals.

The *Djief* uses generally its hind legs as its only power, its front legs only for special activities. That is the result of this fight where the *Djief* was very hurt and jumped around on its hind legs. And up till now the *Djief* uses mostly its hind legs.

2. The *Swi*.
Swi was also present at the fight. *Swi* also wanted to run or jump to look for a safe place, as the *Swi* was also very frightened. Other animals jumped in the ash from the kitchen and the ash flew everywhere. Most of the ash got attached to *Swi* and *Swi* turned all white, and today the *Swi* is white. Because the *Swi* is white it goes and hides high up in the tree, *Swi* climbs up the tree and sits there quietly on a high branch so that the dog cannot see him. During the day *Swi* cannot climb from one tree to another or crawl on the forest floor because *Swi* can easily be spotted by the enemy. The *Swi* eats in the dark or during the night.

3. The *Ames*.
Ames is also very frightened and jumps quickly and runs outside into the forest. *Ames* does not want to climb into a tree or else it will be seen by the dog. Therefore the *Ames* crawls into a hole or crawls into a cave. The family of *Ames* was *Tifdjah*. *Tifdjah* has a hat on his head, which makes it difficult to crawl into a cave. The *Ames* says to *Tifdjah*:

Jij gaat klimmen de boom in en ik kruip de grot in. Ik verberg mij in de grot en jij verberg zich in de boom of in het gat van de boom. *Ames* en *Tifdjah* samen hebben de enige vijand is de hond. *Ames* en *Tifdjah* komen niet op de overdag voor, alleen maar in de nacht doen ze zijn activiteiten voor eten.

You climb into a tree and I will crawl into a cave. I will hide in the cave and you hide in a tree or in a tree hollow. *Ames* and *Tifdjah* together have only one enemy which is the dog. *Ames* and *Tifdjah* do not appear during the day, they only eat during the night.

4. De *Kadjo* :
De *Kadjo* was ook heel bang. De *Kadjo* was ook slim. De *Kadjo* klimt op en zit boven in de plaats waar de brand houten zijn en die plaats ligt boven de keuken. De *Kadjo* was zo heel bang voor de hond en gaat huilen. De *Kadjo* was verdriet want zijn familie de *Djief* zijn testis / scrutum was weg gehakt door de hond. Daarom gaat de *Kadjo* huilen en de *Kadjo* gebruikt zijn voor poten om de tranen te drogen. En tot nu toe de *Kadjo* gaat huilen als de *Kadjo* de vijand ontmoet (de mensen of de hond). Tot nu toe herinert de *Kadjo* heel goed deze vecht partij en de woestheid van de hond voor zijn familie de *Djief.*

4. The *Kadjo*.
The *Kadjo* was also very frightened. The *Kadjo* was also very smart. He climbs up and sits up at the location above the firewood, which is above the kitchen. The *Kadjo* was very afraid of the dog and starts to cry. The *Kadjo* was very sad because his family the *Djief* had his testes / scrotum cut off by the dog. That is why the *Kadjo* cries and the *Kadjo* uses his front legs to wipe off the tears. And up till now the *Kadjo* cries when it meets its enemy (the humans or the dog). *Kadjo* remembers this big fight very well and the viciousness of the dog to its family the *Djief.*

Dat is een beetje over de zogende dieren. De andere dieren waren ook er bij, en dat waren de verschillende vogels. Alle vogels waren ook er bij toen gebeurde de vecht partij. Alle vogels die een rode vlek op zijn borst, op zijn snavel, op zijn poten, op zijn vleugel hebben, dat kwam van de bloed van de *Djief.* Ook vogels die helemaal rood zijn dat kleur kwam van de *Djief* bloed. Toen de vecht partij, willen deze vogels weg vliegen maar vliegen ze tegen of door in, en vallen ze in de bloed van de *Djief.* Daarom worden ze (de vogels) helemaal rood, of met rode vlekken in de verschillende gebied / gedeelte van sommige vogels. De vogels hebben ook vijand en dat is de hond.

That is a little story about the mammals. Other animals were there too, which were various birds. All birds were also present when the fight began. All birds that have a red spot on the chest, on the beak, on the legs or on the wings, this comes from the blood of the *Djief.* Also birds that are all-red are this colour because of the blood of the *Djief.* During the fight these birds wanted to fly away but they all flew against each other and fell in the blood of the *Djief.* That is why they (the birds) are completely red, or have red spots on various parts of the body. The birds also have an enemy, which is the dog.

Index

INDEX – bold= illustration or table

MODERN QUATERNARY RESEARCH IN SOUTHEAST ASIA (MQR)

Annual – ISSN 0168 - 6151

Editors 1975 – 1999: G.-J.Bartstra & W.A.Casparie
Biological-Archaeological Institute,
University of Groningen,
Groningen, The Netherlands

Editors since 2000: J.M. Pasveer & K. Aplin
Department of Archaeology and Natural History,
Australian National University,
Canberra, Australia

VOLUMES PUBLISHED WITHIN THE MQR SERIES:

Volume 1
Contains: *Papers read at a colloquium on research in Indonesia, Groningen.*
Bartstra, G.-J. / Casparie, W.A. (eds)
1975, 23 cm, 108 pp., ISBN: 90 6191 006 4

Volume 2
Contains: *Research reports from Indonesia. Solo Terraces; Man in Indonesia; Hoabinhian in Sumatra; Ulu Leang I Cave, incl. 7 loose maps.*
Bartstra, G.-J. / Casparie, W.A. (eds)
1976, 23 cm, ISBN: 90 6191 013 7, Out of print

Volume 3
Contains: *Phu Wiang pottery; Prehistory NE Thailand andamp; Cambodia; Maritime adaptations andamp; Resources in Sundaic SE Asian prehistory.*
Bartstra, G.-J. / Casparie, W.A. (eds)
1977, 23 cm, 171 pp., ISBN: 90 6191 016 1, Out of print

Volume 4
Contains: *Javan affinities of the tula 'adzeflake' from Australia; The site of Homo erectus mandible F; Age of Pithecanthropus mandible C; C-dating of sea levels N Java; Age of deglaciation of Mt Kinabalu.*
Bartstra, G.-J. / Casparie, W.A. (eds)
1978, 23 cm, 80 pp., ISBN: 90 6191 031 5

Volume 5
Contains: *Stone-Age sites in Cagayan Valley, Philippines; Bifacial stone industry from Thailand; Migration routes of aborigines to Tasmania*, and more.
Bartstra, G.-J. / Casparie, W.A. (eds)
1979, 23 cm, 110 pp., ISBN: 90 6191 083 8

Volume 6

Contains: *The archaeology of the Leang Burung Cave 2, S Sulawesi; Pollen diagram, Tao Sipinggan, N Sumatra; Megalithic remains, Sumberjaya; Geochelone faunas Indonesia*, and more.
Bartstra, G.-J. / Casparie, W.A. (eds)
1981, 23 cm, 128 pp., ISBN: 90 6191 219 9

Volume 7

Contains: *In memoriam Prof.Dr. G.H.R.von Koenigswald. Prehistorians in pre-war Indonesia; The 'Dubois collection'; The Trinil rats; Chronostratigraphy of the Plio-Pleistocene fossil hominids, Java; Mammal fauna of Wajak Cave.*
Bartstra, G.-J. / Casparie, W.A. (eds)
1982, 23 cm, 246 pp., ISBN: 90 6191 200 8

Volume 8

Contains: *Papers on palynology, paleontology, geology, & prehistoric archaeology*
Bartstra, G.-J. / Casparie, W.A. (eds)
1984, 23 cm, 188 pp., ISBN: 90 6191 540 6

Volume 9

Contains: *Papers read at Symposium I, 12th Congress of the Indo-Pacific Prehistory Association, Philippines, 26 January-2 February 1985.*
Bartstra, G.-J. / Casparie, W.A. (eds)
1985, 23 cm, 166 pp., ISBN: 90 6191 605 4

Volume 10: *The Kettledrums of Southeast Asia: A Bronze Age World and its Aftermath*
Bernet Kempers, A.J. (aut) / Bartstra, G.-J. / Casparie, W.A. (eds)
1988, 23 cm, 505 pp., ISBN: 90 6191 541 4

Volume 11

Contains: *Newly found bronze drum from Bali; Sanskrit inscriptions recently discovered in Kedah, Malaysia; Sibisa Swamp, N.Sumatra,* and more.
Bartstra, G.-J. / Casparie, W.A. (eds):
1989, 23 cm, 173 pp., ISBN: 90 6191 883 9, Out of print

Volume 12: *Late Pleistocene and Holocene vegetation of West Java, Indonesia*
Stuijts, I.-L.M.
1993, 23 cm, 186 pp., ISBN: 90 5410 148 2, Out of print

Volume 13: *Dvarapalas in Indonesia: Temple guardians and acculturation*
Bemmel, H.A.van (aut) / Bartstra, G.-J. / Casparie, W.A. (eds)
1994, 23 cm, 265 pp., ISBN: 90 5410 155 5

Volume 14: *Pollen rain, vegetation sand climate in lowland East Java, Indonesia*
Beuning, K.R.M. (aut) / Bartstra, G.-J. / Casparie, W.A. (eds)
1996, 23 cm, 112 pp., ISBN: 90 5410 629 8

Volume 15: *Bird's Head Approaches: Papers Collected for JISR, Irian Jaya Studies – a Programme for Interdisciplinary research*
Bartstra, G.-J. (ed)
1998, 23 cm, 286 pp., ISBN: 90 5410 683 2

Volume 16: *East of Wallace's Line: Studies of Past and Present Maritime Cultures of the Indo-Pacific Region*
O'Connor, S. / Veth, P. (eds)
2000, 23 cm, 380 pp., ISBN: 90 5809 319 0

Volume 17: *The Djief Hunters – 26,000 Years of Rainforest Exploitation on the Bird's Head of Papua, Indonesia*
Pasveer, J.M.
2004, 24 cm, 448 pp., ISBN: 90 5809 663 7

Volume 18: *Quaternary Research in Indonesia*
S.G. Keates / J.M. Pasveer (eds)
2004, 24 cm, 394 pp., ISBN: 90 5809 674 2